SAFETY SYMBOLS

	HAZARD	EXAMPLES	PRECAUTION	REMEDY
DISPOSAL	Special disposal procedures need to be followed.	certain chemicals, living organisms	Do not dispose of these materials in the sink or trash can.	Dispose of wastes as directed by your teacher.
BIOLOGICAL	Organisms or other biological materials that might be harmful to humans	bacteria, fungi, blood, unpreserved tissues, plant materials	Avoid skin contact with these materials. Wear mask or gloves.	Notify your teacher if you suspect contact with material. Wash hands thoroughly.
EXTREME TEMPERATURE	Objects that can burn skin by being too cold or too hot	boiling liquids, hot plates, dry ice, liquid nitrogen	Use proper protection when handling.	Go to your teacher for first aid.
SHARP OBJECT	Use of tools or glassware that can easily puncture or slice skin	razor blades, pins, scalpels, pointed tools, dissecting probes, broken glass	Practice common-sense behavior and follow guidelines for use of the tool.	Go to your teacher for first aid.
FUME	Possible danger to respiratory tract from fumes	ammonia, acetone, nail polish remover, heated sulfur, moth balls	Make sure there is good ventilation. Never smell fumes directly. Wear a mask.	Leave foul area and notify your teacher immediately.
ELECTRICAL	Possible danger from electrical shock or burn	improper grounding, liquid spills, short circuits, exposed wires	Double-check setup with teacher. Check condition of wires and apparatus.	Do not attempt to fix electrical problems. Notify your teacher immediately.
IRRITANT	Substances that can irritate the skin or mucous membranes of the respiratory tract	pollen, moth balls, steel wool, fiberglass, potassium permanganate	Wear dust mask and gloves. Practice extra care when handling these materials.	Go to your teacher for first aid.
CHEMICAL	Chemicals that can react with and destroy tissue and other materials	bleaches such as hydrogen peroxide; acids such as sulfuric acid, hydrochloric acid; bases such as ammonia, sodium hydroxide	Wear goggles, gloves, and an apron.	Immediately flush the affected area with water and notify your teacher.
TOXIC	Substance may be poisonous if touched, inhaled, or swallowed	mercury, many metal compounds, iodine, poinsettia plant parts	Follow your teacher's instructions.	Always wash hands thoroughly after use. Go to your teacher for first aid.
OPEN FLAME	Open flame may ignite flammable chemicals, loose clothing, or hair	alcohol, kerosene, potassium permanganate, hair, clothing	Tie back hair. Avoid wearing loose clothing. Avoid open flames when using flammable chemicals. Be aware of locations of fire safety equipment.	Notify your teacher immediately. Use fire safety equipment if applicable.

 Eye Safety Proper eye protection should be worn at all times by anyone performing or observing science activities.

 Clothing Protection This symbol appears when substances could stain or burn clothing.

 Animal Safety This symbol appears when safety of animals and students must be ensured.

 Radioactivity This symbol appears when radioactive materials are used.

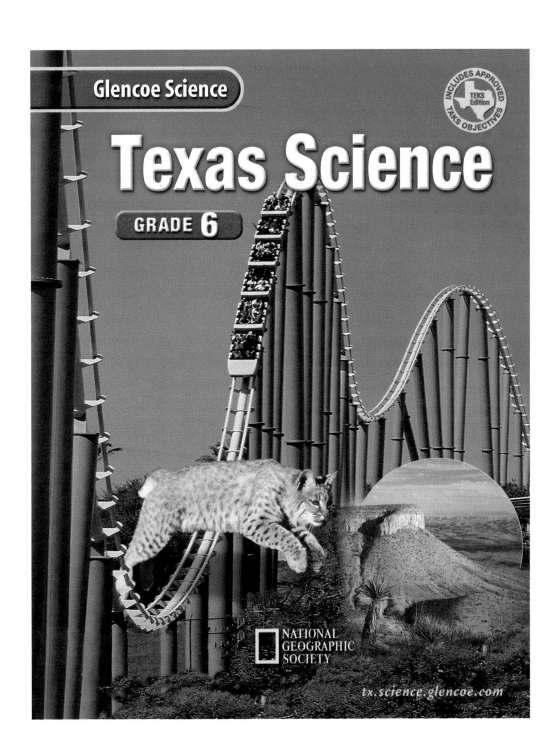

Glencoe Science

Texas Science
GRADE 6

New York, New York   Columbus, Ohio   Woodland Hills, California   Peoria, Illinois

Glencoe Science

TEXAS SCIENCE GRADE 6

Student Edition
Teacher Wraparound Edition
Interactive Teacher Edition CD-ROM
Interactive Lesson Planner CD-ROM
Texas Lesson Plans
Content Outline for Teaching
Dinah Zike's Teaching Science with Foldables
Directed Reading for Content Mastery
Foldables: Improving Reading and Study Skills
Assessment
    Chapter Review
    Chapter Tests
    ExamView Pro Test Bank Software
    Assessment Transparencies
    Performance Assessment in the Science Classroom
    The Princeton Review TAKS Practice Booklet
Directed Reading for Content Mastery in Spanish
Spanish Resources
English/Spanish Guided Reading Audio Program

Reinforcement
Enrichment
Activity Worksheets
Section Focus Transparencies
Teaching Transparencies
Laboratory Activities
Science Inquiry Labs
Critical Thinking/Problem Solving
Reading and Writing Skill Activities
Mathematics Skill Activities
Cultural Diversity
Texas Laboratory Management and Safety in the Science Classroom
MindJogger Videoquizzes and Teacher Guide
Interactive CD-ROM with Presentation Builder
Vocabulary PuzzleMaker Software
Cooperative Learning
Environmental Issues in the Science Classroom
Home and Community Involvement
Using the Internet in the Science Classroom

"Study Tip," "Test-Taking Tip," and "TAKS Practice" features in this book were written by The Princeton Review, the nation's leader in test preparation. Through its association with McGraw-Hill, The Princeton Review offers the best way to help students excel on standardized assessments.

The Princeton Review is not affiliated with Princeton University or Educational Testing Service.

A Division of The McGraw·Hill Companies

Copyright ©2002 by the McGraw-Hill Companies, Inc. All rights reserved. Except as permitted under the United States Copyright Act, no part of this publication may be reproduced or distributed in any form or by any means, or stored in a database or retrieval system, without the prior written permission of the publisher.

The "Visualizing" features found in each chapter of this textbook were designed and developed by the National Geographic Society's Education Division, copyright ©2002 National Geographic Society. The name "National Geographic Society" and the yellow border rectangle are trademarks of the Society, and their use, without prior written permission, is strictly prohibited. All rights reserved.

The "Science and Society" and the "Science and History" features that appear in this book were designed and developed by TIME School Publishing, a division of TIME Magazine. TIME and the red border are trademarks of Time, Inc. All rights reserved.

Cover Images: a roller coaster in one of Texas's amusement parks; a springing bobcat; Sierra Diablo with yucca plants in foreground

Send all inquiries to:
Glencoe/McGraw-Hill
8787 Orion Place
Columbus, OH 43240

ISBN 0-07-825458-2
Printed in the United States of America.
5 6 7 8 9 10 071/055 13 12 11 10 09 08

Authors

National Geographic Society
Education Division
Washington, D.C.

Alton Biggs
Biology Teacher
Allen High School
Allen ISD
Allen, Texas

Lucy Daniel, EdD
Teacher Consultant
Rutherford County Schools
Rutherfordton, North Carolina

Ralph M. Feather Jr., PhD
Science Department Chair,
Earth Science Teacher
Derry Area School District
Derry, Pennsylvania

Susan Leach Snyder
Earth Science Teacher, retired
Jones Middle School
Upper Arlington, Ohio

Dinah Zike
Educational Consultant
Dinah-Might Activities, Inc.
San Antonio, Texas

Contributing Authors

Dan Barrett IV
Science Writer
Chicago Heights, Illinois

Patricia Horton
Mathematics and Science Teacher
Summit Intermediate School
Etiwanda, California

Deborah Lillie
Science Teacher
Sudbury Intermediate School
Sudbury, Massachusetts

Thomas McCarthy, PhD
Science Department Chair
St. Edwards School
Vero Beach, Florida

Texas Science Consultants

José Luis Alvarez, PhD
Math Science Mentor Teacher
TEKS for Leaders Trainer
Ysleta ISD
El Paso, Texas

José Alberto Marquez
TEKS for Leaders Trainer
Ysleta ISD
El Paso, Texas

Sandra West, PhD
Associate Professor of Biology
Southwest Texas State University
San Marcos, Texas

Series Reading Consultants

Elizabeth Babich
Special Education Teacher
Mashpee Public Schools
Mashpee, Massachusetts

Barry Barto
Special Education Teacher
John F. Kennedy Elementary
Manistee, Michigan

Carol A. Senf, PhD
Associate Professor of English
Georgia Institute of Technology
Atlanta, Georgia

Rachel Swaters
Science Teacher
Rolla Middle Schools
Rolla, Missouri

Nancy Woodson, PhD
Professor of English
Otterbein College
Westerville, Ohio

Content Consultants

Leanne Field, PhD
Lecturer Molecular Genetics and Microbiology
University of Texas
Austin, Texas

Jerry Jackson, PhD
Program Director Center for Science, Mathematics and
Technology Education
Florida Gulf Coast University
Fort Meyers, Florida

William C. Keel, PhD
Department of Physics/Astronomy
University of Alabama
Tuscaloosa, Alabama

Linda Knight, EdD
Associate Director
Rice Model Science lab
Houston, Texas

Stephen M. Letro
National Weather Service
Meteorologist in Charge
Jacksonville, Florida

Lisa McGaw
Science Teacher
Hereford High School
Hereford ISD
Hereford, Texas

Madelaine Meek
Physics Consultant
Editor
Lebanon, Ohio

Robert Nierste
Science Department Head
Hendrick Middle School
Plano ISD
Plano, Texas

Sten Odenwald, PhD
NASA, Goddard Space Flight Center
Education and Public Outreach Manager
Greenbelt, Maryland

Connie Rizzo, MD
Professor of Biology
Pace University
New York, New York

Dominic Salinas, PhD
Middle School Science Supervisor
Caddo Parish Schools
Shreveport, Louisiana

Carl Zorn, PhD
Staff Scientist
Jefferson Laboratory
Newport News, Virginia

Series Safety Consultants

Malcolm Cheney, PhD
OSHA Chemical Safety Officer
Hall High School
West Hartford, Connecticut

Aileen Duc, PhD
Science II Teacher
Hendrick Middle School
Plano, Texas

Sandra West, PhD
Associate Professor of Biology
Southwest Texas State University
San Marcos, Texas

Series Math Consultants

Michael Hopper, D.Eng
Manager of Aircraft Certification
Raytheon Company
Greenville, Texas

Teri Willard, EdD
Department of Mathematics
Montana State University
Belgrade, Montana

Reviewers

Sharla Adams
McKinney High School North
McKinney ISD
McKinney, Texas

Maureen Barrett
Thomas E. Harrington Middle School
Mt. Laurel, New Jersey

Desiree Bishop
Baker High School
Mobile, Alabama

William Blair
J. Marshall Middle School
Billerica, Massachusetts

Mary Ferneau
Westview Middle School
Goose Creek, South Carolina

Connie Cook Fontenot
Bethune Academy
Aldine ISD
Houston, Texas

Annette Garcia
Kearney Middle School
Commerce City, Colorado

Nerma Coats Henderson
Pickerington Jr. High School
Pickerington, Ohio

Tammy Ingraham
Westover Park Intermediate School
Canyon ISD
Canyon, Texas

Thomas E. Lynch Jr.
Northport High School
East Northport, New York

Michael Mansour
John Page Middle School
Madison Heights, Michigan

Linda Melcher
Woodmont Middle School
Piedmont, South Carolina

Annette Parrott
Lakeside High School
Atlanta, Georgia

Billye Robbins
Lomax Junior High School
La Porte ISD
La Porte, Texas

Pam Starnes
North Richland Middle School
Birdville ISD
Fort Worth, Texas

Joanne Stickney
Monticello Middle School
Monticello, New York

Series Activity Testers

José Luis Alvarez, PhD
Math/Science Mentor Teacher
Ysleta ISD
El Paso, Texas

Nerma Coats Henderson
Science Teacher
Pickerington Jr. High School
Pickerington, Ohio

Mary Helen Mariscal-Cholka
Science Teacher
William D. Slider Middle School
Socorro ISD
El Paso, Texas

José Alberto Marquez
TEKS for Leaders Trainer
Ysleta ISD
El Paso, Texas

Science Kit and Boreal Laboratories
Tonawanda, New York

Contents in Brief

Grade 6 TEKS

UNIT 1 **The Nature of Science** 2

    Chapter 1 The Nature of Science 4   *(6.5 A, B)*
    Chapter 2 Measurement 40   *(6.6 A)*

UNIT 2 **Interactions of Matter and Energy** 70

    Chapter 3 Properties and Changes of Matter ... 72   *(6.7 A, B; 6.8A)*
    Chapter 4 Forces and Motion 100   *(6.5 A; 6.6 A, B)*
    Chapter 5 Energy 130   *(6.8 A, B; 6.9 A, B, C)*

UNIT 3 **Earth's Systems** 162

    Chapter 6 Rocks and Minerals 164   *(6.6 C; 6.7 A; 6.14 A)*
    Chapter 7 Forces Shaping Earth 196   *(6.6 C; 6.14)*
    Chapter 8 Weathering and Erosion 224   *(6.6 A, C; 6.7A, B)*
    Chapter 9 Groundwater Resources 250   *(6.5 A; 6.6 C; 6.14 B)*
    Chapter 10 The Atmosphere in Motion 282   *(6.2 C; 6.5 A; 6.8 B; 6.14 C)*

UNIT 4 **Solar System** 316

    Chapter 11 Space Technology 318   *(6.6 A; 6.13 A, B)*
    Chapter 12 The Solar System and Beyond 350   *(6.5 A, B; 6.13 A, B)*

UNIT 5 **Living Systems** 384

    Chapter 13 Life's Structure and Function 386   *(6.10 A, B, C; 6.11 B; 6.12 A, B)*
    Chapter 14 The Role of Genes in Inheritance .. 414   *(6.10 A, B; 6.11 A, B, C)*
    Chapter 15 Interactions of Living Things 440   *(6.5 A, B; 6.8 B, C; 6.10 B, C; 6.12 A, B, C)*
    Chapter 16 Animal Behavior 468   *(6.11 A; 6.12 A, B, C)*

GRADE 6 TEKS

(6.1) Scientific processes. The student conducts field and laboratory investigations using safe, environmentally appropriate, and ethical practices. The student is expected to: *(A)* demonstrate safe practices during field and laboratory investigations; and *(B)* make wise choices in the use and conservation of resources and the disposal or recycling of materials.

(6.2) Scientific processes. The student uses scientific inquiry methods during field and laboratory investigations. The student is expected to: *(A)* plan and implement investigative procedures including asking questions, formulating testable hypotheses, and selecting and using equipment and technology; *(B)* collect data by observing and measuring; *(C)* analyze and interpret information to construct reasonable explanations from direct and indirect evidence; *(D)* communicate valid conclusions; and *(E)* construct graphs, tables, maps, and charts using tools including computers to organize, examine, and evaluate data.

(6.3) Scientific processes. The student uses critical thinking and scientific problem solving to make informed decisions. The student is expected to: *(A)* analyze, review, and critique scientific explanations, including hypotheses and theories, as to their strengths and weaknesses using scientific evidence and information; *(B)* draw inferences based on data related to promotional materials for products and services; *(C)* represent the natural world using models and identify their limitations; *(D)* evaluate the impact of research on scientific thought, society, and the environment; and *(E)* connect Grade 6 science concepts with the history of science and contributions of scientists.

(6.4) Scientific processes. The student knows how to use a variety of tools and methods to conduct science inquiry. The student is expected to: *(A)* collect, analyze, and record information using tools including beakers, petri dishes, meter sticks, graduated cylinders, weather instruments, timing devices, hot plates, test tubes, safety goggles, spring scales, magnets, balances, microscopes, telescopes, thermometers, calculators, field equipment, compasses, computers, and computer probes; and *(B)* identify patterns in collected information using percent, average, range, and frequency.

(6.5) Scientific concepts. The student knows that systems may combine with other systems to form a larger system. The student is expected to: *(A)* identify and describe a system that results from the combination of two or more systems such as in the solar system; and *(B)* describe how the properties of a system are different from the properties of its parts.

(6.6) Science concepts. The student knows that there is a relationship between force and motion. The student is expected to: *(A)* identify and describe the changes in position, direction of motion, and speed of an object when acted upon by force; *(B)* demonstrate that changes in motion can be measured and graphically represented; and *(C)* identify forces that shape features of the Earth including uplifting, movement of water, and volcanic activity.

(6.7) Science concepts. The student knows that substances have physical and chemical properties. The student is expected to: *(A)* demonstrate that new substances can be made when two or more substances are chemically combined and compare the properties of the new substances to the original substances; and *(B)* classify substances by their physical and chemical properties.

Grade 6 TEKS

(6.8) Science concepts. The student knows that complex interactions occur between matter and energy. The student is expected to: *(A)* define matter and energy; *(B)* explain and illustrate the interactions between matter and energy in the water cycle and in the decay of biomass such as in a compost bin; and *(C)* describe energy flow in living systems including food chains and food webs.

(6.9) Science concepts. The student knows that obtaining, transforming, and distributing energy affects the environment. The student is expected to: *(A)* identify energy transformations occurring during the production of energy for human use such as electrical energy to heat energy or heat energy to electrical energy; *(B)* compare methods used for transforming energy in devices such as water heaters, cooling systems, or hydroelectric and wind power plants; and *(C)* research and describe energy types from their source to their use and determine if the type is renewable, non-renewable, or inexhaustible.

(6.10) Science concepts. The student knows the relationship between structure and function in living systems. The student is expected to: *(A)* differentiate between structure and function; *(B)* determine that all organisms are composed of cells that carry on functions to sustain life; and *(C)* identify how structure complements function at different levels of organization including organs, organ systems, organisms, and populations.

(6.11) Science concepts. The student knows that traits of species can change through generations and that the instructions for traits are contained in the genetic material of the organisms. The student is expected to: *(A)* identify some changes in traits that can occur over several generations through natural occurrence and selective breeding; *(B)* identify cells as structures containing genetic material; and *(C)* interpret the role of genes in inheritance.

(6.12) Science concepts. The student knows that the responses of organisms are caused by internal or external stimuli. The student is expected to: *(A)* identify responses in organisms to internal stimuli such as hunger or thirst; *(B)* identify responses in organisms to external stimuli such as the presence or absence of heat or light; and *(C)* identify components of an ecosystem to which organisms may respond.

(6.13) Science concepts. The student knows components of our solar system. The student is expected to: *(A)* identify characteristics of objects in our solar system including the Sun, planets, meteorites, comets, asteroids, and moons; and *(B)* describe types of equipment and transportation needed for space travel.

(6.14) Science concepts. The student knows the structures and functions of Earth systems. The student is expected to: *(A)* summarize the rock cycle; *(B)* identify relationships between groundwater and surface water in a watershed; and *(C)* describe components of the atmosphere, including oxygen, nitrogen, and water vapor, and identify the role of atmospheric movement in weather change.

CONTENTS

UNIT 1 — The Nature of Science — 2

CHAPTER 1 — The Nature of Science — 4

- **SECTION 1** What is science? 6
- **SECTION 2** Science in Action 12
- **SECTION 3** Models in Science 21
  - NATIONAL GEOGRAPHIC Visualizing the Modeling of King Tut 24
- **SECTION 4** Evaluating Scientific Explanations 27
  - Activity What is the right answer? 31
  - Activity Identifying Parts of an Investigation 32–33
  - TIME *Science and History*
    Women in Science 34–35

CHAPTER 2 — Measurement — 40

- **SECTION 1** Description and Measurement 42
  - NATIONAL GEOGRAPHIC Visualizing Precision and Accuracy 46
- **SECTION 2** SI Units 50
  - Activity Scale Drawing 55
- **SECTION 3** Drawings, Tables, and Graphs 56
  - Activity: Design Your Own Experiment
    Pace Yourself 60–61
  - Science Stats Biggest, Tallest, Loudest 62–63
  - THE PRINCETON REVIEW 68

Contents

UNIT 2 Interactions of Matter and Energy — 70

CHAPTER 3 — Properties and Changes of Matter — 72

SECTION 1 Physical Properties and Changes 74
    NATIONAL GEOGRAPHIC Visualizing Dichotomous Keys 82

SECTION 2 Chemical Properties and Changes 84
    Activity Liquid Layers 91
    Activity: Design Your Own Experiment
    Fruit Salad Favorites 92–93
    Oops! Accidents in Science Glass from the Past 94–95

CHAPTER 4 — Forces and Motion — 100

SECTION 1 Describing Motion 102
    NATIONAL GEOGRAPHIC Visualizing Earth's Motion 104

SECTION 2 Forces 110
    Activity Toys in Motion 114

SECTION 3 The Laws of Motion 115
    Activity: Use the Internet
    Space Shuttle Speed 122–123
    Science and Language Arts
    "Rayona's Ride" 124–125

CHAPTER 5 — Energy — 130

SECTION 1 What is energy? 132
SECTION 2 Energy Transformations 137
    NATIONAL GEOGRAPHIC Visualizing Energy Transformations 140
    Activity Hearing with Your Jaw 144

SECTION 3 Sources of Energy 145
    Activity: Use the Internet
    Energy to Power Your Life 152–153
    Science Stats Energy to Burn 154–155
    THE PRINCETON REVIEW 160

CONTENTS

UNIT 3 Earth's Systems — 162

CHAPTER 6 Rocks and Minerals — 164

SECTION 1 Minerals—Earth's Jewels 166
SECTION 2 Igneous and Sedimentary Rocks 175
    NATIONAL GEOGRAPHIC Visualizing Igneous Rock Features 178
SECTION 3 Metamorphic Rocks and the Rock Cycle 182
    Activity Gneiss Rice 187
    Activity Classifying Minerals 188–189
    Oops! Accidents in Science Going for the Gold 190–191

CHAPTER 7 Forces Shaping Earth — 196

SECTION 1 Earth's Moving Plates 198
    NATIONAL GEOGRAPHIC Visualizing Rift Valleys 204
    Activity Earth's Moving Plates 208
SECTION 2 Uplift of Earth's Crust 209
    Activity: Model and Invent
      Isostasy 216–217
    Science Stats Mountains 218–219

CHAPTER 8 Weathering and Erosion — 224

SECTION 1 Weathering and Soil Formation 226
    Activity Classifying Soils 232
SECTION 2 Erosion of Earth's Surface 233
    NATIONAL GEOGRAPHIC Visualizing Mass Movement 234
    Activity: Design Your Own Experiment
      Measuring Soil Erosion 242–243
    TIME *Science and History*
      Crumbling Monuments 244–245

Contents

CHAPTER 9
Groundwater Resources — 250
- **SECTION 1** Groundwater 252
  - Activity Artesian Wells 259
- **SECTION 2** Groundwater Pollution and Overuse 260
  - NATIONAL GEOGRAPHIC Visualizing Sources of Groundwater Pollution 261
- **SECTION 3** Caves and Other Groundwater Features 269
  - Activity
    Pollution in Motion 274–275
  - Science Stats Caves 276–277

CHAPTER 10
The Atmosphere in Motion — 282
- **SECTION 1** The Atmosphere 284
  - NATIONAL GEOGRAPHIC Visualizing the Water Cycle 288
- **SECTION 2** Earth's Weather 290
- **SECTION 3** Air Masses and Fronts 298
  - Activity Modeling Air Masses and Fronts 305
  - Activity: Design Your Own Experiment
    Creating Your Own Weather Station 306–307
  - TIME *Science and Society*
    How Zoos Prepare for Hurricanes 308–309
  - THE PRINCETON REVIEW 314

UNIT 4
Solar System — 316

CHAPTER 11
Space Technology — 318
- **SECTION 1** Radiation from Space 320
  - Activity Building a Reflecting Telescope 326
- **SECTION 2** Early Space Missions 327
  - NATIONAL GEOGRAPHIC Visualizing Space Probes 331
- **SECTION 3** Current and Future Space Missions 335
  - Activity: Use the Internet
    Star Sightings 342–343
  - TIME *Science and Society*
    Cities in Space 344–345

xii

Contents

CHAPTER 12

The Solar System and Beyond — 350

SECTION 1 Earth's Place in Space 352
Activity Moon Phases 357
SECTION 2 The Solar System 358
SECTION 3 Stars and Galaxies 366

 Visualizing Galaxies 370

Activity: Design Your Own Experiment
Space Colony 374–375

Science and Language Arts
The Sun and the Moon 376–377

THE PRINCETON REVIEW 382

UNIT 5

Living Systems — 384

CHAPTER 13

Life's Structure and Function — 386

SECTION 1 Cell Structure 388
Activity Comparing Cells 396
SECTION 2 Viewing Cells 397

NATIONAL GEOGRAPHIC Visualizing Microscopes 398

SECTION 3 Viruses 402

Activity: Design Your Own Experiment
Comparing Light Microscopes 406–407

TIME *Science and History*
Cobb Against Cancer 408–409

CHAPTER 14

The Role of Genes in Inheritance — 414

SECTION 1 Continuing Life 416

NATIONAL GEOGRAPHIC Visualizing Human Reproduction 422

Activity Getting DNA from Onion Cells 424
SECTION 2 Genetics—The Study of Inheritance 425

Activity: Use the Internet
Genetic Traits: The Unique You 432–433

TIME *Science and Society*
Separated at Birth 434–435

xiii

Contents

CHAPTER 15

Interactions of Living Things — 440

SECTION 1 The Environment 442
  Activity Delicately Balanced Ecosystems 448
SECTION 2 Interactions Among Living Organisms 449
SECTION 3 Matter and Energy 454
  NATIONAL GEOGRAPHIC Visualizing a Food Chain 455

Activity: Design Your Own Experiment
  Identifying a Limiting Factor 460–461
Science and Language Arts
  The Solace of Open Spaces 462–463

CHAPTER 16

Animal Behavior — 468

SECTION 1 Types of Behavior 470
SECTION 2 Behavioral Interactions 476
  NATIONAL GEOGRAPHIC Visualizing Bioluminescence 481

Activity Observing Earthworm Behavior 485
Activity: Model and Invent
  Animal Habitats 486–487
Oops Accidents in Science Going to the Dogs 488–489
THE PRINCETON REVIEW 494

TEKS Review — 496

Field Guide
  Amusement Park Rides 524
  Building Stones 528
  Living in Space 532
  Insects 536

Skill Handbooks — 540
Reference Handbooks — 565
English Glossary — 579
Spanish Glossary — 589
Index — 601

Interdisciplinary Connections

NATIONAL GEOGRAPHIC Unit Openers

Unit 1 How are arms and centimeters connected? 2
Unit 2 How are charcoal and celebrations connected? 70
Unit 3 How are rocks and fluorescent lights connected? 162
Unit 4 How are Inuit and astronauts connected? 316
Unit 5 How are seaweed and cell cultures connected? 384

NATIONAL GEOGRAPHIC VISUALIZING

1. The Modeling of King Tut 24
2. Precision and Accuracy 46
3. Dichotomous Keys 82
4. Earth's Motion 104
5. Energy Transformations 140
6. Igneous Rock Features 178
7. Rift Valleys 204
8. Mass Movements 234
9. Sources of Groundwater Pollution 261
10. The Water Cycle 288
11. Space Probes 331
12. Galaxies 370
13. Microscopes 398
14. Human Reproduction 422
15. A Food Chain 455
16. Bioluminescence 481

Interdisciplinary Connections

TIME
Science and Society
- 10 How Zoos Prepare for Hurricanes 308
- 11 Cities in Space 344
- 14 Separated at Birth 434

Science and History
- 1   Women in Science 34
- 8   Crumbling Monuments 244
- 13 Cobb Against Cancer 408

Accidents in SCIENCE
- 3   Glass from the Past 94
- 6   Going for the Gold 190
- 16 Going to the Dogs 488

Science and Language Arts
- 4   "Rayona's Ride" 124
- 12 The Sun and the Moon 376
- 15 The Solace of Open Spaces 462

Science Stats
- 2   Biggest, Tallest, Loudest 62
- 5   Energy to Burn 154
- 7   Mountains 218
- 9   Caves 276

Activities

Full Period Labs

1. What is the right answer? ... 31
   Identifying Parts of an Investigation ... 32–33
2. Scale Drawing ... 55
   Design Your Own Experiment:
   Pace Yourself ... 60–61
3. Liquid Layers ... 91
   Design Your Own Experiment: Fruit Salad Favorites ... 92–93
4. Toys in Motion ... 114
   Use the Internet:
   Space Shuttle Speed ... 122–123
5. Hearing with Your Jaw ... 144
   Use the Internet: Energy to Power Your Life ... 152–153
6. Gneiss Rice ... 187
   Classifying Minerals ... 188–189
7. Earth's Moving Plates ... 208
   Model and Invent: Isostasy ... 216–217
8. Classifying Soils ... 232
   Design Your Own Experiment:
   Measuring Soil Erosion ... 242–243
9. Artesian Wells ... 259
   Pollution in Motion ... 274–275
10. Modeling Air Masses and Fronts ... 305
    Design Your Own Experiment:
    Creating Your Own Weather Station ... 306–307
11. Building a Reflecting Telescope ... 326
    Use the Internet: Star Sightings ... 342–343
12. Moon Phases ... 357
    Design Your Own Experiment: Space Colony ... 374–375
13. Comparing Cells ... 396
    Design Your Own Experiment:
    Comparing Light Microscopes ... 406–407
14. Getting DNA from Onion Cells ... 424
    Use the Internet: Genetic Traits: The Unique You ... 432–433
15. Delicately Balanced Ecosystems ... 448
    Design Your Own Experiment:
    Identifying a Limiting Factor ... 460–461
16. Observing Earthworm Behavior ... 485
    Model and Invent: Animal Habitats ... 486–487

Activities

Mini LAB

1. **Try at Home:** Classifying Parts of a System 8
   Try at Home: Forming a Hypothesis 14
   Thinking Like a Scientist 23
2. Measuring Accurately 44
   Try at Home: Measuring Volume 52
3. **Try at Home:** Changing Density 77
   Observing Yeast 88
4. **Try at Home:** Modeling Acceleration 109
   Demonstrating the Third Law of Motion 120
5. **Try at Home:** Analyzing Energy Transformations 139
   Building a Solar Collector 149
6. Classifying Minerals 171
   Try at Home: Modeling How Fossils Form Rocks 180
7. **Try at Home:** Modeling Tension and Compression 205
   Modeling Mountains 211
8. Rock Dissolving Acids 229
   Try at Home: Analyzing Soils 230
9. Measuring Porosity 253
   Try at Home: Modeling Groundwater Pollution 264
10. **Try at Home:** Observing Condensation and Evaporation 289
    Creating a Low-Pressure Center 301
11. **Try at Home:** Observing Effects of Light Pollution 324
    Modeling a Satellite 333
12. Observing Distance and Size 355
    Try at Home: Modeling Constellations 367
13. Modeling Cytoplasm 390
    Try at Home: Observing Magnified Objects 400
14. Observing Yeast Budding 419
    Try at Home: Modeling Probability 427
15. Observing Symbiosis 452
    Try at Home: Modeling the Water Cycle 458
16. Observing Conditioning 474
    Try at Home: Demonstrating Chemical Communication 479

xviii

Activities

Explore Activity

1. Observe how gravity accelerates objects 5
2. Measure length 41
3. Classify coins 73
4. Measure the effect of friction 101
5. Analyze a marble launch 131
6. Observe a rock 165
7. Model Earth's interior 197
8. Model water erosion 225
9. Measuring pore space 251
10. Model the effects of temperature on
    molecules in air 283
11. Model visible light seen through nebulae 319
12. Estimate grains of rice 351
13. Measure a small object 387
14. Compare seeds 415
15. Measure space 441
16. Observe how humans communicate
    without using sound 469

Problem-Solving Activities

1. How can you use a data table to analyze and present data? 17
3. Do light sticks conserve mass? 89
5. Is energy consumption outpacing production? 148
6. How hard are these minerals? 171
7. How can glaciers cause land to rise? 214
8. Can evidence of sheet erosion be seen in a farm field? 239
9. Can stormwater be cleaned and reused for irrigation? 265
12. How can you model distances in the solar system? 363
15. How do changes in Antarctic food webs affect populations? .. 456
16. How can you determine which animals hibernate? 483

Activities

Math Skills Activities

- **2** Rounding 48
- **4** Calculating Average Speed 107
  - Acceleration, Force, and Mass 117
- **10** Calculating Speed 295
- **11** Using a Grid to Draw 328
- **13** Calculate the Ratio of Surface Area to Volume of Cells 394
- **14** Calculating Possible Alleles in Sex Cells 429

Skill Builder Activities

Science

Classifying: 30, 459

Communicating: 11, 49, 113, 136, 181, 231, 268, 297, 341, 373, 395, 431, 459

Comparing and Contrasting: 11, 90, 121, 174, 207, 231, 273, 297, 304, 423

Concept Mapping: 26, 83, 181, 215, 356, 401, 405, 431

Drawing Conclusions: 20, 186, 453

Forming Operational Definitions: 113

Interpreting Data: 136

Interpreting Scientific Illustrations: 395

Making and Using Graphs: 59

Making and Using Tables: 341

Making Models: 373

Measuring in SI: 54

Recognizing Cause and Effect: 241, 258, 268, 289

Recording Data: 109

Recording Observations: 447

Researching Information: 30, 475

Sequencing: 325

Testing a Hypothesis: 143, 484

Science Connections

Math
Converting Units: 54
Solving One-Step Equations: 83, 90, 109, 121, 241, 304, 325, 334, 356, 401, 423, 453, 484
Using Percentages: 174, 258
Using Precision and Significant Digits: 49
Using Proportions: 26, 151

Technology
Developing Multimedia Presentations: 365
Using a Database: 447
Using an Electronic Spreadsheet: 59, 151, 186, 289, 365, 475
Using Graphics Software: 143, 207
Using a Word Processor: 20, 215, 273, 334, 405

Science INTEGRATION

Astronomy: 51
Chemistry: 43, 200, 255, 330, 457, 480
Earth Science: 22, 79, 116, 119, 430, 444
Environmental Science: 394
Health: 85, 263, 286, 321, 421, 471
Life Science: 13, 141, 167, 291, 360
Physics: 177, 235, 372, 400

SCIENCE Online

Collect Data: 76, 118, 184, 332, 353, 368, 404
Data Update: 148, 203, 339
Research: 9, 18, 47, 58, 75, 113, 138, 172, 212, 236, 240, 254, 257, 265, 271, 297, 299, 303, 337, 359, 403, 417, 431, 446, 450, 473, 482

39, 67, 68–69, 99, 129, 159, 160–161, 195, 223, 249, 281, 313, 314–315, 349, 381, 382–383, 413, 439, 467, 493, 494–495

UNIT 1
The Nature of Science

How Are
Arms &
Centimeters
Connected?

About 5,000 years ago, the Egyptians developed one of the earliest recorded units of measurement—the cubit, which was based on the length of the arm from elbow to fingertip. The Egyptian measurement system probably influenced later systems, many of which also were based on body parts such as arms and feet. Such systems, however, could be problematic, since arms and feet vary in length from one person to another. Moreover, each country had its own system, which made it hard for people from different countries to share information. The need for a precise, universal measurement system eventually led to the adoption of the meter as the basic international unit of length. A meter is defined as the distance that light travels in a vacuum in a certain fraction of a second—a distance that never varies. Meters are divided into smaller units called centimeters, which are seen on the rulers here.

SCIENCE CONNECTION

MEASUREMENT SYSTEMS Ancient systems of measurement had their flaws, but they paved the way for the more exact and uniform systems used today. Devise your own measurement system based on parts of your body (for example, the length of your hand or the width of your shoulders) or common objects in your classroom or home. Give names to your units of measurement. Then calculate the width and height of a doorway using one or more of your units.

CHAPTER 1

Science TEKS 6.5, A, B

The Nature of Science

An important part of science is asking questions. Over time, scientists observed an unusual behavior among humpback whales and wondered why they did it. Through scientific investigations, they learned that the humpbacks work together to get food. They swim in circles and blow bubbles. This makes a bubble net that traps small fish and krill—tiny shrimplike animals. Then the whales can swoop up mouthfuls of food.

What do you think?

Science Journal Look at the picture below with a classmate. Discuss what you think this is. Here's a hint: *Dinner is served*. Write your answer or best guess in your Science Journal.

Gravity is a familiar natural force. It keeps you anchored on Earth, but how does it work? Scientists learn about gravity and other concepts by making observations. Noticing things is how scientists start any study of nature. Do the activity below to see how gravity affects objects.

Observe how gravity accelerates objects

1. Collect three identical, unsharpened pencils.
2. Tape two of the pencils together.
3. Hold all the pencils at the same height, as high as you can. Drop them together and observe what happens as they fall.

Observe
Did the single pencil fall faster or slower than the pair? Predict in your Science Journal what would happen if you taped 30 pencils together and dropped them at the same time as you dropped a single pencil.

Before You Read

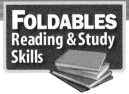

Making a Know-Want-Learn Study Fold Make the following Foldable to help you identify what you already know and what you want to know about science.

1. Stack two sheets of paper in front of you so the short side of both sheets is at the top.
2. Slide the top sheet up so that about 4 cm of the bottom sheet show.
3. Fold both sheets top to bottom to form four tabs and staple along the topfold, as shown.
4. Label the top flap *Science*. Then, label the other flaps *Know*, *Want*, and *Learned*, as shown. Before you read the chapter, write what you know about science on the *Know* tab and what you want to know on the *Want* tab.
5. As you read the chapter, list the things you learn about science on the *Learned* tab.

SECTION 1

What is science?

As You Read

What You'll Learn
- **Define** science and identify questions that science cannot answer.
- **Compare and contrast** theories and laws.
- **Identify** a system and its components.
- **Identify** the three main branches of science.

Vocabulary
science
scientific theory
scientific law
system
life science
Earth science
physical science
technology

Why It's Important
Science can be used to learn more about the world you live in.

Learning About the World

When you think of a scientist, do you imagine a person in a laboratory surrounded by charts, graphs, glass bottles, and bubbling test tubes? It might surprise you to learn that anyone who tries to learn something about the natural world is a scientist. **Science** is a way of learning more about the natural world. Scientists want to know why, how, or when something occurred. This learning process usually begins by keeping your eyes open and asking questions about what you see.

Asking Questions Scientists ask many questions, too. How do things work? What do things look like? What are they made of? Why does something take place? Science can attempt to answer many questions about the natural world, but some questions cannot be answered by science. Look at the situations in **Figure 1.** Who should you vote for? What does this poem mean? Who is your best friend? Questions about art, politics, personal preference, or morality can't be answered by science. Science can't tell you what is right, wrong, good, or bad.

Figure 1
Some questions about topics such as politics, literature, and art cannot be answered by science.

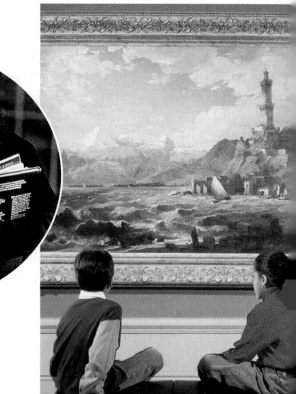

6 CHAPTER 1 The Nature of Science

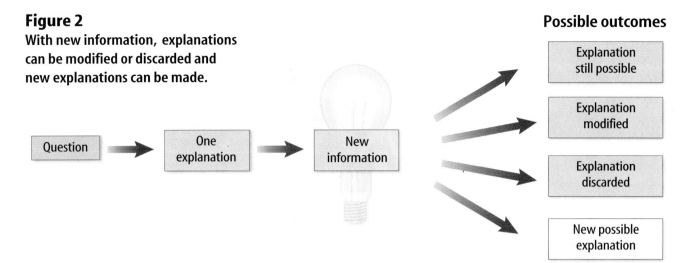

Figure 2 With new information, explanations can be modified or discarded and new explanations can be made.

Possible Explanations If learning about your world begins with asking questions, can science provide answers to these questions? Science can answer a question only with the information available at the time. Any answer is uncertain because people will never know everything about the world around them. With new knowledge, they might realize that some of the old explanations no longer fit the new information. As shown in **Figure 2,** some observations might force scientists to look at old ideas and think of new explanations. Science can only provide possible explanations.

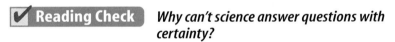 *Why can't science answer questions with certainty?*

Scientific Theories An attempt to explain a pattern observed repeatedly in the natural world is called a **scientific theory.** Theories are not simply guesses or someone's opinions, nor are theories only vague ideas. Theories in science must be supported by observations and results from many investigations. They are the best explanations that have been found so far. However, theories can change. As new data become available, scientists evaluate how the new data fit the theory. If enough new data do not support the theory, the theory can be changed to fit the new observations better.

Scientific Laws A rule that describes a pattern in nature is a **scientific law.** For an observation to become a scientific law, it must be observed repeatedly. The law then stands until someone makes observations that do not follow the law. A law helps you predict that an apple dropped from arm's length will always fall to Earth. The law, however, does not explain why gravity exists or how it works. A law, unlike a theory, does not attempt to explain why something happens. It simply describes a pattern.

Figure 3
Systems are a collection of structures, cycles, and processes. *What systems can you identify in this classroom?*

Classifying Parts of a System

Procedure
Think about how your school's cafeteria is run. Consider the physical structure of the cafeteria. How many people run it? Where does the food come from? How is it prepared? Where does it go? What other parts of the cafeteria system are necessary?

Analysis
Classify the parts of your school cafeteria's system as structures, cycles, or processes.

Systems in Science

Scientists can study many different things in nature. Some might study how the human body works or how planets move around the Sun. Others might study the energy carried in a lightning bolt. What do all of these things have in common? All of them are systems. A **system** is a collection of structures, cycles, and processes that relate to and interact with each other. The structures, cycles, and processes are the parts of a system, just like your stomach is one of the structures of your digestive system.

Reading Check *What is a system?*

Systems are not found just in science. Your school is a system with structures such as the school building, the tables and chairs, you, your teacher, the school bell, your pencil, and many other things. **Figure 3** shows some of these structures. Your school day also has cycles. Your daily class schedule and the calendar of holidays are examples of cycles. Many processes are at work during the school day. When you take a test, your teacher has a process. You might be asked to put your books and papers away and get out a pencil before the test is distributed. When the time is over, you are told to put your pencil down and pass your test to the front of the room.

Parts of a System Interact In a system, structures, cycles, and processes interact. Your daily schedule influences where you go and what time you go. The clock shows the teacher when the test is complete, and you couldn't complete the test without a pencil.

Parts of a Whole All systems are made up of other systems. For example, you are part of your school. The human body is a system—within your body are other systems. Your school is part of a system—district, state, and national. You have your regional school district. Your district is part of a statewide school system. Scientists often break down problems by studying just one part of a system. A scientist might want to learn about how construction of buildings affects the ecosystem. Because an ecosystem has many parts, one scientist might study a particular animal, and another might study the effect of construction on plant life.

The Branches of Science

Science often is divided into three main categories, or branches—life science, Earth science, and physical science. Each branch asks questions about different kinds of systems.

Life Science The study of living systems and the ways in which they interact is called **life science.** Life scientists attempt to answer questions like "How do whales navigate the ocean?" and "How do vaccines prevent disease?" Life scientists can study living organisms, where they live, and how they interact. Dian Fossey, **Figure 4,** was a life scientist who studied gorillas, their habitat, and their behaviors.

People who work in the health field know a lot about the life sciences. Physicians, nurses, physical therapists, dietitians, medical researchers, and others focus on the systems of the human body. Some other examples of careers that use life science include biologists, zookeepers, botanists, farmers, and beekeepers.

SCIENCE Online

Research Visit the Glencoe Science Web site at **tx.science.glencoe.com** for information on Dian Fossey's studies. Write a summary of your research in your Science Journal.

Figure 4
Over a span of 18 years, life scientist Dian Fossey spent much of her time observing mountain gorillas in Rwanda, Africa. She was able to interact with them as she learned about their behavior.

Figure 5 Scientists study a wide range of subjects.

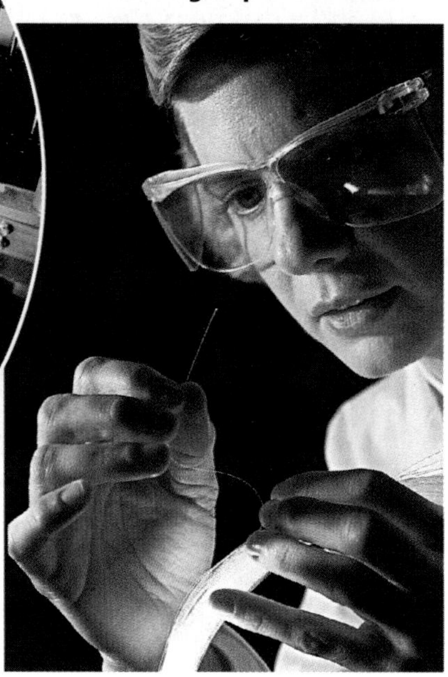

C This physicist is studying light as it travels through optical fibers.

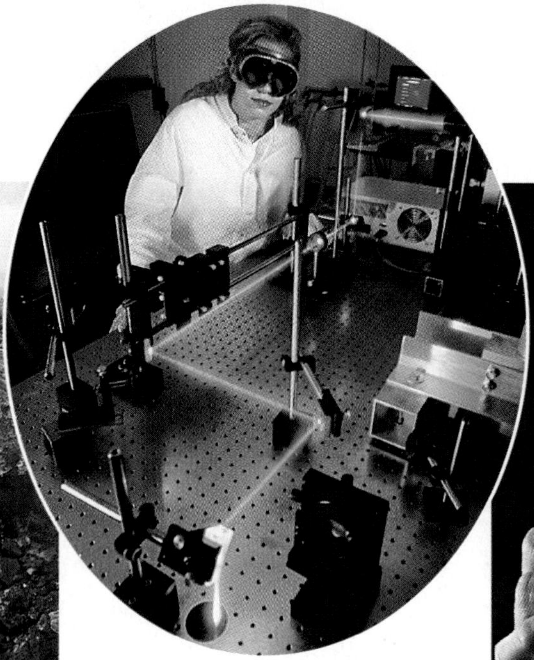

B This chemist is studying the light emitted by certain compounds.

A These volcanologists are studying the temperature of the lava flowing from a volcano.

Earth Science The study of Earth systems and the systems in space is **Earth science.** It includes the study of nonliving things such as rocks, soil, clouds, rivers, oceans, planets, stars, meteors, and black holes. Earth science also covers the weather and climate systems that affect Earth. Earth scientists ask questions like "How can an earthquake be detected?" or "Is water found on other planets?" They make maps and investigate how geologic features formed on land and in the oceans. They also use their knowledge to search for fuels and minerals. Meteorologists study weather and climate. Geologists study rocks and geologic features. **Figure 5A** shows a volcanologist—a person who studies volcanoes—measuring the temperature of lava.

Reading Check *What do Earth scientists study?*

Physical Science The study of matter and energy is **physical science.** Matter is anything that takes up space and has mass. The ability to cause change in matter is energy. Living and nonliving systems are made of matter. Examples include plants, animals, rocks, the atmosphere, and the water in oceans, lakes, and rivers. Physical science can be divided into two general fields—chemistry and physics. Chemistry is the study of matter and the interactions of matter. Physics is the study of energy and its ability to change matter. **Figures 5B** and **5C** show physical scientists at work.

Careers Chemists ask questions such as "How can I make plastic stronger?" or "What can I do to make aspirin more effective?" Physicists might ask other types of questions, such as "How does light travel through glass fibers?" or "How can humans harness the energy of sunlight for their energy needs?"

Many careers are based on the physical sciences. Physicists and chemists are some obvious careers. Ultrasound and X-ray technicians working in the medical field study physical science because they study the energy in ultrasound or X rays and how it affects a living system.

Science and Technology Although learning the answers to scientific questions is important, these answers do not help people directly unless they can be applied in some way. **Technology** is the practical use of science, or applied science, as illustrated in **Figure 6.** Engineers apply science to develop technology. The study of how to use the energy of sunlight is science. Using this knowledge to create solar panels is technology. The study of the behavior of light as it travels through thin, glass, fiber-optic wires is science. The use of optical fibers to transmit information is technology. A scientist uses science to study how the skin of a shark repels water. The application of this knowledge to create a material that helps swimmers slip through the water faster is technology.

Figure 6
Solar-powered cars and the swimsuits worn in the Olympics are examples of technology—the application of science.

Section Assessment

1. What is science?
2. Compare scientific theory and scientific law. Explain how a scientific theory can change.
3. What are the components of a system?
4. Name the three main branches of science.
5. **Think Critically** List two questions that can be answered by science and one that can't be answered by science. Explain.

Skill Builder Activities

6. **Comparing and Contrasting** Compare and contrast life science and physical science. **For more help, refer to the** Science Skill Handbook.
7. **Communicating** In your Science Journal, describe how science and technology are related. **For more help, refer to the** Science Skill Handbook.

SECTION 1 What is science? **11**

SECTION
2 Science in Action

As You Read

What You'll Learn
- **Identify** some skills scientists use.
- **Define** hypothesis.
- **Recognize** the difference between observation and inference.

Vocabulary
hypothesis
infer
controlled experiment
variable
constant

Why It's Important
Science can be used to learn more about the world you live in.

Science Skills

You know that science involves asking questions, but how does asking questions lead to learning? Because no single way to gain knowledge exists, a scientist doesn't start with step one, then go to step two, and so on. Instead, scientists have a huge collection of skills from which to choose. Some of these skills include thinking, observing, predicting, investigating, researching, modeling, measuring, analyzing, and inferring. Science also can advance with luck and creativity.

Science Methods Investigations often follow a general pattern. As illustrated in **Figure 7,** most investigations begin by seeing something and then asking a question about what was observed. Scientists often research by talking with other scientists. They read books and scientific magazines to learn as much as they can about what is already known about their question. Usually, scientists state a possible explanation for their observation. To collect more information, scientists almost always make more observations. They might build a model of what they study or they might perform investigations. Often, they do both. How might you combine some of these skills in an investigation?

Figure 7
Although there are different scientific methods for investigating a specific problem, most investigations follow a general pattern.

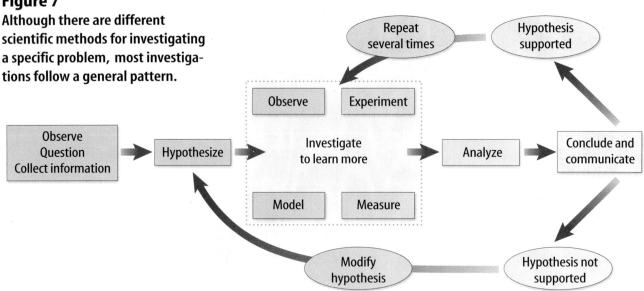

12   CHAPTER 1   The Nature of Science

Figure 8
Investigations often begin by making observations and asking questions.

Questioning and Observing Ms. Clark placed a sealed shoe box on the table at the front of the laboratory. Everyone in the class noticed the box. Within seconds the questions flew. "What's in the box?" "Why is it there?"

Ms. Clark said she would like the class to see how they used some science skills without even realizing it.

"I think that she wants us to find out what's in it," Isabelle said to Marcus.

"Can we touch it?" asked Marcus.

"It's up to you," Ms. Clark said.

Marcus picked up the box and turned it over a few times.

"It's not heavy," Marcus observed. "Whatever is inside slides around." He handed the box to Isabelle.

Isabelle shook the box. The class heard the object strike the sides of the box. With every few shakes, the class heard a metallic sound. The box was passed around for each student to make observations and write them in his or her Science Journal. Some observations are shown in **Figure 8.**

Taking a Guess "I think it's a pair of scissors," said Marcus.

"Aren't scissors lighter than this?" asked Isabelle, while shaking the box. "I think it's a stapler."

"What makes you think so?" asked Ms. Clark.

"Well, staplers are small enough to fit inside a shoe box, and it seems to weigh about the same," said Isabelle.

"We can hear metal when we shake it," said Enrique.

"So, you are guessing that a stapler is in the box?"

"Yes," they agreed.

"You just stated a hypothesis," exclaimed Ms. Clark.

"A what?" asked Marcus.

Life Science
INTEGRATION

Some naturalists study the living world, using mostly their observational skills. They observe animals and plants in their natural environment, taking care not to disturb the organisms they are studying. Make observations of organisms in a nearby park or backyard. Record your observations in your Science Journal.

SECTION 2 Science in Action **13**

Forming a Hypothesis

Procedure
1. Fill a large **pot** with **water**. Drop an unopened **can of diet soda** and an unopened **can of regular soda** into the pot of water and observe what each can does.
2. In your Science Journal, make a list of the possible explanations for your observation. Select the best explanation and write a hypothesis.
3. Read the nutritional facts on the back of each can and compare their ingredients.
4. Revise your hypothesis based on this new information.

Analysis
1. What did you observe when you placed the cans in the water?
2. How did the nutritional information on the cans change your hypothesis?
3. Infer why the two cans behaved differently in the water.

The Hypothesis "A **hypothesis** is a reasonable and educated possible answer based on what you know and what you observe."

"We know that a stapler is small, it can be heavy, and it is made of metal," said Isabelle.

"We observed that what is in the box is small, heavier than a pair of scissors, and made of metal," continued Marcus.

Analyzing Hypotheses "What other possible explanations fit with what you observed?" asked Ms. Clark.

"Well, it has to be a stapler," said Enrique.

"What if it isn't?" asked Ms. Clark. "Maybe you're overlooking explanations because your minds are made up. A good scientist keeps an open mind to every idea and explanation. What if you learn new information that doesn't fit with your original hypothesis? What new information could you gather to verify or disprove your hypothesis?"

"Do you mean a test or something?" asked Marcus.

"I know," said Enrique, "We could get an empty shoe box that is the same size as the mystery box and put a stapler in it. Then we could shake it and see whether it feels and sounds the same." Enrique's test is shown in **Figure 9.**

Making a Prediction "If your hypothesis is correct, what would you expect to happen?" asked Ms. Clark.

"Well, it would be about the same weight and it would slide around a little, just like the other box," said Enrique.

"It would have that same metallic sound when we shake it," said Marcus.

"So, you predict that the test box will feel and sound the same as your mystery box. Go ahead and try it," said Ms. Clark.

Figure 9
Comparing the known information with the unknown information can be valuable even though you cannot see what is inside the closed box.

Testing the Hypothesis Ms. Clark gave the class an empty shoe box that appeared to be identical to the mystery box. Isabelle found a metal stapler. Enrique put the stapler in the box and taped the box closed. Marcus shook the box.

"The stapler does slide around but it feels just a little heavier than what's inside the mystery box," said Marcus. "What do you think?" he asked Isabelle as he handed her the box.

"It is heavier," said Isabelle "and as hard as I shake it, I can't get a metallic sound. What if we find the mass of both boxes? Then we'll know the exact mass difference between the two."

Using a balance, as shown in **Figure 10,** the class found that the test box had a mass of 410 g, and the mystery box had a mass of 270 g.

Figure 10
Laboratory balances are used to find the mass of objects.

Organizing Your Findings "Okay. Now you have some new information," said Ms. Clark. "But before you draw any conclusions, let's organize what we know. Then we'll have a summary of our observations and can refer back to them when we are drawing our conclusions."

"We could make a chart of our observations in our Science Journals," said Marcus.

"We could compare the observations of the mystery box with the observations of the test box," said Isabelle. The chart that the class made is shown in **Table 1.**

Table 1 Observation Chart

Questions	Mystery Box	Our Box
Does it roll or slide?	It slides and appears to be flat.	It slides and appears to be flat.
Does it make any sounds?	It makes a metallic sound when it strikes the sides of the box.	The stapler makes a thudding sound when it strikes the sides of the box.
Is the mass evenly distributed in the box?	No. The object doesn't completely fill the box.	No. The mass of the stapler is unevenly distributed.
What is the mass of the box?	270 g	410 g

Figure 11
Observations can be used to draw inferences. *Looking at both of these photos, what do you infer has taken place?*

Drawing Conclusions

"What have you learned from your investigation so far?" asked Ms. Clark.

"The first thing that we learned was that our hypothesis wasn't correct," answered Marcus.

"Would you say that your hypothesis was entirely wrong?" asked Ms. Clark.

"The boxes don't weigh the same, and the box with the stapler doesn't make the same sound as the mystery box. But there could be a difference in the kind of stapler in the box. It could be a different size or made of different materials."

"So you infer that the object in the mystery box is not exactly the same type of stapler, right?" asked Ms. Clark.

"What does *infer* mean?" asked Isabelle.

"To **infer** something means to draw a conclusion based on what you observe," answered Ms. Clark.

"So we inferred that the things in the boxes had to be different because our observations of the two boxes are different," said Marcus.

"I guess we're back to where we started," said Enrique. "We still don't know what's in the mystery box."

"Do you know more than you did before you started?" asked Ms. Clark.

"We eliminated one possibility," Isabelle added.

"Yes. We inferred that it's not a stapler, at least not like the one in the test box," said Marcus.

"So even if your observations don't support your hypothesis, you know more than you did when you started," said Ms. Clark.

Continuing to Learn "So when do we get to open the box and see what it is?" asked Marcus.

"Let me ask you this," said Ms. Clark. "Do you think scientists always get a chance to look inside to see if they are right?"

"If they are studying something too big or too small to see, I guess they can't," replied Isabelle. "What do they do in those cases?"

"As you learned, your first hypothesis might not be supported by your investigation. Instead of giving up, you continue to gather information by making more observations, making new hypotheses, and by investigating further. Some scientists have spent lifetimes researching their questions. Science takes patience and persistence," said Ms. Clark.

Communicating Your Findings A big part of science is communicating your findings. It is not unusual for one scientist to continue the work of another or to try to duplicate the work of another scientist. It is important for scientists to communicate to others not only the results of the investigation, but also the methods by which the investigation was done. Scientists often publish reports in journals, books, and on the Internet to show other scientists the work that was completed. They also might attend meetings where they make speeches about their work. Scientists from around the world learn from each other, and it is important for them to exchange information freely.

Like the science-fair student in **Figure 12** demonstrates, an important part of doing science is the ability to communicate methods and results to others.

Figure 12
Books, presentations, and meetings are some of the many ways people in science communicate their findings.

 *Why do scientists share information?*

Problem-Solving Activity

MATH TEKS
6.1 A; 6.2 B

How can you use a data table to analyze and present data?

Suppose you were given the average temperatures in a city for the four seasons in 1997, 1998, and 1999: spring 1997 was 11°C; summer 1997 was 25°C; fall 1997 was 5°C; winter 1997 was −5°C; spring 1998 was 9°C; summer 1998 was 36°C; fall 1998 was 10°C; winter 1998 was −3°C; spring 1999 was 10°C; summer 1999 was 30°C; fall 1999 was 9°C; and winter 1999 was −2°C. How can you tell in which of the years each season had its coldest average?

Seasonal Temperatures (°C)			
	1997	1998	1999
Spring			
Summer			
Fall			
Winter			

Identifying the Problem
The information that is given is not in a format that is easy to see at a glance. It would be more helpful to put it in a table that allows you to compare the data.

Solving the Problem
1. Create a table with rows for seasons and columns for the years. Now insert the values you were given. You should be able to see that the four coldest seasons were spring 1998, summer 1997, fall 1997, and winter 1997.
2. Use your new table to find out which season had the greatest difference in temperatures over the three years from 1997 through 1999.
3. What other observations or comparisons can you make from the table you've created on seasonal temperatures?

SCIENCE Online

Research Visit the Glencoe Science Web site at **tx.science.glencoe.com** for information on variables and constants. Make a poster showing the differences between these two parts of a reliable investigation.

Experiments

Different types of questions call for different types of investigations. Ms. Clark's class made many observations about their mystery box and about their test box. They wanted to know what was inside. To answer their question, building a model—the test box—was an effective way to learn more about the mystery box. Some questions ask about the effects of one factor on another. One way to investigate these kinds of questions is by doing a controlled experiment. A **controlled experiment** involves changing one factor and observing its effect on another while keeping all other factors constant.

Variables and Constants Imagine a race in which the lengths of the lanes vary. Some lanes are 102 m long, some are 98 m long, and a few are 100 m long. When the first runner crosses the finish line, is he or she the fastest? Not necessarily. The lanes in the race have different lengths.

Variables are factors that can be changed in an experiment. Reliable experiments, like the race shown in **Figure 13,** attempt to change one variable and observe the effect of this change on another variable. The variable that is changed in an experiment is called the independent variable. The dependent variable changes as a result of a change in the independent variable. It usually is the dependent variable that is observed in an experiment. Scientists attempt to keep all other variables constant—or unchanged.

The variables that are not changed in an experiment are called **constants.** Examples of constants in the race include track material, wind speed, and distance. This way it is easier to determine exactly which variable is responsible for the runners' finish times. In this race, the runners' abilities were varied. The runners' finish times were observed.

Figure 13
The 400-m race is an example of a controlled experiment. The distance, track material, and wind speed are constants. The runners' abilities and their finish times are varied.

Figure 14
Safety is the most important aspect of any investigation.

Laboratory Safety

In your science class, you will perform many types of investigations. However, performing scientific investigations involves more than just following specific steps. You also must learn how to keep yourself and those around you safe by obeying the safety symbol warnings, shown in **Figure 15.**

In a Laboratory When scientists work in a laboratory, as shown in **Figure 14,** they take many safety precautions.

The most important safety advice in a science lab is to think before you act. Always check with your teacher several times in the planning stage of any investigation. Also make sure you know the location of safety equipment in the laboratory room and how to use this equipment, including the eyewashes, thermal mitts, and fire extinguisher.

Good safety habits include the following suggestions. Before conducting any investigation, find and follow all safety symbols listed in your investigation. You always should wear an apron and goggles to protect yourself from chemicals, flames, and pointed objects. Keep goggles on until activity, cleanup, and handwashing are complete. Always slant test tubes away from yourself and others when heating them. Never eat, drink, or apply makeup in the lab. Report all accidents and injuries to your teacher and always wash your hands after working with lab materials.

In the Field Investigations also take place outside the lab, in streams, farm fields, and other places. Scientists must follow safety regulations there, as well, such as wearing eye goggles and any other special safety equipment that is needed. Never reach into holes or under rocks. Always wash your hands after you've finished your field work.

 Eye Safety
 Clothing Protection
 Disposal
 Biological
 Extreme Temperature
 Sharp Object
 Fume
 Irritant
 Toxic
 Animal Safety
 Open Flame
 Electrical
 Chemical
 Radioactivity

Figure 15
Safety symbols are present on nearly every investigation you will do this year. *What safety symbols are on the lab the student is preparing to do in **Figure 14?***

Figure 16
Accidents are not planned. Safety precautions must be followed to prevent injury.

Why have safety rules? Doing science in the class laboratory or in the field can be much more interesting than reading about it. However, safety rules must be strictly followed, so that the possibility of an accident greatly decreases. However, you can't predict when something will go wrong.

Think of a person taking a trip in a car. Most of the time when someone drives somewhere in a vehicle, an accident, like the one shown in **Figure 16,** does not occur. But to be safe, drivers and passengers always should wear safety belts. Likewise, you always should wear and use appropriate safety gear in the lab—whether you are conducting an investigation or just observing. The most important aspect of any investigation is to conduct it safely.

Section 2 Assessment

1. What are four steps scientific investigations often follow?
2. Is a hypothesis as firm as a theory? Explain.
3. What is the difference between an inference and an observation?
4. Why is it important always to use the proper safety equipment?
5. **Think Critically** You are going to use bleach in an investigation. Bleach can irritate your skin, damage your eyes, and stain your clothes. What safety symbols should be listed with this investigation? Explain.

Skill Builder Activities

6. **Drawing Conclusions** While waiting outside your classroom door, the bell rings for school to start. According to your watch, you still have 3 min to get to your classroom. Based on these observations, what can you conclude about your watch? **For more help, refer to the** Science Skill Handbook.
7. **Using a Word Processor** Describe the different types of safety equipment you should use if you are working with a flammable liquid in the lab. **For more help, refer to the** Technology Skill Handbook.

SECTION 3
Models in Science

Why are models necessary?

Just as you can take many different paths in an investigation, you can test a hypothesis in many different ways. Ms. Clark's class tested their hypothesis by building a model of the mystery box. A model is one way to test a hypothesis. In science, a **model** is any representation of an object or an event used as a tool for understanding the natural world.

Models can help you visualize, or picture in your mind, something that is difficult to see or understand. Ms. Clark's class made a model because they couldn't see the item inside the box. Models can be of things that are too small or too big to see. They also can be of things that can't be seen because they don't exist anymore or they haven't been created yet. Models also can show events that occur too slowly or too quickly to see. **Figure 17** shows different kinds of models.

As You Read

What You'll Learn
- **Describe** various types of models.
- **Discuss** limitations of models.

Vocabulary
model

Why It's Important
Models can be used to help understand difficult concepts.

Figure 17
Models help scientists visualize and study complex things and things that can't be seen.

Solar system model

Prototype model

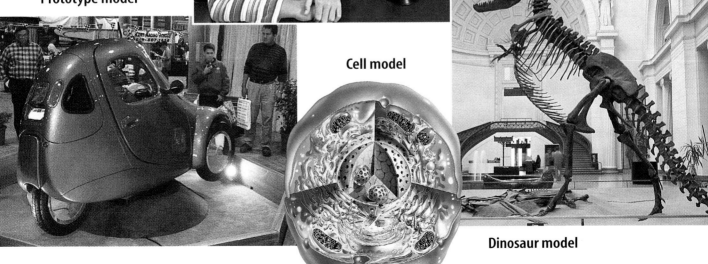

Cell model

Dinosaur model

Types of Models

Most models fall into three basic types—physical models, computer models, and idea models. Depending on the reason that a model is needed, scientists can choose to use one or more than one type of model.

Physical Models Models that you can see and touch are called physical models. Examples include things such as a tabletop solar system, a globe of Earth, a replica of the inside of a cell, or a gumdrop-toothpick model of a chemical compound. Models show how parts relate to one another. They also can be used to show how things appear when they change position or how they react when an outside force acts on them.

Computer Models Computer models are built using computer software. You can't touch them, but you can view them on a computer screen. Some computer models can model events that take a long time or take place too quickly to see. For example, a computer can model the movement of large plates in the Earth and might help predict earthquakes.

Computers also can model motions and positions of things that would take hours or days to calculate by hand or even using a calculator. They can also predict the effect of different systems or forces. **Figure 18** shows how computer models are used by scientists to help predict the weather based on the motion of air currents in the atmosphere.

✓ **Reading Check** *What can computer models do?*

Earth Science INTEGRATION

A basic type of map used to represent an area of land is the topographic map. It shows the natural features of the land, in addition to artificial features such as political boundaries. Draw a topographic map of your classroom in your Science Journal. Indicate the various heights of chairs, desks, and cabinets, with different colors.

Figure 18
A weather map is a computer model showing weather patterns over large areas. Scientists can use this information to predict the weather and to alert people to potentially dangerous weather on the way.

Figure 19
Models can be created using various types of tools.

Idea Models Some models are ideas or concepts that describe how someone thinks about something in the natural world. Albert Einstein is famous for his theory of relativity, which involves the relationship between matter and energy. One of the most famous models Einstein used for this theory is the mathematical equation $E = mc^2$. This explains that mass, m, can be changed into energy, E. Einstein's idea models never could be built as physical models, because they are basically ideas.

Making Models

The process of making a model is something like a sketch artist at work, as shown in **Figure 19.** The sketch artist attempts to draw a picture from the description given by someone. The more detailed the description is, the better the picture will be. Like a scientist who studies data from many sources, the sketch artist can make a sketch based on more than one person's observation. The final sketch isn't a photograph, but if the information is accurate, the sketch should look realistic. Scientific models are made much the same way. The more information a scientist gathers, the more accurate the model will be. The process of constructing a model of King Tutankhamun, who lived more than 3,000 years ago, is shown in **Figure 20.**

✓ **Reading Check** *How are sketches like scientific models?*

Using Models

When you think of a model, you might think of a model airplane or a model of a building. Not all models are for scientific purposes. You use models, and you might not realize it. Drawings, maps, recipes, and globes are all examples of models.

Thinking Like a Scientist

Procedure
1. Pour 15 mL of **water** into a **test tube.**
2. Slowly pour 5 mL of **vegetable oil** into the test tube.
3. Add two drops of **food coloring** and observe the liquid for 5 min.

Analysis
1. Record your observations of the test tube's contents before and after the oil and the food coloring were added to it.
2. Infer a scientific explanation for your observations.

SECTION 3 Models in Science **23**

NATIONAL GEOGRAPHIC VISUALIZING THE MODELING OF KING TUT

Figure 20

More than 3,000 years ago, King Tutankhamun ruled over Egypt. His reign was a short one, and he died when he was just 18. In 1922, his mummified body was discovered, and in 1983 scientists recreated the face of this most famous of Egyptian kings. Some of the steps in building the model are shown here.

This is the most familiar image of the face of King Tut—the gold funerary mask that was found covering his skeletal face.

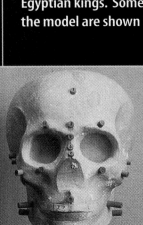

A First, a scientist used measurements and X rays to create a cast of the young king's skull. Depth markers (in red) were then glued onto the skull to indicate the likely thickness of muscle and other tissue.

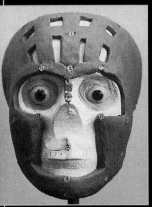

B Clay was applied to fill in the area between the markers.

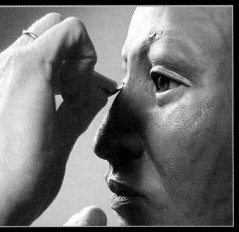

C Next, the features were sculpted. Here, eyelids are fashioned over inlaid prosthetic, or artificial, eyes.

D When this model of King Tut's face was completed, the long-dead ruler seemed to come to life.

24 **CHAPTER 1** The Nature of Science

Models Communicate Some models are used to communicate observations and ideas to other people. Often, it is easier to communicate ideas you have by making a model instead of writing your ideas in words. This way others can visualize them, too.

Models Test Predictions Some models are used to test predictions. Ms. Clark's class predicted that a box with a stapler in it would have characteristics similar to their mystery box. To test this prediction, the class made a model. Automobile and airplane engineers use wind tunnels to test predictions about how air will interact with their products.

Models Save Time, Money, and Lives Other models are used because working with and testing a model can be safer and less expensive than using the real thing. Some of these models are shown in **Figure 21**. For example, crash-test dummies are used in place of people when testing the effects of automobile crashes. To help train astronauts in the conditions they will encounter in space, NASA has built a special airplane. This airplane flies in an arc that creates the condition of weightlessness for 20 to 25 seconds. Making several trips in the airplane is easier, safer, and less expensive than making a trip into space.

Figure 21
Models are a safe and relatively inexpensive way to test ideas.

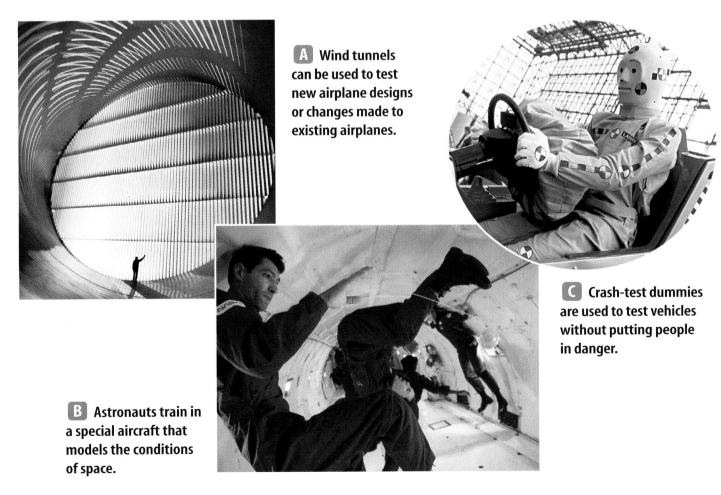

A Wind tunnels can be used to test new airplane designs or changes made to existing airplanes.

C Crash-test dummies are used to test vehicles without putting people in danger.

B Astronauts train in a special aircraft that models the conditions of space.

Figure 22
The model of Earth's solar system changed as new information was gathered.

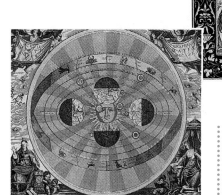

A An early model of the solar system had Earth in the center with everything revolving around it.

B Later on, a new model had the Sun in the center with everything revolving around it.

Limitations of Models

The solar system is too large to be viewed all at once, so models are made to understand it. Many years ago, scientists thought that Earth was the center of the universe and the sky was a blanket that covered the planet.

Later, through observation, it was discovered that the objects you see in the sky are the Sun, the Moon, stars, and other planets. This new model explained the solar system differently. Earth was still the center, but everything else orbited it.

Models Change Still later, through more observation, it was discovered that the Sun is the center of the solar system. Earth, along with the other planets, orbits the Sun. In addition, it was discovered that other planets also have moons that orbit them. A new model, shown in **Figure 22B,** was developed to show this.

Earlier models of the solar system were not meant to be misleading. Scientists made the best models they could with the information they had. More importantly, their models gave future scientists information to build upon. Models are not necessarily perfect, but they provide a visual tool to learn from.

Section 3 Assessment

1. What type of models can be used to model weather? How are they used?
2. How are models used in science?
3. How do consumer product testing services use models to ensure the safety of the final products produced?
4. Make a table describing three types of models, their advantages and limitations.
5. **Think Critically** Explain how some models are better than others for certain situations.

Skill Builder Activities

6. **Concept Mapping** Develop a concept map to explain models and their uses in science. How is this concept map a model? **For more help, refer to the** Science Skill Handbook.
7. **Using Proportions** On a map of a state, the scale shows that 1 cm is approximately 5 km. If the distance between two cities is 1.7 cm on the map, how many kilometers separate them? **For more help, refer to the** Math Skill Handbook.

SECTION 4
Evaluating Scientific Explanation

Believe it or not?

Look at the photo in **Figure 23.** Do you believe what you see? Do you believe everything you read or hear? Think of something that someone told you that you didn't believe. Why didn't you believe it? Chances are you looked at the facts you were given and decided that there wasn't enough proof to make you believe it. What you did was evaluate, or judge the reliability of what you heard. When you hear a statement, you ask the question "How do you know?" If you decide that what you are told is reliable, then you believe it. If it seems unreliable, then you don't believe it.

Critical Thinking When you evaluate something, you use critical thinking. **Critical thinking** means combining what you already know with the new facts that you are given to decide if you should agree with something. You can evaluate an explanation by breaking it down into two parts. First you can look at and evaluate the observations. Based upon what you know, are the observations accurate? Then you can evaluate the inferences—or conclusions made about the observations. Do the conclusions made from the observations make sense?

As You Read

What You'll Learn
- Evaluate scientific explanations.
- Evaluate promotional claims.

Vocabulary
critical thinking

Why It's Important
Evaluating scientific claims can help you make better decisions.

Figure 23
In science, observations and inferences are not always agreed upon by everyone. *Do you see the same things your classmates see in this photo?*

27

Table 2 Favorite Foods

People's Preference	Tally	Frequency
pepperoni pizza	|||| |||| |||| |||| |||| |||| |||| ||	37
hamburgers with ketchup	|||| |||| |||| |||| |||| |||	28

Figure 24
These scientists are writing down their observations during their investigation rather than waiting until they are back on land. *Do you think this will increase or decrease the reliability of their data?*

Evaluating the Data

A scientific investigation always contains observations—often called data. These might be descriptions, tables, graphs, or drawings. When evaluating a scientific claim, you might first look to see whether any data are given. You should be cautious about believing any claim that is not supported by data.

Are the data specific? The data given to back up a claim should be specific. That means they need to be exact. What if your friend tells you that many people like pizza more than they like hamburgers? What else do you need to know before you agree with your friend? You might want to hear about a specific number of people rather than unspecific words like *many* and *more*. You might want to know how many people like pizza more than hamburgers. How many people were asked about which kind of food they liked more? When you are given specific data, a statement is more reliable and you are more likely to believe it. An example of data in the form of a frequency table is shown in **Table 2**. A frequency table shows how many times types of data occur. Scientists must back up their scientific statements with specific data.

Take Good Notes Scientists must take thorough notes at the time of an investigation, as the scientists shown in **Figure 24** are doing. Important details can be forgotten if you wait several hours or days before you write down your observations. It is also important for you to write down every observation, including ones that you don't expect. Often, great discoveries are made when something unexpected happens in an investigation.

Your Science Journal During this course, you will be keeping a science journal. You will write down what you do and see during your investigations. Your observations should be detailed enough that another person could read what you wrote and repeat the investigation exactly as you performed it. Instead of writing "the stuff changed color," you might say "the clear liquid turned to bright red when I added a drop of food coloring." Detailed observations written down during an investigation are more reliable than sketchy observations written from memory. Practice your observation skills by describing what you see in **Figure 25.**

Can the data be repeated? If your friend told you he could hit a baseball 100 m, but couldn't do it when you were around, you probably wouldn't believe him. Scientists also require repeatable evidence. When a scientist describes an investigation, as shown in **Figure 26,** other scientists should be able to do the investigation and get the same results. The results must be repeatable. When evaluating scientific data, look to see whether other scientists have repeated the data. If not, the data might not be reliable.

Evaluating the Conclusions

When you think about a conclusion that someone has made, you can ask yourself two questions. First, does the conclusion make sense? Second, are there any other possible explanations? Suppose you hear on the radio that your school will be running on a two-hour delay in the morning because of snow. You look outside. The roads are clear of snow. Does the conclusion that snow is the cause for the delay make sense? What else could cause the delay? Maybe it is too foggy or icy for the buses to run. Maybe there is a problem with the school building. The original conclusion is not reliable unless the other possible explanations are proven unlikely.

Figure 25
Detailed observations are important in order to get reliable data. *Write down at least five sentences describing what you see in this photo.*

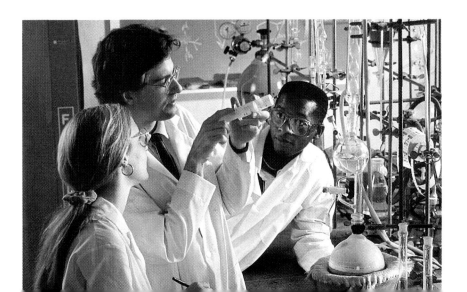

Figure 26
Working together is an important part of science. Several scientists must repeat an experiment and obtain the same results before data are considered reliable.

Figure 27
All material should be read with an analytical mind. *What does this advertisement mean?*

Evaluating Promotional Materials

Scientific processes are not used only in the laboratory. Suppose you saw an advertisement in the newspaper like the one in **Figure 27.** What would you think? First, you might ask, "Does this make sense?" It seems unbelievable. You would probably want to hear some of the scientific data supporting the claim before you would believe it. How was this claim tested? How is the amount of wrinkling in skin measured? You might also want to know if an independent laboratory repeated the results. An independent laboratory is one that is not related in any way to the company that is selling the product or service. It has nothing to gain from the sales of the product. Results from an independent laboratory usually are more reliable than results from a laboratory paid by the selling company. Advertising materials are designed to get you to buy a product or service. It is important that you carefully evaluate advertising claims and the data that support them before making a quick decision to spend your money.

Section 4 Assessment

1. Explain what is meant by critical thinking and give an example.
2. What types of scientific claims should be verified?
3. Name two parts of a scientific explanation. Give examples of ways to evaluate the reliability of each part.
4. How can vague claims in advertising be misleading?
5. **Think Critically** An advertisement on a food package claims it contains Glistain, a safe, taste enhancer. Make a list of at least ten questions you would ask when evaluating the claim.

Skill Builder Activities

6. **Classifying** Watch three television commercials and read three magazine advertisements. In your Science Journal, record the claims that each advertisement made. Classify each claim as being vague, misleading, reliable, and/or scientific. **For more help, refer to the** Science Skill Handbook.

7. **Researching Information** Visit your school library and choose an article from a news magazine. Pick one that deals with a scientific claim. Learn more about the claim and evaluate it using the scientific process. **For more help, refer to the** Science Skill Handbook.

Activity

What is the right answer?

Scientists sometimes develop more than one explanation for observations. Can more than one explanation be correct? Do scientific explanations depend on judgment?

What You'll Investigate
Can more than one explanation apply to the same observation?

Materials
cardboard mailing tubes   length of rope
*empty shoe boxes   scissors
*Alternate Materials

Goals
- **Make a hypothesis** to explain an observation.
- **Construct** a model to support your hypothesis.
- **Refine** your model based on testing.

Safety Precautions
WARNING: *Be careful when punching holes with sharp tools.*

Procedure

1. You will be shown a cardboard tube with four ropes coming out of it, one longer than the others. Your teacher will show you that when any of the three short ropes—A, C, or D—are pulled, the longer rope, B, gets shorter. Pulling on rope B returns the other ropes to their original lengths.

2. Make a hypothesis as to how the teacher's model works.

3. **Sketch** a model of a tube with ropes based on your hypothesis. Check your sketch to be sure that your model will do what you expect. Revise your sketch if necessary.

4. Using a cardboard tube and two lengths of rope, build a model according to your design. Test your model by pulling each of the ropes. If it does not perform as planned, modify your hypothesis and your model to make it work like your teacher's model.

Conclude and Apply

1. **Compare** your model with those made by others in your class.

2. Can more than one design give the same result? Can more than one explanation apply to the same observation? Explain.

3. Without opening the tube, can you tell which model is exactly like your teacher's?

Communicating Your Data

Make a display of your working model. Include sketches of your designs. **For more help, refer to the** Science Skill Handbook.

Activity

Identifying Parts of an Investigation

Science investigations contain many parts. How can you identify the various parts of an investigation? In addition to variables and constants, many experiments contain a control. A control is one test, or trial, where everything is held constant. A scientist compares the control trial to the other trials.

What You'll Investigate
What are the various parts of an experiment to test which fertilizer helps a plant grow best?

Materials
description of fertilizer experiment

Goals
- **Identify** parts of an experiment.
- **Identify** constants, variables, and controls in the experiment.
- **Graph** the results of the experiment and draw appropriate conclusions.

Procedure

1. **Read** the description of the fertilizer experiment.
2. **List** factors that remained constant in the experiment.
3. **Identify** any variables in the experiment.
4. **Identify** the control in the experiment.
5. **Identify** one possible hypothesis that the gardener could have tested in her investigation.
6. **Describe** how the gardener went about testing her hypothesis using different types of fertilizers.
7. **Graph** the data that the gardener collected in a line graph.

32 CHAPTER 1 The Nature of Science

Using Scientific Methods

A gardener was interested in helping her plants grow faster. When she went to the nursery, she found three fertilizers available for her plants. One of those fertilizers, fertilizer A, was recommended to her. However, she decided to conduct a test to determine which of the three fertilizers, if any, helped her plants grow fastest. The gardener planted four seeds, each in a separate pot. She used the same type of pot and the same type of soil in each pot. She fertilized one seed with fertilizer A, one with fertilizer B, and one with fertilizer C. She did not fertilize the fourth seed. She placed the four pots near one another in her garden. She made sure to give each plant the same amount of water each day. She measured the height of the plants each week and recorded her data. After eight weeks of careful observation and record keeping, she had the following table of data.

Plant Height (cm)				
Week	Fertilizer A	Fertilizer B	Fertilizer C	No Fertilizer
1	0	0	0	0
2	2	4	1	1
3	5	8	5	4
4	9	13	8	7
5	14	18	12	10
6	20	24	15	13
7	27	31	19	16
8	35	39	22	20

Conclude and Apply

1. **Describe** the results indicated by your graph. What part of an investigation have you just done?
2. Based on the results in the table and your graph, which fertilizer do you think the gardener should use if she wants her plants to grow the fastest? What part of an investigation have you just done?
3. Suppose the gardener told a friend who also grows these plants about her results. What is this an example of?
4. Suppose fertilizer B is much more expensive than fertilizers A and C. Would this affect which fertilizer you think the gardener should buy? Why or why not?
5. Does every researcher need the same hypothesis for an experiment? What is a second possible hypothesis for this experiment (different from the one you wrote in step 5 in the Procedure section)?
6. Did the gardener conduct an adequate test of her hypothesis? Explain why or why not.

Compare your conclusions with those of other students in your class. **For more help, refer to the** Science Skill Handbook.

TIME SCIENCE AND HISTORY

SCIENCE CAN CHANGE THE COURSE OF HISTORY!

Women in Science

Nobel prizes are given every year in many areas of science.

Is your family doctor a man or a woman? To your great-grandparents, such a question would likely have seemed odd. Why? Because 100 years ago, there were only a handful of women in scientific fields such as medicine. Women then weren't encouraged to study science as they are today. But that does not mean that there were no female scientists back in your great-grandparents' day. Many women managed to overcome great barriers and, like the more recent Nobel prizewinners featured in this article, made discoveries that changed the world.

Maria Goeppert Mayer

Dr. Maria Goeppert Mayer won the Nobel Prize in Physics in 1963 for her work on the structure of an atom. An atom is made up of protons, neutrons, and electrons. The protons and neutrons exist in the nucleus, or center, of the atom. The electrons orbit the nucleus in shells. Mayer proposed a similar shell model for the protons and neutrons inside the nucleus. This model greatly increased human understanding of atoms, which make up all forms of matter. About the Nobel prize, she said, "To my surprise, winning the prize wasn't half as exciting as doing the work itself. That was the fun—seeing it work out."

Rita Levi-Montalcini

In 1986, the Nobel Prize in Medicine went to Dr. Rita Levi-Montalcini, a biologist from Italy, for her discovery of growth factors. Growth factors regulate the growth of cells and organs in the body. Because of her work, doctors are better able to understand why tumors form and wounds heal. Although she was a bright student, Dr. Levi-Montalcini almost did not go to college. Rita's father believed that it was not acceptable for women to have a professional career. Rita asked for her father's permission to prepare for the medical school entrance exams, and he agreed. After less than one year of study, Rita scored highest in the rankings on the entrance exams and was one of seven women in a school of over three hundred medical students.

Rosalyn Sussman Yalow

In 1977, Dr. Rosalyn Sussman Yalow, a nuclear physicist, was awarded the Nobel Prize in Medicine for discovering a way to measure substances in the blood that are present in tiny amounts, such as hormones and drugs.

The discovery made it possible for doctors to diagnose problems that they could not detect before. Upon winning the prize, Yalow spoke out against discrimination of women. She said, "The world cannot afford the loss of the talents of half its people if we are to solve the many problems which beset us."

CONNECTIONS Research Write short biographies about recent Nobel prizewinners in physics, chemistry, and medicine. In addition to facts about their lives, explain why the scientists were awarded the prize. How did their discoveries impact their scientific fields or people in general?

Online For more information, visit tx.science.glencoe.com

Chapter 1 Study Guide

Reviewing Main Ideas

Section 1 What is science?

1. Science is a way of learning more about the natural world. It can provide only possible explanations for questions.

2. A scientific law describes a pattern in nature.

3. A scientific theory attempts to explain patterns in nature.

4. Systems are a collection of structures, cycles, and processes that interact. *Can you identify structures, cycles, and processes in this system?*

5. Science can be divided into three branches—life science, Earth science, and physical science.

6. Technology is the application of science.

Section 2 Science in Action

1. Science involves using a collection of skills.

2. A hypothesis is a reasonable guess based on what you know and observe.

3. An inference is a conclusion based on observation.

4. Controlled experiments involve changing one variable while keeping others constant.

5. You should always obey laboratory safety symbols. You should also wear and use appropriate gear in the laboratory.

Section 3 Models in Science

1. A model is any representation of an object or an event used as a tool for understanding the natural world.

2. There are physical, computer, and idea models.

3. Models can communicate ideas; test predictions; and save time, money, and lives. *How is this model used?*

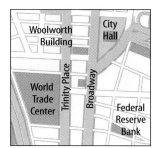

4. Models change as more information is learned.

Section 4 Evaluating Scientific Explanations

1. An explanation can be evaluated by looking at the observations and the conclusions in an experiment.

2. Reliable data are specific and repeatable by other scientists.

3. Detailed notes must be taken *during* an investigation.

4. To be reliable, a conclusion must make sense and be the most likely explanation.

After You Read

Without looking at the chapter or at your Foldable, write what you learned about science on the *Learned* fold of your Know-Want-Learn Study Fold.

Chapter 1 Study Guide

Visualizing Main Ideas

Complete the following concept map.

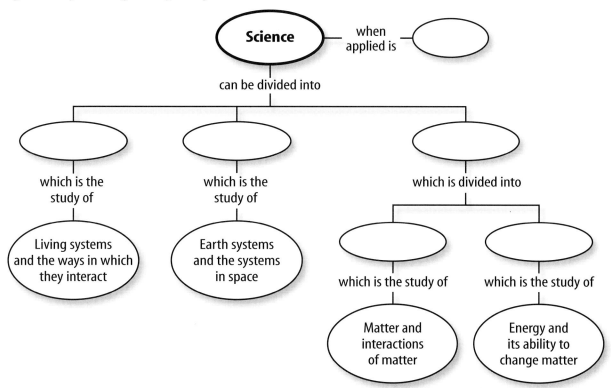

Vocabulary Review

Vocabulary Words

a. constant
b. controlled experiment
c. critical thinking
d. Earth science
e. hypothesis
f. infer
g. life science
h. model
i. physical science
j. science
k. scientific law
l. scientific theory
m. system
n. technology
o. variable

Using Vocabulary

Explain the relationship between the words in the following sets.

1. hypothesis, scientific theory
2. constant, variable
3. science, technology
4. science, system
5. Earth science, physical science
6. critical thinking, infer
7. scientific law, observation
8. model, system
9. controlled experiment, variable
10. scientific theory, scientific law

 **Study Tip**

Make a note of anything you don't understand so that you'll remember to ask your teacher about it.

CHAPTER STUDY GUIDE 37

Chapter 1 Assessment & TEKS Review

Checking Concepts

Choose the word or phrase that best answers the question.

1. What does infer mean?
   A) make observations   C) replace
   B) draw a conclusion   D) test

2. Which is an example of technology?
   A) a squirt bottle   C) a cat
   B) a poem   D) physical science

3. Which branch of science includes the study of weather?
   A) life science   C) physical science
   B) Earth science   D) engineering

4. What explains something that takes place in the natural world?
   A) scientific law   C) scientific theory
   B) technology   D) experiments

5. Which of the following cannot protect you from splashing acid?
   A) goggles   C) fire extinguisher
   B) apron   D) gloves

6. If the results from your investigation do not support your hypothesis, what should you do?
   A) Do nothing.
   B) You should repeat the investigation until it agrees with the hypothesis.
   C) Modify your hypothesis.
   D) Change your data to fit your hypothesis.

7. Which of the following is NOT an example of a scientific hypothesis?
   A) Earthquakes happen because of stresses along continental plates.
   B) Some animals can detect ultrasound frequencies caused by earthquakes.
   C) Paintings are prettier than sculptures.
   D) Lava takes different forms depending on how it cools.

8. An airplane model is an example of what type of model?
   A) physical   C) idea
   B) computer   D) mental

9. Using a computer to make a three-dimensional picture of a building is a type of which of the following?
   A) model   C) constant
   B) hypothesis   D) variable

10. Which of the following increases the reliability of a scientific explanation?
    A) vague statements
    B) notes taken after an investigation
    C) repeatable data
    D) several likely explanations

Thinking Critically

11. Is evaluating a play in English class science? Explain.

12. Why is it a good idea to repeat an experiment a few times and compare results? Explain.

13. How is using a rock hammer an example of technology? Explain.

14. Why is it important to record and measure data accurately during an experiment?

15. What type of model would most likely be used in classrooms to help young children learn science? Explain.

Developing Skills

16. **Comparing and Contrasting** How are scientific theories and laws similar? How are they different?

Chapter 1 Assessment

17. **Drawing Conclusions** When scientists study how well new medicines work, one group of patients receives the medicine. A second group does not. Why?

18. **Forming Hypotheses** Make a hypothesis about the quickest way to get to school in the morning. How could you test your hypothesis?

19. **Making Operational Definitions** How does a scientific law differ from a state law? Give examples of both types of laws.

20. **Making and Using Tables** Mohs hardness scale measures how easily an object can be scratched. The higher the number is, the harder the material is. Use the table below to identify which material is the hardest and which is the softest.

Hardness	
Object	Mohs Scale
copper	3.5
diamond	10
fingernail	2.5
glass	5.5
quartz	7
steel file	6.5

Performance Assessment

21. **Write a Story** Write a story illustrating what science is and how it is used to investigate problems.

TECHNOLOGY

Go to the Glencoe Science Web site at **tx.science.glencoe.com** or use the **Glencoe Science CD-ROM** for additional chapter assessment.

TAKS Practice

Sally and Rafael have just learned about the parts of the solar system in science class. They decided to build a large model to better understand it. *TEKS 6.2 C, D; 6.3 C*

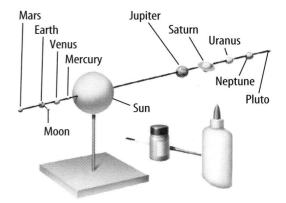

Study the diagram and answer the following questions.

1. According to this information, Rafael and Sally's model of the solar system best represents which kind of scientific model?
   A) idea
   B) computer
   C) physical
   D) realistic

2. According to this model, all of the following are represented EXCEPT _____.
   F) the Sun
   G) the Moon
   H) planets
   J) stars

CHAPTER 2

Science TEKS 6.6 A

Measurement

Does the expression "winning by a nose" mean anything to you? If you have ever "won by a nose," that means the race was close. Sometimes horse races, such as this one, are so close the winner has to be determined by a photograph. But there is more to measure than just how close the race was. How fast did the horse run? Did he break a record? In this chapter, you will learn how scientists measure things like distance, time, volume, and temperature. You also will learn how to use illustrations, pictures, and graphs to communicate measurements.

What do you think?

Science Journal Look at the picture below with a classmate. Discuss what you think this might be. Here's a hint: *How fast did you come up with an answer?* Write your answer or best guess in your Science Journal.

Explore Activity

You make measurements every day. If you want to communicate those measurements to others, how can you be sure that they will understand exactly what you mean? Using vague words without units won't work. Do the Explore Activity below to see the confusion that can result from using measurements that aren't standard.

Measure length

1. As a class, choose six objects to measure in your classroom.
2. Measure each object using the width of your hand and write your measurements in your Science Journal.
3. Compare your measurements to those of your classmates.

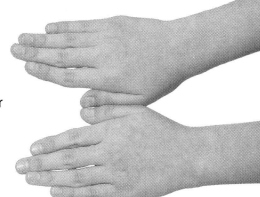

Observe

Is your hand the same width as your classmates' hands? Discuss in your Science Journal why it is better to switch from using hands to using units of measurement that are the same all the time.

Before You Read

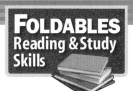

Making an Organizational Study Fold When information is grouped into clear categories, it is easier to understand what you are learning. Before you begin reading, make the following Foldable to help you organize your thoughts about measurements.

1. Place a sheet of paper in front of you so the short side is at the top. Fold the paper in half from the left side to the right side two times. Unfold all the folds.
2. Fold the paper from top to bottom in equal thirds and then in half. Unfold all the folds.
3. Trace over all the fold lines and label the table you created. Label the columns: *Estimate It, Measure It,* and *Round It,* as shown. Label the rows: *Length of* _____ , *Volume of* _____ , *Mass of* _____ , *Temperature of* _____ , and *Rate of* _____ , as shown.
4. Before you read the chapter, select objects to measure and estimate their measurements. As you read the chapter, complete the *Measure It* column.

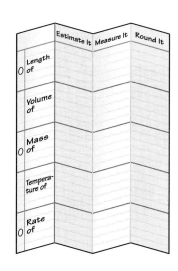

41

SECTION 1

Description and Measurement

As You Read

What You'll Learn
- **Determine** how reasonable a measurement is by estimating.
- **Identify** and use the rules for rounding a number.
- **Distinguish** between precision and accuracy in measurements.

Vocabulary
measurement
estimation
precision
accuracy

Why It's Important
Measurement helps you communicate information and ideas.

Measurement

How would you describe what you are wearing today? You might start with the colors of your outfit, and perhaps you would even describe the style. Then you might mention sizes—size 7 shoes, size 14 shirt. Every day you are surrounded by numbers. **Measurement** is a way to describe the world with numbers. It answers questions such as how much, how long, or how far. Measurement can describe the amount of milk in a carton, the cost of a new compact disc, or the distance between your home and your school. It also can describe the volume of water in a swimming pool, the mass of an atom, or how fast a penguin's heart pumps blood.

The circular device in **Figure 1** is designed to measure the performance of an automobile in a crash test. Engineers use this information to design safer vehicles. In scientific endeavors, it is important that scientists rely on measurements instead of the opinions of individuals. You would not know how safe the automobile is if this researcher turned in a report that said, "Vehicle did fairly well in head-on collision when traveling at a moderate speed." What does "fairly well" mean? What is a "moderate speed?"

Figure 1
This device measures the range of motion of a seat-belted mannequin in a simulated accident.

Figure 2
Accurate measurement of distance and time is important for competitive sports like track and field. *Why wouldn't a clock that measured in minutes be precise enough for this race?*

Describing Events Measurement also can describe events such as the one shown in **Figure 2.** In the 1956 summer Olympics, sprinter Betty Cuthbert of Australia came in first in the women's 200-m dash. She ran the race in 23.4 s. In the 2000 summer Olympics, Marion Jones of the United States won the 100-m dash in a time of 10.75 s. In this example, measurements convey information about the year of the race, its length, the finishing order, and the time. Information about who competed and in what event are not measurements but help describe the event completely.

Estimation

What happens when you want to know the size of an object but you can't measure it? Perhaps it is too large to measure or you don't have a ruler handy. **Estimation** can help you make a rough measurement of an object. When you estimate, you can use your knowledge of the size of something familiar to estimate the size of a new object. Estimation is a skill based on previous experience and is useful when you are in a hurry and exact numbers are not required. Estimation is a valuable skill that improves with experience, practice, and understanding.

 *When should you not estimate a value?*

How practical is the skill of estimation? In many instances, estimation is used on a daily basis. A caterer prepares for each night's crowd based on an estimation of how many will order each entree. A chef makes her prize-winning chili. She doesn't measure the cumin; she adds "just that much." Firefighters estimate how much hose to pull off the truck when they arrive at a burning building.

Chemistry INTEGRATION

A description of matter that does not involve measurement is *qualitative*. For example, water is composed of hydrogen and oxygen. A *quantitative* description uses numbers to describe. For example, one water molecule is composed of one oxygen atom and two hydrogen atoms. Research another compound containing hydrogen and oxygen—hydrogen peroxide. Infer a qualitative and quantitative description of hydrogen peroxide in your Science Journal.

Figure 3
This student is about 1.5 m tall. *Estimate the height of the tree in the photo.*

Using Estimation You can use comparisons to estimate measurements. For example, the tree in **Figure 3** is too tall to measure easily, but because you know the height of the student next to the tree, you can estimate the height of the tree. When you estimate, you often use the word *about*. For example, doorknobs are about 1 m above the floor, a sack of flour has a mass of about 2 kg, and you can walk about 5 km in an hour.

Estimation also is used to check that an answer is reasonable. Suppose you calculate your friend's running speed as 47 m/s. You are familiar with how long a second is and how long a meter is. Think about it. Can your friend really run a 50-m dash in 1 s? Estimation tells you that 47 m/s is unrealistically fast and you need to check your work.

Precision and Accuracy

One way to evaluate measurements is to determine whether they are precise. **Precision** is a description of how close measurements are to each other. Suppose you measure the distance between your home and your school five times with an odometer. Each time, you determine the distance to be 2.7 km. Suppose a friend repeated the measurements and measured 2.7 km on two days, 2.8 km on two days, and 2.6 km on the fifth day. Because your measurements were closer to each other than your friend's measurements, yours were more precise. The term *precision* also is used when discussing the number of decimal places a measuring device can measure. A clock with a second hand is considered more precise than one with only an hour hand.

Mini LAB

Measuring Accurately

Procedure
1. Fill a **400-mL beaker** with **crushed ice**. Add enough **cold water** to fill the beaker.
2. Make three measurements of the temperature of the ice water using a **computer temperature probe.** Remove the computer probe and allow it to warm to room temperature between each measurement. Record the measurements in your **Science Journal.**
3. Repeat step two using an **alcohol thermometer.**

Analysis
1. Average each set of measurements.
2. Which measuring device is more precise? Explain. Can you determine which is more accurate? How?

44 CHAPTER 2 Measurement

Degrees of Precision The timing for Olympic events has become more precise over the years. Events that were measured in tenths of a second 100 years ago are measured to the hundredth of a second today. Today's measuring devices are more precise. **Figure 4** shows an example of measurements of time with varying degrees of precision.

Accuracy When you compare a measurement to the real, actual, or accepted value, you are describing **accuracy.** A watch with a second hand is more precise than one with only an hour hand, but if it is not properly set, the readings could be off by an hour or more. Therefore, the watch is not accurate. On the other hand, measurements of 1.03 m, 1.04 m, and 1.06 m compared to an actual value of 1.05 m is accurate, but not precise. **Figure 5** illustrates the difference between precision and accuracy.

Reading Check *What is the difference between precision and accuracy?*

Figure 4
Each of these clocks provides a different level of precision. *Which of the three could you use to be sure to make the 3:35 bus?*

A Before the invention of clocks, as they are known today, a sundial was used. As the Sun passes through the sky, a shadow moves around the dial.

B For centuries, analog clocks—the kind with a face—were the standard.

C Digital clocks are now as common as analog ones.

SECTION 1 Description and Measurement **45**

NATIONAL GEOGRAPHIC VISUALIZING PRECISION AND ACCURACY

Figure 5

From golf to gymnastics, many sports require precision and accuracy. Archery—a sport that involves shooting arrows into a target—clearly shows the relationship between these two factors. An archer must be accurate enough to hit the bull's-eye and precise enough to do it repeatedly.

A The archer who shot these arrows is neither accurate nor precise—the arrows are scattered all around the target.

C Here we have a winner! All of the arrows have hit the bull's-eye, a result that is both precise and accurate.

B This archer's attempt demonstrates precision but not accuracy—the arrows were shot consistently to the left of the target's center.

 **Health INTEGRATION** Precision and accuracy are important in many medical procedures. One of these procedures is the delivery of radiation in the treatment of cancerous tumors. Because radiation damages cells, it is important to limit the radiation to only the cancerous cells that are to be destroyed. A technique called Stereotactic Radiotherapy (SRT) allows doctors to be accurate and precise in delivering radiation to areas of the brain. The patient makes an impression of his or her teeth on a bite plate that is then attached to the radiation machine. This same bite plate is used for every treatment to position the patient precisely the same way each time. A CAT scan locates the tumor in relation to the bite plate, and the doctors can pinpoint with accuracy and precision where the radiation should go.

Rounding a Measurement Not all measurements have to be made with instruments that measure with great precision like the scale in **Figure 6.** Suppose you need to measure the length of the sidewalk outside your school. You could measure it to the nearest millimeter. However, you probably would need to know the length only to the nearest meter or tenth of a meter. So, if you found that the length was 135.841 m, you could round off that number to the nearest tenth of a meter and still be considered accurate. How would you round this number? To round a given value, follow these steps:

1. Look at the digit to the right of the place being rounded to.
   - If the digit to the right is 0, 1, 2, 3, or 4, the digit being rounded to remains the same.
   - If the digit to the right is 5, 6, 7, 8, or 9, the digit being rounded to increases by one.
2. The digits to the right of the digit being rounded to are deleted if they are also to the right of a decimal. If they are to the left of a decimal, they are changed to zeros.

Look back at the sidewalk example. If you want to round the sidewalk length of 135.841 to the tenths place, you look at the digit to the right of the 8. Because that digit is a 4, you keep the 8 and round it off to 135.8 m. If you want to round to the ones place, you look at the digit to the right of the 5. In this case you have an 8, so you round up, changing the 5 to a 6, and your answer is 136 m.

Research Visit the Glencoe Science Web site at **tx.science.glencoe.com** for more information about measurement. Communicate to your class what you learn.

Figure 6
This laboratory scale measures to the nearest hundredth of a gram.

SECTION 1 Description and Measurement **47**

Precision and Number of Digits When might you need to round a number? Suppose you want to divide a 2-L bottle of soft drink equally among seven people. When you divide 2 by 7, your calculator display reads as shown in **Figure 7.** Will you measure exactly 0.285 714 285 L for each person? No. All you need to know is that each person gets about 0.3 L of soft drink.

Using Precision and Significant Digits The number of digits that truly reflect the precision of a number are called the significant digits or significant figures. They are figured as follows:

- Digits other than zero are always significant.
- Final zeros after a decimal point (6.545600 g) are significant.
- Zeros between any other digits (507.0301 g) are significant.
- Initial zeros (0.0002030 g) are NOT significant.
- Zeros in a whole number (1650) may or may not be significant.
- A number obtained by counting instead of measuring, such as the number of people in a room or the number of meters in a kilometer, has infinite significant figures.

Math Skills Activity

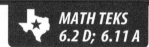

MATH TEKS
6.2 D; 6.11 A

Rounding

Example Problem

The mass of one object is 6.941 g. The mass of a second object is 20.180 g. You need to know these values only to the nearest whole number to solve a problem. What are the rounded values?

Solution

1. *This is what you know:*   mass of first object = 6.941 g
   mass of second object = 20.180 g

2. *This is what you need to know:* the number to the right of the one's place
   first object: 9, second object: 1

3. *This is what you need to use:* digits 0, 1, 2, 3, 4 remain the same
   for digits 5, 6, 7, 8, 9, round up

4. *Solution:* first object: 9 makes the 6 round up = 7
   second object: 1 makes the 0 remain the same = 20

Practice Problem

What are the rounded masses of the objects to the nearest tenth of a unit?

For more help, refer to the Math Skill Handbook.

Following the Rules In the soda example you have an exact number, seven, for the number of people. This number has infinite significant digits. You also have the number two, for how many liters of soda you have. This has only one significant digit.

There are also rules to follow when deciding the number of significant digits in the answer to a calculation. They depend on what kind of calculation you are doing.

- For multiplication and division, you determine the number of significant digits in each number in your problem. The significant digits of your answer are determined by the number with fewer digits.

$$6.14 \times 5.6 = \boxed{34}.384$$
3 digits   2 digits   2 digits

- For addition and subtraction, you determine the place value of each number in your problem. The significant digits of the answer is determined by the number that is least precise.

```
  6.14    to the hundredths
+ 5.6     to the tenths
─────
 11.7 4   to the tenths
```

Therefore, in the soda example you are dividing and the limiting number of digits is determined by the amount of soda, 2 L. There is one significant digit there; therefore, your answer has one.

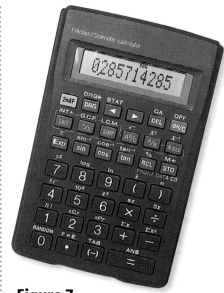

Figure 7
Sometimes considering the size of each digit will help you realize they are unneeded. In this calculation, the seven ten-thousandths of a liter represents just a few drops of soda.

 Reading Check *What determines the number of significant digits in the answer to an addition problem?*

Section 1 Assessment

1. Estimate the distance between your desk and your teacher's desk. Explain the method you used.
2. Measure the height of your desk to the nearest half centimeter.
3. Sarah's garden is 11.72 m long. Round to the nearest tenth of a meter.
4. John's puppy has chewed on his ruler. Will John's measurements be accurate or precise?
5. **Think Critically** Would the sum of 5.7 and 6.2 need to be rounded? Why or why not? Would the sum of 3.28 and 4.1 need to be rounded? Why or why not?

Skill Builder Activities

6. **Using Precision and Significant Digits** Perform the following calculations and express the answer using the correct number of significant digits: 42.35 + 214; 225/12. **For more help, refer to the** Math Skill Handbook.

7. **Communicating** Describe your backpack in your Science Journal. Include in your description one set of qualities that have no measurements, such as color and texture, and one set of measured quantities, such as width and mass. **For more help, refer to the** Science Skill Handbook.

SECTION 1 Description and Measurement **49**

SECTION 2 SI Units

As You Read

What You'll Learn
- Identify the purpose of SI.
- Identify the SI units of length, volume, mass, temperature, time, and rate.

Vocabulary
SI
meter
mass
kilogram
kelvin
rate

Why It's Important
The SI system is used throughout the world, allowing you to measure quantities in the exact same way as other students around the world.

The International System

Can you imagine how confusing it would be if people in every country used different measuring systems? Sharing data and ideas would be complicated. To avoid confusion, scientists established the International System of Units, or **SI**, in 1960 as the accepted system for measurement. It was designed to provide a worldwide standard of physical measurement for science, industry, and commerce. SI units are shown in **Table 1**.

Reading Check Why was SI established?

The SI units are related by multiples of ten. Any SI unit can be converted to a smaller or larger SI unit by multiplying by a power of 10. For example, to rewrite a kilogram measurement in grams, you multiply by 1,000. The new unit is renamed by changing the prefix, as shown in **Table 2**. For example, one millionth of a meter is one *micro*-meter. One thousand grams is one *kilo*gram. **Table 3** shows some common objects and their measurements in SI units.

Table 1 SI Base Units

Quantity	Unit	Symbol
length	meter	m
mass	kilogram	kg
temperature	kelvin	K
time	second	s
electric current	ampere	A
amount of substance	mole	mol
intensity of light	candela	cd

Table 2 SI Prefixes

Prefix	Multiplier
giga-	1,000,000,000
mega-	1,000,000
kilo-	1,000
hecto-	100
deka-	10
[unit]	1
deci-	0.1
centi-	0.01
milli-	0.001
micro-	0.000 001
nano-	0.000 000 001

Length

Length is defined as the distance between two points. Lengths measured with different tools can describe a range of things from the distance from Earth to Mars to the thickness of a human hair. In your laboratory activities, you usually will measure length with a metric ruler or meterstick.

The **meter** (m) is the SI unit of length. One meter is about the length of a baseball bat. The size of a room or the dimensions of a building would be measured in meters. For example, the height of the Washington Monument in Washington, D.C. is 169 m.

Smaller objects can be measured in centimeters (cm) or millimeters (mm). The length of your textbook or pencil would be measured in centimeters. A twenty-dollar bill is 15.5 cm long. You would use millimeters to measure the width of the words on this page. To measure the length of small things such as blood cells, bacteria, or viruses, scientists use micrometers (millionths of a meter) and nanometers (billionths of a meter).

A Long Way Sometimes people need to measure long distances, such as the distance a migrating bird travels or the distance from Earth to the Moon. To measure such lengths, you use kilometers. Kilometers might be most familiar to you as the distance traveled in a car or the measure of a long-distance race, as shown in **Figure 8.** The course of a marathon is measured carefully so that the competitors run 42.2 km. When you drive from New York to Los Angeles, you cover 4,501 km.

Figure 8
These runners have just completed a 10-kilometer race—known as a 10K. *About how many kilometers is the distance between your home and your school?*

Astronomy
INTEGRATION

How important are accurate measurements? In 1999, the *Mars Climate Orbiter* disappeared as it was to begin orbiting Mars. NASA later discovered that a unit system error caused the flight path to be incorrect and the orbiter to be lost. Research the error and determine what systems of units were involved. How can using two systems of units cause errors?

Table 3 Common Objects in SI Measurements

Object	Type of Measurement	Measurement
can of soda	volume	355 mL
bag of potatoes	mass	4.5 kg
fluorescent tube	length	1.2 m
refrigerator	temperature	276 K

Figure 9
A cubic meter equals the volume of a cube 1 m by 1 m by 1 m. *How many cubic centimeters are in a cubic meter?*

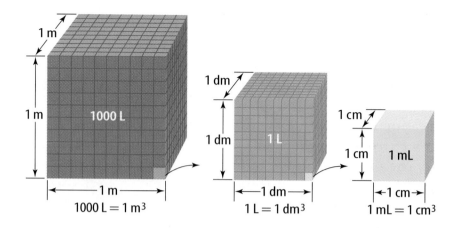

Volume

The amount of space an object occupies is its volume. The cubic meter (m^3), shown in **Figure 9,** is the SI unit of volume. You can measure smaller volumes with the cubic centimeter (cm^3 or cc). To find the volume of a square or rectangular object, such as a brick or your textbook, measure its length, width, and height and multiply them together. What is the volume of a compact disc case?

You are probably familiar with a 2-L bottle. A liter is a measurement of liquid volume. A cube 10 cm by 10 cm by 10 cm holds 1 L (1,000 cm^3) of water. A cube 1 cm on a side holds 1 mL (1 cm^3) of water.

Volume by Immersion Not all objects have an even, regular shape. How can you find the volume of something irregular like a rock or a piece of metal?

Have you ever added ice cubes to a nearly full glass of water only to have the water overflow? Why did the water overflow? Did you suddenly have more water? The volume of water did not increase at all, but the water was displaced when the ice cubes were added. Each ice cube takes up space or has volume. The difference in the volume of water before and after the addition of the ice cubes equals the volume of the ice cubes that are under the surface of the water.

The ice cubes took up space and caused the total volume in the glass to increase. When you measure the volume of an irregular object, you do the same thing. You start with a known volume of water and drop in, or immerse, the object. The increase in the volume of water is equal to the volume of the object.

TRY AT HOME Mini LAB

Measuring Volume

Procedure
1. Fill a plastic or glass **liquid measuring cup** until half full with **water.** Measure the volume.
2. Find an **object,** such as a rock, that will fit in your measuring cup.
3. Carefully lower the object into the water. If it floats, push it just under the surface with a **pencil.**
4. Record in your **Science Journal** the new volume of the water.

Analysis
1. How much space does the object occupy?
2. If 1 mL of water occupies exactly 1 cm^3 of space, what is the volume of the object in cm^3?

52 CHAPTER 2 Measurement

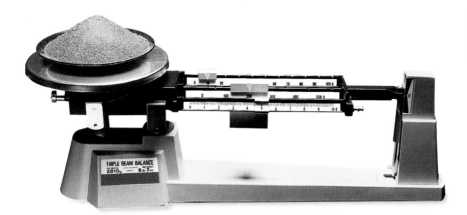

Figure 10
A triple beam balance compares an unknown mass to known masses.

Mass

The **mass** of an object measures the amount of matter in the object. The **kilogram** (kg) is the SI unit for mass. One liter of water has a mass of about 1 kg. Smaller masses are measured in grams (g). One gram is about the mass of a large paper clip.

You can determine mass with a triple beam balance, shown in **Figure 10**. The balance compares an object to a known mass. It is balanced when the known standard mass of the slides on the balance is equal to the object on the pan.

Why use the word *mass* instead of *weight*? Weight and mass are not the same. Mass depends only on the amount of matter in an object. If you ride in an elevator in the morning and then ride in the space shuttle later that afternoon, your mass is the same. Mass does not change when only your location changes.

Weight Weight is a measurement of force. The SI unit for weight is the newton (N). Weight depends on gravity, which can change depending on where the object is located. A spring scale measures how a planet's gravitational force pulls on objects. Several spring scales are shown in **Figure 11.**

If you were to travel to other planets, your weight would change, even though you would still be the same size and have the same mass. This is because gravitational force is different on each planet. If you could take your bathroom scale, which uses a spring, to each of the planets in this solar system, you would find that you weigh much less on Mars and much more on Jupiter. A mass of 75 pounds, or 34 kg, on Earth is a weight of 332 N. On Mars, the same mass is 126 N, and on Jupiter it is 782 N.

 *What does weight measure?*

Figure 11
A spring scale measures an object's weight by how much it stretches a spring.

Figure 12
The kelvin scale starts at 0 K. In theory, 0 K is the coldest temperature possible in nature.

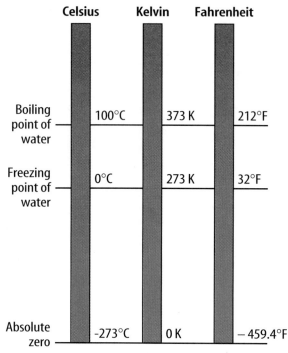

Temperature

The physical property of temperature is related to how hot or cold an object is. Temperature is a measure of the kinetic energy, or energy of motion, of the particles that make up matter.

The Fahrenheit and Celsius temperature scales are two common scales used on thermometers. Temperature is measured in SI with the **kelvin (K)** scale. A 1-K difference in temperature is the same as a 1°C difference in temperature, as shown in **Figure 12**. However, the two scales do not start at zero.

Time and Rates

Time is the interval between two events. The SI unit of time is the second (s). Time also is measured in hours (h). Although the hour is not an SI unit, it is easier to use for long periods of time. Can you imagine hearing that a marathon was run in 7,620 s instead of 2 h and 7 min?

A **rate** is the amount of change of one measurement in a given amount of time. One rate you are familiar with is speed, which is the distance traveled in a given time. Speeds often are measured in kilometers per hour (km/h).

The unit that is changing does not necessarily have to be an SI unit. For example, you can measure the number of cars that pass through an intersection per hour in cars/h. The annual rate of inflation can be measured in percent/year.

Section Assessment

1. Describe a situation in which different units of measure could cause confusion.
2. What type of quantity does the cubic meter measure?
3. How would you change a measurement in centimeters to kilometers?
4. What SI unit replaces the pound? What does this measure?
5. **Think Critically** You are told to find the mass of a metal cube. How will you do it?

Skill Builder Activities

6. **Measuring in SI** Measure the length, volume, and mass of your textbook in SI units. Describe any tools or calculations you use. **For more help, refer to the Science Skill Handbook.**
7. **Converting Units** A block of wood is 0.2 m by 0.1 m by 0.5 m. Find its dimensions in centimeters. Use these to find its volume in cubic centimeters. Show your work. **For more help, refer to the Math Skill Handbook.**

Activity

Scale Drawing

A scale drawing is used to represent something that is too large or too small to be drawn at its actual size. Blueprints for a house are a good example of a scale drawing.

What You'll Investigate
How can you represent your classroom accurately in a scale drawing?

Materials
1-cm graph paper
pencil
metric ruler
meterstick

Goals
- **Measure** using SI.
- **Make** a data table.
- **Calculate** new measurements.
- **Make** an accurate scale drawing.

Procedure
1. Use your meterstick to measure the length and width of your classroom. Note the locations and sizes of doors and windows.
2. **Record** the lengths of each item in a data table similar to the one below.
3. Use a scale of 2 cm = 1 m to calculate the lengths to be used in the drawing. Record them in your data table.
4. **Draw** the floor plan. Include the scale.

Room Dimensions		
Part of Room	Distance in Room (m)	Distance on Drawing (cm)

Conclude and Apply
1. How did you calculate the lengths to be used on your drawing? Did you put a scale on your drawing?
2. What would your scale drawing look like if you chose a different scale?
3. **Sketch** your room at home, estimating the distances. Compare this sketch to your scale drawing of the classroom. When would you use each type of illustration?
4. What measuring tool simplifies this task?

Communicating Your Data

Measure your room at home and compare it to the estimates on your sketch. Explain to someone at home what you did and how well you estimated the measurements. **For more help, refer to the** Science Skill Handbook.

SECTION
3 Drawings, Tables, and Graphs

As You Read

What You'll Learn
- **Describe** how to use pictures and tables to give information.
- **Identify** and use three types of graphs.
- **Distinguish** the correct use of each type of graph.

Vocabulary
table
graph
line graph
bar graph
circle graph

Why It's Important
Illustrations, tables, and graphs help you communicate data about the world around you in an organized and efficient way.

Scientific Illustrations

Most science books include pictures. Photographs and drawings model and illustrate ideas and sometimes make new information more clear than written text can. For example, a drawing of an airplane engine shows how all the parts fit together much better than several pages of text could describe it.

Drawings A drawing is sometimes the best choice to show details. For example, a canyon cut through red rock reveals many rock layers. If the layers are all shades of red, a drawing can show exactly where the lines between the layers are. The drawing can emphasize only the things that are necessary to show.

A drawing also can show things you can't see. You can't see the entire solar system, but drawings show you what it looks like. Also, you can make quick sketches to help model problems. For example, you could draw the outline of two continents to show how they might have fit together at one time.

A drawing can show hidden things, as well. A drawing can show the details of the water cycle, as in **Figure 13.** Architects use drawings to show what the inside of a building will look like. Biologists use drawings to show where the nerves in your arm are found.

Figure 13
This drawing shows details of the water cycle that can't be seen in a photograph.

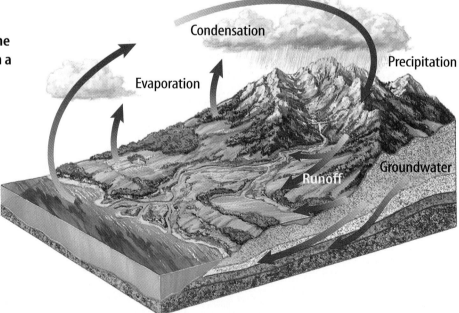

56 CHAPTER 2 Measurement

Photographs A still photograph shows an object exactly as it is at a single moment in time. Movies show how an object moves and can be slowed down or sped up to show interesting features. In your schoolwork, you might use photographs in a report. For example, you could show the different types of trees in your neighborhood for a report on ecology.

Tables and Graphs

Everyone who deals with numbers and compares measurements needs an organized way to collect and display data. A **table** displays information in rows and columns so that it is easier to read and understand, as seen in **Table 4.** The data in the table could be presented in a paragraph, but it would be harder to pick out the facts or make comparisons.

A **graph** is used to collect, organize, and summarize data in a visual way. The relationships between the data often are seen more clearly when shown in a graph. Three common types of graphs are line, bar, and circle graphs.

Line Graph A **line graph** shows the relationship between two variables. A variable is something that can change, or vary, such as the temperature of a liquid or the number of people in a race. Both variables in a line graph must be numbers. An example of a line graph is shown in **Figure 14.** One variable is shown on the horizontal axis, or *x*-axis, of the graph. The other variable is placed along the vertical axis, or *y*-axis. A line on the graph shows the relationship between the two variables.

Table 4 Endangered Animal Species in the United States

Year	Number of Endangered Animal Species
1980	174
1982	179
1984	192
1986	213
1988	245
1990	263
1992	284
1994	321
1996	324
1998	357

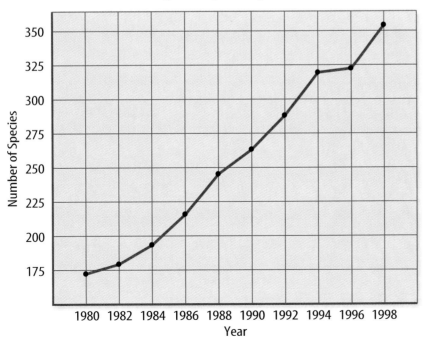

Figure 14
To find the number of endangered animal species in 1988, find that year on the *x*-axis and see what number corresponds to it on the *y*-axis.

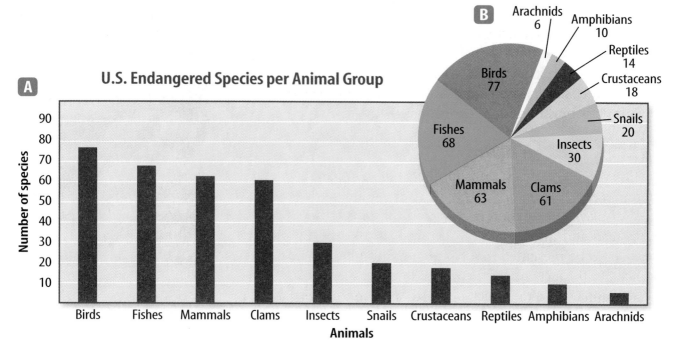

Figure 15

A Bar graphs allow you to picture the results easily. *Which category of animals has the most endangered species?* **B** On this circle graph, you can see what part of the whole each animal represents.

Bar Graph A **bar graph** uses rectangular blocks, or bars, of varying sizes to show the relationships among variables. One variable is divided into parts. It can be numbers, such as the time of day, or a category, such as an animal. The second variable must be a number. The bars show the size of the second variable. For example, if you made a bar graph of the endangered species data from **Figure 14,** the bar for 1990 would represent 263 species. An example of a bar graph is shown in **Figure 15A.**

Circle Graph Suppose you want to show the relationship among the types of endangered species. A **circle graph** shows the parts of a whole. Circle graphs are sometimes called pie graphs. Each piece of pie visually represents a fraction of the total. Looking at the circle graph in **Figure 15B,** you see quickly which animals have the highest number of endangered species by comparing the sizes of the pieces of pie.

A circle has a total of 360°. To make a circle graph, you need to determine what fraction of 360 each part should be. First, determine the total of the parts. In **Figure 15B,** the total of the parts, or endangered species, is 367. One fraction of the total, *Mammals,* is 63 of 367 species. What fraction of 360 is this? To determine this, set up a ratio and solve for x:

$$\frac{63}{367} = \frac{x}{360°} \quad x = 61.8°$$

Mammals will have an angle of 61.8° in the graph. The other angles in the circle are determined the same way.

Research Visit the Glencoe Science Web site at **tx.science.glencoe.com** for more information about scientific illustrations. Communicate to your class what you learn.

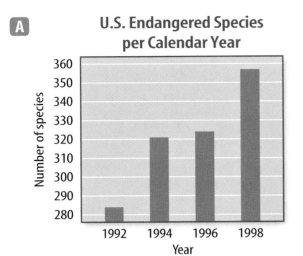
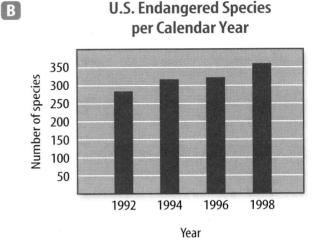

Reading Graphs When you are using or making graphs to display data, be careful—the scale of a graph can be misleading. The way the scale on a graph is marked can create the wrong impression, as seen in **Figure 16A.** Until you see that the *y*-axis doesn't start at zero, it appears that the number of endangered species has quadrupled in just six years.

This is called a broken scale and is used to highlight small but significant changes, just as an inset on a map draws attention to a small area of a larger map. **Figure 16B** shows the same data on a graph that does not have a broken scale. The number of species has only increased 22 percent from 1980 to 1986. Both graphs have correct data, but must be read carefully. Always analyze the measurements and graphs that you come across. If there is a surprising result, look closer at the scale.

Figure 16
Careful reading of graphs is important. **A** This graph does not start at zero, which makes it appear that the number of species has more than quadrupled from 1980 to 1986. **B** The actual increase is about 22 percent as you can see from this full graph. The broken scale must be noted in order to interpret the results correctly.

Section Assessment

1. Describe a time when an illustration would be helpful in everyday activities.
2. Explain how to use **Figure 16** to find the number of endangered species in 1998.
3. Explain the difference between tables and graphs.
4. Suppose your class surveys students about after-school activities. What type of graph would you use to display your data? Explain.
5. **Think Critically** How are line, bar, and circle graphs the same? How are they different?

Skill Builder Activities

6. **Making and Using Graphs** Record the amount of time you spend reading each day for the next week. Then make a graph to display the data. What type of graph will you use? Could more than one kind of graph be used? **For more help, refer to the** Science Skill Handbook.

7. **Using an Electronic Spreadsheet** Use a spreadsheet to display how the total mass of a 500-kg elevator changes as 50-kg passengers are added one at a time. **For more help, refer to the** Technology Skill Handbook.

Activity: Design Your Own Experiment

Pace Yourself

Track meets and other competitions require participants to walk, run, or wheel a distance that has been precisely measured. Officials make sure all participants begin at the same time, and each person's time is stopped at the finish line. If you are practicing for the Houston Marathon or the Austin 10K, you need to know your speed or pace in order to compare it with those of other participants. How can your performance be measured accurately?

Recognize the Problem

How will you measure the speed of each person in your group? How will you display these data?

Form a Hypothesis

Think about the information you have learned about precision, measurement, and graphing. In your group, make a hypothesis about a technique that will provide you with the most precise measurement of each person's pace.

Goals
- **Design** an experiment that allows you to measure speed for each member of your group accurately.
- **Display** data in a table and a graph.

Possible Materials
meterstick
stopwatch
*watch with a second hand
*Alternate materials

Safety Precautions

Work in an area where it is safe to run. Participate only if you are physically able to exercise safely. As you design your plan, make a list of all the specific safety and health precautions you will take as you perform the investigation. Get your teacher's approval of the list before you begin.

Using Scientific Methods

Test Your Hypothesis

Plan

1. As a group, decide what materials you will need.
2. How far will you travel? How will you measure that distance? How precise can you be?
3. How will you measure time? How precise can you be?
4. List the steps and materials you will use to test your hypothesis. Be specific. Will you try any part of your test more than once?
5. Before you begin, create a data table. Your group must decide on its design. Be sure to leave enough room to record the results for each person's time. If more than one trial is to be run for each measurement, include room for the additional data.

Do

1. Make sure that your teacher approves your plan before you start.
2. Carry out the experiment as planned and approved.
3. Be sure to record your data in the data table as you proceed with the measurements.

Analyze Your Data

1. **Graph** your data. What type of graph would be best?
2. Are your data table and graph easy to understand? Explain.
3. How do you know that your measurements are precise?
4. Do any of your data appear to be out of line with the rest?

Draw Conclusions

1. How is it possible for different members of a group to find different times while measuring the same event?
2. What tools would help you collect more precise data?
3. What other data displays could you use? What are the advantages and disadvantages of each?

Communicating Your Data

Make a larger version of your graph to display in your classroom with the graphs of other groups. **For more help, refer to the** Science Skill Handbook.

ACTIVITY 61

Science Stats

Biggest, Tallest, Loudest

Did you know...

... The world's most massive flower belongs to a species called *Rafflesia* (ruh FLEE zhee uh) and has a mass of up to 11 kg. The diameter, or the distance across the flower's petals, can measure up to 1 m.

... The world's tallest building is the Petronus Towers in Kuala Lumpur, Malaysia. It is 452 m tall. The tallest building in the United States is Chicago's Sears Tower, shown here, which measures 442 m.

... The Grand Canyon is so deep— as much as 1,800 m—that it can hold more than four Empire State Buildings stacked on top of one another.

... The world's tallest tree is a coast redwood in the Montgomery Woods State Park in California. The tree stands 112.1 m high.

62 CHAPTER 2 Measurement

Connecting To Math

...The largest animal on Earth is the blue whale. It can grow to be 33.5 m long. If 20 people who are each 1.65 m tall were lying head to toe, it would almost equal this length.

How do they measure up?

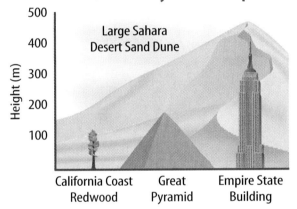

...One of the loudest explosions on Earth was the 1883 eruption of Krakatau (krah kuh TAHEW), an Indonesian volcano. It was heard from more than 3,500 km away.

Do the Math

1. How many of the largest rafflesia petals would you have to place side by side to equal the length of a blue whale?
2. When Krakatau erupted, it ejected 18,000 km^3 of ash and rock. Other large eruptions released the following: Mount Pinatubo—7,000 km^3, Mount Katmai—13,000 km^3, Tambora—30,000 km^3, Vesuvius—5,000 km^3. Make a bar graph to compare the sizes of these eruptions.
3. Use the information provided about the Grand Canyon to calculate how many Sears Towers would have to stand end on end to equal the depth of the canyon.

Go Further

Do research on the Internet at **tx.science.glencoe.com** to find facts that describe some of the shortest, smallest, or fastest things on Earth. Create a class bulletin board with the facts you and your classmates find.

SCIENCE STATS

Chapter 2 Study Guide

Reviewing Main Ideas

Section 1 Description and Measurement

1. Measurements such as length, volume, mass, temperature, and rates are used to describe objects and events.
2. Estimation is used to make an educated guess at a measurement.
3. Accuracy describes how close a measurement is to the true value.
4. Precision describes how close measurements are to each other. *Are the shots accurate or precise on the basketball hoop shown?*

Section 2 SI Units

1. The international system of measurement is called SI. It is used throughout the world for communicating data.
2. The SI unit of length is the meter. Volume—the amount of space an object occupies—can be measured in cubic meters. The mass of an object is measured in kilograms. The SI unit of temperature is the kelvin. *What type of measurement is being made according to the sign shown?*

Section 3 Drawings, Tables, and Graphs

1. Tables, photographs, drawings, and graphs can sometimes present data more clearly than explaining everything in words. Scientists use these tools to collect, organize, summarize, and display data in a way that is easy to use and understand.
2. The three common types of graphs are line graphs, bar graphs, and circle graphs. *Which city on the line graph shown is the coldest in the fifth month?*

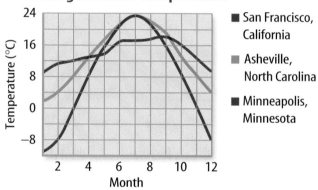

3. Line graphs show the relationship between two variables that are numbers on an *x*-axis and a *y*-axis. Bar graphs divide a variable into parts to show a relationship. Circle graphs show the parts of a whole like pieces of a pie.

After You Read

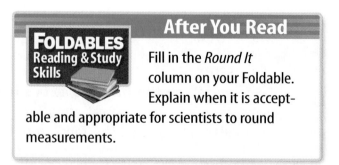

Fill in the *Round It* column on your Foldable. Explain when it is acceptable and appropriate for scientists to round measurements.

Chapter 2 Study Guide

Visualizing Main Ideas

Complete the following concept map.

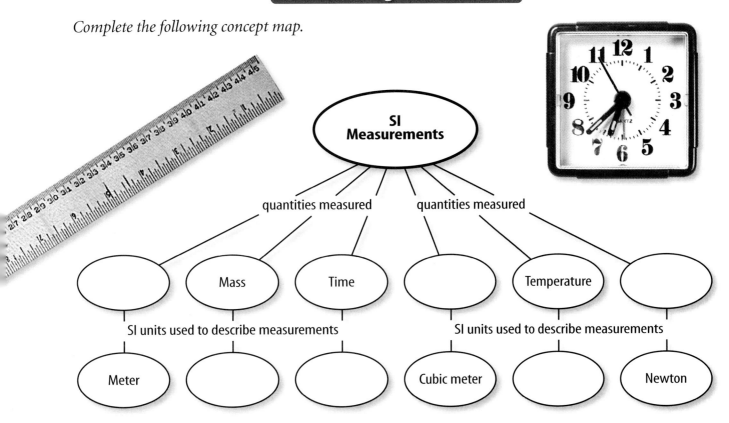

Vocabulary Review

Vocabulary Words

a. accuracy
b. bar graph
c. circle graph
d. estimation
e. graph
f. kelvin
g. kilogram
h. line graph
i. mass
j. measurement
k. meter
l. precision
m. rate
n. SI
o. table

Study Tip

When you encounter new vocabulary, write it down in your Science Journal. This will help you understand and remember them.

Using Vocabulary

Each phrase below describes a vocabulary word. Write the word that matches the phrase describing it.

1. the SI unit for length
2. a description with numbers
3. a method of making a rough measurement
4. the amount of matter in an object
5. a graph that shows parts of a whole
6. a description of how close measurements are to each other
7. the SI unit for temperature
8. an international system of units

Chapter 2 Assessment & TEKS Review

Checking Concepts

Choose the word or phrase that best answers the question.

1. The measurement 25.81 g is precise to the nearest what?
   A) gram
   B) kilogram
   C) tenth of a gram
   D) hundredth of a gram

2. What is the SI unit of mass?
   A) kilometer   C) liter
   B) meter   D) kilogram

3. What would you use to measure length?
   A) graduated cylinder
   B) balance
   C) meterstick
   D) spring scale

4. The cubic meter is the SI unit of what?
   A) volume   C) mass
   B) weight   D) distance

5. Which term describes how close measurements are to each other?
   A) significant digits   C) accuracy
   B) estimation   D) precision

6. Which is a temperature scale?
   A) volume   C) Celsius
   B) mass   D) mercury

7. Which is used to organize data?
   A) table   C) precision
   B) rate   D) meterstick

8. To show the number of wins for each football team in your district, which of the following would you use?
   A) photograph   C) bar graph
   B) line graph   D) SI

9. What organizes data in rows and columns?
   A) bar graph   C) line graph
   B) circle graph   D) table

10. To show 25 percent on a circle graph, the section must measure what angle?
    A) 25°   C) 180°
    B) 90°   D) 360°

Thinking Critically

11. How would you estimate the volume your backpack could hold?

12. Why do scientists in the United States use SI rather than the English system (feet, pounds, pints, etc.) of measurement?

13. List the following in order from smallest to largest: 1 m, 1 mm, 10 km, 100 mm.

14. Describe an instance when you would use a line graph. Can you use a bar graph for the same purpose?

15. Computer graphics artists can specify the color of a point on a monitor by using characters for the intensities of three colors of light. Why was this method of describing color invented?

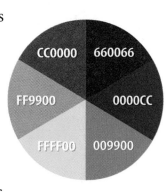

Developing Skills

16. **Measuring in SI** Make a fist. Use a centimeter ruler to measure the height, width, and depth of your fist.

17. **Comparing and Contrasting** How are volume, length, and mass similar? How are they different? Give several examples of units that are used to measure each quantity. Which units are SI?

Chapter 2 Assessment

18. Making and Using Graphs The table shows the area of several bodies of water. Make a bar graph of the data.

Areas of Bodies of Water	
Body of Water	Area (km^2)
Currituck Sound (North Carolina)	301
Pocomoke Sound (Maryland/Virginia)	286
Chincoteague Bay (Maryland/Virginia)	272
Core Sound (North Carolina)	229

19. Interpreting Scientific Illustrations What does the figure show? How has this drawing been simplified?

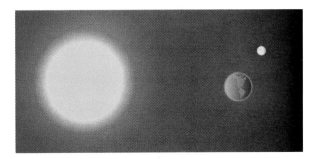

Performance Assessment

20. Poster Make a poster to alert the public about the benefits of using SI units.

21. Newspaper Search Look through a week's worth of newspapers and evaluate any graphs or tables that you find.

Technology

Go to the Glencoe Science Web site at **tx.science.glencoe.com** or use the **Glencoe Science CD-ROM** for additional chapter assessment.

THE PRINCETON REVIEW — TAKS Practice

Some students in Mrs. Olsen's science class measured their masses during three consecutive months. They placed their results in the following table. Study the table and answer the following questions.
TEKS 6.2 C; 6.4 B

Student Masses: Sept. – Nov. 1999			
Student	September	October	November
Domingo	41.13 kg	40.92 kg	42.27 kg
Latoya	35.21 kg	35.56 kg	36.07 kg
Benjamin	45,330 g	45,680 g	45,530 g
Poloma	31.78 kg	31.55 kg	31.51 kg
Frederick	50,870 g	51,880 g	51,030 g
Fiona	37.62 kg	37.71 kg	37.85 kg

1. According to the table, which shows Frederick's weight in kilograms for the three months?
   A) 5.087, 5.118, 5.103
   B) 50.87, 51.88, 51.03
   C) 508.7, 511.8, 510.3
   D) 5,087, 5,118, 5,103

2. According to this information, which lists the students from lightest to heaviest during November?
   F) Poloma, Benjamin, Domingo, Frederick
   G) Domingo, Latoya, Frederick, Benjamin
   H) Fiona, Domingo, Benjamin, Frederick
   J) Frederick, Benjamin, Domingo, Poloma

Reading Comprehension

Read the passage carefully. Then read each question that follows the passage. Decide which is the best answer to each question.

Test-Taking Tip After you read and think about the passage, write one or two sentences that summarize the most important points. Read your sentences out loud.

History of Measurement Units

In modern society, we use units of measurement that have been defined and agreed upon by international scientists. In ancient times, people were just beginning to invent and use units of measurement. For example, thousands of years ago, a cabinetmaker would build one cabinet at a time and measure the pieces of wood needed relative to the size of the other pieces of that cabinet. Today, factories manufacture many of the same products. Ancient cabinetmakers rarely made two cabinets that were exactly the same. Eventually, it became obvious that units of measurement had to mean the same thing to everybody.

Measurements, such as the inch, foot, and yard, began many years ago as fairly crude units. For example, the modern-day inch began as "the width of one's thumb." The foot was originally defined as "the length of one's foot." The yard was defined as "the distance from the tip of one's nose to the end of one's arm."

Although using these units of measurement was easier than not using any units of measurement, these ancient units were confusing. Human beings come in many different sizes and shapes, and one person's foot can be much larger than another person's foot. So, whose foot defines a foot? Whose thumb width defines an inch? Ancient civilizations used these kinds of measurements for thousands of years. Over time, these units were redefined and standardized, eventually becoming the exact units of measurement that you know today.

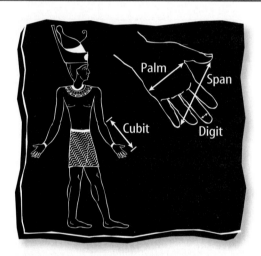

Ancient people created units of measurement.

1. Based on the passage, the reader can conclude that _____. *Reading TEKS 6.10 H*
   A) ancient cultures had no concept of measurement
   B) standards of measurement developed over a long period of time
   C) ancient people were probably good at communicating exacts units of measurement
   D) units of measurement are needed only in modern and technologically advanced societies

2. According to the passage, which of these best describes ancient units of measurement? *Reading TEKS 6.11 A*
   F) precise
   G) incorrect
   H) approximate
   J) irresponsible

TAKS Practice

Reasoning and Skills

Read each question and choose the best answer.

1. All of these are things that can be accurately measured EXCEPT ____.
   Science TEKS 6.2 A, 6.2 D
   A) the temperature of a human body
   B) the space that a couch takes up in a living room
   C) the beauty in a piece of artwork
   D) the mass of a rock from the Moon

   Test-Taking Tip Think about the reasons why we use measurement and the kinds of things that can and cannot be measured.

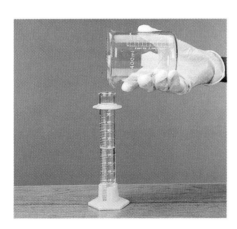

2. Every day at school, Jodie and William record the amount of water left until all of the water has evaporated. When they are finished, they want to clearly represent their data about water and evaporation time to the class. Which of these would best represent their data? *Science TEKS 6.2 E*
   F) A circle graph   H) A bar graph
   G) A line graph   J) A realistic drawing

   Test-Taking Tip Consider the benefits and drawbacks of each answer choice.

Group S
- How does this medication work to reduce a fever?
- Why are there tides?
- What is the melting point of iron?
- What was Earth's climate like in the past?

Group T
- Who would make the best class president?
- Is it right to compliment a friend's new shirt if you don't like the shirt?
- Shouldn't everyone like cauliflower, broccoli, turnips, and spinach?
- Why is a sunset beautiful?

3. The questions in Group S are different from the questions in Group T because only the questions in Group S ____. *Science TEKS 6.2 H*
   A) cannot be answered by science
   B) will have answers that are solely opinions
   C) can be answered with absolute certainty
   D) can be answered by science

   Test-Taking Tip Think about the kinds of questions that scientists try to and are able to answer.

Consider this question carefully before writing your answer on a separate sheet of paper.

4. Suppose that you have just picked up a rock along the river and you want to learn more about it. What steps would you go through to learn about the rock? *Science TEKS 6.2 A*

   Test-Taking Tip Think about the procedures that scientists must go through in order to discover and explain.

STANDARDIZED TEST PRACTICE 69

UNIT 2
Interactions of Matter and Energy

How Are Charcoal & Celebrations Connected?

NATIONAL GEOGRAPHIC

According to one report, one day in the tenth century in China, a cook combined charcoal with two other ingredients that were common in Chinese kitchens. The result was a spectacular explosion of sparks. Whether or not that story is true, most experts agree that fireworks originated in China. The Chinese discovered that if the ingredients were put into a bamboo tube, the force of the reaction would send the tube zooming into the sky. The spectacular light and noise were perfect for celebrations. Traders carried the art of firework making westward to Europe. The Europeans added new colors to the bursts by mixing various chemicals into the explosive powder. Today, people all over the world use colorful fireworks to celebrate special occasions.

SCIENCE CONNECTION

CHEMICAL REACTIONS Inside some fireworks shells, different chemicals are arranged in sequence. As each chemical ignites, a chemical reaction occurs, releasing sparks in particular colors. For example, sodium compounds produce the color yellow, and barium nitrates make green. Research to find out what chemicals would be responsible for a spectacular fireworks display in red, white, and blue?

CHAPTER 3

Science TEKS 6.7 A, B; 6.8 A

Properties and Changes of Matter

Wendy Craig Duncan carried the Olympic flame underwater on the way to the 2000 Summer Olympics in Sydney, Australia. How many different states of matter can you find in this picture? What chemical and physical changes are taking place? In this chapter, you will learn about the four states of matter, the physical and chemical properties of matter and how those properties can change.

What do you think?

Science Journal What is the colored band on this fish tank? Why would it change colors? Here's a hint: *Would you want your fish to be chilly?* Write your thoughts and ideas in your Science Journal.

72

Your teacher has given you a collection of pennies. It is your task to separate these pennies into groups while using words to describe each set. In this chapter, you will learn how to identify things based on their physical and chemical properties. With an understanding of these principles of matter, you will discover how things are classified or put into groups.

Classify coins

1. Observe the collection of pennies.
2. Choose a characteristic that will allow you to separate the pennies into groups.
3. Classify and sort each penny based on the chosen feature. Tally your data in a frequency table.
4. Explain how you classified the pennies. Compare your system of classification with those of others in the classroom.

Observe

Think about how your group classified its pennies. Describe the system your group used in your Science Journal.

Before You Read

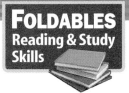

Making an Organizational Study Fold Make the following Foldable to help you organize your thoughts into clear categories about properties of matter.

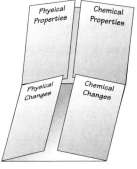

1. Place a sheet of paper in front of you so the long side is at the top. Fold the paper in half from the left side to the right side. Unfold.
2. Fold each side in to the center fold line to divide the paper into fourths. Fold the paper in half from top to bottom. Unfold.
3. Through the top thickness of paper, cut along both of the middle fold lines to form four tabs. Label the tabs *Physical Properties, Physical Changes, Chemical Properties,* and *Chemical Changes.*
4. Before you read the chapter, define each term on the front of the tabs. As you read the chapter, correct your definitions and write about each under the tabs.

SECTION 1

Physical Properties and Changes

As You Read

What You'll Learn
- **Identify** physical properties of matter.
- **Describe** the states of matter.
- **Classify** matter using physical properties.

Vocabulary
physical property
matter
physical change
density
state of matter
melting point
boiling point

Why It's Important
Observing physical properties will help you interpret the world around you.

 Review *Grade 5 TEKS*

For a review of the Grade 5 TEKS *Physical Properties of Matter,* see page 502.

Using Your Senses

As you look in your empty wallet and realize that your allowance isn't coming anytime soon, you decide to get an after-school job. You've been hired at the new grocery store that will open next month. They are getting everything ready for the grand opening, and you will be helping make decisions about where things will go and how they will be arranged.

When you come into a new situation or have to make any kind of decision, what do you usually do first? Most people would make some observations. Observing involves seeing, hearing, tasting, touching, and smelling.

Whether in a new job or in the laboratory, you use your senses to observe materials. Any characteristic of a material that can be observed or measured without changing the identity of the material is a **physical property**. However, it is important to never taste, touch, or smell any of the materials being used in the lab without guidance, as noted in **Figure 1.** For safety reasons you will rely mostly on other observations.

 Watch

 Listen

Do NOT touch

Do NOT smell

Do NOT taste

Figure 1
For safety reasons, in the laboratory you usually use only two of your senses—sight and hearing. Many chemicals can be dangerous to touch, taste, and smell.

Figure 2
The identity of the material does not necessarily depend on its color. Each of these bottles is made of high-density polyethylene (HDPE).

Physical Properties

On the first day of your new job, the boss gives you an inventory list and a drawing of the store layout. She explains that every employee is going to give his or her input as to how the merchandise should be arranged. Where will you begin?

You decide that the first thing you'll do is make some observations about the items on the list. One of the key senses used in observing physical properties is sight, so you go shopping to look at what you will be arranging.

Color and Shape Everything that you can see, touch, smell, or taste is matter. **Matter** is anything that has mass and takes up space. What things do you observe about the matter on your inventory list? The list already is organized by similarity of products, so you go to an aisle and look.

Color is the first thing you notice. The laundry detergent bottles you are looking at come in every color. Maybe you will organize them in the colors of the rainbow. You make a note and look more closely. Each bottle or box has a different shape. Some are square, some rectangular, and some are a free-form shape. You could arrange the packages by their shape.

When the plastic used to make the packaging is molded, it changes shape. However, the material is still plastic. This type of change is called a physical change. It is important to realize that in a **physical change,** the physical properties of a substance change, but the identity of the substance does not change. Notice **Figure 2.** The detergent bottles are made of high-density polyethylene regardless of the differences in the physical properties of color or shape.

Research Visit the Glencoe Science Web site at **tx.science.glencoe.com** for more information about classifying matter by its physical properties. Communicate to your class what you learn.

 *What is matter?*

Length and Mass Some properties of matter can be identified by using your senses, and other properties can be measured. How much is there? How much space does it take up?

One useful and measurable physical property is length. Length is measured using a ruler, meterstick, or tape measure, as shown in **Figure 3.** Objects can be classified by their length. For example, you could choose to organize the French bread in the bakery section of your store by the length of the loaf. But, even though the dough has been shaped in different lengths, it is still French bread.

Back in the laundry aisle, you notice a child struggling to lift one of the boxes of detergent. That raises a question. How much detergent is in each box? Mass is a physical property that describes the amount of material in an object. Some of the boxes are heavy, but, the formula of the detergent hasn't changed from the small box to the large box. Organizing the boxes by mass is another option.

Volume and Density Mass isn't the only physical property that describes how much of something you have. Another measurement is volume. Volume measures the amount of space an object takes up. Liquids usually are measured by volume. The juice bottles on your list could be organized by volume.

Another measurable physical property related to mass and volume is **density**—the amount of mass a material has for a given volume. You notice this property when you try to lift two things of equal volume that have different masses. Density is found by dividing the mass of an object by its volume.

$$\text{density} = \text{mass/volume, or } D = m/V$$

Collect Data Visit the Glencoe Science Web site at **tx.science.glencoe.com** for information about density. Communicate to your class what you learn.

Figure 3
The length of any object can be measured with the appropriate tool.

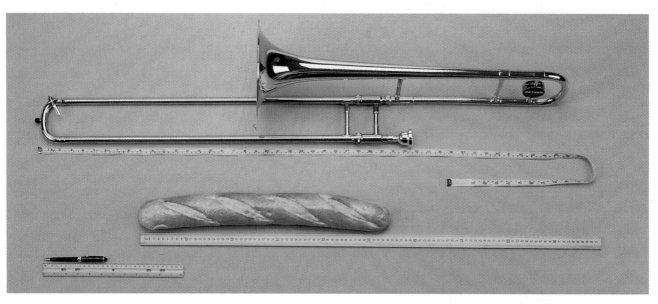

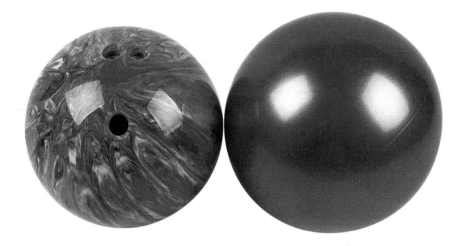

Figure 4
These balls take up about the same space, but the bowling ball on the left has more mass than the kickball on the right. Therefore, the bowling ball is more dense.

Same Volume, Different Mass Figure 4 shows two balls that are the same size but not the same mass. The bowling ball is more dense than the kickball. The customers of your grocery store will notice the density of their bags of groceries if the baggers load all of the canned goods in one bag and put all of the cereal and napkins in the other.

The density of a material stays the same as long as pressure and temperature stay the same. Water at room temperature has a density of 1.00 g/cm³. However, when you do change the temperature or pressure, the density of a material can change. Water kept in the freezer at 0°C is in the form of ice. The density of that ice is 0.9168 g/cm³. Has the identity of water changed? No, but something has changed.

 *What two measurements are related in the measurement of density?*

States of Matter

What changes when water goes from 20°C to 0°C? It changes from a liquid to a solid. The four **states of matter** are solid, liquid, gas, and plasma (PLAZ muh). The state of matter of a substance depends on its temperature and pressure. Three of these states of matter are things you talk about or experience every day, but the term *plasma* might be unfamiliar. The plasma state occurs at very high temperatures and is found in fluorescent (floo RE sunt) lightbulbs, the atmosphere, and in lightning strikes.

As you look at the products to shelve in your grocery store, you might make choices of classification based on the state of matter. The state of matter of a material is another physical property. The liquid juices all will be in one place, and the solid, frozen juice concentrates will be in another.

Changing Density
Procedure
1. Stir three tablespoons of **baking soda** in 3/4 cup of **warm water.** Set aside.
2. Pour **vinegar** into a **clear glass** until it is half full.
3. Slowly pour the baking soda and warm water mixture into the glass of vinegar.
4. After most of the fizzing has stopped, place three halves of **raisins** in the glass.
5. Record your observations in your Science Journal or on a computer.

Analysis
1. How did the bubbles affect the raisins?
2. Infer what changed the density of the raisins.

SECTION 1 Physical Properties and Changes

Moving Particles Matter is made up of moving particles. The state of matter is determined by how much energy the particles have. The particles of a solid vibrate in a fixed position. They remain close together and give the solid a definite shape and volume. The particles of a liquid are moving much faster and have enough energy to slide past one another. This allows a liquid to take the shape of its container. The particles of a gas are moving so quickly that they have enough energy to move freely away from other particles. The particles of a gas take up as much space as possible and will spread out to fill any container. **Figure 5** illustrates the differences in the states of water.

✔ **Reading Check** *Which state of matter doesn't conform to the shape of the container?*

Changes of State You witness a change of state when you put ice cubes in water and they melt. You still have water but in another form. The opposite physical change happens when you put liquid water in ice-cube trays and pop them in your freezer. The water doesn't change identity—only the state it is in.

For your job, you will need to make some decisions based on the ability of materials to change state. You don't want all those frozen items thawing out and becoming slushy liquid. You also don't want some of the liquids to get so cold that they freeze.

Figure 5
Water can be in three different states: solid, liquid, and gas. The molecules in a solid are tightly packed and vibrate in place, but in a liquid they can slip past each other because they have more energy to move. In a gas, they move freely all around the container with even more energy.

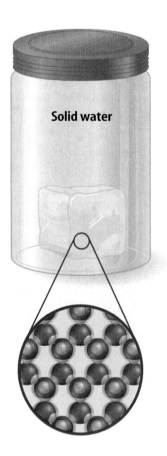
Solid water

Liquid water

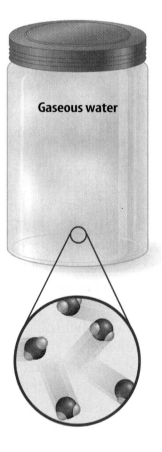
Gaseous water

Melting and Boiling Points At what temperature will water in the form of ice change into a liquid? The temperature at which a solid becomes a liquid is its **melting point**. The melting point of a pure substance does not change with the amount of the substance. This means that a small sliver of ice and a block of ice the size of a house both will melt at 0°C. Lead always melts at 327.5°C. When a substance melts, it changes from a solid to a liquid. This is a physical change, and the melting point is a physical property.

At what temperature will liquid water change to a gas? The **boiling point** is the temperature at which a substance in the liquid state becomes a gas. Each pure substance has a unique boiling point at atmospheric pressure. The boiling point of water is 100°C at atmospheric pressure. The boiling point of nitrogen is −195.8°C, so it changes to a gas when it warms after being spilled into the open air, as shown in **Figure 6**. The boiling point, like the melting point, does not depend on the amount of the substance.

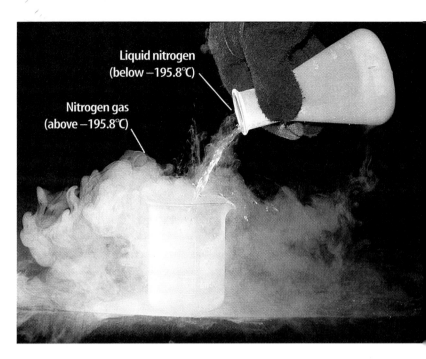

Figure 6
When liquid nitrogen is poured from a flask, you see an instant change to gas because nitrogen's boiling point is −195.8°C, which is much lower than room temperature.

 *What physical change takes place at the boiling point?*

However, the boiling point and melting point can help to identify a substance. If you observe a clear liquid that boils at 56.1°C at atmospheric pressure, it is not water. Water boils at 100°C. If you know the boiling points and melting points of substances, you can classify substances based on those points.

Metallic Properties

Other physical properties allow you to classify substances as metals. You already have seen how you can classify things as solids, liquids, or gases or according to color, shape, length, mass, volume, or density. What properties do metals have?

How do metals look and act? Often the first thing you notice about something that is a metal is its shiny appearance. This is due to the way light is reflected from the surface of the metal. This shine is called luster. New handlebars on a bike have a metallic luster. Other words to describe the appearance of nonmetallic objects are *pearly, milky,* or *dull.*

Earth Science INTEGRATION

When geologists describe rocks, they use specific terms that have meaning to all other scientists who read their descriptions. To describe the appearance of a rock or mineral, they use the following terms: *metallic, adamantine, vitreous, resinous, pearly, silky,* and *greasy.* Research these terms and write a definition and example of each in your Science Journal.

SECTION 1 Physical Properties and Changes **79**

Figure 7
This artist has taken advantage of the ductility of metal by choosing wire as the medium for this sculpture.

Figure 8
This junkyard magnet pulls scrap metal that can be salvaged from the rest of the debris. It is sorting by a physical property.

Uses of Metals Metals can be used in unique ways because of some of the physical properties they have. For example, many metals can be hammered, pressed, or rolled into thin sheets. This property of metals is called malleability (mal lee uh BIH luh tee). The malleability of copper makes it an ideal choice for artwork such as the Statue of Liberty. Many metals can be drawn into wires as shown in **Figure 7.** This property is called ductility (duk TIH luh tee). The wires in buildings and most electrical equipment and household appliances are made from copper. Silver and platinum are also ductile.

You probably observe another physical property of some metals every day when you go to the refrigerator to get milk or juice for breakfast. Your refrigerator door is made of metal. Some metals respond to magnets. Most people make use of that property and put reminder notes, artwork, and photos on their refrigerators. Some metals have groups of atoms that can be affected by the force of a magnet, and they are attracted to the magnet because of that force. The magnet in **Figure 8** is being used to select metallic objects.

At the grocery store, your employer might think about these properties of metals as she looks at grocery carts and thinks about shelving. Malleable carts can be dented. How could the shelf's attraction of magnets be used to post advertisements or weekly specials? Perhaps the prices could be fixed to the shelves with magnetic numbers. After you observe the physical properties of an object, you can make use of those properties.

Using Physical Properties

In the previous pages, many physical properties were discussed. These physical properties—such as appearance, state, shape, length, mass, volume, ability to attract a magnet, density, melting point, boiling point, malleability, and ductility—can be used to help you identify, separate, and classify substances.

For example, salt can be described as a white solid. Each salt crystal, if you look at it under a microscope, could be described as having a three-dimensional cubic structure. You can measure the mass, volume, and density of a sample of salt or find out if it would attract a magnet. These are examples of how physical properties can be used to identify a substance.

Figure 9
Coins can be sorted by their physical properties. Sorting by size is used here.

Sorting and Separating When you do laundry, you sort according to physical properties. Perhaps you sort by color. When you select a heat setting on an iron, you classify the clothes by the type of fabric. When miners during the Gold Rush panned for gold, they separated the dirt and rocks by density of particles. **Figure 9** shows a coin sorter that separates the coins based on their size. Iron filings can be separated from sand by using a magnet.

Scientists who work with animals use physical properties or characteristics to determine the identity of a specimen. They do this by using a tool called a dichotomous (di KAH tuh mus) key. The term *dichotomous* refers to two parts or divisions. Part of a dichotomous key for identifying hard-shelled crabs is shown in **Figure 10.** To begin the identification of your unknown animal, you are given two choices. Your animal will match only one of the choices. In the key in **Figure 10,** you are to determine whether or not your crab lives in a borrowed shell. Based on your answer, you are either directed to another set of choices or given the name of the crab you are identifying.

NATIONAL GEOGRAPHIC VISUALIZING DICHOTOMOUS KEYS

Figure 10

Whether in the laboratory or in the field, scientists often encounter substances or organisms that they cannot immediately identify. One approach to tracking down the identity of such "unknowns" is to use a dichotomous key, such as the one shown. The key is designed so a user can compare physical properties or characteristics of the unknown substance or organism—in this case, a crab—with characteristics of known organisms in a stepwise manner. With each step, a choice must be made. Each choice leads to subsequent steps that guide the user through the key until a positive identification is made.

Dichotomous Key

1. A. Lives in a "borrowed" shell (usually some type of snail shell) — **Hermit Crab**
   B. Does not live in a "borrowed" shell — **go to #2**
2. A. Shell completely overlaps the walking legs — **Box Crab**
   B. Walking legs are exposed — **Kelp Crab**

Can you identify the three crabs shown here by following this dichotomous key?

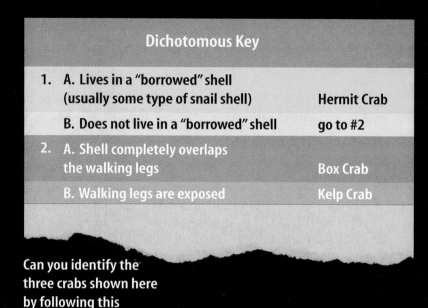

82 CHAPTER 3

Everyday Examples Identification by physical properties is a subject in science that is easy to observe in the real world. Suppose you volunteer to help your friend choose a family pet. While visiting the local animal shelter, you spot a cute dog. The dog looks like the one in **Figure 11.** You look at the sign on the cage. It says that the dog is male, one to two years old, and its breed is unknown. You and your friend wonder what breed of dog he is. What kind of information do you and your friend need to figure out the dog's breed? First, you need a thorough description of the physical properties of the dog. What does the dog look like? Second, you need to know the descriptions of various breeds of dogs. Then you can match up the description of the dog with the correct breed. The dog you found is a white, medium-sized dog with large black spots on his back. He also has black ears and a black mask around his eyes. The manager of the shelter tells you that the dog is close to full-grown. What breed is the dog?

Figure 11
Physical descriptions are used to determine the identities of unknown things. *What physical properties are used to describe this dog?*

Narrowing the Options To find out, you may need to research the various breeds of dogs and their descriptions. Often, determining the identity of something that is unknown is easiest by using the process of elimination. You figure out all of the breeds the dog can't be. Then your list of possible breeds is smaller. Upon looking at the descriptions of various breeds, you eliminate small dog and large dog breeds. You also eliminate breeds that do not contain white dogs. With the remaining breeds, you might look at photos to see which ones most resemble your dog. Scientists use similar methods to determine the identities of living and nonliving things.

Section 1 Assessment

1. What physical properties can you use to describe your science textbook?
2. What property of matter is used to measure the amount of space an object takes?
3. What are the four states of matter? Describe each and give an example.
4. How might a substance such as water have two different densities?
5. **Think Critically** Explain why the boiling point is the same for 1 L and 3 L of water. Will it take the same amount of time for each volume of water to begin to boil?

Skill Builder Activities

6. **Concept Mapping** Using a computer, draw a cycle concept map with the steps for changing an ice cube into steam. Use the terms *melting point* and *boiling point* in your answer. **For more help, refer to the** Science Skill Handbook.

7. **Solving One-Step Equations** Calculate the volume of an object that has a length of 10 cm, a width of 10 cm, and a height of 10 cm. Use the formula for volume: $V = lwh$. Express your answer using cubic centimeters (cm^3). **For more help, refer to the** Math Skill Handbook.

SECTION 2
Chemical Properties and Changes

As You Read

What You'll Learn
- **Recognize** chemical properties.
- **Identify** chemical changes.
- **Classify** matter according to chemical properties.
- **Describe** the law of conservation of mass.

Vocabulary
chemical property
chemical change
conservation of mass

Why It's Important
Knowing the chemical properties will allow you to distinguish differences in matter.

Ability to Change

It is time to celebrate. You and your coworkers have cooperated in classifying all of the products and setting up the shelves in the new grocery store. The store manager agrees to a celebration party and campfire at the nearby park. Several large pieces of firewood and some small pieces of kindling are needed to start the campfire. After the campfire, all that remains of the wood is a small pile of ash. Where did the wood go? What property of the wood is responsible for this change?

All of the properties that you observed and used for classification in the first section were physical properties that you could observe easily. In addition, even when those properties changed, the identity of the object remained the same. Something different seems to have happened in the bonfire example.

Some properties do indicate a change of identity for the substances involved. A **chemical property** is any characteristic that gives a substance the ability to undergo a change that results in a new substance. **Figure 12** shows some properties of substances that can be observed only as they undergo a chemical change.

✓ **Reading Check** *What does a chemical property give a substance the ability to do?*

Figure 12
These are four examples of chemical properties.

Flammability

Reacts with oxygen

Reacts with light

Reacts with water

84 CHAPTER 3 Properties and Changes of Matter

Common Chemical Properties

You don't have to be in a laboratory to see changes that take place because of chemical properties. These are called chemical changes. A **chemical change** is a change in the identity of a substance due to the chemical properties of that substance. A new substance or substances are formed in such a change.

The bonfire you enjoyed to celebrate the opening of the grocery store resulted in chemical changes. The oxygen in the air reacted with the wood and cardboard to form a new substance called ash. Wood and cardboard can burn. This chemical property is called flammability. Some products have warnings on their labels about keeping them away from heat and flame because of the flammability of the materials. Sometimes after a bonfire you see stones that didn't burn around the edge of the ashes. These stones have the chemical property of being incombustible.

Common Reactions An unpainted iron gate, such as the one shown in **Figure 13A,** will rust in time. The rust is a result of oxygen in the air reacting with the iron and causing corrosion. The corrosion produces a new substance called iron oxide. Other chemical reactions occur when metals interact with other elements. **Figure 13B** shows tarnish, the grayish-brown film that develops on silver when it reacts with sulfur in the air. The ability to react with oxygen or sulfur is a chemical property. **Figure 13C** shows another example of this chemical property.

Have you ever sliced an apple or banana and left it sitting on the table? The brownish coloring that you notice is a chemical change that occurs between the fruit and the oxygen in the air. Those who work in the produce department at the grocery store must be careful with any fruit they slice to use as samples. Although nothing is wrong with brown apples, they don't look appetizing.

Figure 13
Many kinds of interactions with oxygen can occur.
A An untreated iron gate will rust. **B** Silver dishes develop tarnish. **C** Copper sculptures develop a green patina, which is a mixture of copper compounds.

Health
INTEGRATION

Researchers have discovered an enzyme in fruit that is involved in the browning process. They are doing experiments to try to grow grapevines in which the level of this enzyme, polyphenol oxidase (PPO), is reduced. This could result in grapes that do not brown as quickly. Write a paragraph in your Science Journal about why this would be helpful to fruit growers, store owners, and customers.

SECTION 2 Chemical Properties and Changes

Heat and Light Vitamins often are dispensed in dark-brown bottles. Do you know why? Many vitamins have the ability to change when exposed to light. This is a chemical property. They are protected in those colored bottles from undergoing a chemical change with light.

Some substances are sensitive to heat and will undergo a chemical change only when heated or cooled. One example is limestone. Limestone is generally thought of as unreactive. Some limestone formations have been around for centuries without changing. However, if limestone is heated, it goes through a chemical change and produces carbon dioxide and lime, a chemical used in many industrial processes. The chemical property in this case is the ability to change when heated.

Another chemical property is the ability to change with electrical contact. Electricity can cause a change in some substances and decompose some compounds. Water is one of those compounds that can be broken down with electricity.

Something New

The important difference in a chemical change is that a new substance is formed. Because of chemical changes, you can enjoy many things in life that you would not have experienced without them. What about that perfect, browned marshmallow you roasted at the bonfire? A chemical change occurred as a result of the fire to make the taste and the appearance different.

Sugar is normally a white, crystalline substance, but after you heat it over a flame, it turns to a dark-brown caramel. A new substance has been formed. Sugar also can undergo a chemical change when sulfuric acid is added to it. The new substance is obviously different from the original, as shown in **Figure 14**.

You could not enjoy a birthday cake if the eggs, sugar, flour, baking powder, and other ingredients didn't change chemically. You would have what you see when you pour the gooey batter into the pan.

Figure 14
When sugar and sulfuric acid combine, a chemical change occurs and a new substance forms. During this reaction, the mixture foams and a toxic gas is released, leaving only water and air-filled carbon behind. (Because a toxic gas is released, students should never perform this as an activity.)

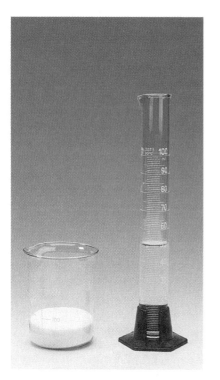

Signs of Change How do you know that you have a new substance? Is it just because it looks different? You could put a salad in a blender and it would look different, but a chemical change would not have occurred. You still would have lettuce, carrots, and any other vegetables that were there to begin with.

You can look for signs when evaluating whether you have a new substance as a result of a chemical change. Look at the piece of birthday cake in **Figure 15.** When a cake bakes, gas bubbles form and grow within the ingredients. Bubbles are a sign that a chemical change has taken place. When you look closely at a piece of cake, you can see the airholes left from the bubbles.

Other signs of change include the production of heat, light, smoke, change in color, and sound. Which of these signs of change would you have seen or heard during the bonfire?

Figure 15
The evidence of a chemical change in the cake is the holes left by the air bubbles that were produced during baking.

Is it reversible? One other way to determine whether a physical change or a chemical change has occurred is to decide whether or not you can reverse the change by simple physical means. Physical changes usually can be reversed easily. For example, melted butter can become solid again if it is placed in the refrigerator. A figure made of modeling clay, like the one in **Figure 16,** can be smashed to fit back into a container. However, chemical changes can't be reversed using physical means. For example, the ashes in a fireplace cannot be put back together to make the logs that you had to start with. Can you find the egg in a cake? Where is the white flour?

Figure 16
A change such as molding clay or changing shape can be undone easily.

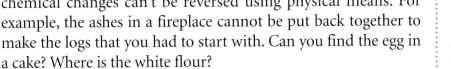 Reading Check *What kind of change can be reversed easily?*

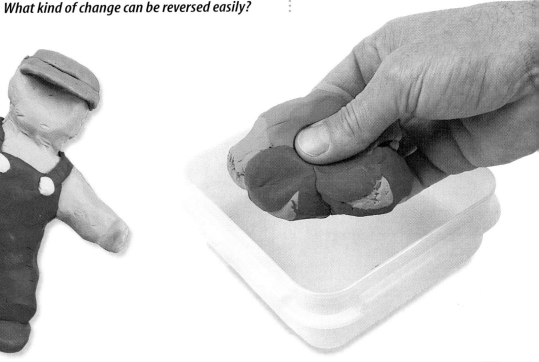

Table 1 Comparing Properties	
Physical Properties	color, shape, length, mass, volume, density, state, ability to attract a magnet, melting point, boiling point, malleability, ductility
Chemical Properties	flammability; ability to react with: oxygen, electricity, light, water, heat, vinegar, bleach, etc.

Classifying According to Chemical Properties Classifying according to physical properties is often easier than classifying according to chemical properties. **Table 1** summarizes the two kinds of properties. The physical properties of a substance are easily observed, but the chemical properties can't be observed without changing the substance. However, once you know the chemical properties, you can classify and identify matter based on those properties. For example, if you try to burn what looks like a piece of wood but find that it won't burn, you can rule out the possibility that it is untreated wood.

In a grocery store, the products sometimes are separated according to their flammability or sensitivity to light or heat. You don't often see the produce section in front of big windows where heat and light come in. The fruit and vegetables would undergo a chemical change and ripen too quickly. You also won't find the lighter fluid and rubbing alcohol near the bakery or other places where heat and flame could be present.

Architects and product designers have to take into account the chemical properties of materials when they design buildings and merchandise. For example, children's sleepwear and bedding can't be made of a flammable fabric. Also, some of the architects designing the most modern buildings are choosing materials like titanium because it does not react with oxygen like many other metals do.

Conservation of Mass

It was so convenient to turn the firewood into the small pile of ash left after the campfire. You began with many kilograms of flammable substances but ended up with just a few kilograms of ash. Could this be a solution to the problems with landfills and garbage dumps? Why not burn all the trash? If you could make such a reduction without creating undesirable materials, this would be a great solution.

Mini LAB

Observing Yeast

Procedure

1. Observe a **tablespoon** of **dry yeast** with a **hand lens.** Draw and describe what you observe.
2. Put the yeast in 50 mL of warm, not hot, **water.**
3. Compare your observations of the dry yeast with those of the wet yeast.
4. Put a pinch of **sugar** in the water and observe for 15 minutes.
5. Record your observations.

Analysis

1. Are new substances formed when sugar is added to the water and yeast? Explain.
2. Do you think this is a chemical change or a physical change? Explain.

Mass Is Not Destroyed Before you celebrate your discovery, think this through. Did mass really disappear during the fire? It appears that way when you compare the mass of the pile of ashes to the mass of the firewood you started with. The law of **conservation of mass** states that the mass of what you end with is always the same as the mass of what you start with.

This law was first investigated about 200 years ago, and many investigations since then have proven it to be true. One experiment done by French scientist Antoine Lavoisier was a small version of a campfire. He determined that a fire does not make mass disappear or truly get rid of anything. The question, however, remains. Where did the mass go? The ashes aren't heavy enough to account for the mass of all of the pieces of firewood.

Where did the mass go? If you look at the campfire example more closely, you see that the law of conservation of mass is true. When flammable materials burn, they combine with oxygen. Ash, smoke, and gases are produced. The smoke and gases escape into the air. If you could measure the mass of the oxygen and all of the original flammable materials that were burned and compare it to the remaining mass of the ash, smoke, and gas, they would be equal.

Problem-Solving Activity

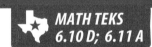
MATH TEKS
6.10 D; 6.11 A

Do light sticks conserve mass?

Light sticks often are used on Halloween to light the way for trick-or-treaters. They make children visible to drivers. They also are used as toys, for camping, marking trails, emergency traffic problems, by the military, and they work well underwater. A light stick contains two chemicals in separate tubes. When you break the inner tube, the two chemicals react producing a greenish light. The chemicals are not toxic, and they will not catch fire.

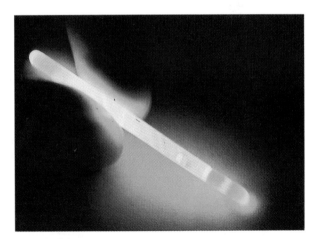

Identifying the Problem

In all reactions that occur in the world, mass is never lost or gained. This is the law of conservation of mass. An example of this phenomenon is the light stick. How can you prove this?

Solving the Problem

Describe how you could show that a light stick does not gain or lose mass when you allow the reaction to take place. Is this reaction a chemical or physical change? What is your evidence?

SECTION 2 Chemical Properties and Changes

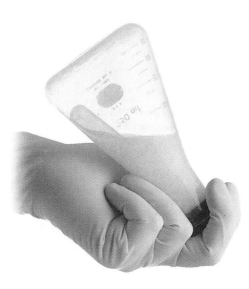
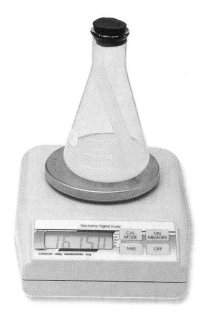

Figure 17
This reaction demonstrates the conservation of mass. Although a chemical change has occurred and new substances were made, the mass remained constant.

Before and After Mass is not destroyed or created during any chemical change. The conservation of mass is demonstrated in **Figure 17.** In the first photo, you see one substance in the flask and a different substance contained in a test tube inside the flask. The total mass is 16.150 g. In the second photo, the flask is turned upside down. This allows the two substances to mix and react. Because the flask is sealed, nothing is allowed to escape. In the third photo, the flask is placed on the balance again and the total mass is determined to be 16.150 g. If no mass is lost or gained, what happens in a reaction? Instead of disappearing or appearing, the particles in the substances rearrange into different combinations with different properties.

Section Assessment

1. What is a chemical property?
2. List four chemical properties.
3. What are some of the signs that a chemical change has occurred?
4. Describe the law of conservation of mass. Give an example.
5. **Think Critically** You see a bright flash and then flames as your teacher performs a demonstration for the class. Is this an example of a physical change or a chemical change? Explain.

Skill Builder Activities

6. **Comparing and Contrasting** Compare and contrast physical change and chemical change. **For more help, refer to the** Science Skill Handbook.
7. **Solving One-Step Equations** A student heats 4.00 g of a blue compound to produce 2.56 g of a white compound and an unknown amount of colorless gas. What is the mass of this gas? **For more help, refer to the** Math Skill Handbook.

Activity

Liquid Layers

Why must you shake up a bottle of Italian salad dressing before using it? Have you observed how the liquids in some dressings separate into two distinct layers? In this activity, you will experiment with creating layers of liquids.

What You'll Investigate
What would several liquids and solids of different densities look like when put into the same container?

Goals
- **Create** layers of liquids using liquids of different densities.
- **Observe** where solids of different densities will rest in the liquid layers.
- **Infer** the densities of the different materials.

Materials
250-mL beaker
graduated cylinder
corn syrup
glycerin
water
corn oil
rubbing alcohol
penny
plastic sphere
rubber ball

Safety Precautions

Procedure
1. Pour 40 mL of corn syrup into your beaker.
2. Slowly pour 40 mL of glycerin into the beaker. Allow the glycerin to trickle down the sides of the container and observe.
3. Slowly pour 40 mL of water into the beaker and observe.
4. Slowly pour 40 mL of corn oil into the beaker and observe.
5. Slowly pour 40 mL of rubbing alcohol into the beaker and observe.
6. Carefully drop the penny, plastic sphere, and rubber ball into the beaker and observe where these items come to a stop.

Conclude and Apply
1. In your Science Journal, draw a picture of the liquids and solids in your flask. Label your diagram.
2. **Describe** what happened to the five liquids when you poured them into the beaker.
3. **Infer** why the liquids behaved this way.
4. **Describe** what happened to the three solids you placed into the beaker.
5. **List** the substances you used in your activity in order from those with the highest density to those with the lowest density.
6. If water has a density of 1 g/cm^3, what can you infer about the densities of the solids and other liquids?

Communicating Your Data
Draw a labeled poster of the substances you placed in your beaker. Research the densities of each substance and include these densities on your poster. **For more help, refer to the** Science Skill Handbook.

ACTIVITY 91

Activity: Design Your Own Experiment

Fruit Salad Favorites

When you are looking forward to enjoying a tasty, sweet fruit salad at a picnic, the last thing you want to see is brown fruit in the bowl. What can you do about this problem? Your teacher has given you a few different kinds of fruit. It is your task to perform a test in which you will observe a physical change and a chemical change.

Recognize the Problem

Can a chemical change be controlled?

Form a Hypothesis

Based on your reading and observations, state a hypothesis about whether you can control a chemical change.

Goals

- **Design** an experiment that identifies physical changes and chemical changes in fruit.
- **Observe** whether chemical changes can be controlled.

Possible Materials

bananas
apples
pears
plastic or glass mixing bowls (2)
lemon/water solution (500 mL)
paring knife

Safety Precautions

WARNING: *Be careful when working with sharp objects. Always keep hands away from sharp blades. Never eat anything in the laboratory.*

92 CHAPTER 3

Using Scientific Methods

Test Your Hypothesis

Plan

1. As a group, agree upon the hypothesis and decide how you will test it. Identify what results will confirm the hypothesis.
2. **List** each of the steps you will need in order to test your hypothesis. Be specific. Describe exactly what you will do in each step. List all of your materials.
3. Prepare a data table in your Science Journal or on a computer for your observations.
4. Read the entire investigation to make sure all steps are in logical order.
5. **Identify** all constants, variables, and controls of the investigation.

Do

1. Ask your teacher to approve your plan and choice of constants, variables and controls before you start.
2. Perform the investigation as planned.
3. While doing the investigation, record your observations and complete the data table you prepared in your Science Journal.

Analyze Your Data

1. **Compare and contrast** the changes you observe in the control and the test fruit.
2. **Compare** your results with those of other groups.
3. What was your control in this investigation?
4. What are your variables?
5. Did you encounter any problems carrying out the investigation?
6. Do you have any suggestions for changes in a future investigation?

Draw Conclusions

1. Did the results support your hypothesis? Explain.
2. **Describe** what effect refrigerating the two salads would have on the fruit.
3. What will you do with the fruit from this experiment? Could it be eaten?

*C*ommunicating Your Data

Write a page for an illustrated cookbook explaining the benefits you found in this experiment. Include drawings and a step-by-step procedure. **For more help, refer to the Science Skill Handbook.**

ACTIVITY 93

Oops! Accidents in SCIENCE
SOMETIMES GREAT DISCOVERIES HAPPEN BY ACCIDENT!

Glass from the Past
THE PROCESS OF MAKING GLASS PROBABLY STARTED BY CHANCE

These 2,000-year-old glass containers were found in Syria.

Syria, 2500 B.C. The cooking fire had blazed on the sandy beach for days. The people of ancient Syria danced around it, cooked meat over it, and roasted grains in it as they celebrated their good harvest. The next year, they returned to the same beach, again to celebrate. In digging out the old firepit, they came across something strange—the sand under last year's fire had changed into a hard, greenish object that broke like obsidian when it was dropped. Over years of observation, the Syrians realized that fire melted the sand into a different material with different properties. They had discovered glass.

The story is fictional. But, glass-making did begin in Syria and Egypt about 4,500 years ago and it is most likely that cooking or signal fires lit on sandy surfaces led to its discovery. Glass beads have been found from that time. These early glassmakers melted sand in clay pots over hot fires, and began experimenting with adding different materials to the mixture. Only small amounts of glass could be made at one time—and only small objects. From 2500 to 1500 B.C., glass objects were made by pressing molten glass into a mold and letting it cool into the desired shape. Beads, bowls, and jewelry were the most frequently made items.

What's the recipe?

Even in ancient times, "recipes" existed for different kinds and colors of glass. For example, glass for cooking and glass for jewelry probably had different recipes.

The Rose Planetarium in New York City, is made mostly of glass.

But some ingredients stayed the same. Sand, made mostly of silica, was—and still remains—the primary ingredient of glass. Silica can be melted into glass, but it takes an extremely hot fire and breaks easily. By trial and error, glass-makers found that adding other materials to the sand changed the sand's properties. They also discovered that different materials added to the sand would make it easier to produce certain objects.

Sodium carbonate from wood ash is added to lower the boiling temperature of silica. Calcium carbonate from limestone is added to stabilize the glass. Without the limestone, glass would dissolve in water!

In more recent times, the technology of glass-making has developed rapidly, with new techniques and new types of glass following quickly after one another. For example, in 1938, an experiment with soap film led to the invention of glare-proof glass that is used to make camera and telescope lenses. Physicist Katharine Burr Blodgett was researching ways to apply a very thin film of soap to glass—a film that was only one molecule thick. As she built up the layers molecule by molecule, she found that at exactly 44 layers—or 44 molecules—thick, the light rays moving into the glass were cancelled out by the light rays moving out of the glass. "You keep barking up so many wrong trees in research," Blodgett said. "This time I … barked up one that held what I was looking for."

Sculptures made of glass are considered works of art.

CONNECTIONS **List** Make a list of all the objects made of glass that you see around you at school and at home. Are there any surprises? Research any items you're not sure of, then classify the items as primarily decorative, structural, or useful.

SCIENCE *Online*
For more information, visit
tx.science.glencoe.com

Chapter 3 Study Guide

Reviewing Main Ideas

Section 1 Physical Properties and Changes

1. Any characteristic of a material that can be observed or measured is a physical property. *What can you observe or measure about the baby in the photo?*

2. The four states of matter are solid, liquid, gas, and plasma. The state of matter is a physical property that is dependent on temperature and pressure.

3. State, color, shape, length, mass, volume, attraction by a magnet, density, melting point, boiling point, luster, malleability, and ductility are common physical properties.

4. In a physical change the properties of a substance change but the identity of the substance always stays the same.

5. You can classify materials according to their physical properties. *What different physical properties could you use to classify the objects in the photo?*

Section 2 Chemical Properties and Changes

1. Chemical properties give a substance the ability to undergo a chemical change. *What chemical property is being displayed in the photo?*

2. Common chemical properties include: ability to burn, reacts with oxygen, reacts with heat or light, and breaks down with electricity.

3. In a chemical change substances combine to form a new material. *Is the balloon about to undergo a chemical or physical change?*

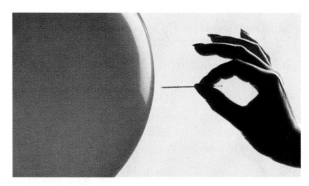

4. The mass of the products of a chemical change is always the same as the mass of what you started with.

5. A chemical change results in a substance with a new identity, but matter is not created or destroyed.

After You Read

Use the information in your Foldable to compare and contrast physical and chemical properties of matter. Write about each on the back of the tabs.

Chapter 3 Study Guide

Visualizing Main Ideas

Complete the following table comparing properties of different objects.

Properties of Matter

Type of Matter	Physical Properties	Chemical Properties
log		
pillow		
bowl of cookie dough		
book		
glass of orange juice		

Vocabulary Review

Vocabulary Words

a. boiling point
b. chemical change
c. chemical property
d. conservation of mass
e. density
f. matter
g. melting point
h. physical change
i. physical property
j. states of matter

Using Vocabulary

The sentences below include terms that have been used incorrectly. Change the incorrect terms so the sentence reads correctly.

1. The <u>boiling point</u> is the temperature at which matter in a solid state changes to a liquid.
2. <u>Matter</u> is a measure of the mass of an object in a given volume.
3. A <u>chemical change</u> is easily observed or measured without changing the object.
4. <u>Physical changes</u> result in a new substance and cannot be reversed by physical means.
5. The four states of matter are solid, <u>volume</u>, liquid, and <u>melting point</u>.

Study Tip

Make a plan. Before you start your homework, write out a checklist of what you need to do for each subject. As you finish each item, check it off.

CHAPTER STUDY GUIDE 97

Chapter 3 Assessment & TEKS Review

Checking Concepts

Choose the word or phrase that best answers the question.

1. What statement describes the physical property of density?
   A) the distance between two points
   B) how light is reflected from an object's surface
   C) the amount of mass for a given volume
   D) the amount of space an object takes up

2. Which of the following is an example of a physical change?
   A) tarnishing   C) burning
   B) rusting   D) melting

3. Which of the choices below describes a boiling point?
   A) a chemical property
   B) a chemical change
   C) a physical property
   D) a color change

4. Which of the following is a sign that a chemical change has occurred?
   A) smoke   C) change in shape
   B) broken pieces D) change in state

5. Which describes what volume is?
   A) the area of a square
   B) the amount of space an object takes up
   C) the distance between two points
   D) the temperature at which boiling begins

6. What property is described by the ability of metals to be hammered into sheets?
   A) mass   C) volume
   B) density   D) malleability

7. Which of these is a chemical property?
   A) size
   B) density
   C) flammability
   D) volume

8. When iron reacts with oxygen, what substance is produced?
   A) tarnish
   B) rust
   C) patina
   D) ashes

9. What kind of change results in a new substance being produced?
   A) chemical   C) physical
   B) mass   D) change of state

10. What is conserved during any change?
    A) color   C) identity
    B) volume   D) mass

Thinking Critically

11. Use the law of conservation of matter to explain what happens to atoms when they combine to form a new substance.

12. Describe the four states of matter. How are they different?

13. A globe is placed on your desk and you are asked to identify its physical properties. How would you describe the globe?

14. What information do you need to know about a material to find its density?

Developing Skills

15. **Classifying** Classify the following as a chemical or physical change: an egg breaks, a newspaper burns in the fireplace, a dish of ice cream is left out and melts, and a loaf of bread is baked.

Chapter 3 Assessment

16. **Measuring in SI** Find the density of the piece of lead that has a mass of 49.01 g and a volume of 4.5 mL.

17. **Concept Mapping** Use the spider-mapping skill to organize and define physical properties of matter. Include the concepts of color, shape, length, density, mass, states of matter, volume, density, melting point, and boiling point.

18. **Making and Using Tables** Complete the table by supplying the missing information.

States of Matter		
Type	**Definition**	**Examples**
solid		books, desk, chair, ice cubes
liquid	Particles do not stay in one position. They move past each other.	
gas		oxygen, helium, vapor

19. **Drawing Conclusions** List the physical and chemical properties and changes that describe the process of scrambling eggs.

Performance Assessment

20. **Comic Strip** Create a comic strip demonstrating a chemical change in a substance. Include captions and drawings that demonstrate your understanding of conservation of matter.

TECHNOLOGY

Go to the Glencoe Science Web site at **tx.science.glencoe.com** or use the **Glencoe Science CD-ROM** for additional chapter assessment.

THE PRINCETON REVIEW — TAKS Practice

Unknown matter can be identified by taking a sample of it and comparing its physical properties to those of already identified substances. *TEKS 7.1 A; 7.2 C; 7.7 B*

Physical Properties		
Substance	**Density (g/mL)**	**Color**
Gasoline	0.703	Clear
Aluminum	2.700	Silver
Methane	0.466	Colorless
Water	1.000	Clear

Study the table and answer the following questions.

1. A scientist has a sample of a substance with a density greater than 1 g/mL. According to the table, which of these substances has a density greater than 1 g/mL?
   A) gasoline
   B) water
   C) aluminum
   D) methane

2. A physical change occurs when the form or appearance of a substance is changed. All of the following are examples of physical changes EXCEPT _____.
   F) the vaporization of gasoline
   G) corrosion of aluminum
   H) methane diluted by air
   J) water freezing into ice

CHAPTER

Science TEKS 6.5 A; 6.6 A, B

Forces and Motion

In the span of a few extremely exhilarating seconds, this bungy jumper has experienced a crash course in forces and motion—as well as speed, velocity, acceleration, gravity, mass, weight, and inertia. After she passes the course, she would probably love to explain these concepts to you—from somewhere on solid ground. In this chapter, you'll study them from the safety of a chair. You'll also learn how Newton's laws of motion describe what happened to the bungy jumper after she fell.

What do you think?

Science Journal Look at this photo with a friend. Discuss what this might be or what is happening. Here's a hint: *In a few more seconds, a drastic reduction in motion will be welcomed.* Write your answer in your Science Journal.

 **W**hy can you slide down pavement when it is covered with ice, but not when it is dry? In both cases, the force of friction brings you to a stop, but ice has much less friction than concrete allowing you to glide farther. The force of friction usually gets larger as a surface becomes rougher. Explore how the roughness of different surfaces affects the motion of a moving object.

Measure the effect of friction

1. Lay your textbook flat on the floor. Place the clip end of a clipboard on the edge of your textbook.
2. Lay a path of fine-grained sandpaper on the floor leading away from the clipboard.
3. Roll a toy car down the board onto the path.
4. Measure the distance the car traveled on the sandpaper path.
5. Repeat the same procedure using medium-grained and then coarse-grained sandpaper.

Observe

Record your measurements in your Science Journal. Write a paragraph explaining the relationship between the roughness of each type of sandpaper and the distance that the car traveled on the sandpaper.

Before You Read

Making a Cause and Effect Study Fold Make the following Foldable to help you understand the cause and effect relationship of forces and motion.

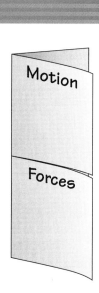

1. Place a sheet of paper in front of you so the long side is at the top. Fold the paper in half from the left side to the right side. Fold top to bottom and crease. Then unfold.
2. Through the top thickness of paper, cut along the middle fold line to form two tabs as shown.
3. Label the tabs *Motion* and *Forces* as shown.
4. As you read the chapter, list information about what causes forces and how forces cause motion under the tabs.

SECTION 1

Describing Motion

As You Read

What You'll Learn
- **Identify** when motion occurs.
- **Explain** relative motion.
- **Compare** speed, velocity, and acceleration.

Vocabulary
displacement
speed
velocity
acceleration

Why It's Important
You must interpret movement every day.

 Grade 5 TEKS Review

For a review of the Grade 5 TEKS *Change Occurs in Cycles,* see page 499.

Figure 1
Two cars are sitting side by side. One of them moves. *How can you tell which one has moved?*

Motion

Every day you are surrounded by motion. You recognize motion when you see cars and people move around you. You sense that air moves as it blows against your face—or that pieces of Earth's crust move if you have ever felt the shaking of an earthquake.

Even as you sit in a quiet country field, Earth is moving around the Sun, and the Moon is moving around Earth. As you sit on a bus that is too crowded for the passengers to move, you see buildings and other objects seemingly slide past you. Are you moving or are they moving? Can you tell? So how do you know when something is moving?

Motion Is Relative

This might have happened to you. You are sitting in a parked car and are startled to notice that the car is rolling backwards. After you look around, you realize that it is the car next to you that is moving forward. Why does your car seem to be moving backwards? At first you could see that your car was getting farther away from the other car. As shown in **Figure 1,** if you thought that the other car was not moving, then it was your car that had to be moving. Yet after you looked around, you could tell that your car was in the same place in the parking lot. You knew then that your car hadn't moved, and it was the other car that had moved forward.

102 CHAPTER 4 Forces and Motion

Reference Points You used the parking lot as a reference point to determine which car moved. You realized that where you were in the parking lot had not changed. In other words, your position relative to the parking lot didn't change. You knew then that relative to the parking lot, you were not moving, so it was the other car that moved.

Motion always is described relative to some reference point. **Figure 2** shows an in-line skater in a park. Suppose the tree in the figure is the reference point. The skater is in motion relative to the tree because the position of the skater has changed relative to the tree.

Reference points are needed to describe the motion of larger objects as well—such as Earth. Look at **Figure 3.** Earth is moving around the Sun. If the Sun is the reference point, Earth moves in a nearly circular path around the Sun. However, the Sun is moving relative to the Milky Way Galaxy and the Milky Way is in motion relative to other galaxies in the universe. The motion of Earth through space depends on the reference point that is chosen.

Choosing a Reference Point Suppose you were to describe the motion of a baseball as it sped past a batter. You wouldn't worry about the motion of the ball relative to the solar system or the galaxy. You would be concerned only with its motion relative to the player waiting to hit the ball. In other words, you would choose the batter as the reference point. You select the reference point to describe the motion that is important to you and the situation you are in. If you are walking on a bus that is moving, you could describe your motion relative to the bus or relative to the ground. Depending on the reference point you choose, how would your speed be different?

Figure 2
Sometimes it is obvious which object has moved. The tree doesn't move, so the skater must have moved.

SECTION 1 Describing Motion **103**

NATIONAL GEOGRAPHIC VISUALIZING EARTH'S MOTION

Figure 3

In the vastness of space, Earth's motion can be described only in relation to other objects, such as stars and galaxies. The illustration here shows how Earth moves relative to the Sun and to the Milky Way Galaxy, which is part of a cluster of galaxies called the Local Group.

A Imagine you are looking down on the Sun's North Pole. From this perspective, Earth traces out a nearly circular path, moving counterclockwise in its orbit around the Sun.

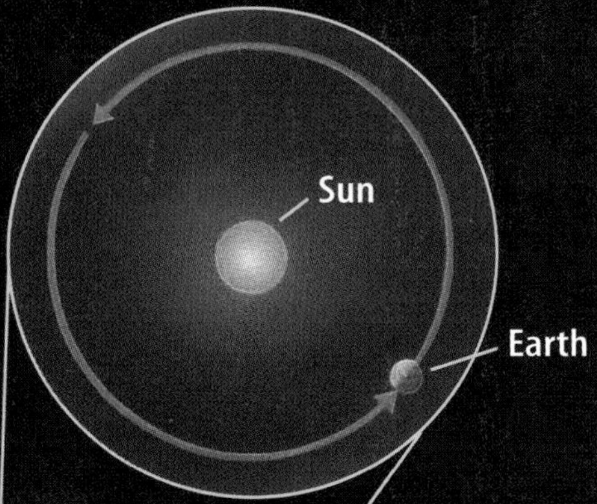

B The Sun belongs to a collection of several billion stars that make up the Milky Way Galaxy. Viewed from the "top" of the galaxy, the Sun moves clockwise in a nearly circular orbit around the galaxy's center. Earth's orbit around the Sun is not in the same plane as the galaxy. As a result, Earth's motion traces out a corkscrew path* as it moves with the Sun relative to the center of the galaxy.

*Earth's corkscrew path not shown to scale.

C The Milky Way Galaxy is moving relative to the center of a cluster of galaxies called the Local Group. So you can think of Earth's motion this way: Earth orbits the Sun, which moves around the Milky Way Galaxy, which in turn is moving around the center of the Local Group.

Changing Position

When you ride a bike, you are in motion relative to the ground. That motion can be described in many ways. How far did you travel? Where did you go? How long did it take? How fast were you riding? Did you speed up or slow down?

When something is moving, its location relative to a reference point is changing. The location of an object is how far the object is from a reference point. This also is called the position of the object. All motion involves a change of position. The position is measured relative to a reference point, as shown in **Figure 4.**

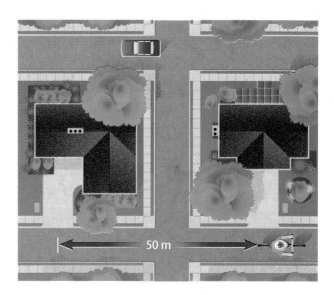

Figure 4
Suppose the house is the reference point. The position of the bike rider is 50 m away from the house.

Distance One way to describe your change in position is to tell how far you went, or the distance you traveled. Suppose you travel in a straight line, and don't change direction. Then you can calculate the distance you traveled by subtracting your final position from your starting position.

For example, suppose you are 50 m from your house and you want to ride to a friend's house. You start pedaling your bike and ride in a straight line. You stop when you are 150 m from your house, as shown in **Figure 5.** Your final position is 150 m, and your starting position is 50 m. So the distance you traveled is: 150 m − 50 m = 100 m.

Reading Check *What does distance measure?*

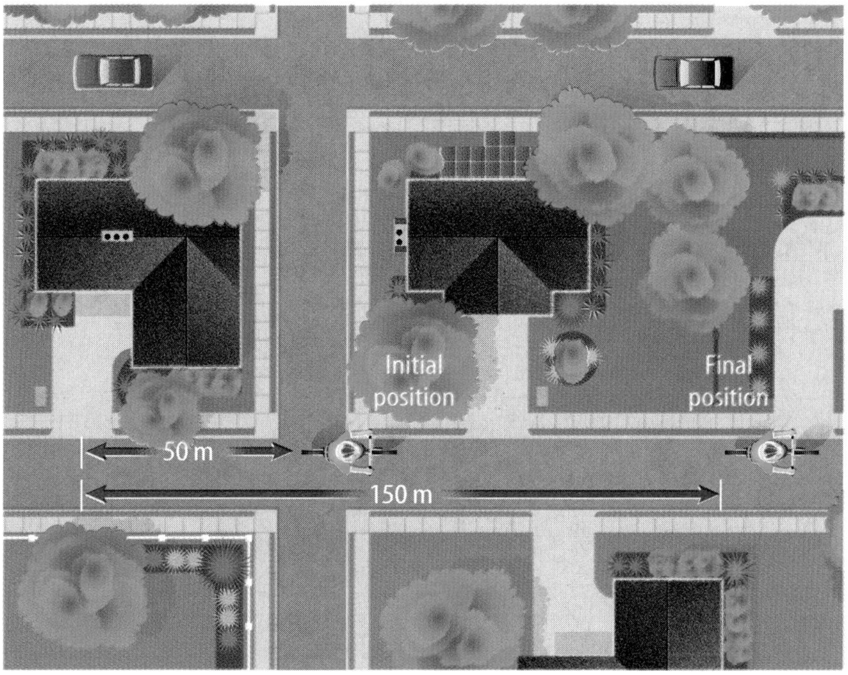

Figure 5
The position of the bike rider has changed. The distance traveled is the difference between the final position and the initial position. In this example the distance traveled is 100 m.

SECTION 1 Describing Motion **105**

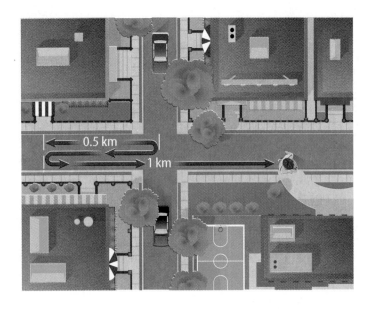

Figure 6
Your travel distance and displacement can be different. *How is the displacement different from the distance traveled in this example?*

Displacement Sometimes motion is not in a single direction. For example, suppose you are walking to school, which is a 1-km trip. You get halfway to school when you realize you forgot your lunch. You turn around, walk home, get your lunch, and then walk to school. How far did you travel?

Look at **Figure 6.** Your trip to school involves three parts. In the first part, you get halfway to school. The distance you travel is 0.5 km. In the second part, you walk back home. The distance you travel is 0.5 km. In the third part, you walk all the way to school and travel a distance of 1 km. To find the total distance you travel, add the distances traveled for all three parts: 0.5 km + 0.5 km + 1.0 km. The total distance you walked is 2.0 km.

Even though you walked a distance of 2.0 km, you are only 1.0 km from you house when you get to school. In other words, the distance between your starting position and your final position is 1.0 km. The direction and the distance between the final position and the starting position is the **displacement.** Your displacement is 1 km east. Displacement and distance traveled describe motion in different ways. Distance traveled depends on the entire path you walked. Displacement depends on only your starting position and your final position.

Distance and displacement are useful in different ways for describing motion. A biologist studying behavior in bees might record the entire path a bee took when it searched for food, located it, collected it, returned to the hive, and interacted with other bees. On the other hand, a biologist studying the migration of monarch butterflies might record only where a butterfly stops each night. This biologist is interested only in the butterflies' displacement after each day.

What is speed?

When you ride your bike to your friend's house, how fast do you ride? Suppose Monday it takes 30 min to get there, but on Friday it takes only 20 min. On which day were you moving faster? You were moving faster on Friday, because it took less time. When you move faster, it takes less time to travel the same distance. A way to describe how fast you are moving is by measuring your speed. **Speed** is the distance traveled divided by the time needed to travel that distance.

106 CHAPTER 4 Forces and Motion

Average Speed As you biked to your friend's house, you probably slowed down as you had to climb a hill and speeded up as you came down the other side. Perhaps you had to stop at an intersection. Your speed probably changed several times for various reasons.

If someone were to ask you what your speed was for your ride, what would you say? If your speed changed, what speed would describe how fast you traveled for the entire trip? One way to answer this is to determine your average speed. Average speed is found by dividing the total distance traveled by the total time.

$$\text{average speed} = \frac{\text{total distance}}{\text{total time}}$$

If you bike 5 km in half an hour, you can calculate your average speed like this:

$$\text{average speed} = \frac{5 \text{ km}}{0.5 \text{ h}} = 10 \text{ km/h}$$

Math Skills Activity

Calculating Average Speed

Example Problem

On a short bike ride, you ride 20 km. It takes you an hour and a half to complete your ride. What was your average speed?

Solution

1. *This is what you know:* distance = 20 km
   Time = 1.5 h
2. *This is what you need to know:* average speed
3. *This is the equation you need to use:* average speed = total distance/total time
4. *Substitute the known values:* average speed = 20 km/1.5 h = 13.3 km/h

Check your answer by multiplying by the time. Do you calculate the same distance that was given?

Practice Problem

On a long cross-country trip, you cover 3,600 km in about 50 h. What was your average speed for the trip?

For more help, refer to the Math Skill Handbook.

SECTION 1 Describing Motion

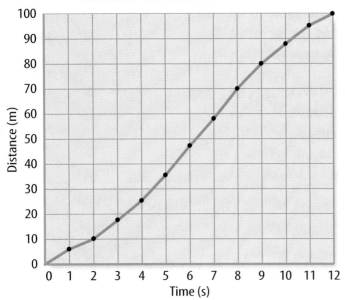

Graphing Motion The way in which something has moved can be shown with a distance-time graph. On a distance-time graph, time is plotted along the horizontal axis. The distance traveled is plotted on the vertical axis. Each point plotted on the graph shows how far something has moved, and how much time was used to travel that distance.

Figure 7 shows the distance-time graph for a sprinter running a 100-m dash. The distance the sprinter moved was plotted on the graph for each second of the race. According to the graph, after 2 s she has traveled 10 m. How far did the sprinter run between 8 s and 9 s? What is her average speed during this time interval?

Figure 7
This distance-time graph shows how the distance traveled by a sprinter changed during a 100-m race. *What was the sprinter's average speed over the entire race?*

Velocity

When school is over for the day, you and a classmate walk home. Both of you leave at the same time, but walk in different directions. If you walk at the same speed, after 10 min you have traveled the same distance. However, you are not at the same place after walking for 10 min because you are walking in different directions. Even though your speed and the distance traveled are the same, your displacement is different. Sometimes the direction of motion is important. The direction and speed of an object are described by the object's velocity. **Velocity** is the speed in a particular direction.

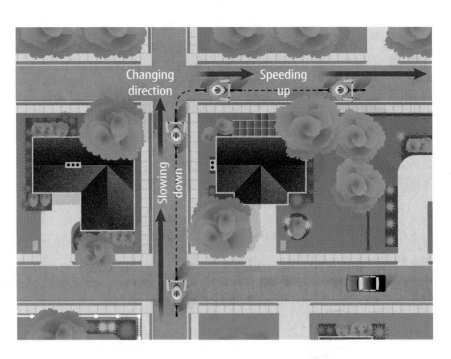

Figure 8
The arrow shows the bike rider's velocity. The direction of the arrow shows the direction of motion. The arrow's length shows the bike rider's speed. The velocity of an object changes when its speed or direction changes.

Changing Velocity The velocity of an object can change as it moves along. For example, look at **Figure 8.** As you rode to your friend's house, your beginning velocity might have been 15 km/h north. When you slowed and turned the corner, your velocity changed to 8 km/h east. When you turned, your speed and direction changed. Then you pedaled harder to speed up, and your velocity was 15 km/h east. This time your direction stayed the same but your speed changed. Is your velocity now the same as when you began? No, because you are now going east instead of north. Your speed is the same, but your direction has changed.

 *What two things does a velocity measurement describe?*

Acceleration

Distance, displacement, speed, and velocity tell where, how far, how fast, and in what direction. Sometimes it is important to know how motion is changing. Is the approaching train going faster and faster or slowing to a stop? Is the in-line skater changing direction? The acceleration of an object describes how its velocity is changing. **Acceleration** is a change in velocity divided by the amount of time over which the change occurs. The change in velocity can be due to a change in speed, a change in direction, or both. An object is accelerating if it is speeding up, slowing down, or turning. A figure skater moving in a circle at constant speed is still accelerating because the direction of motion is changing.

TRY AT HOME Mini LAB

Modeling Acceleration

Procedure
1. Use **masking tape** to lay a course on the floor. Mark a start, and place marks along a straight path 10 cm, 40 cm, 90 cm, 160 cm, and 250 cm from the start.
2. Clap a steady beat. On the first beat, the person walking the course is at Start. On the second beat, the walker is on the first mark and so on. The walker is moving at a constant acceleration.

Analysis
1. Describe what happens to your speed as you move along the course. Infer what would happen if the course were extended farther.
2. Repeat step 2, starting at the other end of the course. Are you still accelerating? Explain.

Section Assessment

1. Compare and contrast velocity and acceleration.
2. You walk 200 m north, then 275 m south. What total distance did you walk?
3. In question 2, how far are you from your starting position? In what direction?
4. Why is a reference point needed to describe motion?
5. **Think Critically** What does it mean about your trip if your displacement is equal to the distance you traveled?

Skill Builder Activities

6. **Recording Data** Mark off a short distance on the floor and use a stopwatch to time yourself walking slowly, walking at a moderate pace, and walking quickly. Calculate your speed in each case. **For more help, refer to the** Science Skill Handbook.
7. **Solving One-Step Equations** A car is moving forward at a rate of 30 m/s. How far will the car have traveled after 4 s? **For more help, refer to the** Math Skill Handbook.

SECTION 1 Describing Motion

SECTION 2 Forces

As You Read

What You'll Learn
- **Identify** force, balanced forces, and net force.
- **Explain** friction.

Vocabulary
force
inertia
balanced forces
friction
gravity

Why It's Important
Forces cause every change in motion you can see around you.

What is a Force?

When you pedal your bike to speed up, you push against pedals. When you want to slow down, you pull on the brake levers. In both cases you exerted a force on the bike. A **force** is a push or a pull.

A force has size and direction. For example, pushing this book from the side will cause it to slide across the desk. However, pushing downward on the book will not cause it to move. The direction of a force is important.

Force and Change in Motion

When you push or pull on something you change its position and its motion. Look at **Figure 9.** When a book is sitting on the desk, pushing on it starts it moving. By applying a force, you changed the position of the book. You push or pull on a door to make it open or close. You bring a moving shopping cart to a stop by pushing or pulling on it. In all these cases, the position and motion of something changed when a force was applied.

In all of these cases, the motion of an object changes when a force acts on it. A force makes an object speed up, slow down, or change direction. Recall that an object accelerates when it speeds up, slows down, or changes direction. So, when a force acts on something, the force causes it to accelerate.

Figure 9
A force causes the motion of each of these objects to change. *How is each object accelerating?*

110 CHAPTER 4 Forces and Motion

Inertia and Mass How does an object respond to a force? If you give this book a push, you can slide it off your desk. However, if you give a car the same push, you know the car won't move. Some moving objects are easy to stop. It's easy to stop a moving basketball but hard to stop a rolling car. Compared to the book or the basketball, a car has more inertia. **Inertia** measures an object's tendency to resist changing its motion. The more inertia an object has, the harder it is to start the object moving or to slow it down.

Inertia depends on the amount of matter in an object, or its mass. The more inertia an object has, the less effect a force has on its motion. Imagine trying to push a toy truck compared with pushing a full-sized automobile, as shown in **Figure 10**. As the object becomes more massive, it becomes harder to move.

Figure 10
The more mass a vehicle has, the more inertia it has. It will take more force to start it moving.
Which vehicle has more inertia?

Balanced Forces Sometimes more than one force acts on an object. Suppose you and a friend are pushing on a chest of drawers, as shown in **Figure 11**. If you push on opposite sides, but with the same force, the chest doesn't move. It's as if no forces were acting on the chest. The two equal forces acting in opposite directions have canceled each other. Forces that cancel each other are called **balanced forces.** If the forces acting on something are balanced, its motion doesn't change.

Figure 11
The motion of an object depends on the direction of the forces applied, as well as how strong the forces are.

A If the forces are opposite and equal, the chest won't move.

B If they are opposite but unequal, the chest will move in the direction of the stronger force.

C If the forces are in the same direction, the forces add together.

SECTION 2 Forces

Figure 12
Friction occurs between any two surfaces that slide past each other.

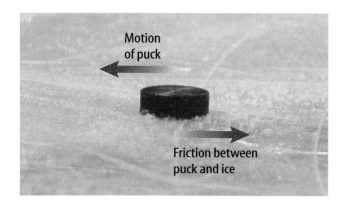

A Friction slows the puck and brings it to a stop.

B Friction between your shoes and the skateboard keeps you from sliding off as you slow down.

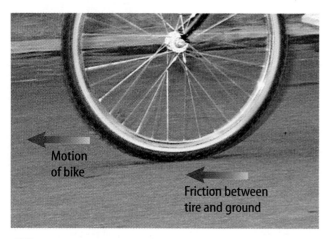

C Friction between the tire and the ground helps wheels turn without slipping.

Unbalanced Forces If both of you push on the same side of the chest, it moves faster than if only one of you pushed. The forces exerted by each of you were in the same direction. These forces added together to form the net force. If you push on one side of the chest and your friend pushes on the other side, how will the chest move? If you push harder, the chest will move in the direction you are pushing. However, the chest will move more slowly than if you were pushing alone. When the two forces are in opposite directions, the net force is the difference between the forces. Unbalanced forces do not cancel each other. If more than one force acts on an object, its motion will change only if the forces acting on it are unbalanced.

Reading Check *What is the difference between balanced and unbalanced forces?*

Friction

Rub your hand on the top of your desk. You can feel a force on your hand, which slows it. This force is **friction,** which resists motion between two touching surfaces. The rougher the surfaces are, the greater the friction is. Friction can be reduced by making the surfaces smoother. This is sometimes done by applying oil or grease to the surfaces.

Friction is present in almost all motion. **Figure 12** shows some examples of friction. When a hockey puck slides across the ice, friction between the puck and the ice slows the puck down and makes it stop. When you ride on a skateboard, friction between your feet and the board keeps you from sliding off when you start and stop. When you ride a bike, friction keeps the wheels from slipping on the ground.

Air Resistance Place your hand in the air rushing past a moving car and you can feel a force pushing on it. This force is air resistance, which is a type of friction. Air resistance slows objects moving through the air. For example, air resistance on an open parachute slows a falling skydiver.

112 CHAPTER 4 Forces and Motion

Gravity

If you hold a tennis ball and then let it go, the tennis ball falls to the ground. The motion of the tennis ball changed after you let it go, so a force was acting on it. The force pulling it down to the ground is gravity. **Gravity** is the pull that all matter exerts on other matter. When you dropped the tennis ball, Earth pulled the tennis ball downward.

Earth and the tennis ball exerted a gravitational pull on each other, even though they weren't touching. However, the force of gravity between two objects becomes weaker as the objects get farther apart. Also, the gravitational force is weaker if the mass of the objects is less. For example, gravitational pull exists between you and the chair across the room. However, the mass of the chair is much less than Earth's mass so the gravitational pull between you and the chair is so small that you don't even notice it.

Research Visit the Glencoe Science Web site at **tx.science.glencoe.com** for more information about how gravity acts between astronomical objects such as the sun and planets. Communicate to your class what you learn.

Mass and Weight Do you know how much you weigh? Is your weight the same thing as your mass? The answer is no. Weight and mass are different. Weight is the gravitational force between an object and the planet or moon where the object is. On Earth, your weight is the strength of the gravitational pull on you due to Earth's gravity. When you stand on a scale, you are measuring the pull of Earth's gravity. Because weight is a force, it is measured in newtons. Recall that mass is the amount of matter in an object. Mass is measured in kilograms. How would your weight change if you were far from Earth?

Section Assessment

1. What is a force?
2. As you sit, the force of gravity is pulling you toward the ground. Are the forces acting on you balanced or unbalanced? Explain.
3. When you stand on a slope, why don't you slide?
4. Explain the differences between inertia, mass, and weight.
5. **Think Critically** A skater pushes off with one foot and then glides in a circle. Is a net force acting on the skater? How can you tell?

Skill Builder Activities

6. **Forming Operational Definitions** You stand on a rickety box. If it is not strong enough to hold you, it collapses. Use this example to form operational definitions of the terms *balanced* and *unbalanced forces*. **For more help, refer to the** Science Skill Handbook.

7. **Communicating** In your Science Journal, describe all the forces that act on a ball when you toss it into the air and then catch it. When are forces balanced? When are they unbalanced? **For more help, refer to the** Science Skill Handbook.

SECTION 2 Forces

Activity

Toys in Motion

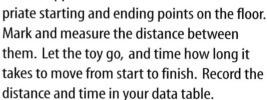

All around you are examples of moving objects. To determine an object's speed, two things must be measured. One is the distance the object moved. The other is the time it took to move that distance.

What You'll Investigate
How can you calculate the speed of moving toys?

Materials
meterstick
stopwatch
clock with a second hand
toys that move—spring-wound, battery-operated, or push toys
graph paper
calculator

Alternate materials

Goals
- **Demonstrate** that changes in motion can be measured and represented graphically.
- **Measure** distance and time using SI units.
- **Calculate** and graph speeds.

Safety Precautions

Procedure
1. On your paper, or with a computer, prepare a data table like the one shown below.

Toy Speeds			
Toy	Total Distance (m)	Time (s)	Speed (m/s)

2. Make a couple of test runs with the toy. Choose appropriate starting and ending points on the floor. Mark and measure the distance between them. Let the toy go, and time how long it takes to move from start to finish. Record the distance and time in your data table.
3. Repeat step 2 for four more toys.
4. With a calculator, **calculate** the speeds of the different toys and record them in your data table.
5. Create a bar graph showing the speeds of the toys. Put the names of the toys on the *x*-axis and the speeds on the *y*-axis.

Conclude and Apply
1. Does a relationship exist between the size of the toy and its speed? Explain. Compare your answers to your classmates' answers. Did they find the same relationship? Why or why not?
2. What forces made the different toys slow down or stop moving?
3. Is your speed calculation the same whether you measure distance traveled or displacement? Which should you measure?

Communicating Your Data

Based on your results, label each toy according to its behavior; for example, fastest toy, or curviest path. **For more help, refer to the Science Skill Handbook.**

SECTION 3 The Laws of Motion

Newton's Laws of Motion

What kinds of things have you seen in motion today? Maybe you saw cars starting, stopping, and turning on busy streets. Perhaps you saw airplanes flying beneath clouds that seemed to drift along. Maybe you watched the wind scramble a pile of leaves or saw the Moon rise. Cars, clouds, leaves, wind, and the Moon—all these things seem so different. However, their motion can be explained using only three rules.

These three rules are known as *Newton's laws of motion.* The laws were presented together for the first time by Isaac Newton in 1687. They explain how forces can change the motion of an object. These rules apply to all objects on Earth and in space. The same laws that describe the motion of a skateboard or the curved path of a batted ball also can predict the motion of the planets.

The First Law of Motion

When your friend holds the football on the tee and you give it a kick, the ball is put into motion as your foot hits it. In a basketball game, a teammate throws the ball to you, but you can't reach it, and the ball sails out of bounds. In both cases the motion of the balls can be explained by Newton's first law of motion. According to the **first law of motion,** an object will remain at rest, or keep moving in a straight line with constant speed, unless an unbalanced force acts on it.

Suppose you take a shot on the basketball court. After the ball leaves your hand, it doesn't travel in a straight line. Instead, its path curves downward and goes through the hoop, as shown in **Figure 13.** Why didn't the ball obey the first law of motion and travel in a straight line with constant speed? The force of gravity pulled the ball downward. Objects that are rolling or sliding are slowed by the force of friction. Here on Earth, gravity and friction keep objects from moving in a straight line with constant speed.

As You Read

What You'll Learn
- **Analyze** motion using Newton's laws.
- **Calculate** acceleration using force.

Vocabulary
first law of motion
second law of motion
third law of motion

Why It's Important
Newton's three laws will help you understand motion.

Figure 13
As a result of gravity, the ball doesn't move in a straight line with constant speed. Without gravity, the ball would follow the path shown by the dashed line.

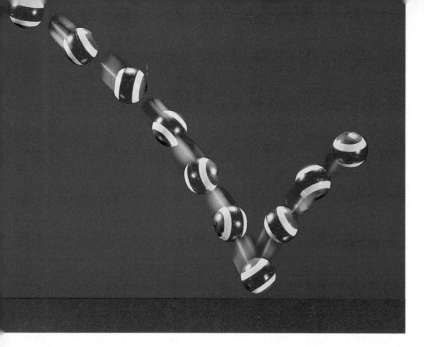

Figure 14
The ball was moving downward until it struck the floor and changed direction. According to Newton's first law of motion, the floor exerted a force on the ball.

Earth Science
INTEGRATION

The May 18, 1980 eruption of Mount Saint Helens in Washington included a lateral blast of volcanic material. Research the events that led to this famous explosion of ash and steam. Identify the force behind the eruption and the direction of the force that propelled ash and gas.

Understanding Motion It took a long time to understand motion. One reason was the people did not understand the behavior of friction, or that friction is a force. Because friction causes moving objects to stop, people thought the natural state of an object was to be at rest. For an object to be in motion, something had to be continuously pushing or pulling it. As soon as the force stopped, nature would bring the object to rest.

The sixteenth-century Italian scientist Galileo was one of the first to understand that an object in constant motion is as natural as an object at rest. It was usually the force of friction that made moving objects come to a stop. To keep an object moving, a force had to be applied to overcome the effects of friction. If friction could be removed, once an object was in motion, it would continue to move in a straight line with constant speed.

Newton's first law means that an object can speed up, slow down or change direction only if a force acts on it as shown in **Figure 14.** Only forces can cause changes in motion.

The Second Law of Motion

How do forces cause motion to change? Suppose you pick up your backpack when it's full of books. As you lift it, you change its motion. The backpack was at rest, and you caused it to move when you pulled it upward. Recall that anytime the motion of something changes, it is accelerating. The backpack went from being at rest to moving. When you pulled the backpack upward, you caused it to accelerate. In other words, the acceleration of the backpack was caused by the force you exerted on it. Newton's **second law of motion** says that an unbalanced force acting on an object causes the object to accelerate in the direction of the force.

Force, Mass, and Acceleration According to the second law of motion, acceleration can be calculated by dividing the unbalanced force exerted on an object by the mass of the object. This relationship between force, mass, and acceleration can be written as this equation:

$$a = F/m$$

Calculating Acceleration In this equation, a stands for the acceleration, F is the unbalanced force, and m is the mass of the object. Recall that force, mass, and acceleration have units. Force is measured in newtons (N), acceleration is measured in units of meters per second squared (m/s^2), and mass has units of kilograms (kg). The newton is a unit equal to the combined units kg · m/s^2.

You can calculate the acceleration of an object when you know the force exerted on the object and the mass of the object. For example, suppose you exert a force of 10 N on a basketball. The basketball has a mass of 1.0 kg. The acceleration can be calculated as shown below:

$$a = F/m$$
$$= 10 \text{ N}/1.0 \text{ kg}$$
$$= 10 \text{ m/s}^2$$

How does acceleration affect how you feel on a roller coaster? To find out more about acceleration and amusement park rides, see the **Amusement Park Rides Field Guide** at the back of the book.

Math Skills Activity

MATH TEKS
6.2 C; 6.8 B

Acceleration, Force and Mass

Example Problem

You have to push your 60-kg basketball hoop off of the driveway. The unbalanced force on the hoop due to your push and friction is 6 N. Find the acceleration.

1. This is what you know:
   force: $F = 6$ N
   mass: $m = 60$ kg

2. This is what you need to know:
   acceleration: a

3. This is the equation you need to use:
   $a = F/m$

4. Substitute the known values:
   $a = (6 \text{ N}) / (60 \text{ kg}) = 0.1 \text{ m/s}^2$

Check your answer by multiplying by the mass. Do you calculate the same force that was given?

Practice Problems

You are pushing a 3-kg lawn chair across the patio. You exert a force of 12 N. Find the acceleration.

For more help, refer to the **Math Skill Handbook.**

SECTION 3 The Laws of Motion

Figure 15 A tennis player's serve changes the motion of the tennis ball.

A The ball moves in a straight line when tossed upward.

B The motion of the ball changes direction when the racket exerts a force on it. *What other force is acting on the ball while it is in the air?*

Collect Data Visit the Glencoe Science Web site at **tx.science.glencoe.com** for data about accelerations for different vehicles. Communicate to your class what you learn.

Acceleration and Direction According to the second law of motion, when a force acts on an object, its acceleration is in the same direction as the force. If you pull on a wagon that is at rest, the wagon starts moving in the same direction as your pull. When a tennis player serves the ball, the ball changes direction and moves in the direction of the force exerted by the racket, as shown in **Figure 15.**

What if a soccer ball comes rolling toward you and you put out your foot to stop it? The force of your foot was opposite to the motion of the ball, so it slowed to a stop. The ball has acceleration, but it is in a direction opposite to its motion.

Acceleration and Force Acceleration is related to change in motion. The greater the acceleration is, the faster the motion of an object changes. Look at the equation for acceleration. What happens to the acceleration of a 1-kg basketball if the force is doubled from 10 N to 20 N? Then the acceleration is doubled to 20 m/s^2. If the force exerted on the object becomes larger, the acceleration becomes larger. How does the speed of a baseball change if you hit it hard or if you just nudge it with the bat? The ball moves faster when you hit it hard. According to the second law of motion, a larger force on an object causes a larger acceleration.

 **Reading Check** *How are acceleration and force related?*

Acceleration and Mass How does acceleration depend on the mass of an object? In the previous example, what happens to the acceleration if the ball is a softball instead of a baseball? A softball has a greater mass than a baseball. If the force that is used to hit the ball remains the same, the acceleration is less for the more massive softball. Likewise, if your backpack is full of books, you have to pull hard to lift it off your desk. If it's empty, the same force causes it to move more quickly. When the backpack is full of books, its mass is larger. If the same amount of force is exerted, the object with the smaller mass has a larger acceleration.

The Third Law of Motion

How high can you jump? Think about the forces acting on you when you jump. Gravity is pulling you downward, so an upward force must be exerted on you that is greater than the force of gravity. Where does this force come from? Maybe you think it comes from your legs and feet, pushing you upward. You're partly right.

Look at **Figure 16.** Your legs and feet push downward on the ground. In response, the ground pushes upward on you. This is the force that enables you to leave Earth for an instant. Newton's third law of motion describes how objects exert forces on each other. According to the **third law of motion,** when a force is applied on an object, an equal force is applied by the object in the opposite direction. When you pushed down on the ground, the ground pushed back up on you.

Earth Science INTEGRATION

The forces that move the continental plates are large. The motion produced, however, is small because the masses are so large. Research how fast Earth's plates are moving and write a paragraph in your Science Journal describing what you find.

Figure 16
When you jump, you are using the third law of motion. You exert a force on the ground, and the ground exerts an equal force on you.

Mini LAB

Demonstrating the Third Law of Motion

Procedure
1. Blow up a **balloon** to maximum size. Hold its end to prevent air from escaping.
2. **Tape** the center of the balloon to a small, **toy car** so the opening of the balloon points backward.
3. Set the car on the floor and release the air in the balloon.

Analyze
1. Describe how the toy car moved when you let go of the balloon.
2. How does Newton's third law of motion explain what you observe?

Figure 17
The swimmer exerts a force on the water, and the water exerts a reaction force on the swimmer.

Force Pairs The forces exerted on an object and by the object are called a force pair. The force pairs act in opposite directions and are always equal in size. Therefore, to jump higher, you must push harder on the ground. Then the ground pushes back harder on you.

You might think that if force pairs are equal in size and in opposite directions, they must cancel out. But remember that force pairs act on different objects. When you push on the ground, you exert a force on Earth, but because Earth has so much mass, your force has no noticeable effect on its motion. Your mass is much smaller than that of Earth, so the same size force causes you to spring into the air. Only if equal and opposite forces act on the same object do they cancel out.

Action and Reaction According to the third law of motion forces always act in pairs. When you push on a wall, the wall pushes back on you. One force of the force pair is called the action force, and the other is called the reaction force. Your push on the wall is the action force, and the wall pushing back on you is the reaction force. For every action force, there is a reaction force that is equal in size but opposite in direction. Every time you push on an object, the object pushes back on you. Every time you pull on an object, the object pulls back on you.

Examples of Action and Reaction If you swing a hammer and hit a nail, you know what happens. The hammer was moving before it hit the nail but stopped moving after it hit the nail. The change in motion of the hammer indicates that a force acted on it. The hammer exerted a force on the nail. At the same time, the nail exerted a force on the hammer, which caused it to stop moving. The action force was the force the hammer exerted on the nail. The reaction force was the force the nail exerted on the hammer.

Sometimes the action and reaction forces are hard to identify. In **Figure 17,** when the swimmer pushes his hands through the water, he exerts a force on the water. Even though the water isn't a solid, such as a wall, the water still exerts a reaction force on the swimmer. This is the force that causes the swimmer to move forward.

120 **CHAPTER 4** Forces and Motion

Applications of Newton's Laws of Motion

Newton's ideas about motion have been tested many times and have been found to apply to objects everywhere. This is why they are called laws. They apply in outer space and to large objects like planets as well as they do to small objects here on Earth. You experience examples of Newton's laws every day. When you walk, your feet push on the ground. According to the third law of motion, the ground then pushes on you, and you move forward, as shown in **Figure 18.** When you bang your shin on a piece of furniture, it hurts. You pushed on the furniture with your shin, and the furniture pushed back on your shin. Can you see other examples of Newton's laws of motion in your everyday life?

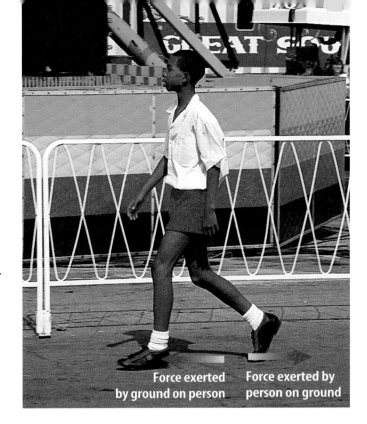

Force exerted by ground on person   Force exerted by person on ground

Figure 18
When you walk, the force exerted on you by the ground causes you to move forward.

✔ **Reading Check** *Under what circumstances do the laws of motion apply?*

Section 3 Assessment

1. The forces acting on an object are balanced. What do Newton's first and second laws say about the motion of the object?
2. Suppose you are sitting still in a chair. According to Newton's second law of motion, are the forces acting on you balanced or unbalanced? Explain.
3. Compare the acceleration of a 5-kg object acted on by forces of 1 N and 2 N.
4. When you jump upward, you slow down and eventually fall back to Earth. Which law of motion explains this?
5. **Think Critically** Use Newton's laws of motion to explain why a large predator might have trouble pursuing a smaller prey animal that changes direction quickly during the chase.

Skill Builder Activities

6. **Comparing and Contrasting** Compare and contrast Newton's second law of motion with his third law of motion. Use at least two examples of objects in motion from everyday experience to demonstrate the similarities and differences in how the two laws describe motion. Build or display a model to show one of your examples. **For more help, refer to the** Science Skill Handbook.
7. **Solving One-Step Equations** A student standing on an icy parking lot throws a ball with a force of 20 N. If friction is ignored, and the student has a mass of 60 kg, what is the student's acceleration? **For more help, refer to the** Math Skill Handbook.

SECTION 3 The Laws of Motion

Activity
Use the Internet

Space Shuttle Speed

The space shuttle is a fast-moving machine. When it's in orbit, the Mission Control Center at the Johnson Space Center in Houston, Texas, keeps track of the shuttle's motion. The first shuttle mission in 1981 traveled more than 1 million km during a flight lasting more than 54 h. That's an average speed of almost 32,000 km/h. Investigate space shuttle data to find how far the shuttle has flown on each mission as well as the amount of time each mission lasted. With that data, you'll be able to calculate the average speed at which the space shuttle flies.

Recognize the Problem

How fast does the space shuttle fly?

Form a Hypothesis

What is the average speed of each mission? Which mission had the fastest average speed? Which mission had the slowest average speed? Why was there a difference in speeds among the different missions?

Goals
- **Identify** distance and flight duration for five space shuttle missions.
- **Calculate** average speed for each of the five space shuttle missions.
- **Compare** average speed data for five space shuttle missions.

Data Source
Go to the Glencoe Science Web site at **tx.science.glencoe.com** to get more information about calculating average speed, and for data collected by other students.

Shuttle Speed Data					
	Shuttle Missions				
	Mission 1	Mission 2	Mission 3	Mission 4	Mission 5
Launch Date					
Total Distance traveled (km)					
Total Time (h)					
Average Speed (km/h)					

122 CHAPTER 4 Forces and Motion

Using Scientific Methods

Test Your Hypothesis

Plan

1. In order to compare average speed data for five space shuttle missions, find out the total distance traveled and the time duration for each mission.

2. Review the time duration data. Make sure that the data represents the number of hours of the mission for each mission. You will need to convert values expressed in days, minutes, and seconds to hours in order to calculate the average speed.

3. Find out the space shuttle's average speed for each of the missions. To do this, divide the distance (in kilometers) by the time in hours. This will give you the average speed for the mission in km/h.

Do

1. Make sure your teacher approves your plan before you start.

2. Go to the Glencoe Science Web site at **tx.science.glencoe.com** to post your data.

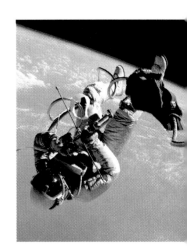

Analyze Your Data

1. For each shuttle mission you investigated, determine the average speed for that mission.

2. Which of the missions had the fastest average speed?

3. Which of the missions had the slowest average speed?

4. What is the difference between the fastest average speed and the slowest average speed?

Draw Conclusions

1. Find this *Use the Internet* activity on the Glencoe Science Web site at **tx.science.glencoe.com.** Post your data in the table provided. Study other students' data and compare them to your data. Review data that other students entered.

2. Does the total mission time affect the average speed? Why or why not?

3. What factors affect the average speed for each mission?

4. **Calculate** the range in the average speeds by subtracting the smallest average speed from the largest average speed. What factors could make the average speeds different?

Once you've compared data with others on the web site, communicate to your class as a whole—especially any major differences. Discuss those differences.

Science and Language Arts

"Rayona's Ride"
from the novel *A Yellow Raft in Blue Water*
by Michael Dorris

Respond to the Reading

1. If the horse and Rayona are similar to a "wind-up toy that has been turned one twist too many," what kind of motion can the reader expect them to have?
2. To what animal does Rayona compare herself?
3. Where does the story take place?

The sounds of the rodeo around me fade in my concentration. There's a drone in my ears that blocks out everything else, pasts and futures and long-range worries. The horse and I are held in a vise[1], a wind-up toy that has been turned one twist too many, a spring coiled beyond its limit.

"Now!" I cry, aloud or to myself I don't know. Everything has boiled down to this instant. There's nothing in the world except in the hand of the gate judge, lowering in slow motion to the catch that contains us....

Wheeling and spinning, tilting and beating, my breath the song, the horse the dance. Time is gone. All the ordinary ways of things, the gettings from here to there, the one and twos, forgot. The crowd is color, the whirl of a spun top. The noises blend into a waving band that flies around us like a ribbon on a string. Beneath me four feet dance, pounding and leaping and turning and stomping. My legs flap like wings. I sail above, first to one side, then the other, remembering more than feeling the slaps of our bodies together. Things happen faster than understanding, faster than ideas. I'm a bird coasting, shot free into the music, spiraling into a place without bones or weight. I'm on the ground. Unmoving. The heels of my hands sunk in the dust of the arena....

[1] a device used to hold something tightly in place

Coming-of-Age Story

Understanding Literature

Coming-of-Age Story "Rayona's Ride" is a chapter in the book *A Yellow Raft in Blue Water*, which is a coming-of-age story. A coming-of-age story describes a young person's growing up and maturing. The passage you just read is a turning point or climax for Rayona, the 15-year-old narrator. A climax is the point in the story with the greatest interest or suspense. The climax signals the moment of Rayona's growing up. What signs in the passage show that this is an important moment?

Science Connection Reference points are needed to describe the motion of objects. In *Rayona's Ride*, Rayona's motion on her horse is described in relation to different reference points throughout the ride. As the ride begins, Rayona uses herself as the reference point and sees the crowd as moving like a ribbon on a string around her. At the end of her ride, Rayona uses Earth as a reference point to describe her motion when she spirals through the air and lands motionless on the ground. The reference points during Rayona's ride change depending on her movement on or off the horse.

Linking Science and Writing

Write a Story Write the climax of your own coming-of-age story. Write about a time in your life that represents the moment when you went from being a young person to a young adult.

Career Connection

Test Pilot and Flight Test Engineer

Nadia Roberts is a lecturer and engineering instructor at the National Test Pilot School in Mojave, California. There, she teaches courses for test pilots and flight test engineers. A test pilot or flight engineer must make sure that airplanes can safely do what they were designed to do. They also study the interaction between humans and aircraft. This means that the test pilot spends a lot of time flying new or updated airplanes. If something does not work the way it should, the test pilot must determine the problem.

What might pilots use as reference points?

SCIENCE *Online* To learn more about careers in aeronautics, visit the Glencoe Science Web site at **tx.science.glencoe.com**.

Chapter 4 Study Guide

Reviewing Main Ideas

Section 1 Describing Motion

1. Motion is described relative to a reference point. *What reference points could you use to describe the motion of the lead goose?*

2. Motion can be measured in two ways. Distance is the total path traveled. Displacement includes the distance and direction from the starting point.

3. Speed measures how fast you cover a given distance.

$$\text{average speed} = \frac{\text{total distance}}{\text{total time}}$$

Velocity includes speed and direction.

4. An object that accelerates speeds up, slows down, or turns.

Section 2 Forces

1. A force is a push or a pull. It has a size and a direction. An object acted on by an unbalanced force will change its motion and accelerate.

2. The tendency of an object to resist changing its motion is called inertia. Inertia depends on the amount of mass an object has. *Which dumbbell in the photo has the most inertia?*

3. Balanced forces cancel each other. When forces are unbalanced, a change in motion will occur.

4. Friction is the force that resists motion between two surfaces that are in contact. Gravity is a force that acts between all objects that have mass.

Section 3 The Laws of Motion

1. Newton's first law of motion states that an object will remain at rest or move at constant speed until an unbalanced force acts on it. *How does Newton's first law explain the motion of this baseball?*

2. Newton's second law of motion states that an object acted on by an unbalanced force will accelerate in the direction of the force according to the following formula.

$$\text{acceleration} = \frac{\text{force}}{\text{mass}}$$

3. Newton's third law of motion states that objects exert forces on each other that are equal, but in opposite directions.

FOLDABLES
Reading & Study Skills

After You Read

On the front of the tabs of your Cause and Effect Study Fold describe how the size and direction of a force affect motion.

Chapter 4 Study Guide

Visualizing Main Ideas

Complete the following concept map about motion.

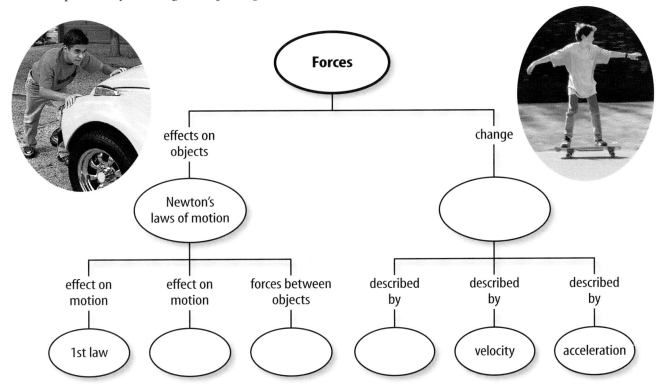

Vocabulary Review

Vocabulary Words

a. acceleration
b. balanced forces
c. displacement
d. first law of motion
e. force
f. friction
g. gravity
h. inertia
i. second law of motion
j. speed
k. third law of motion
l. velocity

 **Study Tip**

Make flashcards for new vocabulary words. Put the word on one side and the definition on the other. Use them to quiz yourself.

Using Vocabulary

Give the vocabulary word that answers each question.

1. What can be stated as "for every action, there is an equal and opposite reaction"?
2. What is a measure of speed and direction?
3. What acts on a motionless object?
4. What describes how an object speeds up, slows down, or turns?
5. What force resists motion between two surfaces?
6. What property explains why a heavy object is harder to move than a light one?
7. What causes motion to change?

CHAPTER STUDY GUIDE 127

Chapter 4 Assessment & TEKS Review

Checking Concepts

Choose the word or phrase that best answers the question.

1. A push is an example of what?
   A) acceleration   C) displacement
   B) velocity   D) force

2. What force slows a sliding box?
   A) acceleration   C) velocity
   B) friction   D) inertia

3. Which of these is not a unit for speed?
   A) m/s   C) cm/s^2
   B) km/h   D) m/day

4. Which of the following would NOT include a direction?
   A) force   C) velocity
   B) acceleration   D) distance

5. What causes a dropped coin to fall?
   A) inertia   C) gravity
   B) velocity   D) friction

6. What is an object's resistance to changing its motion called?
   A) force
   B) gravity
   C) acceleration
   D) inertia

7. An unbalanced force to the left accelerates an object in what direction?
   A) right
   B) left
   C) forward
   D) depends on initial velocity of object

8. A 10-N force is exerted on an object with a mass of 2 kg. What is the acceleration of the object?
   A) 20 m/s^2
   B) 10 m/s^2
   C) 5 m/s^2
   D) depends on the direction of the force

9. You are riding a bike. Which of the following is an example of balanced forces?
   A) You pedal to speed up.
   B) You turn at constant speed.
   C) You coast to slow down.
   D) You pedal at constant velocity.

10. You push against a stationary wall with a force of 20 N forward. What is the force the wall exerts on you?
    A) 20 N backward   C) 10 N backward
    B) 20 N forward   D) 10 N forward

Thinking Critically

11. A baseball is pitched east at 40 km/h. A batter hits it so it moves west at 40 km/h. Did the ball accelerate? Explain.

12. Maureen walked 3 km east in 0.75 h. It took 0.3 h to jog back to her starting point. What was her velocity in each direction?

13. A 200-N net force is applied to a 40-kg object and a 10-kg object. Which one accelerates more? By how much? Explain.

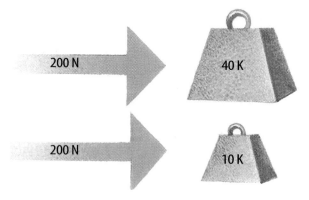

14. A 20-N force is applied to a 5-kg object. It does not change its motion. What could explain this?

15. A box with a weight of 500 N is placed on a table that can support 450 N. Are the forces balanced? What happens?

Chapter 4 Assessment

Developing Skills

16. Comparing and Contrasting Compare and contrast displacement and distance.

17. Making and Using Graphs Make a distance-time graph for a person walking at 1.5 m/s for 10 s.

18. Identifying and Manipulating Variables and Controls You use a spring scale to apply the same force to pull 1-kg, 2-kg, and 4-kg objects. What remains constant? What is the variable?

19. Making Models Use spring scales to demonstrate Newton's third law of motion.

20. Making and Using Graphs Analyze the graph below. What is the speed between 20 s and 40 s? What is the acceleration between 20 s and 40 s?

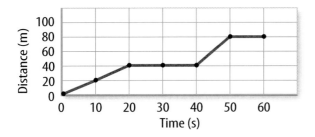

Performance Assessment

21. Oral Presentation Research the science of motion used by manufacturers of sports equipment. What variables affect the motion? Present your findings in a speech to your class.

Technology

Go to the Glencoe Science Web site at **tx.science.glencoe.com** or use the **Glencoe Science CD-ROM** for additional chapter assessment.

TAKS Practice

A map is a tool that can be used to find your way or to trace a route that you have taken. The map below shows the pathway a sixth-grade student took as she walked from her home to a friend's house after school one afternoon. Study the map to find out about the total distance the student walked and to find out about her displacement. *TEKS 6.6 B*

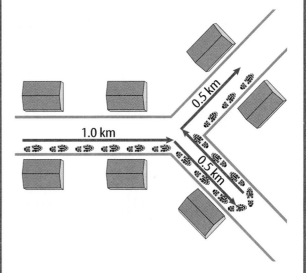

Study the illustration and answer the following questions.

1. According to the map, what is the total distance traveled by the student?
 A) 0.5 km **C)** 1.5 km
 B) 1.0 km **D)** 2.5 km

2. If the whole walk took the student 1 h, what was her average speed?
 F) 1.0 km/h **H)** 2.5 km/h
 G) 1.5 km/h **J)** 3.5 km/h

3. If the student walks back home later, what is her total displacement?
 A) 0.0 km **C)** 1.5 km
 B) 1.0 km **D)** 2.5 km

CHAPTER ASSESSMENT **129**

CHAPTER 5

Science TEKS 6.8 A, B; 6.9 A, B, C

Energy

Volcanoes, earthquakes, lightning, and hurricanes produce some of the most powerful forces in nature. Every one of these phenomena contains a tremendous amount of energy. The river of lava shown in this picture flowing from Mount Etna in Italy has heat energy, light energy, and energy of motion. In this chapter, you will learn about different forms and sources of energy. You also will learn how energy can be transformed from one form into another, and how some forms of energy can be used.

What do you think?

Science Journal Look at the picture below with a classmate. Discuss what is happening. Here's a hint: *Concentrating energy is the key to what is happening here.* Write your answer or best guess in your Science Journal.

 A marble and a piece of wood are on a countertop. If nothing disturbs them, they will remain there. However, if you tilt the wood and roll the marble down the slope, the marble acquires a new property—the ability to do something.

Analyze a marble launch

1. Make a track by slightly separating two metersticks placed side by side.
2. On a table, raise one end of the track slightly and measure the height.
3. Roll a marble down the track. Measure the distance from its starting point to where it hits the floor. Repeat. Calculate the average of the two measurements.
4. Repeat steps 2 and 3 for three different heights. Predict what will happen if you use a heavier marble. Test your prediction and record your observations.

Observe

In your Science Journal, describe your experiment and what you discovered. How did the different heights cause the distance to change?

Before You Read

FOLDABLES Reading & Study Skills

Making a Know-Want-Learn Study Fold Make the following Foldable to help identify what you already know and what you want to know about energy.

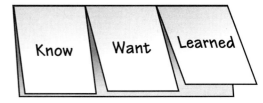

1. Place a sheet of paper in front of you so the long side is at the top. Fold the paper in half from top to bottom.
2. Fold both sides in. Unfold the paper so three sections show.
3. Through the top thickness of paper, cut along each of the fold lines to the topfold, forming three tabs. Label the tabs *Know*, *Want*, and *Learned*.
4. Before you read the chapter, write what you know and what you want to know under the tabs. As you read the chapter, correct what you have written and add more questions.

SECTION 1 What is energy?

As You Read

What You'll Learn
- **Explain** what energy is.
- **Distinguish** between kinetic energy and potential energy.
- **Identify** the various forms of energy.

Vocabulary
energy
kinetic energy
potential energy
thermal energy
chemical energy
radiant energy
electrical energy
nuclear energy

Why It's Important
Energy is the source of all activity.

Figure 1
Energy is the ability to cause change. *How can these objects cause change?*

 Grade 5 TEKS Review
For a review of the Grade 5 TEKS *Forms of Energy* see page 505.

The Nature of Energy

What comes to mind when you hear the word *energy?* Do you picture running, leaping, and spinning like a dancer or a gymnast? How would you define energy? When an object has energy, it can make things happen. In other words, **energy** is the ability to cause change. What do the items shown in **Figure 1** have in common?

Look around and notice the changes that are occurring—someone walking by or a ray of sunshine that is streaming through the window and warming your desk. Maybe you can see the wind moving the leaves on a tree. What changes are occurring?

Transferring Energy You might not realize it, but you have a large amount of energy. In fact, everything around you has energy, but you notice it only when a change takes place. Anytime a change occurs, energy is transferred from one object to another. You hear a footstep because energy is transferred from a foot hitting the ground to your ears. Leaves are put into motion when energy in the moving wind is transferred to them. The spot on the desktop becomes warmer when energy is transferred to it from the sunlight. In fact, all objects, including leaves and desktops, have energy.

132 CHAPTER 5 Energy

Energy of Motion

Things that move can cause change. A bowling ball rolls down the alley and knocks down some pins, as in **Figure 2A.** Is energy involved? A change occurs when the pins fall over. The bowling ball causes this change, so the bowling ball has energy. The energy in the motion of the bowling ball causes the pins to fall. As the ball moves, it has a form of energy called kinetic energy. **Kinetic energy** is the energy an object has due to its motion. If an object isn't moving, it doesn't have kinetic energy.

Kinetic Energy and Speed If you roll the bowling ball so it moves faster, what happens when it hits the pins? It might knock down more pins, or it might cause the pins to go flying farther. A faster ball causes more change to occur than a ball that is moving slowly. Look at **Figure 2B.** The professional bowler rolls a fast-moving bowling ball. When her ball hits the pins, pins go flying faster and farther than for a slower-moving ball. All that action signals that her ball has more energy. The faster the ball goes, the more kinetic energy it has. This is true for all moving objects. Kinetic energy increases as an object moves faster.

Kinetic Energy and Mass Suppose, as shown in **Figure 2C,** you roll a volleyball down the alley instead of a bowling ball. If the volleyball travels at the same speed as a bowling ball, do you think it will send pins flying as far? The answer is no. The volleyball might not knock down any pins. Does the volleyball have less energy than the bowling ball even though they are traveling at the same speed? An important difference between the volleyball and the bowling ball is that the volleyball has less mass. Even though the volleyball is moving at the same speed as the bowling ball, the volleyball has less kinetic energy because it has less mass. Kinetic energy also depends on the mass of a moving object. Kinetic energy increases as the mass of the object increases.

✔ **Reading Check** *Why does a volleyball knock over fewer pins than a bowling ball?*

Figure 2
The kinetic energy of an object depends on two quantities.
What are those quantities?

A This ball has kinetic energy because it is rolling down the alley.

B This ball has more kinetic energy because it has more speed.

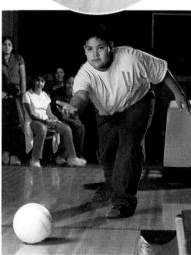

C This ball has less kinetic energy because it has less mass.

Figure 3
The potential energy of an object depends on its mass and height above the ground. *Which vase has more potential energy, the red one or the blue one?*

Energy of Position

An object can have energy even though it is not moving. For example, a glass of water sitting on the kitchen table doesn't have any kinetic energy because it isn't moving. If you accidentally nudge the glass and it falls on the floor, changes occur. Gravity pulls the glass downward, and the glass has energy of motion as it falls. Where did this energy come from?

When the glass was sitting on the table, it had potential (puh TEN chul) energy. **Potential energy** is the energy stored in an object because of its position. In this case, the position is the height of the glass above the floor. The potential energy of the glass changes to kinetic energy as the glass falls. The potential energy of the glass is greater if it is higher above the floor. Potential energy also depends on mass. The more mass an object has, the more potential energy it has. Which object in **Figure 3** has the most potential energy?

Forms of Energy

Food, sunlight, and wind have energy, yet they seem different because they contain different forms of energy. Food and sunlight contain forms of energy different from the kinetic energy in the motion of the wind. The warmth you feel from sunlight is another type of energy that is different from the energy of motion or position.

Thermal Energy The feeling of warmth from sunlight signals that your body is acquiring more thermal energy. All objects have **thermal energy** that increases as its temperature increases. A cup of hot chocolate has more thermal energy than a cup of cold water, as shown in **Figure 4.** Similarly, the cup of water has more thermal energy than a block of ice of the same mass. Your body continually produces thermal energy. Many chemical reactions that take place inside your cells produce thermal energy. Where does this energy come from? Thermal energy released by chemical reactions comes from another form of energy called chemical energy.

Figure 4
The hotter an object is, the more thermal energy it has. A cup of hot chocolate has more thermal energy than a cup of water, which has more thermal energy than a block of ice with the same mass.

Figure 5
Complex chemicals in food store chemical energy. During activity, the chemical energy transforms into kinetic energy.

Chemical Energy When you eat a meal, you are putting a source of energy inside your body. Food contains chemical energy that your body uses to provide energy for your brain, to power your movements, and to fuel your growth. As in **Figure 5,** food contains chemicals, such as sugar, which can be broken apart in cells. These chemicals are made of atoms that are bonded together, and energy is stored in the bonds between atoms. **Chemical energy** is the energy stored in chemical bonds. When chemicals are broken apart and new chemicals are formed, some of this energy is released. The flame of a candle is the result of chemical energy stored in the wax. When the wax burns, chemical energy is transformed into thermal energy and light energy.

Light Energy Light from the candle flame travels through the air at an incredibly fast speed of 300,000 km/s. This is fast enough to circle Earth almost eight times in 1 s. When light strikes something, it can be absorbed, transmitted, or reflected. If the light is absorbed, it will cause the object to warm up. In other words, the thermal energy of the object has increased because light transferred energy to it. The type of energy light carries is called **radiant energy**. **Figure 6** shows a coil of wire that produces radiant energy when it is heated. To heat the metal, another type of energy can be used—electrical energy.

Figure 6
Electrical energy is transformed into thermal energy in the metal heating coil. As the metal becomes hotter, it emits more radiant energy.

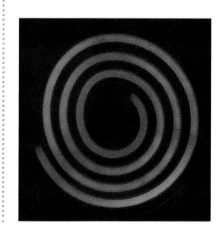

✓ **Reading Check** *How do you know that light has energy?*

SECTION 1 What is energy? **135**

Electrical Energy Electrical lighting is one of the many ways electrical energy is used. Look around at all the devices that use electrical energy. The electric current that comes out of batteries and wall sockets carries **electrical energy.** The amount of electrical energy depends on the voltage. The current out of a 120-V wall socket carries more energy than the current out of a 1.5-V battery. To produce the enormous quantities of electrical energy consumed each day, large power plants are needed. In the United States, about 20 percent of the electrical energy that is generated comes from nuclear power plants.

Figure 7
Complex power plants are required to obtain useful energy from the nucleus of an atom.

Nuclear Energy Nuclear power plants use the energy stored in the nucleus of an atom to generate electricity. Every atomic nucleus contains energy—**nuclear energy**—that can be transformed into other forms of energy. However, releasing the nuclear energy is a difficult process. It involves the construction of complex power plants, as shown in **Figure 7.** In contrast, all that is needed to release chemical energy from wood is a match.

Section Assessment

1. How do you know if an object has energy? Do you have energy? Does a rock?
2. Contrast chemical and nuclear energy.
3. How can chemical energy transform into thermal energy? Into light energy?
4. If two vases are side by side on a high shelf, could one have more potential energy than the other? Explain.
5. **Think Critically** A golf ball and a bowling ball have the same kinetic energy. Which one is moving faster? Explain your answer using what you know about kinetic energy. Suppose the golf ball and the bowling ball have the same speed. Which of the two has more kinetic energy?

Skill Builder Activities

6. **Interpreting Data** Review your results from the Explore Activity. Where did the marble have the most kinetic energy? Where did the marble have the most potential energy? Can you infer a relationship between kinetic energy and potential energy based on your observations? **For more help, refer to the Science Skill Handbook.**

7. **Communicating** The term *energy* is used in everyday language. In your Science Journal, record different expressions and ways of using the word *energy.* Decide which ones match the definition of energy presented in this section. **For more help, refer to the Science Skill Handbook.**

SECTION 2

Energy Transformations

Changing Forms of Energy

Chemical, thermal, radiant, and electrical are some of the forms that energy can have. In the world around you, energy is transforming continually between one form and another. You observe some of these transformations by noticing a change in your environment. Forest fires are a dramatic example of an environmental change that can occur naturally as a result of lightning strikes. Another type of change, shown in **Figure 8,** is a mountain biker pedaling to the top of a hill. What energy transformations occur as he moves up the hill?

Tracking Energy Transformations As the mountain biker pedals, many energy transformations are taking place. In his leg muscles, chemical energy is transforming into kinetic energy. The kinetic energy of his leg muscle transforms into kinetic energy of the bicycle. Some of this energy transforms into potential energy as he moves up the hill. Also, some energy is transformed into thermal energy. His body is warmer because chemical energy is being released. Because of friction, the mechanical parts of the bicycle are warmer, too.

As You Read

What You'll Learn
- **Apply** the law of conservation of energy to energy transformations.
- **Identify** how energy changes form.
- **Describe** how electric power plants produce energy.

Vocabulary
law of conservation of energy
generator
turbine

Why It's Important
Many devices you use every day change energy from one form to another.

Figure 8
The ability to transform energy allows the biker to climb the hill.
Identify all the forms of energy that are represented in the photograph.

SECTION 2 Energy Transformations **137**

SCIENCE Online

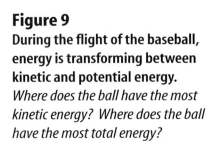

Research Visit the Glencoe Science Web site at **tx.science.glencoe.com** for more information about how energy changes form when it is transformed from one form to another. Use a spreadsheet program to summarize what you've learned.

The Law of Conservation of Energy

It can be a challenge to track energy as it moves from object to object. However, one extremely important principle can serve as a guide as you trace the flow of energy. According to the **law of conservation of energy,** energy is never created or destroyed. The only thing that changes is the form in which energy appears. When the biker is resting at the summit, all his original energy is still around. Some of the energy is in the form of potential energy, which he will use as he coasts down the hill. Some of this energy was changed to thermal energy by friction in the bike. Chemical energy was also changed to thermal energy in the biker's muscles, making him feel hot. As he rests, this thermal energy moves from his body to the air around him. No energy is missing—it can all be accounted for.

✓ **Reading Check** *Can energy ever be lost? Why or why not?*

Changing Kinetic and Potential Energy

The law of conservation of energy can be used to identify the energy changes in a system, especially if the system is not too complicated. For example, tossing a ball into the air and catching it is a simple system. As shown in **Figure 9,** as the ball leaves your hand, most of its energy is kinetic. As the ball rises, it slows and loses kinetic energy. But, the total energy of the ball hasn't changed. The loss of kinetic energy equals the gain of potential energy as the ball flies higher in the air. The total amount of energy always remains constant. Energy moves from place to place and changes form, but it never is created or destroyed.

Figure 9
During the flight of the baseball, energy is transforming between kinetic and potential energy. *Where does the ball have the most kinetic energy? Where does the ball have the most total energy?*

Figure 10
Hybrid cars that use an electric motor and a gasoline engine for power are now available. These cars get up to 70 mpg. Inventions such as the hybrid car make energy transformations more efficient.

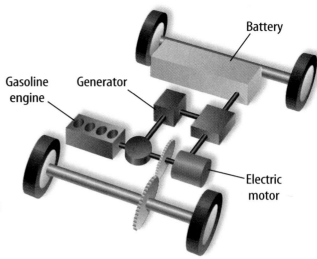

Energy Changes Form

Energy transformations occur constantly all around you. Many machines are devices that transform energy from one form to another. For example, an automobile engine transforms the chemical energy in gasoline into energy of motion. However, not all of the chemical energy is converted into kinetic energy. Instead, some of the chemical energy is converted into thermal energy, and the engine becomes hot. An engine that converts chemical energy into more kinetic energy is a more efficient engine. New types of cars, like the one shown in **Figure 10**, use an electric motor along with a gasoline engine. These engines are more efficient so the car can travel farther on a gallon of gas.

Transforming Chemical Energy
Inside your body, chemical energy also is transformed into kinetic energy. Look at **Figure 11**. The transformation of chemical to kinetic energy occurs in muscle cells. There, chemical reactions take place that cause certain molecules to change shape. Your muscle contracts when many of these changes occur, and a part of your body moves.

The matter contained in living organisms, or biomass, contains chemical energy. When organisms die, chemical compounds in their biomass break down, or decompose. Bacteria, fungi and other organisms help convert these chemical compounds to simpler chemicals that can be used by other living things.

Thermal energy also is released as these changes occur. For example, a compost pile can contain plant matter, such as grass clippings and leaves. As the compost pile decomposes, chemical energy is converted into thermal energy. This can cause the temperature of a compost pile to reach 60°C.

Analyzing Energy Transformations

Procedure
1. Place soft **clay** on the floor and smooth out its surface.
2. Hold a **marble** 1.5 m above the clay and drop it. Measure the depth of the crater made by the marble.
3. Repeat this procedure using a **steel ball**, a **rubber ball**, and a **table-tennis ball**.

Analysis
1. Compare the depths of the craters to determine which ball had the most kinetic energy as it hit the clay. Why did this ball have the most kinetic energy?
2. Explain how potential energy was transformed into kinetic energy during your activity.

NATIONAL GEOGRAPHIC VISUALIZING ENERGY TRANSFORMATIONS

Figure 11

Paddling a raft, throwing a baseball, playing the violin — your skeletal muscles make these and countless other body movements possible. Muscles work by pulling, or contracting. At the cellular level, muscle contractions are powered by reactions that transform chemical energy into kinetic energy.

▶ Energy transformations taking place in your muscles provide the power to move.

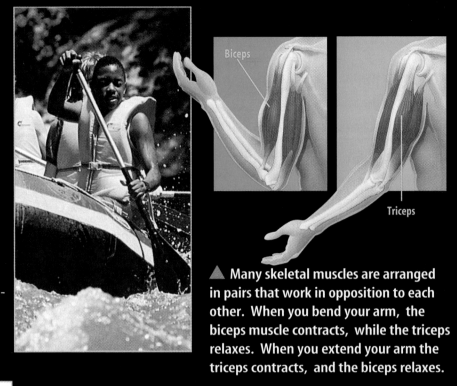

▲ Many skeletal muscles are arranged in pairs that work in opposition to each other. When you bend your arm, the biceps muscle contracts, while the triceps relaxes. When you extend your arm the triceps contracts, and the biceps relaxes.

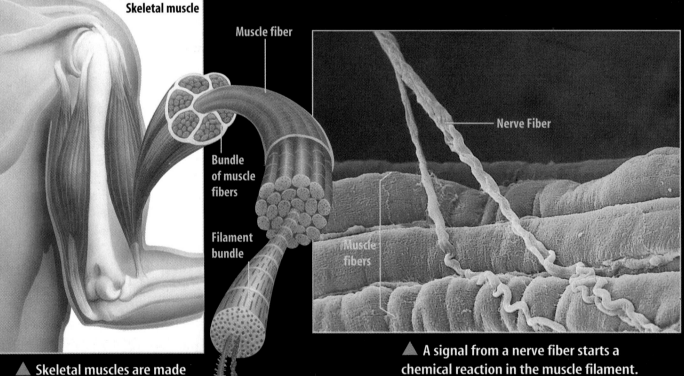

▲ Skeletal muscles are made up of bundles of muscle cells, or fibers. Each fiber is composed of many bundles of muscle filaments.

▲ A signal from a nerve fiber starts a chemical reaction in the muscle filament. This causes molecules in the muscle filament to gain energy and move. Many filaments moving together cause the muscle to contract.

Figure 12
The simple act of listening to a radio involves many energy transformations. A few are diagrammed here.

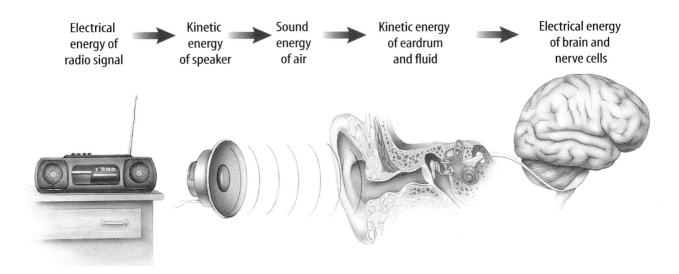

Transforming Electrical Energy Every day you use electrical energy. When you flip a light switch, or turn on a radio or television, or use a hair drier, you are transforming electrical energy to other forms of energy. Every time you plug something into a wall outlet, or use a battery, you are using electrical energy. **Figure 12** shows how electrical energy is transformed into other forms of energy when you listen to a radio. A loudspeaker in the radio converts electrical energy into sound waves that travel to your ear—energy in motion. The energy that is carried by the sound waves causes parts of the ear to move also. This energy of motion is transformed again into chemical and electrical energy in nerve cells, which send the energy to your brain. After your brain interprets this energy as a voice or music, where does the energy go? The energy finally is transformed into thermal energy.

Transforming Thermal Energy Different forms of energy can be transformed into thermal energy. For example, chemical energy changes into thermal energy when something burns. Electrical energy changes into thermal energy when a wire that is carrying an electric current gets hot. Thermal energy can be used to heat buildings and keep you warm. Thermal energy also can be used to heat water. If water is heated to its boiling point, it changes to steam. This steam can be transformed to kinetic energy by steam engines, like the steam locomotives that used to pull trains. Thermal energy also can be transformed into radiant energy. For example, when a bar of metal is heated to a high temperature, it glows and gives off light.

Life Science INTEGRATION

Most organisms have some adaptation for maintaining the correct amount of thermal energy in their bodies. Those living in cooler climates have thick fur coats to keep heat in, and those living in desert regions have skin that reflects the rays of the Sun to keep heat out. Research some of the adaptations different organisms have for controlling the heat in their bodies.

How Thermal Energy Moves Thermal energy can move from one place to another. Look at **Figure 13.** The hot chocolate has thermal energy that moves from the cup to the cooler air around it and to the cooler spoon. Thermal energy only moves from something at a higher temperature to something at a lower temperature.

Generating Electrical Energy

The enormous amount of electrical energy that is used every day is too large to be stored in batteries. The electrical energy that is available for use at any wall socket must be generated continually by power plants. Every power plant works on the same principle—energy is used to turn a large generator. A **generator** is a device that transforms kinetic energy into electrical energy. In fossil fuel power plants, coal, oil, or natural gas is burned to boil water. As the hot water boils, the steam rushes through a **turbine,** which contains a set of narrowly spaced fan blades. The steam pushes on the blades and turns the turbine, which in turn rotates a shaft in the generator to produce the electrical energy, as shown in **Figure 14.**

Figure 13
Thermal energy moves from the hot chocolate to the cooler surroundings. *What happens to the hot chocolate as it loses thermal energy?*

Figure 14
A coal-burning power plant transforms the chemical energy in coal into electrical energy. *What are some of the other energy sources that power plants use?*

Reading Check *What does a generator do?*

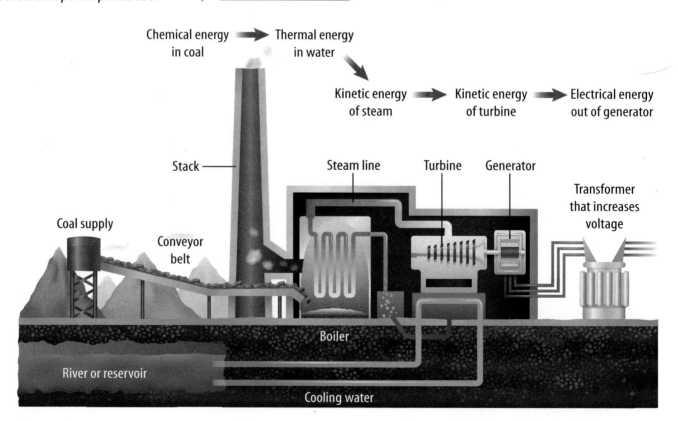

142 CHAPTER 5 Energy

Power Plants Almost 90 percent of the electrical energy generated in the United States is produced by nuclear and fossil fuel power plants, as shown in **Figure 15**. Other types of power plants include hydroelectric (hi droh ih LEK trihk) and wind. Hydroelectric power plants transform the kinetic energy of moving water into electrical energy. Wind power plants transform the kinetic energy of moving air into electrical energy. In these power plants, a generator converts the kinetic energy of moving water or wind to electrical energy.

To analyze the energy transformations in a power plant, you can diagram the energy changes using arrows. A coal-burning power plant generates electrical energy through the following series of energy transformations.

chemical energy of coal → thermal energy of water → kinetic energy of steam → kinetic energy of turbine → electrical energy out of generator

Nuclear power plants use a similar series of transformations. Hydroelectric plants, however, skip the steps that change water into steam because the water strikes the turbine directly.

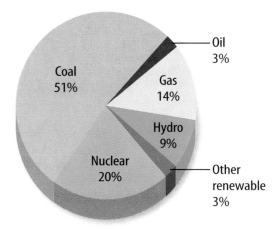

Figure 15
The graph shows sources of electrical energy in the United States. *Which energy source do you think is being used to provide the electricity for the lights overhead?*

Section 2 Assessment

1. What is the law of conservation of energy?
2. If your body temperature is 37°C and you are sitting in a room which has a temperature of 25°C, does your body gain or lose thermal energy? Explain.
3. What are the basic steps involved in generating electricity at a power plant?
4. Diagram the energy transformations that take place at a hydroelectric power plant.
5. **Think Critically** You begin pedaling your bicycle, making it move faster and faster. You notice that at first it is easy to speed up, but then it becomes difficult. You pedal with all your strength, yet you cannot go any faster. Use energy concepts to explain what is happening.

Skill Builder Activities

6. **Testing a Hypothesis** If you drop a rubber ball onto a hard surface, the first bounce will be the highest. How much lower will the second bounce be? If you drop the ball on the top of a shoe box, will it bounce as high? Make a hypothesis. Design and conduct an experiment to test your hypothesis. **For more help, refer to the Science Skill Handbook.**

7. **Using Graphics Software** Use graphics software to diagram all the energy transformations that take place during a conversation. What forms of energy are in the sequence from one person making a sound to a second person hearing that sound? **For more help, refer to the Technology Skill Handbook.**

SECTION 2 Energy Transformations

Activity

Hearing with Your Jaw

You probably have listened to music using speakers or headphones. Have you ever considered how energy is transferred to get the energy from the radio or CD player to your brain? What type of energy is needed to power the radio or CD player? Where does this energy come from? How does that energy become sound? How does the sound get to you? In this activity, the sound from a radio or CD player is going to travel through a motor before entering your body through your jaw instead of your ears.

What You'll Investigate
How can energy be transferred from a radio or CD player to your brain?

Materials
radio or CD player
small electrical motor
headphone jack

Goals
- **Identify** energy transfers and transformations.
- **Explain** your results in terms of transformations of energy and conservation of energy.

Safety Symbols

Procedure
1. Go to one of the places in the room with a motor/radio assembly.
2. Turn on the radio or CD player so that you hear the music.
3. Push the headphone jack into the headphone plug on the radio or CD player.
4. Press the axle of the motor against the side of your jaw.

Conclude and Apply
1. **Describe** what you heard in your Science Journal.
2. What type of energy did you have in the beginning? In the end?
3. **Draw** a diagram to show all the energy transformations taking place.
4. Did anything get hotter as a result of this activity? Explain.
5. **Explain** your results using the law of conservation of energy.

Communicating Your Data
Compare your conclusions with those of other students in your class. **For more help, refer to the Science Skill Handbook.**

144   CHAPTER 5   Energy

SECTION 3: Sources of Energy

Using Energy

Press a button on the remote control and your favorite program appears on television. Open your refrigerator and pull out something cold to drink. Ride to the mall in a car. For any of these things to occur, a transfer of energy must take place. Radiant energy is transferred to your television, electrical energy is transferred to your refrigerator, and the chemical energy in gasoline is transferred to the engine of the car.

Every day energy is used to provide light and to heat and cool homes, schools, and workplaces. Energy is used to run cars, buses, trucks, trains, and airplanes that transport people and materials from one place to another. Energy also is used to make clothing and other materials and to cook food.

According to the law of conservation of energy, energy can't be created or destroyed. Energy only can change form. If a car or refrigerator can't create the energy they use, then where does this energy come from?

Energy Resources

Energy cannot be made, but must come from the natural world. As you can see in **Figure 16,** the surface of Earth receives energy from two sources—the Sun and radioactive atoms in Earth's interior. Of these two energy sources, the energy from the Sun has much more impact on your life. Nearly all the energy you used today can be traced to the Sun, even the gasoline used to power the car or school bus you came to school in.

As You Read

What You'll Learn
- **Explain** what renewable, nonrenewable, and alternative resources are.
- **Describe** the advantages and disadvantages of using various energy sources.

Vocabulary
nonrenewable resource
alternative resource
inexhaustible resource
renewable resource
photovoltaic

Why It's Important
Energy is vital for survival and making life comfortable. Developing new energy sources will improve modern standards of living.

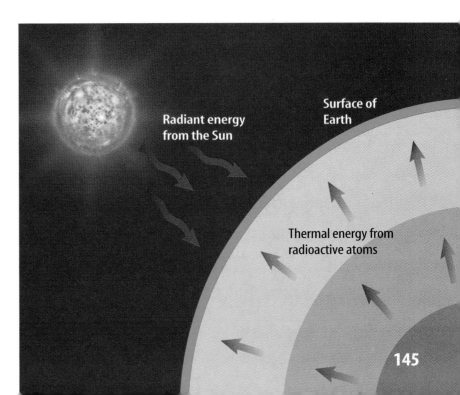

Figure 16
All the energy you use can be traced to one of two sources—the Sun or radioactive atoms in Earth's interior.

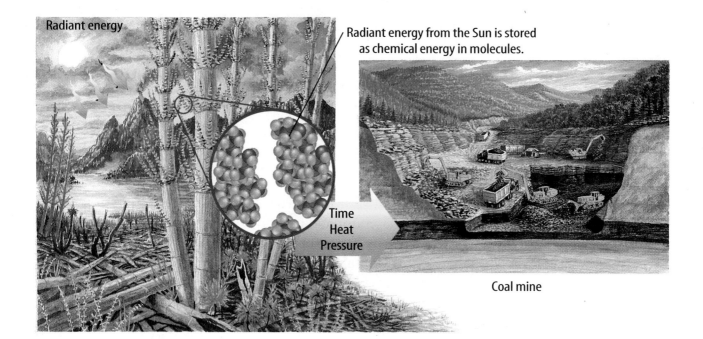

Figure 17
Coal is formed after the molecules in ancient plants are heated under pressure for millions of years. The energy stored by the molecules in coal originally came from the Sun.

Fossil fuels occur only in certain places around the world and must be transported to where they are used. For example, pipelines and supertankers are used to transport oil over long distances. Research the effects on the environment of obtaining and transporting fossil fuels.

Fossil Fuels

Fossil fuels are coal, oil, and natural gas. Oil and natural gas were made from the remains of microscopic organisms that lived in Earth's oceans millions of years ago. Heat and pressure gradually turned these ancient organisms into oil and natural gas. Coal was formed by a similar process from the remains of ancient plants that once lived on land, as shown in **Figure 17.**

Through the process of photosynthesis, ancient plants converted the radiant energy in sunlight to chemical energy stored in various types of molecules. Heat and pressure changed these molecules into other types of molecules as fossil fuels formed. Chemical energy stored in these molecules is released when fossil fuels are burned.

Using Fossil Fuels The energy used when you ride in a car, turn on a light, or use an electric appliance usually comes from burning fossil fuels. However, it takes millions of years to replace each drop of gasoline and each lump of coal that is burned. This means that the supply of oil on Earth will continue to decrease as oil is used. An energy source that is used up much faster than it can be replaced is a **nonrenewable resource.** Fossil fuels are nonrenewable resources.

Burning fossil fuels to produce energy also generates chemical compounds that cause pollution. Each year billions of kilograms of air pollutants are produced by burning fossil fuels. These pollutants can cause respiratory illnesses and acid rain. Also, the carbon dioxide gas formed when fossil fuels are burned might cause Earth's climate to warm.

Nuclear Energy

Can you imagine running an automobile on 1 kg of fuel that releases almost 3 million times more energy than 1 L of gas? What could supply so much energy from so little mass? The answer is the nuclei of uranium atoms. Some of these nuclei are unstable and break apart, releasing enormous amounts of energy in the process. This energy can be used to generate electricity by heating water to produce steam that spins an electric generator, as shown in **Figure 18.** Because no fossil fuels are burned, generating electricity using nuclear energy helps make the supply of fossil fuels last longer. Also, unlike fossil fuel power plants, nuclear power plants produce almost no air pollution. In one year, a typical nuclear power plant generates enough energy to supply 600,000 homes with power and produces only 1 m^3 of waste.

Nuclear Wastes Like all energy sources, nuclear energy has its advantages and disadvantages. One disadvantage is the amount of nonrenewable uranium in Earth's crust. Another is that the waste produced by nuclear power plants is radioactive and can be dangerous to living things. Some of the materials in the nuclear waste will remain radioactive for many thousands of years. As a result the waste must be stored so no radioactivity is released into the environment for a long time. One method is to seal the waste in a ceramic material, place the ceramic in protective containers, and then bury the containers far underground. However, the burial site would have to be chosen carefully so underground water supplies aren't contaminated. Also, the site would have to be safe from earthquakes and other natural disasters that might cause radioactive material to be released.

Figure 18
To obtain electrical energy from nuclear energy, a series of energy transformations must occur.

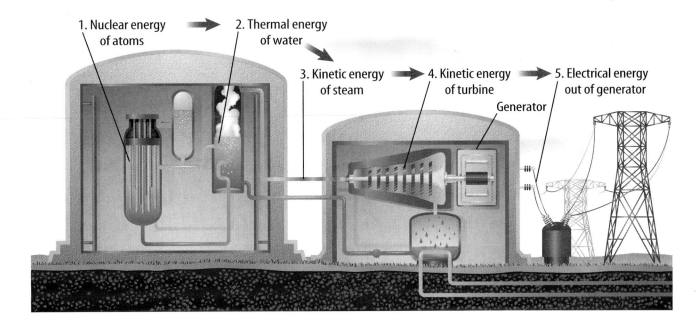

SECTION 3 Sources of Energy **147**

Hydroelectricity

Currently, transforming the potential energy of water that is trapped behind dams supplies the world with almost 20 percent of its electrical energy. Hydroelectricity is the largest renewable source of energy. A **renewable resource** is an energy source that is replenished continually. As long as enough rain and snow fall to keep rivers flowing, hydroelectric power plants can generate electrical energy, as shown in **Figure 19**.

Although production of hydroelectricity is largely pollution free, it has one major problem. It disrupts the life cycle of aquatic animals, especially fish. This is particularly true in the Northwest where salmon spawn and run. Because salmon return to the spot where they were hatched to lay their eggs, the development of dams has hindered a large fraction of salmon from reproducing. This has greatly reduced the salmon population. Efforts to correct the problem have resulted in plans to remove a number of dams. In an attempt to help fish bypass some dams, fish ladders are being installed. Like most energy sources, hydroelectricity has advantages and disadvantages.

SCIENCE Online

Data Update Visit the Glencoe Science Web site at **tx.science.glencoe.com** for data about the use of hydroelectricity in various parts of the world. Using a classroom map, present your finds to your class.

Problem-Solving Activity

Is energy consumption outpacing production?

MATH TEKS 6.5, 6.10 D

You use energy every day—to get to school, to watch TV, and to heat or cool your home. The amount of energy consumed by an average person has increased over the last 50 years. Consequently, more energy must be produced.

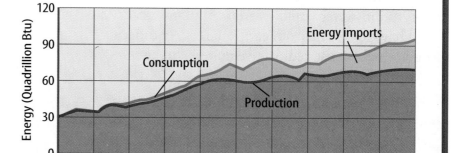

Identifying the Problem

The following graph shows the energy produced and consumed in the United States from 1949 to 1999. How does energy that is consumed by Americans compare with energy that is produced in the United States?

Solving the problem

1. Determine the approximate amount of energy produced in 1949 and in 1999 and how much it has increased in 50 years. Has it doubled or tripled?
2. Do the same for consumption. Has it doubled or tripled?
3. Using your answers for steps 1 and 2, and the graph, where does the additional energy that is needed come from? Give some examples.

148 CHAPTER 5 Energy

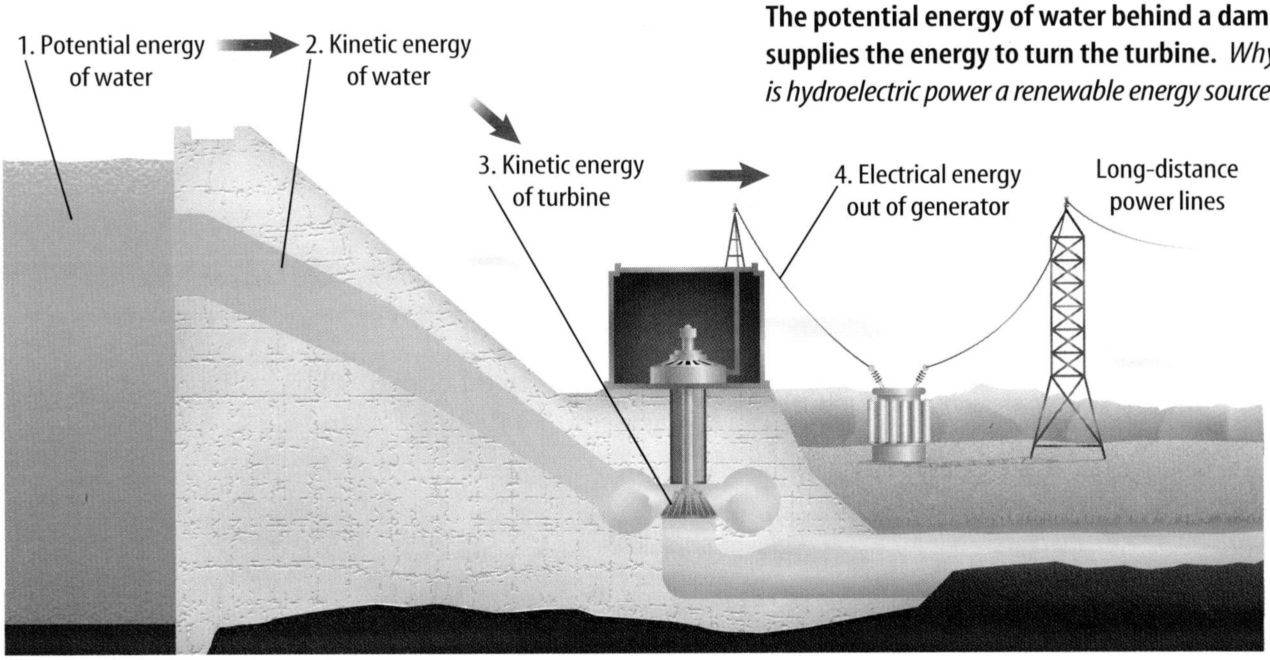

Figure 19
The potential energy of water behind a dam supplies the energy to turn the turbine. *Why is hydroelectric power a renewable energy source?*

1. Potential energy of water
2. Kinetic energy of water
3. Kinetic energy of turbine
4. Electrical energy out of generator
Long-distance power lines

Alternative Sources of Energy

Electrical energy can be generated in several ways. However, each has disadvantages that can affect the environment and the quality of life for humans. Research is being done to develop new sources of energy that are safer and cause less harm to the environment. These sources often are called **alternative resources.** These alternative resources include solar energy, wind, and geothermal energy.

Solar Energy

The Sun is the origin of almost all the energy that is used on Earth. Because the Sun will go on producing an enormous amount of energy for billions of years, the Sun is an inexhaustible source of energy. An **inexhaustible resource** is an energy source that can't be used up by humans.

Each day, on average, the amount of solar energy that strikes the United States is more than the total amount of energy used by the entire country in a year. However, less than 0.1 percent of the energy used in the United States comes directly from the Sun. One reason is that solar energy is more expensive to use than fossil fuels. However, as the supply of fossil fuels decreases, the cost of finding and mining these fuels may increase. Then, it may be cheaper to use solar energy or other energy sources to generate electricity and heat buildings than to use fossil fuels.

 *What is an inexhaustible energy source?*

Building a Solar Collector

Procedure
1. Line a **large pot** with **black plastic** and fill with **water.**
2. Stretch **clear-plastic wrap** over the pot and tape it taut.
3. Make a slit in the top and slide a **thermometer** or a **computer probe** into the water.
4. Place your solar collector in direct sunlight and monitor the temperature change every 3 min for 15 min.
5. Repeat your experiment without using any black plastic.

Analysis
1. Graph the temperature changes in both setups.
2. Explain how your solar collector works.

SECTION 3 Sources of Energy

Collecting the Sun's Energy Two types of collectors capture the Sun's rays. If you look around your neighborhood, you might see large, rectangular panels attached to the roofs of buildings or houses. If, as in **Figure 20A,** pipes come out of the panel, it is a thermal collector. Using a black surface, a thermal collector heats water by directly absorbing the Sun's radiant energy. Water circulating in this system can be heated to about 70°C. If the panel has no pipes, it is a photovoltaic (foh toh vol TAY ihk) collector, like the one pictured in **Figure 20B.** A **photovoltaic** is a device that transforms radiant energy directly into electrical energy. Photovoltaics are used to power calculators and satellites, including the *International Space Station.*

✓ Reading Check   *What does a photovoltaic do?*

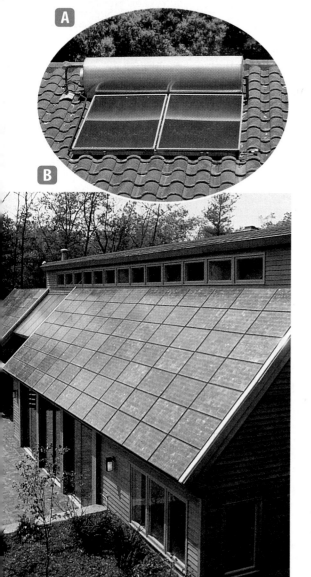

Figure 20
Solar energy can be collected and utilized by individuals using A thermal collectors or B photovoltaic collectors.

Geothermal Energy

Imagine taking a journey to the center of Earth—down to about 6,400 km below the surface. As you went deeper and deeper, you would find the temperature increasing. At a depth of only 100 km, the temperature would be over 900°C. Because Earth's interior is hotter than its surface, heat flows from inside Earth. This heat is called geothermal energy.

In some places cracks in Earth's crust enable molten rock to rise close to the surface. This molten rock can heat underground water. Geothermal power plants change this hot water to steam that spins a turbine and generates electricity. Geothermal energy is an inexhaustible energy source. California currently generates about 5 percent of its electricity using this energy source.

Heat pumps Geothermal energy enables heat pumps to heat and cool buildings. At a depth of several meters, geothermal energy helps keep the temperature of the ground nearly constant at about 10°C to 20°C. A heat pump contains a water-filled loop of pipe that is buried to a depth where the temperature is constant. In summer the air is warmer than this underground temperature. Warm water from the building is pumped through the pipe down into the ground, where the water is cooled. The water cools and is pumped back to the building where it absorbs more heat, and the cycle is repeated. In the wintertime, the air is cooler than the ground below. Then cool water absorbs heat from the ground and releases it into the building.

Wind

Wind is another inexhaustible supply of energy. Modern windmills, like the ones in **Figure 21,** convert the kinetic energy of the wind to electrical energy. The propeller is connected to a generator so that electrical energy is generated when wind spins the propeller. These windmills produce almost no pollution. Some disadvantages are that windmills produce noise and that large areas of land are needed. Also, studies have shown that birds sometimes are killed by windmills.

Conserving Energy

Fossil fuels are a valuable resource. Not only are they burned to provide energy, but oil and coal also are used to make plastics and other materials. One way to make the supply of fossil fuels last longer is to use less energy. Reducing the use of energy is called conserving energy.

You can conserve energy and also save money by turning off lights and appliances such as televisions when you are not using them. Also keep doors and windows closed tightly when it's cold or hot to keep heat from leaking out of or into your house. Energy could also be conserved if buildings are properly insulated, especially around windows. The use of oil could be reduced if cars were used less and made more efficient, so they went farther on a L of gas. Recycling materials such as aluminum cans and glass also helps conserve energy.

Figure 21
Windmills work on the same basic principles as a power plant. Instead of steam turning a turbine, wind turns the rotors. *What are some of the advantages and disadvantages of using windmills?*

Section Assessment

1. What is the ultimate source of most of the energy stored on Earth?
2. What is a renewable resource? Give an example of a renewable and nonrenewable resource and explain the difference.
3. Explain why a heat pump is able to cool a building in the summer and heat the same building in the winter.
4. What are the disadvantages of using hydro-electricity and solar energy?
5. **Think Critically** Explain whether or not the following statement is true: All energy on Earth can be traced back to the Sun.

Skill Builder Activities

6. **Using an Electronic Spreadsheet** Use a spreadsheet to compare the effects on the environment of using fossil fuels, nuclear energy, and dams to produce electricity. Include in your spreadsheet the environmental effects of obtaining, transforming, and distributing the energy. **For more help, refer to the** Technology Skill Handbook.

7. **Using Proportions** As you go deeper into Earth, it becomes hotter. Using the information from this section, calculate the increase in temperature at a depth of 200 m below Earth's surface. **For more help, refer to the** Math Skill Handbook.

Activity: Use the Internet

Energy to Power Your Life

Over the past 100 years, the amount of energy used in the United States and elsewhere has greatly increased. Today, a number of energy sources are available, such as coal, oil, natural gas, nuclear energy, hydroelectric power, wind, and solar energy. Some of these energy sources are being used up and are nonrenewable, but others are replaced as fast as they are used and, therefore, are renewable. Some energy sources are so vast that human usage has almost no effect on the amount available. These energy sources are inexhaustible.

Think about the types of energy you use at home and school every day. In this activity, you will investigate how and where energy is produced, and how it gets to you. You will also investigate alternative ways energy can be produced, and whether these sources are renewable, nonrenewable, or inexhaustible.

Recognize the Problem

What are the sources of the energy you use every day?

Form a Hypothesis

When you wake up in the morning and turn on a light, you use electrical energy. When you ride to school in a car or bus, its engine consumes chemical energy. What other types of energy do you use? Where is that energy produced? Which energy sources are nonrenewable, which are renewable, and which are inexhaustible? What are other sources of energy that you could use instead?

Local Energy Information	
Energy Type	
Where is that energy produced?	
How is that energy produced?	
How is that energy delivered to you?	
Is the energy source renewable, nonrenewable, or inexhaustible?	
What type of alternative energy source could you use instead?	

Goals

- **Identify** how energy you use is produced and delivered.
- **Investigate** alternative sources for the energy you use.
- **Outline** a plan for how these alternative sources of energy could be used.

Data Source

SCIENCEOnline Go to the Glencoe Science Web site at **tx.science.glencoe.com** for more information about sources of energy and for data collected by other students.

152 CHAPTER 5 Energy

Using Scientific Methods

Test Your Hypothesis

Plan

1. Think about the activities you do every day and the things you use. When you watch television, listen to the radio, ride in a car, use a hair drier, or turn on the air conditioning, you use energy. Select one activity or appliance that uses energy.
2. **Identify** the type of energy that is used.
3. **Investigate** how that energy is produced and delivered to you.
4. **Determine** if the energy source is renewable, nonrenewable, or inexhaustible.
5. If your energy source is nonrenewable, describe how the energy you use could be produced by renewable sources.

Do

1. Make sure your teacher approves your plan before you start.
2. Organize your findings in a data table, similar to the one that is shown.
3. Go to the Glencoe Science Web site at **tx.science.glencoe.com** to post your data.

Analyze Your Data

1. **Describe** the process for producing and delivering the energy source you researched. How is it created, and how does it get to you?
2. How much of the energy you use every day comes from the energy source you investigated?
3. Is the energy source you researched renewable, nonrenewable, or inexhaustible? Why?
4. What other renewable or inexhaustible energy sources are used, or could be used, to generate electricity in your area?

Draw Conclusions

1. If the energy source you investigated is nonrenewable, describe how you could reduce your use of this energy source.
2. What alternative sources of energy could you use for everyday energy needs? On the computer, create a plan for using renewable or inexhaustible sources.

Communicating Your Data

SCIENCE Online Find this *Use the Internet* activity on the Glencoe Science Web site at **tx.science.glencoe.com**. Post your data in the table that is provided. **Compare** your data to those of other students. **Combine** your data with those of other students and make inferences using the combined data.

ACTIVITY 153

Science Stats

Energy to Burn

Did you know...

...Garbage—paper, vegetation, animal waste, and more—could be a huge source of energy. Garbage converted to fuel can be used for heating, cooking, transportation, and electricity production. There are approximately 140 garbage-burning power plants in the United States. One truckload of garbage could produce energy equal to about 21 barrels of oil.

...The energy released by the average hurricane is equal to about 200 times the total energy produced by all the world's electrical power plants. Almost all of this energy is released as heat when raindrops are formed.

...The energy Earth gets each half hour from the Sun is enough to meet the world's demands for a year. Renewable and inexhaustible resources, including the Sun, account for only 18 percent of the energy that is used worldwide.

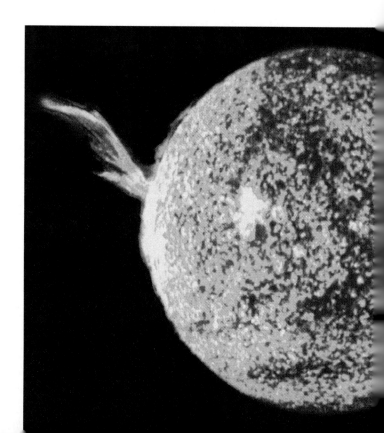

154 **CHAPTER 5** Energy

Connecting To Math

...Gas isn't the only source of energy for cars. Hybrid cars that use both gasoline engines and electric motors are being developed. These cars get higher gas mileage than cars powered by gasoline engines alone—over 60 miles per gallon.

...The largest energy failure in history occurred in 1965 and left 30 million people without electricity at the same time. The problem began because of a faulty relay switch in Ontario, Canada. An area from Ontario through most of New England was affected.

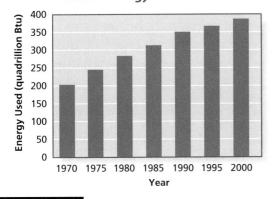

...The Calories in one medium apple will give you enough energy to walk for about 15 min, swim for about 10 min, or jog for about 9 min.

Do the Math

1. If 180 million people were living in the United States in 1965 and 24 million of them lost power, approximately what percentage of U.S. residents lost power?
2. How many liters of gas would you save by taking the hybrid car on a 174-km trip rather than taking the vehicle with the lowest gas mileage?
3. If walking for 15 min requires 80 Calories of fuel (from food), how many Calories would someone need to consume to walk for 1 h?

Go Further

Where would you place solar collectors in the United States? Why? For more information on solar energy, go to **tx.science.glencoe.com**.

SCIENCE STATS 155

Chapter 5 Study Guide

Reviewing Main Ideas

Section 1 What is energy?

1. Energy is the ability to cause change. Energy is found in many forms.
2. Moving objects have kinetic energy. The potential energy of an object depends on its height and mass.
3. Potential energy is the energy of position. Radiant energy is the energy of light.
4. Electric current carries electrical energy, and atomic nuclei contain nuclear energy. *What are all the forms of energy that are represented in this picture?*

Section 2 Energy Transformations

1. Energy can be transformed from one form to another. Energy transformations cause changes to occur.
2. All energy transformations obey the law of conservation of energy, which means no energy is ever created or destroyed.
3. Chemical and electrical energy can be converted into other forms of energy such as radiant energy and thermal energy. Thermal energy moves from warm to cool objects.
4. Power plants convert a source of energy into electrical energy. The kinetic energy of steam spins a turbine which causes a generator to spin. The spinning generator produces electricity.

Section 3 Sources of Energy

1. Fossil fuels and nuclear energy are nonrenewable energy sources. The use of each of these energy sources produces waste products.
2. Renewable and inexhaustible energy sources include hydroelectric, solar, geothermal, and wind energy. *Why is the energy source used by this hydroelectric plant a renewable source of energy?*
3. Energy shortages might be prevented by conserving energy. Each energy source has advantages and disadvantages. *Look at the two photographs below of the same view of New York but at different times. Explain the difference.*

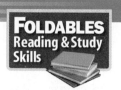

After You Read

Write what you learned about the types, sources and transformation of energy under the Learned tab of your Know-Want-Learn Study Fold.

Chapter 5 Study Guide

Visualizing Main Ideas

Use the following terms and phrases to complete the concept map about energy sources: fossil fuels, hydroelectric, solar, wind, oil, coal, photovoltaic, *and* nonrenewable resources.

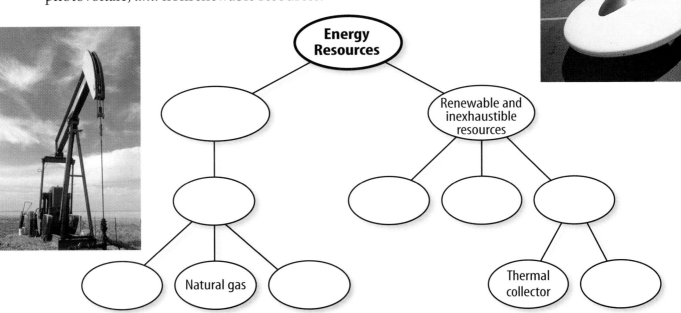

Vocabulary Review

Vocabulary Words

a. alternative resource
b. chemical energy
c. electrical energy
d. energy
e. generator
f. inexhaustible resource
g. kinetic energy
h. law of conservation of energy
i. nonrenewable resource
j. nuclear energy
k. photovoltaic
l. potential energy
m. radiant energy
n. renewable resource
o. thermal energy
p. turbine

Using Vocabulary

For each set of terms below, explain the relationship that exists.

1. electrical energy, nuclear energy
2. turbine, generator
3. photovoltaic, radiant energy, electrical energy
4. renewable resource, inexhaustible resource
5. potential energy, kinetic energy
6. kinetic energy, electrical energy, generator
7. thermal energy, radiant energy
8. law of conservation of energy, energy transformations
9. nonrenewable resource, chemical energy

 **Study Tip**

Practice reading graphs and charts. Make a table that contains the same information as a graph does.

CHAPTER STUDY GUIDE **157**

Chapter 5 Assessment & TEKS Review

Checking Concepts

1. Objects that are able to fall have what type of energy?
   A) kinetic C) potential
   B) radiant D) electrical

2. Which form of energy does light have?
   A) electrical C) kinetic
   B) nuclear D) radiant

3. Muscles perform what type of energy transformation?
   A) kinetic to potential
   B) kinetic to electrical
   C) thermal to radiant
   D) chemical to kinetic

4. Photovoltaics perform what type of energy transformation?
   A) thermal to radiant
   B) kinetic to electrical
   C) radiant to electrical
   D) electrical to thermal

5. Which form of energy does food have?
   A) chemical C) radiant
   B) potential D) electrical

6. Solar energy, wind, and geothermal are what type of energy resource?
   A) inexhaustible C) nonrenewable
   B) inexpensive D) chemical

7. Which of the following is a nonrenewable source of energy?
   A) hydroelectricity C) wind
   B) nuclear D) solar

8. Which of the following does NOT require a generator to generate electricity?
   A) solar C) hydroelectric
   B) wind D) nuclear

9. Which of the following are fossil fuels?
   A) gas C) oil
   B) coal D) all of these

10. From where does the surface of Earth acquire most of its energy?
    A) radioactivity C) chemicals
    B) Sun D) wind

Thinking Critically

11. Explain how the motion of a swing illustrates the transformation between potential and kinetic energy.

12. A skateboard that is coasting along a flat surface will slow down and come to a stop. Explain what happens to the kinetic energy of the skateboard.

13. Describe the energy transformations that occur in the process of toasting a bagel in an electric toaster.

14. In what ways is the formation of coal like the formation of oil and natural gas? How is it different?

15. Explain the difference between the law of conservation of energy and conserving energy. How can conserving energy help prevent energy shortages?

Developing Skills

16. **Researching Information** Find out how spacecraft, such as *Galileo,* obtain the energy they need to operate as they travel through the solar system.

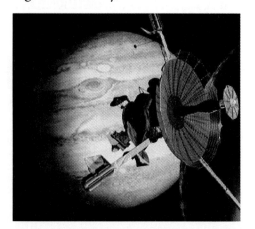

Chapter 5 Assessment

17. Concept Mapping Complete this concept map about energy.

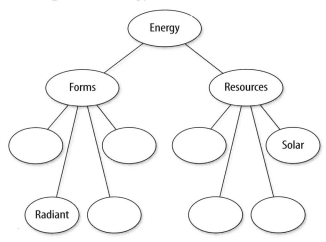

18. Classifying A proposal has been made to use wheat as a source of biomass energy. It can be made into alcohol, which can be burned in engines. How would you classify this source of energy? What are some advantages and disadvantages in using this as a source of energy? Explain.

Performance Assessment

19. Multimedia Presentation Alternative sources of energy that weren't discussed include tidal energy, biomass energy, wave energy, and hydrogen fuel cells. Research an alternative energy source and then prepare a digital slide show about the information you found. Use the concepts you learned from this chapter to inform your classmates about the future prospects of such an energy source.

TECHNOLOGY

Go to the Glencoe Science Web site at **tx.science.glencoe.com** or use the **Glencoe Science CD-ROM** for additional chapter assessment.

TAKS Practice

Throughout the course of one day, you engage in dozens of energy transformations. The table below gives some examples of different energy transformations.
TEKS 6.9 A, B

Types of Energy Transformation	
Energy Transformation	**Example**
Potential → Kinetic	Ball rolling down a hill
Kinetic → Potential	A pebble tossed upward
Electrical → Radiant	A desk lamp
Chemical → Thermal	Burning fossil fuels
	Music from a radio

1. According to the table, burning coal in a stove is an example of
   A) potential → kinetic
   B) kinetic → potential
   C) electrical → radiant
   D) chemical → thermal

2. Which of these energy transformations will complete the table?
   F) electrical → radiant
   G) sound → electrical
   H) electrical → sound
   J) electrical → chemical

3. An image displayed on a computer screen most closely matches which example in terms of energy transformation?
   A) ball rolling down a hill
   B) a desk lamp
   C) burning fossil fuels
   D) music from a radio

Reading Comprehension

Read the passage. Then read each question that follows the passage. Decide which is the best answer to each question.

Electric Cars: The cars of the future?

Have you ever wondered how a car is able to move? In car engines, gasoline is burned to convert chemical energy into thermal energy. The engine then changes some of this thermal energy into kinetic energy that causes the wheels to turn. However, some car manufacturers are also exploring whether cars can be developed that will run on electrical energy rather than gasoline.

Electric cars would use electrical energy to power an electric motor that turns the car's wheels. The electrical energy would be provided by a battery. In a battery chemical reactions occur that convert chemical energy into electrical energy. Eventually, the chemicals in the battery are changed into other chemical compounds and the battery can no longer produce electrical energy. When <u>rechargeable</u> batteries are recharged, the chemical reactions in the battery are reversed. Then the chemicals in the battery that produce electrical energy are restored.

While electric cars would produce no pollution while they are driven, there are potential environmental problems. The rechargeable batteries used by electric cars are heavy, expensive, and contain hazardous materials such as lead. As a result, the manufacture and disposal of these batteries can create environmental problems. Also, the electricity used to charge these batteries usually is generated by power plants that can produce air pollution and other by-products.

Other types of electric cars are being developed that use a hydrogen fuel cell instead of batteries. In this fuel cell hydrogen gas reacts with oxygen to produce electricity. The hydrogen gas can be obtained from water. Although the fuel cell produces almost no pollution, electricity from power plants still is needed to generate the hydrogen gas the fuel cell uses. Research is being done to find other ways to produce hydrogen gas that would result in less pollution.

Several carmakers have developed hybrid cars that combine an internal combustion engine, a battery, and an electric motor. The electric motor assists the internal combustion engine in providing power. During braking, about half of the engine's kinetic energy is used to charge the battery.

Test-Taking Tip Read the passage slowly to make sure you don't miss any important details.

1. From the story, you can tell that the word <u>rechargeable</u> means _____. *Reading TEKS 6.9 B*
   A) brand new
   B) reusable
   C) paid by credit card
   D) disposable

2. The hydrogen gas used in a hydrogen fuel cell can be obtained from _____. *Reading TEKS 6.9 B*
   F) water
   G) electric cars
   H) power plants
   J) hybrid cars

TAKS Practice

Reasoning and Skills

Read each question and choose the best answer.

1. The picture shows an experiment to see which liquid will boil first. Which of the following would make it a better-designed experiment? *Science TEKS 6.2 A*
   A) put a thermometer in one container
   B) use a different amount of liquid in each container
   C) use the same size container for each liquid
   D) use solids instead of liquids

Test-Taking Tip It is important in experiments that only one factor is being tested.

2. What probably is being measured in the picture? *Science TEKS 6.4 A*
   F) the volume of the rock
   G) the mass of the rock
   H) the length of the rock
   J) the texture of the rock

Test-Taking Tip Think about what characteristic a scale measures.

3. Objects moving across slippery surfaces have less friction than objects moving across rough surfaces. On which surface would it be easiest to pull a sled?
   Science TEKS 6.6 A
   A) rug
   B) cement
   C) dirt
   D) ice

Test-Taking Tip Which surface would have the least amount of friction for the sled to slide over?

4. A marshmallow is held over a fire too long and it burns. Which of the following is this an example of? *Science TEKS 6.7 B*
   F) physical property
   G) chemical property
   H) physical change
   J) chemical change

Consider this question carefully before writing your answer on a separate sheet of paper.

5. Usually there is more than one force acting on an object. These forces can be balanced or unbalanced. Describe a situation where the forces acting on an object are balanced. Be sure to explain what the forces are.
   Science TEKS 6.6 A

Test-Taking Tip When the forces acting on an object are unbalanced, that object will move.

STANDARDIZED TEST PRACTICE 161

UNIT 3
Earth's Systems

How Are Rocks & Fluorescent Lights Connected?

NATIONAL GEOGRAPHIC

Around 1600, an Italian cobbler found a rock that contained a mineral that could be made to glow in the dark. The discovery led other people to seek materials with similar properties. Eventually, scientists identified many fluorescent and phosphorescent (fahs fuh RE sunt) substances—substances that react to certain forms of energy by giving off their own light. As seen above, a fluorescent mineral may look one way in ordinary light (front), but may give off a strange glow (back) when exposed to ultraviolet light. In the 1850s, a scientist wondered whether the fluorescent properties of a substance could be harnessed to create a new type of lighting. The scientist put a fluorescent material inside a glass tube and sent an electric charge through the tube, creating the first fluorescent lamp. Today, fluorescent light bulbs are widely used in office buildings, schools, and factories.

SCIENCE CONNECTION

FLUORESCENT MINERALS Some minerals fluoresce—give off visible light in various colors—when exposed to invisible ultraviolet (UV) light. Using library resources or the Glencoe Science Web site at tx.science.glencoe.com, find out more about fluorescence in minerals. Write a paragraph that answers the following questions: Would observing specimens under UV light be a reliable way for a geologist to identify minerals? Why or why not?

CHAPTER 6

Science TEKS 6.6 C; 6.7 A; 6.14 A

Rocks and Minerals

Spectacular natural scenes like this one at Pikes Peak in Colorado are often shaped by rock formations. How did rocks form? What are they made of? In this chapter you will learn the answers. In addition you will find out where gemstones and valuable metals such as gold and copper come from. Rocks and minerals are the basic materials of Earth's surface. Read on to discover how they are classified and how they are related.

What do you think?

Science Journal Look at the picture below with a classmate. Discuss what you think this might be or what is happening. Here's a hint: *It is far cooler now than it used to be.* Write your answer or best guess in your Science Journal.

The view is spectacular! You and a friend have successfully scaled Pikes Peak. Now that you have reached the top, you also have a chance to look more closely at the rock you've been climbing. First, you notice that it sparkles in the Sun because of the silvery specks that are stuck in the rock. Looking closer, you also see clear, glassy pieces and pink, irregular chunks. What is the rock made of? How did it get here?

Observe a rock

1. Obtain a sparkling rock from your teacher. You also will need a hand lens.
2. Observe the rock with the hand lens. Your job is to observe and record as many of the features of the rock as you can.
3. Return the rock to your teacher.
4. Describe your rock so other students could identify it from a variety of rocks.

Observe

How do the parts of the rock fit together to form the whole thing? Describe this in your Science Journal and make a drawing. Be sure to label the colors and shapes in your drawing.

Before You Read

Making a Venn Diagram Study Fold Make the following Foldable to compare and contrast the characteristics of rocks and minerals.

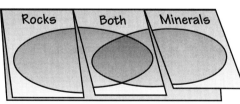

1. Place a sheet of paper in front of you so the long side is at the top. Fold the paper in half from top to bottom.
2. Fold both sides in. Unfold the paper so three sections show.
3. Through the top thickness of paper, cut along each of the fold lines to the top fold, forming three tabs. Label each tab *Rocks, Both,* and *Minerals* and draw ovals across the front of the paper as shown.
4. As you read the chapter, write what you learn about rocks and minerals under the left and right tabs.

165

SECTION 1

Minerals—Earth's Jewels

As You Read

What You'll Learn
- **Identify** the difference between a mineral and a rock.
- **Describe** the properties that are used to identify minerals.

Vocabulary
mineral   gem
rock   ore
crystal

Why It's Important
Minerals are the basic substances of nature that humans use for a variety of purposes.

What is a mineral?

Suppose you were planning an expedition to find minerals (MIHN uh ruhlz). Where would you look? Do you think you'll have to crawl into a cave or brave the depths of a mine? Well, put away your flashlight. You can find minerals in your own home—in the salt shaker and in your pencil. Metal pots, glassware, and ceramic dishes are products made from minerals. Minerals and products from them, shown in **Figure 1,** surround you.

Minerals Defined **Minerals** are inorganic, solid materials found in nature. Inorganic means they usually are not formed by plants or animals. You could go outside and find minerals that occur as gleaming crystals—or as small grains in ordinary rocks. X-ray patterns of a mineral show an orderly arrangement of atoms that looks something like a garden trellis. Evidence of this orderly arrangement is the beautiful crystal shape often seen in minerals. The particular chemical makeup and arrangement of the atoms in the crystal is unique to each mineral. **Rocks,** such as the one used in the Explore Activity, usually are made of two or more minerals. Each mineral has unique characteristics you can use to identify it. So far, more than 4,000 minerals have been identified.

Figure 1
You use minerals every day without realizing it. Minerals are used to make many common objects.

A The "lead" in a pencil is not lead. It is the mineral graphite.

B The mineral quartz is used to make the glass that you use every day.

How do minerals form? Minerals form in several ways. One way is from melted rock inside Earth called magma. As magma cools, atoms combine in orderly patterns to form minerals. Minerals also form from melted rock that reaches Earth's surface. Melted rock at Earth's surface is called lava.

Evaporation can form minerals. Just as salt crystals appear when seawater evaporates, other dissolved minerals, such as gypsum, can crystallize. A process called precipitation (prih sih puh TAY shun) can form minerals, too. Water can hold only so much dissolved material. Any extra separates and falls out as a solid. Large areas of the ocean floor are covered with manganese nodules that formed in this way. These metallic spheres average 25 cm in diameter. They crystallized directly from seawater containing metal atoms.

Figure 2
This cluster of fluorite crystals formed from a solution rich in dissolved minerals.

Formation Clues Sometimes, you can tell how a mineral formed by how it looks. Large mineral grains that fit together like a puzzle seem to show up in rocks formed from slow-cooling magma. If you see large, perfectly formed crystals, it means the mineral had plenty of space in which to grow. This is a sign they may have formed in open pockets within the rock.

The crystals you see in **Figure 2** grew this way from a solution that was rich in dissolved minerals. To figure out how a mineral was formed, you have to look at the size of the mineral crystal and how the crystals fit together.

Grade 5 TEKS Review

For a review of the Grade 5 TEKS *The Natural World of Earth and Sky,* see page 517.

Properties of Minerals

The cheers are deafening. The crowd is jumping and screaming. From your seat high in the bleachers, you see someone who is wearing a yellow shirt and has long, dark hair in braids, just like a friend you saw this morning. You're only sure it's your friend when she turns and you recognize her smile. You've identified your friend by physical properties that set her apart from other people—her clothing, hair color and style, and facial features. Each mineral, too, has a set of physical properties that can be used to identify it. Most common minerals can be identified with items you have around the house and can carry in your pocket, such as a penny or a steel file. With a little practice you soon can recognize mineral shapes, too. Next you will learn about properties that will help you identify minerals.

Life Science INTEGRATION

Bones, such as those found in humans and horses, contain tiny crystals of the mineral apatite. Research apatite and report your findings to your class.

SECTION 1 Minerals—Earth's Jewels **167**

Figure 3
The mineral pyrite often forms crystals with six faces. *Why do you think pyrite also is called "fool's gold"?*

Crystals All minerals have an orderly pattern of atoms. The atoms making up the mineral are arranged in a repeating pattern. Solid materials that have such a pattern of atoms are called **crystals.** Sometimes crystals have smooth surfaces called crystal faces. The mineral pyrite commonly forms crystals with six crystal faces, as shown in **Figure 3.**

Reading Check *What distinguishes crystals from other types of solid matter?*

Cleavage and Fracture Another clue to a mineral's identity is the way it breaks. Minerals that split into pieces with smooth, regular planes that reflect light are said to have cleavage (KLEE vihj). The mineral mica in **Figure 4A** shows cleavage by splitting into thin sheets. Splitting one of these minerals along a cleavage surface is something like peeling off a piece of presliced cheese. Cleavage is caused by weaknesses within the arrangement of atoms that make up the mineral.

Not all minerals have cleavage. Some break into pieces with jagged or rough edges. Instead of neat slices, these pieces are shaped more like hunks of cheese torn from an unsliced block. Materials that break this way, such as quartz, have what is called fracture (FRAK chur). **Figure 4C** shows the fracture of flint.

Figure 4
Some minerals have one or more directions of cleavage. If minerals do not break along flat surfaces, they have fracture.

A Mica has one direction of cleavage and can be peeled off in sheets.

B The mineral halite, also called rock salt, has three directions of cleavage at right angles to each other. *Why might grains of rock salt look like little cubes?*

C Fracture can be jagged and irregular or smooth and curvy like in flint.

168 CHAPTER 6 Rocks and Minerals

Figure 5
The mineral calcite can form in a wide variety of colors. The colors are caused by slight impurities.

Color The reddish-gold color of a new penny shows you that it contains copper. The bright yellow color of sulfur is a valuable clue to its identity. Sometimes a mineral's color can help you figure out what it is. But color also can fool you. The common mineral pyrite (PI rite) has a shiny, gold color similar to real gold—close enough to disappoint many prospectors during the California Gold Rush in the 1800s. Because of this, pyrite also is called fool's gold. While different minerals can look similar in color, the same mineral can occur in a variety of colors. The mineral calcite, for example, can be many different colors, as shown in **Figure 5.**

Streak and Luster Scraping a mineral sample across an unglazed, white tile, called a streak plate, produces a streak of color, as shown in **Figure 6.** Oddly enough, the streak is not necessarily the same color as the mineral itself. This streak of powdered mineral is more useful for identification than the mineral's color. Gold prospectors could have saved themselves a lot of heartache if they had known about the streak test. Pyrite makes a greenish-black or brownish-black streak, but real gold makes a yellow streak.

Is the mineral shiny? Dull? Pearly? Words like these describe another property of minerals called luster. Luster describes how light reflects from a mineral's surface. If it shines like a metal, the mineral has metallic (muh TA lihk) luster. Nonmetallic minerals can be described as having pearly, glassy, dull, or earthy luster. You can use color, streak, and luster to help identify minerals.

Figure 6
Streak is the color of the powdered mineral. The mineral hematite has a characteristic reddish-brown streak. *How do you obtain a mineral's streak?*

SECTION 1 Minerals—Earth's Jewels

Table 1 Mohs Scale

Mineral	Hardness	Hardness of Common Objects
Talc	1 (softest)	
Gypsum	2	fingernail (2.5)
Calcite	3	copper penny (3.5)
Fluorite	4	iron nail (4.5)
Apatite	5	glass (5.5)
Feldspar	6	steel file (6.5)
Quartz	7	streak plate (7)
Topaz	8	
Corundum	9	
Diamond	10 (hardest)	

Hardness As you investigate different minerals, you'll find that some are harder than others. Some minerals, like talc, are so soft that they can be scratched with a fingernail. Others, like diamond, are so hard that they can be used to cut almost anything else.

In 1822, an Austrian geologist named Friedrich Mohs also noticed this property. He developed a way to classify minerals by their hardness. The Mohs scale, shown in **Table 1,** classifies minerals from 1 (softest) to 10 (hardest). You can determine hardness by trying to scratch one mineral with another to see which is harder. For example, fluorite (4 on the Mohs scale) will scratch calcite (3 on the scale), but fluorite cannot scratch apatite (5 on the scale). You also can use a homemade mineral identification kit—a penny, a nail, and a small glass plate with smooth edges. Simply find out what scratches what. Is the mineral hard enough to scratch a penny? Will it scratch glass?

Specific Gravity Some minerals are heavier for their size than others. Specific gravity compares the weight of a mineral with the weight of an equal volume of water. Pyrite—or fool's gold—is about five times heavier than water. Real gold is more than 15 times heavier than water. You easily could sense this difference by holding each one in your hand. Measuring specific gravity is another way you can identify minerals.

Figure 7
Calcite has the unique property of double refraction.

Other Properties Some minerals have other unusual properties that can help identify them. The mineral magnetite will attract a magnet. The mineral calcite has two unusual properties. It will fizz when it comes into contact with an acid like vinegar. Also, if you look through a clear calcite crystal, you will see a double image, as shown in **Figure 7.** Scientists taste some minerals to identify them, but you should not try this yourself. Halite, also called rock salt, has a salty taste.

Together, all of the properties you have read about are used to identify minerals. Learn to use them and you can be a mineral detective.

Common Minerals

In the Chapter Opener, the rocks making up Pikes Peak were made of minerals. But only a small number of the more than 4,000 minerals make up most rocks. These minerals often are called the rock-forming minerals. If you can recognize these minerals, you will be able to identify most rocks. Other minerals are much rarer. However, some of these rare minerals also are important because they are used as gems or are ore minerals, which are sources of valuable metals.

Most of the rock-forming minerals are silicates, which contain the elements silicon and oxygen. The mineral quartz is pure silica (SiO_2). More than half of the minerals in Earth's crust are forms of a silicate mineral called feldspar. Other important rock-forming minerals are carbonates—or compounds containing carbon and oxygen. The carbonate mineral calcite makes up the common rock limestone.

Reading Check *Why is the silicate mineral feldspar important?*

Other common minerals can be found in rocks that formed at the bottom of ancient, evaporating seas. Rock comprised of the mineral gypsum is abundant in many places, and rock salt, made of the mineral halite, underlies large parts of the Midwest.

Mini LAB

Classifying Minerals

Procedure
1. Touch a **magnet** to samples of **quartz, calcite, hornblende,** and **magnetite.** Record which mineral attracts the magnet.
2. Place each sample in a small **beaker** that is half full of **vinegar.** Record what happens.
3. Rinse samples with **water.**

Analysis
1. Describe how each mineral reacted to the tests in steps 1 and 2.
2. Describe in a data table the other physical properties of the four minerals.

Problem-Solving Activity

How hard are these minerals?

MATH TEKS
6.10 D; 6.13 A

Some minerals, like diamonds, are hard. Others, like talc, are soft. How can you determine the hardness of a mineral?

Identifying the Problem

The table at the right shows the results of a hardness test done using some common items as tools (a fingernail, penny, nail, and steel file) to scratch certain minerals (halite, turquoise, an emerald, a ruby, and graphite). The testing tools are listed at the top from softest (fingernail) to hardest (steel file). The table shows which minerals were scratched by which tools. Examine the table to determine the relative hardness of each mineral.

Hardness Test

Mineral	Fingernail	Penny	Nail	Steel File
Turquoise	N	N	Y	Y
Halite	N	Y	Y	Y
Ruby	N	N	N	N
Graphite	Y	Y	Y	Y
Emerald	N	N	N	N

Solving the problem
1. Is it possible to rank the five minerals from softest to hardest using the data in the table above? Why or why not?
2. What method could you use to determine whether the ruby or the emerald is harder?

Figure 8
A This garnet crystal is encrusted with other minerals but still shines a deep red. **B** Cut garnet is a prized gemstone.

Gems Which would you rather win, a diamond ring or a quartz ring? A diamond ring would be more valuable. Why? The diamond in a ring is a kind of mineral called a gem. **Gems** are minerals that are rare and can be cut and polished, giving them a beautiful appearance, as shown in **Figure 8.** This makes them ideal for jewelry. To be gem quality, most minerals must be clear with no blemishes or cracks. A gem also must have a beautiful luster or color. Few minerals meet these standards. That's why the ones that do are rare and valuable.

The Making of a Gem One reason why gems are so rare is that they are produced under special conditions. Diamond, for instance, is a form of the element carbon. Scientists can make artificial diamonds in laboratories, but they must use extremely high pressures. These pressures are greater than any found within Earth's crust. Therefore, scientists suggest that diamond forms deep in Earth's mantle. It takes a special kind of volcanic eruption to bring a diamond close to Earth's surface, where miners can find it. This type of eruption forces magma from the mantle toward the surface of Earth at high speeds, bringing diamond right along with it. This type of magma is called kimberlite magma. **Figure 9** shows a rock from a kimberlite deposit in South Africa that is mined for diamond. Kimberlite deposits are found in the necks of ancient volcanoes.

SCIENCE Online

Research Visit the Glencoe Science Web site at **tx.science.glencoe.com** for more information about gems. Communicate to your class what you learn.

Figure 9
Diamonds sometimes are found in kimberlite deposits.

Figure 10
To be profitable, ores must be found in large deposits or rich veins. Mining is expensive. Copper ore is obtained from this mine in Arizona.

Ores A mineral is called an **ore** if it contains enough of a useful substance that it can be sold for a profit. Many of the metals that humans use come from ores. For example, the iron used to make steel comes from the mineral hematite, lead for batteries is produced from galena, and the magnesium used in vitamins comes from dolomite. Ores of these useful metals must be extracted from Earth in a process called mining. A copper mine is shown in **Figure 10.**

Ore Processing After an ore has been mined, it must be processed to extract the desired mineral or element. **Figure 11** shows a copper smelting plant that melts the ore and then separates and removes most of the unwanted materials. After this smelting process, copper can be refined, which means that it is purified. Then it is processed into many materials that you use every day. Examples of useful copper products include sheet metal products, electrical wiring in cars and homes, and just about anything electronic. Some examples of copper products are shown in **Figure 12.**

Early settlers in Jamestown, Virginia, produced iron by baking moisture out of ore they found in salt marshes. Today much of the iron produced in the United States is highly processed from ore found near Lake Superior.

Figure 11
This smelter in Montana heats and melts copper ore. *Why is smelting necessary to process copper ore?*

SECTION 1 Minerals—Earth's Jewels **173**

Figure 12
Many metal objects you use every day are made with copper. *What other metals are used to produce everyday objects?*

Minerals Around You Now you have a better understanding of minerals and their uses. Can you name five things in your classroom that come from minerals? Can you go outside and find a mineral right now? You will find that minerals are all around you and that you use minerals every day. Next, you will look at rocks, which are Earth materials made up of combinations of minerals.

Section Assessment

1. Explain the difference between a mineral and a rock. Name five common rock-forming minerals.
2. List five properties that are used most commonly to identify minerals.
3. Where in Earth is diamond formed? Describe an event that must occur in order for diamond to reach Earth's surface.
4. When is a mineral considered to be an ore? Describe the steps of mining, smelting, and refining that are used to extract minerals or elements from ores.
5. **Think Critically** Would you want to live close to a working gold mine? Explain.

Skill Builder Activities

6. **Comparing and Contrasting** Gems and ores are some of Earth's rarer minerals. Compare and contrast gems and ores. Why are they so valuable? Explain the importance of both in society today. **For more help, refer to the Science Skill Handbook.**

7. **Using Percentages** In 1996, the United States produced approximately 2,340,000 metric tons of refined copper. In 1997, about 2,440,000 metric tons of refined copper were produced. Compared to the 1996 amount, copper production increased by what percentage in 1997? **For more help, refer to the Math Skill Handbook.**

174 CHAPTER 6 Rocks and Minerals

SECTION 2: Igneous and Sedimentary Rocks

Igneous Rock

A rocky cliff, a jagged mountain peak, and a huge boulder probably all look solid and permanent to you. Rocks seem as if they've always been here and always will be. But little by little, things change constantly on Earth. New rocks form, and old rocks wear away. Such processes produce three main kinds of rocks—igneous, sedimentary, and metamorphic.

The deeper you go into the interior of Earth, the higher the temperature is and the greater the pressure is. Deep inside Earth, it is hot enough to melt rock. **Igneous** (IHG nee us) **rocks** form when melted rock from inside Earth cools. The cooling and hardening that result in igneous rock can occur on Earth, as seen in **Figure 13,** or underneath Earth's surface. When melted rock cools on Earth's surface, it makes an **extrusive** (ehk STREW sihv) igneous rock. Melted rock that cools below Earth's surface forms **intrusive** (ihn trew sihv) igneous rock.

As You Read

What You'll Learn
- **Explain** how extrusive and intrusive igneous rocks are different.
- **Describe** how different types of sedimentary rocks form.

Vocabulary
igneous rock
extrusive
intrusive
sedimentary rock

Why It's Important
Rocks form the land all around you.

Chemical Composition The chemicals in the melted rock determine the color of the resulting rock. If it contains a high percentage of silica and little iron, magnesium, or calcium, the rock will be light in color. Light-colored igneous rocks are called granitic (gra NIH tihk) rocks. If the silica content is far less, but it contains more iron, magnesium, or calcium, a dark-colored or basaltic (buh SAWL tihk) rock will result. Intrusive igneous rocks often are granitic, and extrusive igneous rocks often are basaltic. These two categories are important in classifying igneous rocks.

Figure 13
Sakurajima is a volcano in Japan. During the 1995 eruption, molten rock and solid rock were thrown into the air.

Rocks from Lava Extrusive igneous rocks form when melted rock cools on Earth's surface. Liquid rock that reaches Earth's surface is called lava. Lava cools quickly before large mineral crystals have time to form. That's why extrusive igneous rocks usually have a smooth, sometimes glassy appearance.

Extrusive igneous rocks can form in two ways. In one way, volcanoes erupt and shoot out lava and ash. Also, large cracks in Earth's crust, called fissures (FIH shurz), can open up. When they do, the lava oozes out onto the ground or into water. Oozing lava from a fissure or a volcano is called a lava flow. In Hawaii, lava flows are so common that you can observe one almost every day. Lava flows quickly expose melted rock to air or water. The fastest cooling lava forms no grains at all. This is how obsidian, a type of volcanic glass, forms. Lava trapping large amounts of gas can cool to form igneous rocks containing many holes.

✔ Reading Check *What is a fissure?*

Figure 14
Extrusive igneous rocks form at Earth's surface. Intrusive igneous rocks form inside Earth. Wind and water can erode rocks to expose features such as dikes, sills, and volcanic necks.

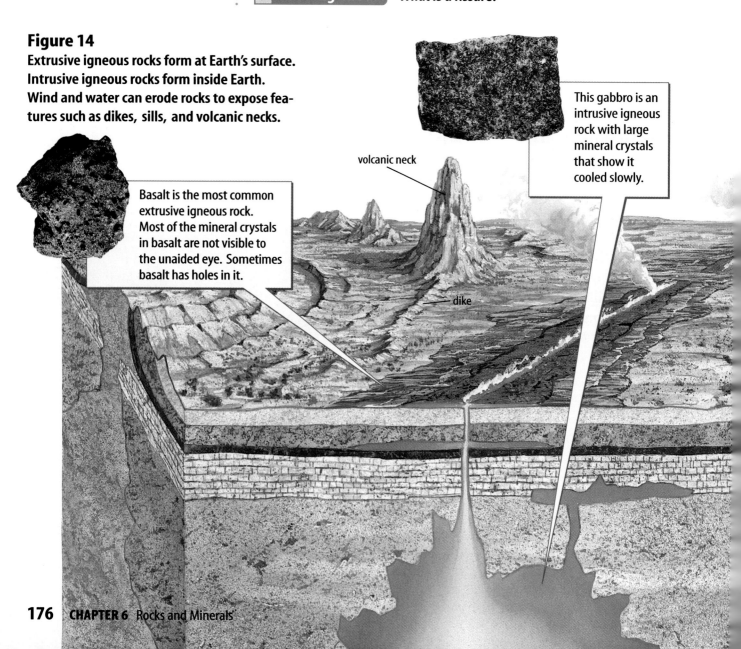

Basalt is the most common extrusive igneous rock. Most of the mineral crystals in basalt are not visible to the unaided eye. Sometimes basalt has holes in it.

This gabbro is an intrusive igneous rock with large mineral crystals that show it cooled slowly.

Rocks from Magma Some melted rock never reaches the surface. Such underground melted rock is called magma. Intrusive igneous rocks are produced when magma cools below the surface of Earth, as shown in **Figure 14.**

Intrusive igneous rocks form when a huge glob of magma from inside Earth rises toward the surface but never reaches it. It's similar to when a helium balloon rises and gets stopped by the ceiling. This hot mass of rock sits under the surface and cools slowly over millions of years until it is solid. The cooling is so slow that the minerals in the magma have time to form large crystals. The size of the mineral crystals is the main difference between intrusive and extrusive igneous rocks. Intrusive igneous rocks have large crystals that are easy to see. Extrusive igneous rocks do not have large crystals that you can see easily. **Figure 15** shows some igneous rock features.

✔ **Reading Check** *How do intrusive and extrusive rocks appear different?*

Physics
INTEGRATION

The extreme heat found inside Earth has several sources. Some is left over from Earth's formation, and some comes from radioactive isotopes that constantly emit heat while they decay deep in Earth's interior. Research to find detailed explanations of these heat sources. Use your own words to explain them in your Science Journal.

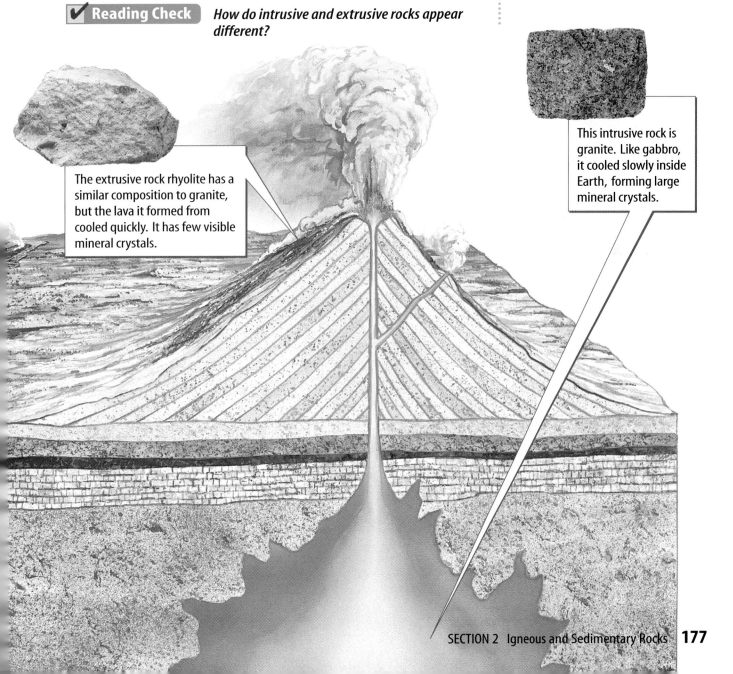

The extrusive rock rhyolite has a similar composition to granite, but the lava it formed from cooled quickly. It has few visible mineral crystals.

This intrusive rock is granite. Like gabbro, it cooled slowly inside Earth, forming large mineral crystals.

SECTION 2 Igneous and Sedimentary Rocks

NATIONAL GEOGRAPHIC VISUALIZING IGNEOUS ROCK FEATURES

Figure 15

Intrusive igneous rocks are formed when a mass of liquid rock, or magma, rises toward Earth's surface and then cools before emerging. The magma cools in a variety of ways. Eventually the rocks may be uplifted and erosion may expose them at Earth's surface. A selection of these formations is shown here.

▶ This dike in Israel's Negev Desert formed when magma squeezed into cracks that cut across rock layers.

▶ A batholith is a very large igneous rock body that forms when rising magma cools below the ground. Towering El Capitan, right, is just one part of a huge batholith. It looms over the entrance to the Yosemite Valley.

▲ Sills such as this one in Death Valley, California, form when magma is forced into spaces that run parallel to rock layers.

▶ Volcanic necks like Shiprock, New Mexico, form when magma hardens inside the vent of a volcano. Because the volcanic rock in the neck is harder than the volcanic rock in the volcano's cone, only the volcanic neck remains after erosion wears the cone away.

178 CHAPTER 6 Rocks and Minerals

Sedimentary Rocks

Pieces of broken rock, shells, mineral grains, and other materials make up what is called sediment (SE duh munt). The sand you squeeze through your toes at the beach is one type of sediment. As shown in **Figure 16,** sediment can collect in layers to form rocks. These are called **sedimentary** (sed uh MEN tuh ree) **rocks.** Rivers, ocean waves, mud slides, and glaciers can carry sediment. Sediment also can be carried by the wind. When sediment is dropped, or deposited, by wind, ice, gravity, or water, it collects in layers. After sediment is deposited, it begins the long process of becoming rock. Most sedimentary rocks take thousands to millions of years to form. The changes that form sedimentary rocks occur continuously. As with igneous rock, there are several kinds of sedimentary rocks. They fall into three main categories.

Figure 16
The layers in these rocks are the different types of sedimentary rocks that have been exposed at Sedona, in Arizona. *What causes the layers seen in sedimentary rocks?*

 **Reading Check** *How is sediment transported?*

Detrital Rocks When you mention sedimentary rocks, most people think about rocks like sandstone, which is a detrital (dih TRI tuhl) rock. Detrital rocks, shown in **Figure 17,** are made of grains of minerals or other rocks that have moved and been deposited in layers by water, ice, gravity, or wind. Other minerals dissolved in water act to cement these particles together. The weight of sediment above them also squeezes or compacts the layers into rock.

Figure 17
Four types of detrital sedimentary rocks include shale, siltstone, sandstone, and conglomerate.

SECTION 2 Igneous and Sedimentary Rocks **179**

Modeling How Fossils Form Rocks

Procedure

1. Fill a small **aluminum pie pan** with pieces of broken **macaroni**. These represent various fossils.
2. Mix 50 mL of **white glue** into 250 mL of **water**. Pour this solution over the macaroni and set it aside to dry.
3. When your fossil rock sample has set, remove it from the pan and compare it with an actual **fossil limestone** sample.

Analysis

1. Explain why you used the glue solution and what this represents in nature.
2. Using whole macaroni samples as a guide, match the macaroni "fossils" in your "rock" to the intact macaroni. Draw and label them in your **Science Journal.**

Identifying Detrital Rocks To identify a detrital sedimentary rock, you use the size of the grains that make up the rock. The smallest, clay-sized grains feel slippery when wet and make up a rock called shale. Silt-sized grains are slightly larger than clay. These make up the rougher-feeling siltstone. Sandstone is made of yet larger, sand-sized grains. Pebbles are larger still. Pebbles mixed and cemented together with other sediment make up rocks called conglomerates (kun GLAHM ruts).

Chemical Rocks Some sedimentary rocks form when seawater, loaded with dissolved minerals, evaporates. Chemical sedimentary rock also forms when mineral-rich water from geysers, hot springs, or salty lakes evaporates, as shown in **Figure 18**. As the water evaporates, layers of the minerals are left behind. If you've ever sat in the Sun after swimming in the ocean, you probably noticed salt crystals on your skin. The seawater on your skin evaporated, leaving behind deposits of halite. The halite was dissolved in the water. Chemical rocks form this way from evaporation or other chemical processes.

Organic Rocks Would it surprise you to know that the chalk your teacher is using on the chalkboard might also be a sedimentary rock? Not only that, but coal, which is used as a fuel to produce electricity, is also a sedimentary rock.

Chalk and coal are examples of the group of sedimentary rocks called organic rocks. Organic rocks form over millions of years. Living matter dies, piles up, and then is compressed into rock. If the rock is produced from layers of plants piled on top of one another, it is called coal. Organic sedimentary rocks also form in the ocean and usually are classified as limestone.

Figure 18
The minerals left behind after a geyser erupts form layers of chemical rock.

Figure 19
There are a variety of organic sedimentary rocks.

A The pyramids in Egypt are made from fossiliferous limestone.

B A thin slice through the limestone shows that it contains many small fossils.

Fossils Chalk and other types of fossiliferous limestone are made from the fossils of millions of tiny organisms, as shown in **Figure 19**. A fossil is the remains or trace of a once-living plant or animal. A dinosaur bone and footprint are both fossils.

Section 2 Assessment

1. Contrast the ways in which extrusive and intrusive igneous rocks are formed.
2. Infer why igneous rocks that solidify underground cool so slowly.
3. Diagram how each of the three kinds of sedimentary rocks forms. List one example of each kind of rock: detrital, chemical, and organic.
4. List in order from smallest to largest the grain sizes used to identify detrital rocks.
5. **Think Critically** If someone handed you a sample of an igneous rock and asked you whether it is extrusive or intrusive, what would you look for first? Explain.

Skill Builder Activities

6. **Concept Mapping** Coal is an organic sedimentary rock that can be used as fuel. Research to find out how coal forms. On a computer, develop an events-chain concept map showing the steps in its formation. **For more help, refer to the** Science Skill Handbook.
7. **Communicating** Research a national park or monument where volcanic activity has taken place. Read about the park and the features that you'd like to see. Then describe the features in your Science Journal. Be sure to explain how each feature formed. **For more help, refer to the** Science Skill Handbook.

SECTION 2 Igneous and Sedimentary Rocks **181**

SECTION 3

Metamorphic Rocks and the Rock Cycle

As You Read

What You'll Learn
- **Describe** the conditions needed for metamorphic rocks to form.
- **Explain** how all rocks are linked by the rock cycle.

Vocabulary
metamorphic rock
foliated
nonfoliated
rock cycle

Why It's Important
Metamorphic rocks and the rock cycle show that Earth is a constantly changing planet.

New Rock from Old Rock

The land around you changed last night—perhaps not measurably, but it changed. Even if you can't detect it, Earth is changing constantly. Wind relocates soil particles. Layers of sediment are piling up on lake bottoms where streams carrying sediment flow into them. Wind and rain are gradually wearing away cliffs. Landmasses are moving at a rate of a few centimeters per year. Rocks are disappearing slowly below Earth's surface. Some of these changes can cause existing rocks to be heated and squeezed, as shown in **Figure 20**. In the process, new rocks form.

It can take millions of years for rocks to change. That's the amount of time that often is necessary for extreme pressure to build while rocks are buried deeply or continents collide. Sometimes existing rocks are cooked when magma moves upward into Earth's crust, changing their mineral crystals. All these events can make new rocks out of old rocks.

✔ **Reading Check** *What events can change rocks?*

Figure 20
The rocks of the Labrador Peninsula in Canada were squeezed into spectacular folds. This photo was taken during the space shuttle *Challenger* mission *STS-41G* in 1984.

182 CHAPTER 6 Rocks and Minerals

Figure 21
High pressure and temperature can cause existing rocks to change into new metamorphic rocks. **A** Granite can change to gniess. **B** The sedimentary rock sandstone can become quartzite, and **C** limestone can change to marble.

Metamorphic Rocks Do you recycle your plastic milk jugs? After the jugs are collected, sorted, and cleaned, they are heated and squeezed into pellets. The pellets later can be made into useful new products. It takes millions of years, but rocks get recycled, too. This process usually occurs thousands of meters below Earth's surface where temperatures and pressures are high. New rocks that form when existing rocks are heated or squeezed are called **metamorphic** (me tuh MOR fihk) **rocks.** The word *metamorphic* means "change of form." This describes well how some rocks take on a whole new look when they are under great temperatures and pressures.

What rocks are used to construct buildings? To find out about building stones, see the **Building Stones Field Guide** at the back of the book.

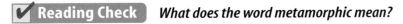 *What does the word metamorphic mean?*

Figure 21 shows three kinds of rocks and what they change into when they are subjected to the forces involved in metamorphism. Not only do the resulting rocks look different, they have recrystallized and might be chemically changed, too. The minerals often align in a distinctive way.

SECTION 3 Metamorphic Rocks and the Rock Cycle **183**

Figure 22
There are many different types of metamorphic rocks.

A This statue is made from marble, a nonfoliated metamorphic rock.

B The roof of this house is made of slate, a foliated metamorphic rock.

Science Online

Collect Data Visit the Glencoe Science Web site at **tx.science.glencoe.com** for data about the rock cycle. Make your own diagram of the rock cycle.

Types of Changed Rocks New metamorphic rocks can form from any existing type of rock—igneous, sedimentary, or metamorphic. A physical characteristic helpful for classifying all rocks is the texture of the rocks. This term refers to the general appearance of the rock. Texture differences in metamorphic rocks divide them into two main groups—foliated (FOH lee ay tud) and nonfoliated, as shown in **Figure 22.**

Foliated rocks have visible layers or elongated grains of minerals. The term *foliated* comes from the Latin *foliatus*, which means "leafy." These minerals have been heated and squeezed into parallel layers, or leaves. Many foliated rocks have bands of different-colored minerals. Slate, gneiss (NISE), phyllite (FIHL ite), and schist (SHIHST) are all examples of foliated rocks.

Nonfoliated rocks do not have distinct layers or bands. These rocks, such as quartzite, marble, and soapstone, often are more even in color than foliated rocks. If the mineral grains are visible at all, they do not seem to line up in any particular direction. Quartzite forms when the quartz sand grains in sandstone fuse after they are squeezed and heated. You can fuse ice crystals in a similar way if you squeeze a snowball. The presssure from your hands creates grains of ice inside.

184 CHAPTER 6 Rocks and Minerals

The Rock Cycle

Rocks are changing constantly from one type to another. If you wanted to describe these processes to someone, how would you do it? Would you use words or pictures? Scientists have created a model in diagram form called the **rock cycle** to show how different kinds of rock are related to one another and how rocks change from one type to another. Each rock is on a continuing journey through the rock cycle, as shown in **Figure 23**. A trip through the rock cycle takes millions of years.

Figure 23
This diagram of the rock cycle shows how rocks are recycled constantly from one kind of rock to another.

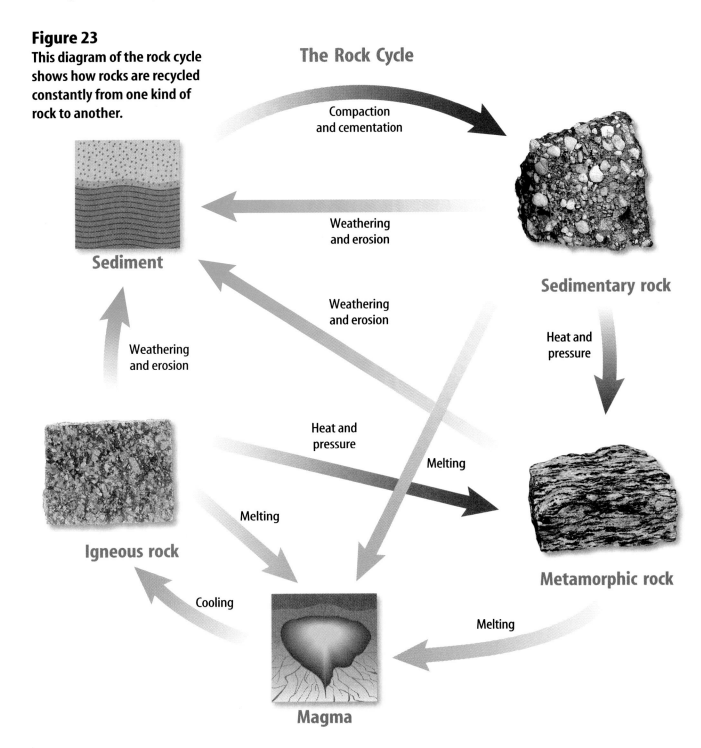

SECTION 3 Metamorphic Rocks and the Rock Cycle **185**

The Journey of a Rock Pick any point on the diagram of the rock cycle in **Figure 23,** and you will see how a rock in that part of the cycle could become any other kind of rock. Start with a blob of lava that oozes to the surface and cools, as shown in **Figure 24.** It forms an igneous rock. Wind, rain, and ice wear away at the rock, breaking off small pieces. These pieces are now called sediment. Streams and rivers carry the sediment to the ocean where it piles up over time. The weight of sediment above compresses the pieces below. Mineral-rich water seeps through the sediment and glues, or cements, it together. It becomes a sedimentary rock. If this sedimentary rock is buried deeply, pressure and heat inside Earth can change it into a metamorphic rock. Metamorphic rock deep inside Earth can melt and begin the cycle again. In this way, all rocks on Earth are changed over millions and millions of years. This process is taking place right now.

Figure 24
This lava in Hawaii is flowing into the ocean and cooling rapidly.

Reading Check *Describe how a metamorphic rock might change into an igneous rock.*

Section 3 Assessment

1. Identify two factors that can produce metamorphic rocks.
2. List examples of foliated and nonfoliated rocks. Explain the difference between the two types of metamorphic rocks.
3. Igneous rocks and metamorphic rocks can form at high temperatures and pressures. Explain the difference between these two rock types.
4. Scientists have diagrammed the rock cycle. Explain what this diagram shows.
5. **Think Critically** Trace the journey of a piece of granite through the rock cycle. Explain how this rock could be changed from an igneous rock to a sedimentary rock and then to a metamorphic rock.

Skill Builder Activities

6. **Drawing Conclusions** Describe an event that is a part of the rock cycle you can observe occurring around you or that you see on television news. Explain the steps leading up to this part of the rock cycle, and the steps that could follow to continue the cycle. **For more help, refer to the** Science Skill Handbook.
7. **Using an Electronic Spreadsheet** Using a spreadsheet program, create a data table to list the properties of different rocks and minerals that you have studied in this chapter. After you've made your table, cut and paste the different rows so as to group like rocks and minerals together. **For more help, refer to the** Technology Skill Handbook.

Activity

Gneiss Rice

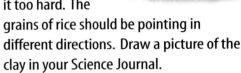

You know that metamorphic rocks often are layered. But did you realize that individual mineral grains can change in orientation? This means that the grains can line up in certain directions. You'll experiment with rice grains in clay to see how foliation is produced.

What You'll Investigate
What conditions will cause an igneous rock to change into a metamorphic rock?

Materials
rolling pin
lump of modeling clay
uncooked rice (wild rice, if available) (200 g)
granite sample
gneiss sample

Goals
- **Investigate** ways rocks are changed.
- **Model** a metamorphic rock texture.

Safety Precautions

WARNING: *Do not taste, eat, or drink any materials used in the lab.*

Procedure

1. **Sketch** the granite specimen in your Science Journal. Be sure that your sketch clearly shows the arrangement of the mineral grains.
2. Pour the rice onto the table. Roll the ball of clay in the rice. Some of the rice will stick to the outside of the ball. Knead the ball until the rice is spread out fairly evenly. Roll and knead the ball again, and repeat until your clay sample has lots of "minerals" distributed throughout it.
3. Using the rolling pin, roll the clay so it is about 0.5 cm thick. Don't roll it too hard. The grains of rice should be pointing in different directions. Draw a picture of the clay in your Science Journal.
4. Take the edge of the clay closest to you and fold it toward the edge farthest from you. Roll the clay in the direction you folded it. Fold and roll the clay in the same direction several more times. Flatten the lump to 0.5 cm in thickness again. Draw what you observe in your "rock" and in the gneiss sample in your Science Journal.

Conclude and Apply

1. What features did the granite and the first lump of clay have in common?
2. What force caused the positions of rice grains in the lump of clay to change? How is this process similar to and different from what happens in nature?

Communicating Your Data

Refer to your Science Journal diagrams and the rock samples provided for you in this activity and make a poster relating this activity to processes in the rock cycle. Be sure to include diagrams of what you did, as well as information on how similar events occur in nature. **For more help, refer to the** Science Skill Handbook.

ACTIVITY 187

Activity

Classifying Minerals

Texas is a great place for collecting minerals. You can find gypsum in the southeast and great quartz crystals in south and west Texas. If you live in Travis County, you might find some strange, gold-colored fossils that have been changed to pyrite or marcasite. If you spend time in Presidio County, you could find geodes, which are hollow rocks filled with mineral crystals. This activity will help you recognize some common minerals.

What You'll Investigate
How to classify a set of minerals.

Materials
set of minerals
hand lens
putty knife
streak plate
Mohs scale
minerals field guide

Safety Precautions

WARNING: *Be careful when using a knife. Never taste any materials used in a lab.*

Goals
- **Test** and observe important mineral characteristics.
- **List** the characteristics of each mineral in a data table.
- **Identify** the minerals based on their characteristics.

188 CHAPTER 6 Rocks and Minerals

Using Scientific Methods

Procedure

1. Copy the data table below into your Science Journal. Based on your observations and streak and hardness tests, fill in columns 2 to 6. In the sixth column—"Scratches which samples?"—list the number of each mineral sample that this sample was able to scratch. Use this information to rank each sample from softest to hardest. Compare these ranks to Mohs scale to help identify the mineral. Consult the rocks and minerals field guide to fill in the last column after compiling all the characteristics.

2. Obtain a classroom set of minerals.

3. **Observe** each sample and conduct appropriate tests to complete as much of your data table as possible.

Mineral Characteristics							
Sample Number	Crystal Shape	Cleavage/ Fracture	Color	Streak and Luster	Scratches which samples?	Hardness Rank	Mineral Name
1							
2							
…							
No. of samples							

Conclude and Apply

1. Based on the information in your data table, identify each mineral.

2. Did you need all of the information in the table to identify each mineral? Explain why or why not.

3. Which characteristics were easy to determine? Which were somewhat more difficult? Explain.

4. Were some characteristics more useful as indicators than others?

5. Would you be able to identify minerals in the field after doing this activity? Which characteristics would be easy to determine on the spot? Which would be difficult?

6. **Describe** how your actions in this activity are similar to those of a scientist. What additional work might a scientist have done to identify these unknown minerals?

Create a visually appealing poster showing the minerals in this activity and the characteristics that were useful for identifying each one. Be sure to include informative labels on your poster.

Oops! Accidents in SCIENCE
SOMETIMES GREAT DISCOVERIES HAPPEN BY ACCIDENT!

Going for the
A time line history of the accidental discovery of gold in California

1840
California is a quiet place. Only a few hundred people live in the small town of San Francisco.

1847
John Sutter hires James Marshall to build a sawmill on his ranch. Marshall and local Native Americans work quickly to harness the water power of the American River. They dig a channel from the river to run the sawmill. The water is used to make the water-wheel work.

Sutter's Mill

1848
On January 24, Marshall notices something glinting in the water. Is it a nugget of gold? Aware that all that glitters is not gold, Marshall hits it with a rock. Marshall knows that "fool's gold" shatters when hit. But this shiny metal bends. More nuggets are found. Marshall shows the gold nugget to Sutter. After some more tests, they decide it is gold! Sutter and Marshall try to keep the discovery a secret, but word leaks out.

Miners hope to strike it rich.

1850
California becomes the thirty-first state.

1849
The Gold Rush hits! A flood of people from around the world descends on northern California. They're dubbed "forty-niners" because they leave home in 1849 to seek their fortunes. San Francisco's population grows to 25,000. Many people become wealthy—but not Marshall or Sutter. Since Sutter doesn't have a legal claim to the land, the U.S. government claims it.

Gold

1854
A giant nugget of gold, the largest known to have been discovered in California, is found in Calaveras County.

1872
As thanks for his contribution to California's growth, the state legislature awards Marshall a pension of $200 a month for two years. The pension is renewed until 1878.

1885
James Marshall dies with barely enough money to cover his funeral.

1864
California's gold rush ends. The rich surface and river placers are largely exhausted. Hydraulic mines are the chief source of gold for the next 20 years.

1880
His pension ended, Marshall is forced to earn a living through various odd jobs, receiving charity, and by selling his autograph. He attempts a lecture tour, but is unsuccessful.

1890
California builds a bronze statue of Marshall in honor of his discovery.

CONNECTIONS Research Trace the history of gold from ancient civilizations to the present. How was gold used in the past? How is it used in the present? What new uses for gold have been discovered? Report to the class.

SCIENCE *Online*
For more information, visit
tx.science.glencoe.com

Chapter 6 Study Guide

Reviewing Main Ideas

Section 1 Minerals—Earth's Jewels

1. Minerals are inorganic solid materials found in nature. They generally have the same chemical makeup, and the atoms always are arranged in an orderly pattern. Rocks are combinations of two or more minerals. *In what way does this amethyst reflect the orderly pattern of its atoms?*

2. These properties can be used to identify minerals—crystal shape, cleavage and fracture, color, streak, luster, hardness, and specific gravity.

3. Gems are minerals that are rare. When cut and polished they are beautiful.

4. Ores of useful minerals must be extracted from Earth in a process called mining. Ores usually must be processed to produce metals.

Section 2 Igneous and Sedimentary Rocks

1. Igneous rocks form when melted rock from inside Earth cools and hardens.

2. Extrusive igneous rocks are formed on Earth's surface and have small or no crystals. Intrusive igneous rocks harden underneath Earth's surface and have large crystals.

3. Sedimentary rocks are made from pieces of other rocks, minerals, or plant and animal matter that collect in layers.

4. There are three groups of sedimentary rocks. Rocks formed from grains of minerals or other rocks are called detrital rocks. *What two processes occurred to change sand into this sandstone?*

5. Rocks formed from a chemical process such as evaporation of mineral-rich water are called chemical rocks. Rocks formed from fossils or plant remains are organic rocks.

Section 3 Metamorphic Rocks and the Rock Cycle

1. Existing rocks change into metamorphic rocks after becoming heated or squeezed inside Earth. The result is a rock with an entirely different appearance.

2. Foliated metamorphic rocks are easy to spot due to the layers of minerals. Nonfoliated metamorphic rocks lack distinct layers.

3. The rock cycle shows how all rocks are related and the processes that change them from one type to another. *What rock type would result from this volcanic eruption?*

After You Read

Use the information in your Venn Diagram Study Fold to compare and contrast rocks and minerals. Write common characteristics under the *Both* tab.

Chapter 6 Study Guide

Visualizing Main Ideas

Complete the concept map using the following terms and phrases: extrusive, organic, foliated, intrusive, chemical, nonfoliated, detrital, metamorphic, *and* sedimentary.

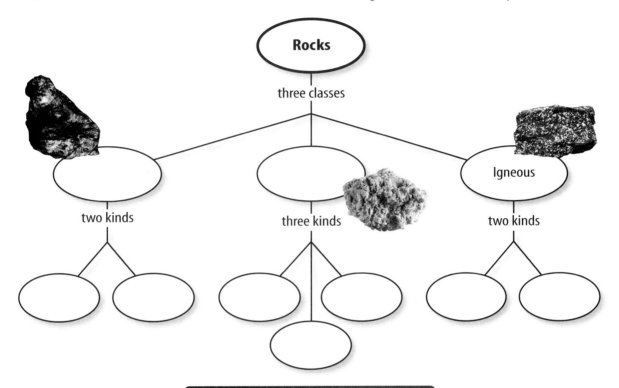

Vocabulary Review

Vocabulary Words

a. crystal
b. extrusive
c. foliated
d. gem
e. igneous rock
f. intrusive
g. metamorphic rock
h. mineral
i. nonfoliated
j. ore
k. rock
l. rock cycle
m. sedimentary rock

Using Vocabulary

Explain the difference between each pair of vocabulary words.

1. mineral, rock
2. crystal, gem
3. cleavage, fracture
4. hardness, streak
5. rock, rock cycle
6. intrusive, extrusive
7. igneous rock, metamorphic rock
8. foliated, nonfoliated
9. rock, ore
10. metamorphic rock, sedimentary rock

 **Study Tip**

Make sure to read over your class notes after each lesson. Reading them will help you better understand what you've learned, as well as prepare you for the next day's lesson.

Chapter 6 Assessment & TEKS Review

Checking Concepts

Choose the word or phrase that best answers the question.

1. Which of the following describes what rocks usually are composed of?
   A) pieces
   B) minerals
   C) fossil fuels
   D) foliations

2. When do metamorphic rocks form?
   A) when layers of sediment are deposited
   B) when lava solidifies in seawater
   C) when particles of rock break off at Earth's surface
   D) when heat and pressure change rocks

3. How can sedimentary rocks be classified?
   A) foliated or nonfoliated
   B) organic, chemical, or detrital
   C) extrusive or intrusive
   D) gems or ores

4. What kind of rocks are produced by volcanic eruptions?
   A) detrital
   B) foliated
   C) organic
   D) extrusive

5. Which of the following must be true for a substance to be considered a mineral?
   A) It must be organic.
   B) It must be glassy.
   C) It must be a gem.
   D) It must be naturally occurring.

6. Which of the following describes grains in igneous rocks that form slowly from magma below Earth's surface?
   A) no grains
   B) visible grains
   C) sedimentary grains
   D) foliated grains

7. How do sedimentary rocks form?
   A) They are deposited on Earth's surface.
   B) They form from magma.
   C) They are squeezed into foliated layers.
   D) They form deep in Earth's crust.

8. Which of these is NOT a physical property of a mineral?
   A) cleavage
   B) organic
   C) fracture
   D) hardness

9. Which is true of all minerals?
   A) They are inorganic solids.
   B) They have a glassy luster.
   C) They have a conchoidal fracture.
   D) They are harder than a penny.

10. Which is true about how all detrital rocks form?
    A) form from grains of preexisting rocks
    B) form from lava
    C) form by evaporation
    D) form from plant remains

Thinking Critically

11. Is a sugar crystal a mineral? Explain.

12. Metal deposits in Antarctica are not considered to be ores. List some reasons for this.

13. How is it possible to find pieces of gneiss, granite, and basalt in a single conglomerate?

14. Would you expect to find a well-preserved dinosaur bone in a metamorphic rock like schist? Explain.

15. Explain how the mineral quartz could be in an igneous rock and in a sedimentary rock.

Developing Skills

16. **Communicating** You are hiking in the mountains and as you cross a shallow stream, you see an unusual rock. You notice that it is full of fossil shells. Your friend asks you what it is. What do you say and why?

Chapter 6 Assessment

17. Classifying Your teacher gives you two clear minerals. What quick test could you do in order to determine which is halite and which is calcite?

18. Concept Mapping Complete this concept map about minerals.

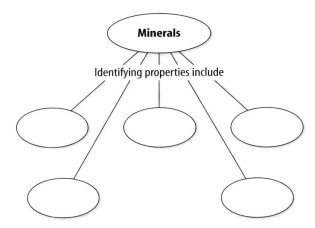

19. Testing a Hypothesis Your teacher gives you a glass plate, a nail, a penny, and a bar magnet. On a computer, describe how you would use these items to determine the hardness and special property of the mineral magnetite. Refer to **Table 1** for help.

Performance Assessment

20. Making models Determine what materials and processes you would need to use to set up a working model of the rock cycle. Describe the ways in which your model is accurate and the ways in which it falls short. Present your model to the class.

TECHNOLOGY

Go to the Glencoe Science Web site at **tx.science.glencoe.com** or use the **Glencoe Science CD-ROM** for additional chapter assessment.

 TAKS Practice

A student was studying for an Earth science test and made the following table to keep track of information. *TEKS 6.14 A*

Type of Rocks	
Type of Rock	Characteristics
Igneous	Form from lava or magma
Sedimentary	Layers of sediment or organic remains
Metamorphic	Result of high temperature or pressure

Examine the table and answer the questions

1. According to the chart, igneous rocks form from _____.
   A) sediment
   B) lava or magma
   C) plant/animal matter
   D) high pressures

2. A rock made up of fossil shells would be a _____.
   F) sedimentary rock
   G) igneous rock
   H) metamorphic rock
   J) fissure.

3. Rocks that form from high heat and pressure are _____.
   A) sedimentary rocks
   B) chemical rocks
   C) metamorphic rocks
   D) organic rocks

CHAPTER 7

Science TEKS 6.6 C; 6.14

Forces Shaping Earth

These majestic, snow-capped mountains in Annapurna, Nepal look rugged and indestructible, but in geologic terms, they are in their infancy. It would take a few hundred million years of erosion for their sharp, jagged peaks to become smooth. They then would resemble more mature mountains, like those found in the eastern United States. In this chapter, you'll learn how the movement of plates formed these mountains and about other Earth forces that shape mountains. You also will study layers of the Earth's interior.

What do you think?

Science Journal Look at the picture below with a classmate. Discuss what this might be or what is happening. Here's a hint: *During earthquakes rocks can grind and scrape.* Write your answer or best guess in your Science Journal.

Geologists know many things about the interior of Earth even though its center is over 6,000 km deep. Use modeling clay to make a model of Earth's interior.

Model Earth's interior

1. Obtain four pieces of clay that are different colors.
2. Roll one piece of clay into a ball. This clay represents the inner core.
3. Wrap another piece of clay around the first ball of clay, making an even bigger ball. This clay represents the outer core.
4. Repeat step 3 with the third piece of clay, which represents Earth's mantle. Wrap your model with a thin layer of the fourth piece of clay to represent the crust.
5. Use a plastic knife to cut the ball of clay in half.

Observe
In your Science Journal, make a sketch of your model and label each of Earth's layers.

Before You Read

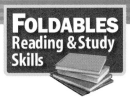

Making a Cause and Effect Study Fold Make the following Foldable to help you understand the cause and effect relationship of Earth's interior and surface.

1. Stack two sheets of paper in front of you so the short side of both sheets is at the top.
2. Slide the top sheet up so about four centimeters of the bottom sheet show.
3. Fold both sheets top to bottom to form four tabs and staple along the topfold. Turn the Foldable so the staples are at the bottom.
4. Label each flap *Inner Core, Outer Core, Mantle,* and *Crust.* Draw in the layers as shown.
5. As you read the chapter, write about each layer under the tabs.

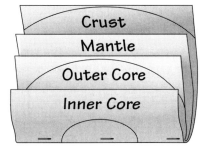

SECTION 1

Earth's Moving Plates

As You Read

What You'll Learn

- **Describe** how Earth's interior is divided into layers.
- **Explain** how plates of Earth's lithosphere move.
- **Discuss** why Earth's plates move.

Vocabulary

inner core
outer core
mantle
crust
plate
fault
subduction
lithosphere

Why It's Important

Forces that cause Earth's plates to move apart, together, or past each other form features on Earth's surface, such as mountains, volcanoes, and mid-ocean ridges.

Clues to Earth's Interior

If someone gives you a wrapped present, how could you figure out what was in it? You might hold it, shake it gently, or weigh it. You'd look for clues that could help you identify the contents of the box. Even though you can't see what's inside the package, these types of clues can help you figure out what it might be. Because you can't see what's inside, the observations you make are known as indirect observations.

Geologists do the same thing when they try to learn about Earth's interior. Although the best way to find out what's inside Earth might be to dig a tunnel to its center, that isn't possible. The deepest mines in the world only scratch Earth's surface. A tunnel would need to be more than 6,000 km deep to reach the center, so geologists must use indirect observations to gather clues about what Earth's interior is made of and how it is structured. This indirect evidence includes information learned by studying earthquakes and rocks that are exposed at Earth's surface.

Waves When you throw a rock into a calm puddle or pond, you observe waves like those shown in **Figure 1.** Waves are disturbances that carry energy through matter or space. When a rock hits water, waves carry some of the rock's kinetic energy, or energy of motion, away from where it hit the water. When an earthquake occurs, as shown in **Figure 2,** energy is carried through objects by waves. The speed of these waves depends on the density and nature of the material they are traveling through. For example, a wave travels faster in solid rock than it does in a liquid. By studying the speed of these waves and the paths they take, geologists uncover clues as to how the planet is put together. In fact, these waves, called seismic waves, speed up in some areas, slow down in other areas, and can be bent or stopped.

Figure 1
Waves carry energy across water just like seismic waves carry energy through Earth.

Figure 2
As seismic waves travel across Earth's surface, the ground shakes and damage occurs.

Rock Clues Another clue to what's inside Earth comes in the form of certain rocks found in different places on Earth's surface. These rocks are made of material similar to what is thought to exist deep inside Earth. The rocks formed far below the surface. Forces inside Earth pushed them closer to the surface, where they eventually were exposed by erosion. The seismic clues and the rock clues suggest that Earth is made up of layers of different kinds of materials.

Earth's Layers

Based on evidence from earthquake waves and exposed rocks, scientists have produced a model of Earth's interior. The model shows that Earth's interior has at least four distinct layers—the inner core, the outer core, the mantle, and the crust. Earth's structure is similar in some ways to the structure of a peach, shown in **Figure 3.** A peach has a thin skin covering the thick, juicy part that you eat. Under that is a large pit that surrounds a seed.

Inner Core The pit and seed are similar to Earth's core. Earth's core is divided into two distinct parts—one that is liquid and one that is solid. The innermost layer of Earth's interior is the solid **inner core.** This part of the core is dense and composed mostly of solid iron. When seismic waves produced by earthquakes reach this layer they speed up, indicating that the inner core is solid.

Conditions in the inner core are extreme compared to those at the surface. At about 5,000°C, the inner core is the hottest part of Earth. Also, because of the weight of the surrounding rock, the core is under tremendous pressure. Pressure, or the force pushing on an area, increases the deeper you go beneath Earth's surface. Pressure increases because more material is pushing toward Earth's center as a result of gravity. The inner core, at the center of Earth, experiences the greatest amount of pressure.

Figure 3
The structure of Earth can be compared to a peach. *If the part of Earth that you live on is like the skin of the peach, what does that tell you about this layer of Earth?*

Outer Core The **outer core** lies above the inner core and is thought to be composed mostly of molten metal. The outer core stops one type of seismic wave and slows down another. Because of this, scientists have concluded that the outer core is a liquid. The location of the outer core is similar to the location of the pit in the peach model. Even the wrinkled surface of the pit resembles the uneven nature of the boundary between Earth's outer core and its mantle as indicated by seismic studies.

✓ **Reading Check** *What peach layer is similar to the outer core?*

Mantle The layer in Earth's interior above the outer core is the mantle. In the peach model, the mantle would be the juicy part of the peach that you would eat. The **mantle** is the largest layer of Earth's interior. Even though it's solid, the mantle flows slowly, similar to putty.

Crust Earth's outermost layer is the **crust.** In the model of the peach, this layer would be the fuzzy skin of the peach. Earth's crust is thin when compared to the other layers, though its thickness does vary. It is thinnest under the oceans and thickest through the continents. All features on Earth's surface are part of the crust.

Chemistry INTEGRATION

Earth's crust is composed of about five percent iron. However, geologists theorize that Earth's core is composed mostly of iron. Research the theory that Earth's core is composed mostly of iron. Analyze, review, and critique the strengths and weaknesses of this theory using scientific evidence and information.

Figure 4
Earth is made up of many layers.
What geologic events have allowed scientists to study Earth's interior?

A The lithosphere is composed of crust and uppermost mantle.

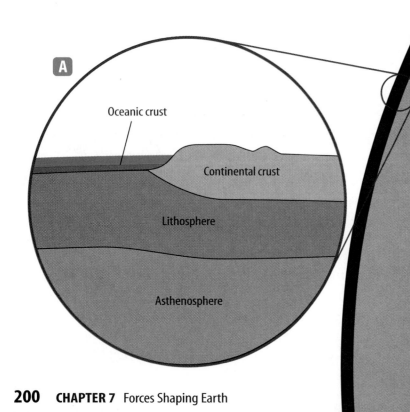

Earth's Structure

Although Earth's structure can be divided into four basic layers, it also can be divided into other layers based on physical properties that change with depth beneath the surface. **Figure 4** shows the structure of Earth and describes some of the properties of its layers. Density, temperature, and pressure are properties that are lowest in the crust and greatest in the inner core.

B The mantle makes up the majority of Earth's mass.

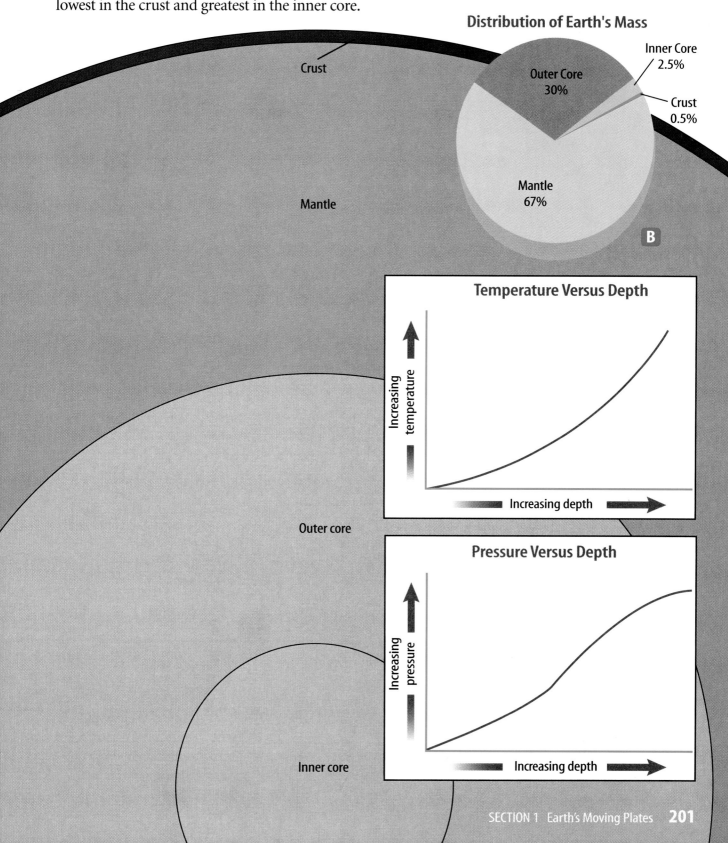

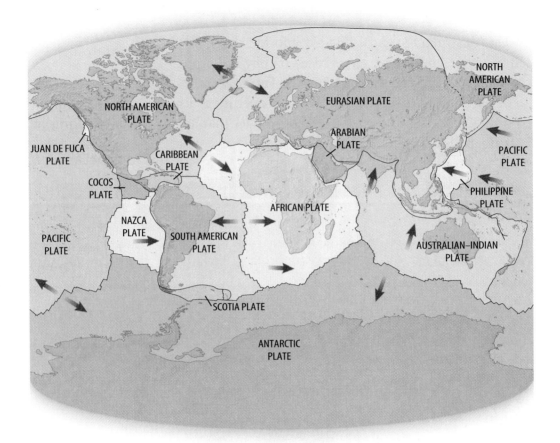

Figure 5
Earth's plates fit together like the pieces of a jigsaw puzzle. *If the plates are moving, what do you suppose happens at the plate boundaries?*

Earth's Plates

Although the crust is separated from the mantle, the uppermost, rigid layer of the mantle moves as if it were part of Earth's crust. The rigid, upper part of Earth's mantle and the crust is called the **lithosphere.** It is broken into about 30 sections or **plates** that move around on the plasticlike asthenosphere, which also is part of the mantle. Earth's major plates vary greatly in size and shape, as shown in **Figure 5.**

Reading Check *What parts of Earth make up the lithosphere?*

The movements of the plates are fairly slow, often taking more than a year to creep a few centimeters. This means that they have not always looked the way they do in **Figure 5.** The plates have not always been their current size and shape, and continents have moved great distances. Antarctica, which now covers the south pole, was once near the equator, and North America was once connected to Africa and Europe.

Lasers and satellites are used to measure the small plate movements, which can add up to great distances over time. If a plate is found to move at 2 centimeters per year on average, how far will it move in 1,000 years? What about in 10 million years?

Plate Boundaries

The places where the edges of different plates meet are called plate boundaries. The constant movement of plates creates forces that affect Earth's surface at the boundaries of the plates. At some boundaries, these forces are large enough to cause mountains to form. Other boundaries form huge rift valleys with active volcanoes. At a third type of boundary, huge faults form. **Faults** are large fractures in rocks along which movement occurs. The movement can cause earthquakes. **Figure 6** shows the different plate motions.

Plates That Move Apart Plates move apart as a result of pulling forces that act in opposite directions on each plate. This pulling force is called tension. **Figure 7** shows what happens as tension continues to pull two plates apart.

One important result of plates separating is the formation of new lithosphere. New lithosphere forms in gaps where the plates pull apart. As tension continues along these boundaries, new gaps form and are filled in by magma that is pushed up from the mantle. Over time, the magma in the gaps cools to become new lithosphere. This process of plate separation and lithosphere formation takes place under the oceans at places called mid-ocean ridges. As new lithosphere moves away from the mid-ocean ridges, it cools and becomes denser.

Data Update Visit the Glencoe Science Web site at **tx.science.glencoe.com** for recent news or magazine articles about Earth's plates and the different boundaries that they form. Communicate to your class what you learn.

Figure 6
Earth's plates can collide, move away from each other, or slide past each other.

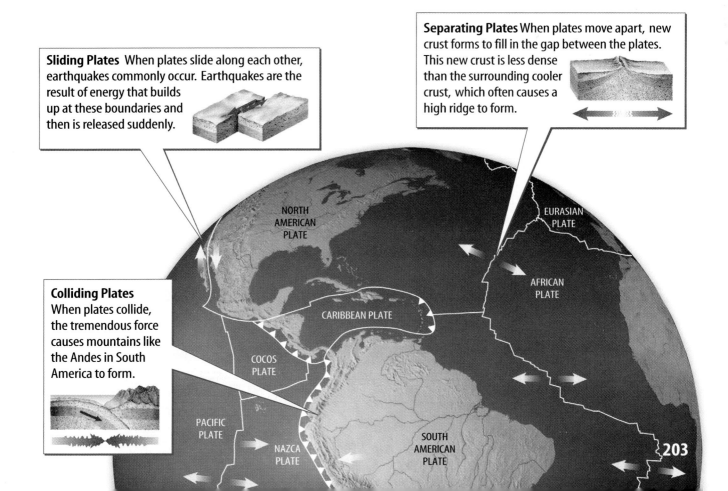

Sliding Plates When plates slide along each other, earthquakes commonly occur. Earthquakes are the result of energy that builds up at these boundaries and then is released suddenly.

Separating Plates When plates move apart, new crust forms to fill in the gap between the plates. This new crust is less dense than the surrounding cooler crust, which often causes a high ridge to form.

Colliding Plates When plates collide, the tremendous force causes mountains like the Andes in South America to form.

NATIONAL GEOGRAPHIC VISUALIZING RIFT VALLEYS

Figure 7

When two continental plates pull apart, rift valleys may form. If spreading continues and the growing rift reaches a coastline, seawater floods in. Beneath the waves, molten rock, or magma, oozes from the weakened and fractured valley floor. In time, the gap between the two continental slabs may widen into a full-fledged ocean. The four steps associated with this process are shown here. Africa's Great Rift Valley, which cuts across the eastern side of Africa for 5,600 km (right), represents the second of these four steps. If rifting processes continue in the Great Rift Valley, East Africa eventually will part from the mainland.

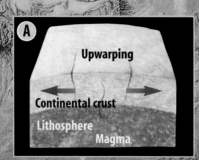

Rising magma forces the crust upward, causing numerous cracks in the rigid crust.

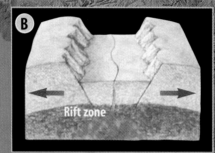

As the crust is pulled apart, large slabs of rock sink, generating a rift zone.

Further spreading generates a narrow sea or lake.

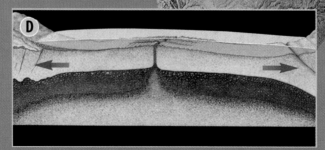

Eventually, an expansive ocean basin and ridge system are created.

CHAPTER 7 Forces Shaping Earth

Plates That Collide When plates move toward each other, they collide, causing several different things to occur. As you can see in **Figure 8,** the outcome depends on the density of the two plates involved. The crust that forms the ocean floors, called oceanic crust, is more dense than the continental crust, which forms continents.

If two continental plates collide, they have a similar density so the collision causes the crust to pile up. When rock converges like this, the force is called compression. Compression causes the rock layers on both plates to crumple and fold. Imagine laying a piece of fabric flat on your desk. If you push the edges of the cloth toward each other, the fabric will crumple and fold over on itself. A similar process occurs when plates crash into each other, causing mountains to form.

Flat rock layers are pushed up into folds. Sometimes the folding is so severe that rock layers bend completely over on themselves, turning upside down. As rock layers are folded and faulted, they pile up and form mountains. The tallest mountains in the world, the Himalaya in Asia, are still rising as two continental plates collide.

Plate Subduction When an oceanic plate collides with another oceanic plate or a continental plate, the more dense one plunges underneath the other, forming a deep trench. When one plate sinks underneath another plate, it's called **subduction.** When a plate subducts, it sinks into the mantle. In this way, Earth's crust does not continue to grow larger. As new crust material is generated at a rift, older crustal material subducts into the mantle.

Modeling Tension and Compression

Procedure
1. Obtain two bars of **taffy.**
2. Hold one bar of taffy between your hands and push your hands together.
3. Record your observations in your **Science Journal.**
4. Hold the other bar of taffy between your hands and pull gently on both ends.
5. Record your observations in your Science Journal.

Analysis
1. On which bar of taffy did you apply tension? Compression?
2. Explain how this applies to plate boundaries.

Figure 8
There are three types of convergent plate boundaries.

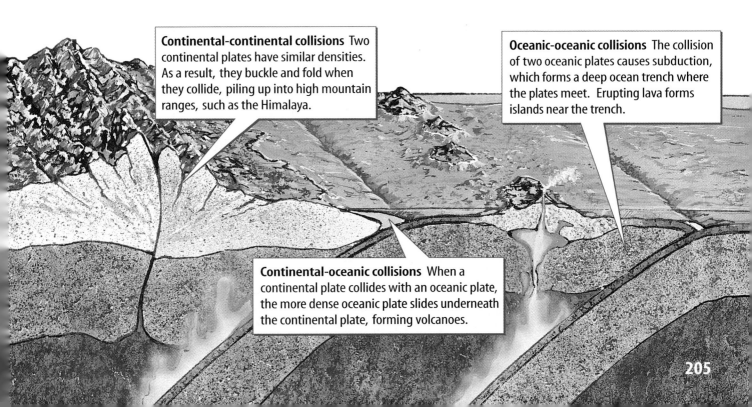

Continental-continental collisions Two continental plates have similar densities. As a result, they buckle and fold when they collide, piling up into high mountain ranges, such as the Himalaya.

Oceanic-oceanic collisions The collision of two oceanic plates causes subduction, which forms a deep ocean trench where the plates meet. Erupting lava forms islands near the trench.

Continental-oceanic collisions When a continental plate collides with an oceanic plate, the more dense oceanic plate slides underneath the continental plate, forming volcanoes.

Figure 9
As two plates slide past each other, their edges grind and scrape. The jerky movement that results causes earthquakes like those frequently felt in California along the San Andreas Fault.

Plates That Slide Past In addition to moving toward and away from one another, plates also can slide past one another. For example, one plate might be moving north while the plate next to it is moving south. The boundary where these plates meet is called a transform boundary. When a force pushes something in two different directions, it's called shearing. Shearing causes the area between the plates to form faults and experience many earthquakes. **Figure 9** shows part of the San Andreas Fault near Taft, California, which is an example of the features that form along a transform boundary.

Why do plates move?

As you can see, Earth's plates are large. To move something so massive requires a tremendous amount of energy. Where does the energy that drives plate movement come from? The reason plates move is complex, and geologists still are trying to understand it fully. So far, scientists have come up with several possible explanations about what is happening inside Earth to cause plate movement. Most of these theories suggest that gravity is the driving force behind it. However, gravity pulls things toward the center of Earth, and plates move sideways across the globe. How does gravity make something move across the surface of Earth?

One theory that could explain plate movement is convection of the mantle. Convection in all materials is driven by differences in density. In the mantle, density differences are caused by uneven heating, which results in a cycling of material, as shown in **Figure 10.** The theory suggests that the plates move as part of this circulation of mantle material.

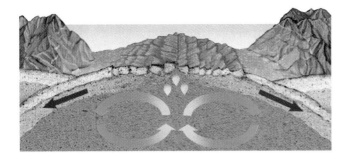

Figure 10
Convection, ridge-push, and slab-pull might all contribute to the motions of Earth's plates.

A Uneven heating of the upper mantle could cause convection.

B Ridge-push could occur at mid-ocean ridges.

Ridge-Push and Slab-pull Other factors, as shown in **Figure 10,** that could play a role in plate movement are ridge-push and slab-pull. Ridge-push occurs at mid-ocean ridges, which are higher than surrounding ocean floor. The plates respond to gravity by sliding down the slope. Slab-pull occurs as the plates move away from the mid-ocean ridges and become cooler, which makes them more dense. A plate can get so dense that it sinks when it collides with another plate. When the more dense plate begins to sink, it becomes easier for it to move across Earth's surface because resistance to movement is reduced.

C Slab-pull could occur where oceanic plates meet other oceanic or continental plates.

Section Assessment

1. How is the speed of earthquake waves used by scientists to provide information about Earth's interior?
2. Name the three types of plate movements and give examples of where they occur.
3. Which layer of Earth's interior is the largest?
4. List the layers of Earth's interior in order of least to greatest density.
5. **Think Critically** How can slab-pull and ridge-push contribute to the movement of a plate at the same time?

Skill Builder Activities

6. **Comparing and Contrasting** Compare and contrast the following pairs of terms: inner core, outer core; ridge-push, slab-pull. **For more help, refer to the** Science Skill Handbook.
7. **Using Graphics Software** Use the graphics capabilities of a computer to produce illustrations of plate movement at the three types of plate boundaries. **For more help, refer to the** Technology Skill Handbook.

SECTION 1 Earth's Moving Plates

Activity

Earth's Moving Plates

You have learned that Earth's surface is separated into plates that move apart, move together, or slide past each other. In this activity, you will observe a process that is thought to cause this plate movement.

What You'll Investigate
What process inside Earth provides the energy for plate motion?

Materials
1-L beaker (2)
food coloring
aluminum foil
pencil
rubber band
water (warm and cold)
2-cm paper squares (3)
small, clear-plastic cup

Goals
- **Observe** movement of solid plates on a liquid.
- **Identify** the cause of plate movement on Earth's surface.

Safety Precautions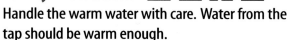
Handle the warm water with care. Water from the tap should be warm enough.

Procedure
1. Fill one of the 1-L beakers with cold water.
2. Fill the small cup with warm water.
3. Add four drops of food coloring to the cup of warm water and cover the top with aluminum foil. Secure the aluminum foil with a rubber band. No air should be underneath the foil.
4. Carefully place the cup of colored, warm water in the bottom of the second 1-L beaker.
5. Carefully pour the cold water from the first 1-L beaker into the second 1-L beaker. Take care not to disturb the cup of colored water.
6. Place the pieces of paper on the surface of the water in the second 1-L beaker.
7. Use a long pencil to make two small holes in the aluminum foil covering the cup.
8. **Observe** what happens to the contents of the cup and to the pieces of paper. Record your observations in your Science Journal.

Conclude and Apply
1. What happened to the colored, warm water originally located in the cup?
2. What effect, if any, does the warm water have on the positions of the floating paper?
3. How is what happens to the warm water similar to processes that occur inside Earth? How is it different?
4. After observing the pieces of paper floating on the cold water, explain what features on Earth's surface they are similar to.

Communicating Your Data
Compare your conclusions with those of other students in your class. **For more help, refer to the** Science Skill Handbook.

208 **CHAPTER 7** Forces Shaping Earth

SECTION 2
Uplift of Earth's Crust

Building Mountains

One popular vacation that people enjoy is a trip to the mountains. Mountains tower over the surrounding land, often providing spectacular views from their summit or from surrounding areas. The highest mountain peak in the world is Mount Everest in the Himalaya in Tibet. Its elevation is more than 8,800 m above sea level. In the United States, the highest mountains reach an elevation of more than 6,000 m. There are four main types of mountains—fault-block, folded, upwarped, and volcanic. Each type forms in a different way and can produce mountains that vary greatly in size.

Age of a Mountain As you can see in **Figure 11,** mountains can be rugged with high, snowcapped peaks, or they can be rounded and forested with gentle valleys and babbling streams. The ruggedness of a mountain chain depends largely on whether or not it is still forming. Mountains like the Himalaya are currently forming at a rate of several centimeters per year, while much older mountains like the Ouachita Mountains in Arkansas stopped forming millions of years ago and are now being eroded by geological processes.

As You Read

What You'll Learn
- **Describe** how Earth's mountains form and erode.
- **Compare** types of mountains.
- **Identify** the forces that shape Earth's mountains.

Vocabulary
fault-block mountain
folded mountain
upwarped mountain
volcanic mountain
isostasy

Why It's Important
Forces inside Earth that cause Earth's plates to move around also are responsible for forming Earth's mountains.

Figure 11
A Mountains can be high and rugged like the mountains of the Himalaya, or **B** they can be large, gently rolling hills like the Ouachita Mountains in Arkansas. *What determines how rugged and high a mountain chain is?*

SECTION 2 Uplift of Earth's Crust **209**

Figure 12
Before tension is applied, the layers of rock are even and fairly level. After tension is applied, huge blocks of rock separate and slip downward. This leaves large, tilted blocks that become mountains.

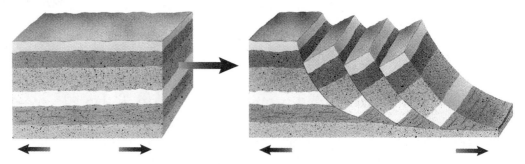

Fault-Block Mountains The first mountains you'll study are fault-block mountains. Some examples are the Sierra Nevada in California and the Teton Range in Wyoming. Recall that pulling forces that occur at the boundaries of plates moving apart, work to create surface features such as rift valleys and faults. Fault-block mountains also form from pulling forces. **Fault-block mountains** are made of huge, tilted blocks of rock that are separated from surrounding rock by faults. When rock layers are pulled from opposite directions, large blocks slide downward, creating peaks and valleys, as shown in **Figure 12.**

Figure 13
The Teton Range in the Grand Teton National Park has sharp, jagged peaks that are characteristic of fault-block mountains.

Models of Mountain Building If you hold a candy bar between your hands and then begin to pull it apart, cracks might form within the chocolate. Similarly, when rocks are pulled apart, faults form. Unlike rocks deep in Earth, rocks at Earth's surface are hard and brittle. When they are pulled apart, large blocks of rock can move along the faults. The Teton Range of Wyoming formed when a block of crust was tilted as one side of the range was uplifted above the neighboring valley. As shown in **Figure 13,** if you travel to the Grand Teton National Park, you will see sharp, jagged peaks that are characteristic of fault-block mountains.

Now, hold a flat piece of clay between your hands and then push your hands together gently. What happens? As you push your hands together, the clay begins to bend and fold over on itself. A similar process causes rocks to fold and bend, causing folded mountains to form on Earth's surface.

Folded Mountains Traveling along a road that is cut into the side of the Appalachian Mountains, you can see that rock layers were folded just as the clay was when it was squeezed, or compressed. Tremendous pushing forces exerted by two of Earth's plates moving together squeezed rock layers from opposite sides. This caused the rock layers to buckle and fold, forming folded mountains. **Folded mountains** are mountains formed by the folding of rock layers caused by compressive forces.

Reading Check *What type of force causes folded mountains to form?*

The Appalachian Mountains are folded mountains that formed about 250 million to 300 million years ago. A small part of the folded Appalachians is shown in **Figure 14.** The compression occurred as the North American Plate and the African Plate moved together. The Appalachians are the oldest mountain range in North America, and also one of the longest. They extend from Alabama northward to Quebec, Canada. Erosion has been acting on these mountains since they were formed. As a result, the Appalachians are small compared to other mountain ranges. At one time, the Appalachian Mountains were higher than the Rocky Mountains are today.

Upwarped Mountains The Adirondack Mountains in New York, the southern Rocky Mountains in Colorado and New Mexico, and the Black Hills in South Dakota are examples of upwarped mountains. **Upwarped mountains** form when forces inside Earth push up the crust. With time, sedimentary rock layers on top will erode, exposing the igneous or metamorphic rocks underneath. The igneous and metamorphic rocks can erode further to form sharp peaks and ridges.

Mini LAB

Modeling Mountains

Procedure
1. Use layers of **clay** to build a model of each major type of mountain.
2. For fault-block mountains, cut the layers of clay with a **plastic knife** to show how one block moves upward and another moves downward.
3. For folded mountains, push on the layers of clay from directly opposite directions.
4. For upwarped mountains, push a large, round object, such as a **ball,** upward from below, forcing the layers of clay to warp.
5. For volcanic mountains, place layer upon layer of clay to form a cone-shaped feature.

Analysis
1. Do any of the mountains you have modeled look similar? Explain.
2. How could you recognize the different types of mountains?

Figure 14
This roadcut in Maryland exposes folded rock layers that formed when the North American Plate and the African Plate collided.

SCIENCE Online

Research Visit the Glencoe Science Web site at **tx.science.glencoe.com** for more information on and photographs of volcanic mountains. Communicate to your class what you learn.

Volcanic Mountains Occasionally, magma from inside Earth reaches the surface. When this happens, the magma is called lava. When hot, molten lava flows onto Earth's surface, volcanic mountains can form. Over time, layer upon layer of lava piles up until a cone-shaped feature called a **volcanic mountain** forms. Washington's Mount St. Helens and Mexico's Mount Popocateptl, shown in **Figure 15,** are examples. Next, you will take a closer look at how volcanic mountains form.

Some volcanic mountains form when large plates of Earth's lithosphere sink into Earth's mantle at subduction zones. As the plates sink deeper into the mantle, they cause melting to occur. The magma produced is less dense than the surrounding rock, so it is forced slowly upward to Earth's surface. If the magma reaches the surface, it can erupt as lava and ash. Layers of these materials can pile up over time to form volcanic mountains.

Figure 15
Volcanic mountains form when lava and ash build up in one area over time.

Crater This bowl-shaped part of the volcano surrounds the vent. Lava often collects here before it flows down the slope.

Vent As magma flows up the pipe, it reaches the surface at an opening called the vent. Side vents often branch off of the main pipe.

Pipe Magma flows through this nearly vertical crack in the rock called the pipe.

Magma Chamber Rising magma forms and fills a large pocket underneath the volcano. This pocket is called the magma chamber. In some cases, one magma chamber feeds several volcanoes.

Magma The hot, molten mixture of rock material and gases is called magma.

Figure 16
The Hawaiian Islands are a series of volcanic mountains that have been built upward from the seafloor.

A Mauna Kea, shown here, forms part of the island of Hawaii.

Underwater Volcanic Mountains

You know that volcanic mountains form on land, but did you know that these mountains also form on the ocean floor? Underwater eruptions can produce mountains beneath the sea. Eventually, if enough lava is erupted, these mountains grow above sea level. For example, Hawaii, shown in **Figure 16A,** is the peak of a huge volcanic mountain that extends above the surface of the water of the Pacific Ocean. **Figure 16B** illustrates how the Hawaiian Islands formed.

Volcanic mountains like the Hawaiian Islands are different from the volcanic mountains that form where one plate subducts beneath another. The Hawaiian Islands formed from material that came from near the boundary between Earth's core and mantle. Hot rock travels through the mantle as a plume and melts to form a hot spot in Earth's crust. As plates travel over the hot spot, a series of volcanoes, as seen in Hawaii, forms. Magma from subduction volcanoes forms much closer to Earth's surface. Hot spot volcanoes also are much larger and have more gently sloping sides than subduction volcanoes.

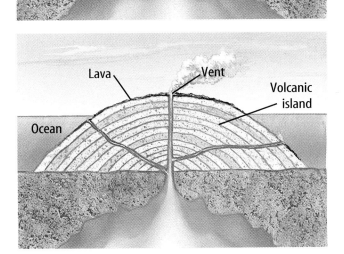

B Hawaii began to form as lava erupted onto the ocean floor. Over time, the mountain grew so large that it rose above sea level.

Reading Check *What type of mountains make up the Hawaiian Islands?*

Other Types of Uplift

You have learned about the origin of the pushing forces that bend crustal rocks during mountain-building processes. However, another force also works to keep mountains elevated above the surrounding land. If you place wooden blocks of various thicknesses in a container of water, you will notice that different blocks of wood float in the water at different heights. Also, the thicker blocks of wood float higher in the water than the thinner blocks do. The buoyant force of the water is balancing the force of gravity. A similar process called isostasy occurs in Earth. According to the principle of **isostasy**, Earth's crust and lithosphere float on the upper part of the mantle.

The effects of isostasy were first noticed near large mountain ranges. Earth's crust is thicker under mountains than it is elsewhere. Also, if mountains continue to get uplifted, the crust under the mountains will become thicker and will extend farther down into the mantle. This is similar to the floating wooden blocks. If you pile another wooden block on a block that is already floating in the water, you will see that the new, larger block will sink down into the water farther than before. You also will see that the new block floats higher than it did before.

Physics INTEGRATION

Using the principle of isostasy, explain in your Science Journal why large features on Earth's surface, such as mountains, float on the layers of Earth beneath them.

Problem-Solving Activity

How can glaciers cause land to rise?

About 20,000 years ago, much of North America was covered by a large glacial ice sheet. How do you think an ice sheet can affect Earth's crust? What do you think happens when the ice melts?

Identifying the Problem

More than 100 years ago, people living in areas that once had been covered by glaciers noticed that features such as old beaches had been tilted. The beaches had a higher elevation in some places and a lower elevation in others. How do you think old beaches could be tilted?

Solving the Problem

1. The weight of glaciers pushes down Earth's crust. What do you think happens after the glacier melts?

2. How could rising crust cause beaches to be tilted? Do you think the crust would rise the same amount everywhere? Explain.

Figure 17
Isostasy makes Earth's crust behave in a similar way to these icebergs. As an iceberg melts and becomes smaller, ice from below the water's surface is forced up.

Adjusting to Gravity Similar to the wooden blocks, if mountains continue to grow larger, they will sink even farther into the mantle. Once mountains stop forming, erosion lowers the mountains and the crust rises again because weight has been removed. If the process continues, the once-thick crust under the mountains will be reduced to the thickness of the crust where no mountains exist.

Icebergs behave in much the same way, as shown in **Figure 17.** The iceberg is largest when it first breaks off of a glacier. As the iceberg floats, it melts and starts to lose mass. This causes the iceberg to rise in the water. Eventually, the iceberg will be much smaller and will not extend as deeply into the water. How is this similar to what happens to mountains?

Section Assessment

1. What are the four main types of mountains found on Earth?
2. If compression were exerted on rock layers, what type of mountains would form?
3. Describe how fault-block mountains form.
4. How does a volcano form?
5. **Think Critically** Put the Appalachian, Himalaya, and Rocky Mountains in order from youngest to oldest knowing that the Himalaya are most rugged and the Appalachians are the least rugged.

Skill Builder Activities

6. **Concept Mapping** Make a chain of events concept map that describes how folded mountains form. **For more help, refer to the** Science Skill Handbook.
7. **Using a Word Processor** On a computer, prepare a table with descriptions of how the four different types of mountains form. Be sure to show any similarities and differences in their formation. **For more help, refer to the** Technology Skill Handbook.

SECTION 2 Uplift of Earth's Crust

Activity
Model and Invent

Isostasy

The principle of isostasy states that Earth's crust floats on the more dense mantle beneath. This is similar to the way objects float in water. What do you think will happen when you add mass to a floating object? What if you take away mass?

Recognize the Problem

How does adding or removing mass affect the way an object floats in a fluid?

Thinking Critically

How could you add mass to or remove mass from a floating object? How will adding or removing mass change the way the object floats?

Goals
- **Observe** the results of isostasy.
- **Predict** what will happen to floating objects when mass is removed or added.

Possible Materials
5-cm × 5-cm × 3-cm wooden blocks (3)
10-cm × 35-cm × 15-cm clear-plastic storage box or other bin
water
permanent marker
ruler

Safety Precautions

Data Source
SCIENCEOnline You can learn more about isostasy and the related concept of buoyancy by visiting the Glencoe Science Web site at **tx.science.glencoe.com.**

216 CHAPTER 7 Forces Shaping Earth

Using Scientific Methods

Planning the Model

1. **Decide** what object(s) you will float in the water initially. How will you remove mass from that object? How will you add mass?
2. What will you observe as the mass changes. How will you record the effects of adding or removing mass?
3. How much water will you use? What problems might you encounter if you have too much or too little water?
4. Will you make any additional measurements or record any other data?
5. **List** all the steps that you plan to do in this activity. Are the steps in a logical order?

Check the Model Plans

1. **Compare** your model plans to those of other students.
2. Make sure your teacher approves your plans before you start.

Making the Model

1. Fill the storage box or bin with an appropriate amount of water.
2. Start by floating the initial object you planned to use in the water. Observe and record relevant data.
3. Follow the list of steps you planned in order to obtain data for removing and adding mass. Observe your model and record all relevant data in your Science Journal.

Analyzing and Applying Results

1. What did your initial object look like? What level did the water rise to when your initial object was placed in the bin? How did you add and remove mass?
2. What happened to the amount of the object that was submerged and the amount sticking out of the water when mass was removed from the object?
3. What happened to the amount of the object that was submerged and the amount sticking out of the water when mass was added?
4. How can you explain your observations about how much of the object was submerged and how much was sticking out of the water? How is this similar to processes that occur in Earth?

Make a poster that illustrates what you have learned about isostasy. **For more help, refer to the** Science Skill Handbook.

ACTIVITY 217

Science Stats

Mountains

Did you know...

...Mount Monadnock is the most-often climbed mountain in the world. Located in New Hampshire, it's climbed by about 125,000 people every year. That's the same amount of people as the population of Topeka, Kansas. Mount Fuji, in Japan, previously claimed this honor.

...The world's longest mountain range is underwater. The mid-ocean ridge that winds around Earth beneath the Arctic, Atlantic, and Pacific Oceans is 65,000 km long. That's four times longer than the combined lengths of the Andes Mountains, the Rocky Mountains, and the Himalaya.

...The beautiful Appalachian Mountains are among the oldest in the world. By 250 million years ago, their formation was complete. Today, the mountains aren't among the tallest because they have been worn down by many millions of years of erosion.

Connecting To Math

...In 1963, Surtsey, a small island, formed when an underwater volcano erupted off the coast of Iceland. The 1.6-km-long island rose to the height of 183 m—about as tall as a 55-story building.

...Many people live on or near mountains, but only about 134 people live in Mountain—Mountain, North Dakota, that is.

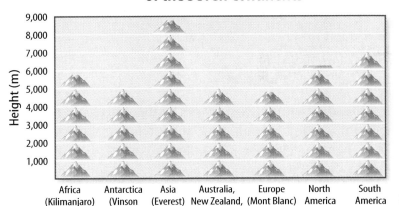

Highest Mountain Peaks on Each of the Seven Continents

Do the Math

1. One trail to the top of Mount Monadnock is about 3 km long. If it takes a climber 4 h to reach the top, how many kilometers does this climber hike in 1 h?
2. Draw a picture to scale that accurately represents the sizes of Mount McKinley (6,194 m), Mount Rainier (4,392 m), and Mount Everest (8,850 m).
3. If the mid-ocean ridge is four times the combined length of the Andes Mountains, the Rocky Mountains, and the Himalaya, what is the combined length of these mountain ranges?

Go Further

Research a mountain on **tx.science.glencoe.com.** Pinpoint its location on a map, and then accurately draw the mountain and the view from its top.

SCIENCE STATS

Chapter 7 Study Guide

Reviewing Main Ideas

Section 1 Earth's Moving Plates

1. Earth's interior is divided into four layers, the inner core, the outer core, the mantle, and the crust. *Can you identify these layers in the diagram below?*

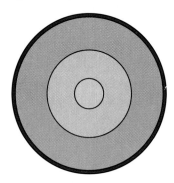

2. Earth's inner and outer cores are thought to be composed mostly of iron. The outer core is thought to be liquid and the inner core is solid.

3. Plates composed of sections of Earth's crust and rigid upper mantle move around on the plasticlike asthenosphere.

4. Earth's plates move together, move apart, and slide past each other. *Which type of movement occurs at the San Andreas Fault in California, shown below?*

5. Evidence suggests that plates move because of various effects of gravity. Convection in Earth's mantle, ridge-push, and slab-pull might all contribute to plate movement.

Section 2 Uplift of Earth's Crust

1. Uplift, as a result of different processes, causes mountains to form. Faulting, folding, upwarping, and volcanic eruptions are all processes that build mountains above the surrounding land.

2. The four main types of mountains are fault-block mountains, folded mountains, upwarped mountains, and volcanic mountains. *Which type of mountain is shown in the photo below?*

3. Compression and tension affect the thickness of Earth's crust.

4. As erosion removes material from the tops of mountains, the mass of the mountains is reduced. As a result of isostasy, the crust is then forced upward.

After You Read

FOLDABLES Reading & Study Skills

To help you review the four layers of Earth's interior, use the Foldable you made at the beginning of the chapter.

Chapter 7 Study Guide

Visualizing Main Ideas

Fill in the following table comparing examples and causes of the four types of mountains.

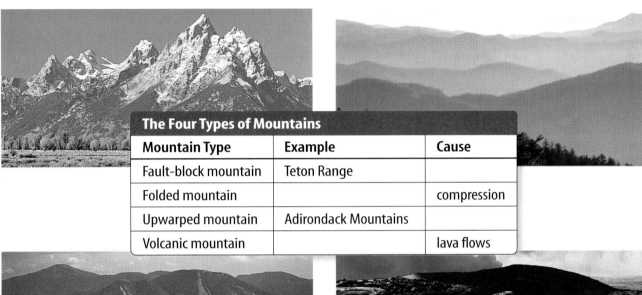

The Four Types of Mountains		
Mountain Type	Example	Cause
Fault-block mountain	Teton Range	
Folded mountain		compression
Upwarped mountain	Adirondack Mountains	
Volcanic mountain		lava flows

Vocabulary Review

Vocabulary Words

a. crust
b. fault
c. fault-block mountain
d. folded mountain
e. inner core
f. isostasy
g. lithosphere
h. mantle
i. outer core
j. plate
k. subduction
l. upwarped mountain
m. volcanic mountain

Using Vocabulary

Answer the following questions with complete sentences.

1. Which part of Earth's core do scientists think is liquid?
2. The Sierra Nevada mountains in California are which type of mountain?
3. What type of mountains form in areas where rocks are being pushed together?
4. What process occurs when a more dense plate sinks beneath a less dense plate?
5. Which type of mountain forms when magma is forced upward and flows onto Earth's surface?

 **Study Tip**

Don't just memorize definitions. Write complete sentences using new vocabulary words to be certain you understand what they mean.

Chapter 7 Assessment & TEKS Review

Checking Concepts

Choose the word or phrase that best answers the question.

1. Which part of Earth is largest?
   A) crust
   B) mantle
   C) outer core
   D) inner core

2. Earth's plates are pieces of which of the following layer of Earth?
   A) lithosphere
   B) asthenosphere
   C) inner core
   D) mantle

3. Which force pushes plates together?
   A) tension
   B) compression
   C) shearing
   D) isostasy

4. Which force occurs where Earth's plates are moving apart?
   A) tension
   B) compression
   C) shear
   D) isostasy

5. Which layer of Earth is thought to be solid and composed mostly of the metal iron?
   A) crust
   B) mantle
   C) outer core
   D) inner core

6. Which of the following suggests that Earth's crust floats on the upper mantle?
   A) tension
   B) compression
   C) shear
   D) isostasy

7. Which type of mountain forms because of compressional forces?
   A) fault-block mountains
   B) folded mountains
   C) upwarped mountains
   D) volcanic mountains

8. Which type of mountain forms because forces inside Earth push up overlying rock layers?
   A) fault-block mountains
   B) folded mountains
   C) upwarped mountains
   D) volcanic mountains

9. Which type of plate movement occurs at transform boundaries?
   A) plates moving together
   B) plates moving apart
   C) plates sinking
   D) plates sliding past each other

10. Which type of plate movement produces deep rifts such as the mid-ocean rift?
    A) plates moving together
    B) plates moving apart
    C) plates sliding past each other
    D) plates sinking

Thinking Critically

11. Which is older, the Great Rift Valley in East Africa, or the Mid-Atlantic Ridge in the Atlantic Ocean? Explain.

12. How can you determine whether or not a mountain is still forming?

13. Seismic waves slow down when entering the asthenosphere. What does this tell you about the nature of the asthenosphere?

14. What would happen to the elevation of the island of Greenland if the ice sheet were to melt away?

15. If you wanted to know whether a certain mountain was formed by compression, what would you look for?

Developing Skills

16. **Comparing and Contrasting** Compare and contrast volcanic and folded mountains. Draw a diagram of each type of mountain. Label important features.

17. **Making Models** Use layers of clay to make a model of fault-block mountains. Draw a diagram of your model.

Chapter 7 Assessment

18. Drawing Conclusions The speed of seismic waves suddenly increases when they go from the upper mantle into the lower mantle. What does this indicate about the comparative densities of the rock in both layers?

19. Using Graphics Software Use the graphics capabilities of a computer to generate a scale illustration of Earth's interior. Include the thickness of each layer in kilometers.

20. Recognizing Cause and Effect What is the effect of subduction at the boundary of two plates?

Performance Assessment

21. Poem Write a poem in a style of your choosing about the spectacular view often associated with mountains. You may wish to write about the scene from the top of a mountain or the one you see from the bottom of the mountain looking up to its peak.

TECHNOLOGY

Go to the Glencoe Science Web site at **tx.science.glencoe.com** or use the **Glencoe Science CD-ROM** for additional chapter assessment.

TAKS Practice

Germaine's homework assignment was to do some research about Earth's plates. In an encyclopedia, he saw a reference to the San Andreas Fault in California. The diagram he found is shown below.
TEKS 6.6 C, TEKS 6.2 C

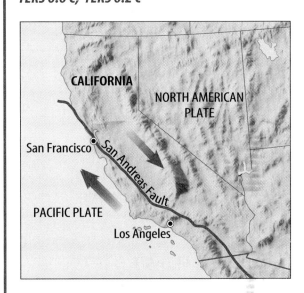

Study the diagram and answer the following questions.

1. Which of the following is the most likely cause of frequent earthquake activity along the San Andreas Fault?
   A) unusually high tides
   B) movement of plates
   C) increased sunspot activity
   D) excessive oil drilling

2. According to the diagram, which statement best describes the motion of the plates on either side of the San Andreas Fault?
   F) They are moving apart.
   G) They are moving together.
   H) They are sliding past each other.
   J) They are not moving.

CHAPTER 8

Science TEKS 6.6 A, C; 6.7 A, B

Weathering and Erosion

Erosion can be a devastating problem in many places in the world, especially in hilly or mountainous regions. Rock and sediment tend to move downhill under the influence of gravity. This mudflow in California threatened lives and destroyed a house. In this chapter, you will learn how weathering and erosion affect rocks. You also will learn how soil develops from weathered rock.

What do you think?

Science Journal Look at the picture below with a classmate. Discuss what this might be. Here's a hint: *Glaciers don't always flow in the same direction.* Write your answer or best guess in your Science Journal.

224

The Grand Canyon is 440 km long, up to 24 km wide, and up to 1,800 m deep. The water of the Colorado River carved the canyon out of rock by wearing away particles and carrying them away for millions of years. The process of wearing away rock is called erosion. Over time, erosion has shaped and reshaped Earth's surface many times. In this activity, you will explore how running water formed the Grand Canyon.

Model water erosion

1. Fill a bread pan with packed sand and form a smooth, even surface.
2. Place the bread pan in a plastic wash tub. Position one end of the wash tub in a sink under the faucet.
3. Place a brick or wood block under the end of the bread pan beneath the faucet.
4. Turn on the water to form a steady trickle of water falling into the pan and observe for 10 min. The wash tub should catch the eroded sand.

Observe

In your Science Journal, draw a top view picture of the erosion pattern formed in the sand by the running water. Write a paragraph describing what the sand would look like if you had left the water running overnight.

Before You Read

Making a Compare and Contrast Study Fold Make the following Foldable to help you see how weathering and erosion are similar and different.

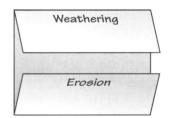

1. Place a sheet of paper in front of you so the short side is at the top. Fold the paper in half from top to bottom and then unfold.
2. Fold in to the centerfold line to divide the paper into fourths.
3. Label the flaps *Weathering* and *Erosion*. Label the middle portion inside your Foldable *Both*. Before you read the chapter, write the definition of each on the front of the flaps.
4. As you read the chapter, write information you learn on the back of the two flaps.

225

SECTION 1

Weathering and Soil Formation

As You Read

What You'll Learn
- **Identify** processes that break rock apart.
- **Describe** processes that chemically change rock.
- **Explain** how soil evolves.

Vocabulary
weathering
mechanical weathering
chemical weathering
soil
topography

Why It's Important
Soil forms when rocks break apart and change chemically. Soil is home to many organisms and most plants need soil in order to grow.

Weathering

Have you noticed potholes in roadways and broken concrete in sidewalks and curbs? When a car rolls over a pothole in the road in late winter or when you step over a broken sidewalk, you know things aren't as solid or permanent as they look. Holes in roads and broken sidewalks show that solid materials can be changed by nature. **Weathering** is a natural process that causes rocks to change, breaks them down, and causes them to crumble. Freezing and thawing, oxygen in the air, and even plants and animals can affect the stability of rock. These are some of the things that cause rocks on Earth's surface to weather, and in some cases, to become soils.

Mechanical Weathering

When a sidewalk breaks apart, a large slab of concrete is broken into many small pieces. The concrete looks the same. It's just broken apart. This is similar to mechanical weathering. **Mechanical weathering** breaks rocks into smaller pieces without changing them chemically. The small pieces are identical in composition to the original rock, as shown in **Figure 1.** Two of the many causes of mechanical weathering are ice wedging and living organisms.

Figure 1
The forces of mechanical weathering break apart rocks. *How do you know that the smaller pieces of granite were produced by mechanical weathering?*

226

Figure 2
Over time, freezing water can break apart rock.

A Water seeps into cracks. The deeper the cracks are, the deeper water can seep in.

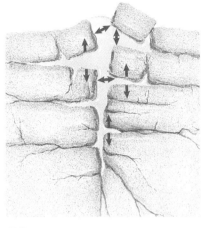
B The water freezes and expands forcing the cracks to open further.

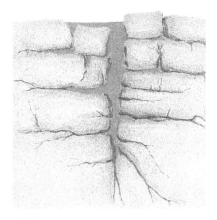
C The ice melts. If the temperature falls below freezing again, the process will repeat itself.

Ice Wedging In some areas of the world, air temperature drops low enough to freeze water. Then, when the temperature rises, the ice thaws. This freezing and thawing cycle breaks up rocks. How can this happen? When it rains or snow melts, water seeps into cracks in rocks. If the temperature drops below freezing, ice crystals form. As the crystals grow, they take up more space than the water did because ice is less dense than water. This expansion exerts pressure on the rocks. With enough force, the rocks will crack further and eventually break apart, as shown in **Figure 2.** Ice wedging also causes potholes to form in roadways.

Reading Check *Explain how ice wedging can break rock apart.*

Plants and Animals Plants and animals also cause mechanical weathering. As shown in **Figure 3,** plants can grow in what seem to be the most inconvenient places. Their roots grow deep into cracks in rock where water collects. As they grow, roots become thicker and longer, slowly exerting pressure and wedging rock apart.

Gophers and prairie dogs also weather rock—as do other animals that burrow in the ground. As they burrow through sediment or soft sedimentary rock, animals break rock apart. They also push some rock and sediment to the surface where another kind of weathering, called chemical weathering, takes place more rapidly.

Figure 3
Tree roots can break rock apart.

SECTION 1 Weathering and Soil Formation

Figure 4
Chemical weathering changes the chemical composition of minerals and rocks. *How is kaolinite different from feldspar?*

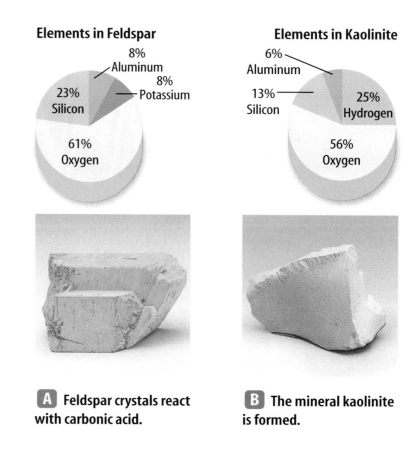

A Feldspar crystals react with carbonic acid.

B The mineral kaolinite is formed.

Chemical Weathering

Chemical weathering occurs when the chemical composition of rock changes. This kind of weathering is rapid in tropical regions where it's moist and warm most of the time. Because desert areas have little rainfall and polar regions have low temperatures, chemical weathering occurs slowly in these areas. **Table 1** summarizes the rates of chemical weathering for different climates. Two important causes of chemical weathering are natural acids and oxygen.

Reading Check *Why is chemical weathering rapid in the tropics?*

Table 1 Rates of Weathering	
Climate	Chemical Weathering
Hot and dry	Slow
Hot and wet	Fast
Cold and dry	Slow
Cold and wet	Slow

Natural Acids Some rocks react chemically with natural acids in the environment. When water mixes with carbon dioxide in air or soil, for example, carbonic acid forms. Carbonic acid can change the chemical composition of minerals in rocks, as shown in **Figure 4**.

Although carbonic acid is weak, it reacts chemically with many rocks. Vinegar reacts with the calcium carbonate in chalk, dissolving it. In a similar way, when carbonic acid comes in contact with rocks like limestone, dolomite, and marble, they dissolve. Other rocks also weather when exposed to carbonic acid.

Plant Acids Plant roots also produce acid that reacts with rocks. Many plants produce a substance called tannin. In solution, tannin forms tannic acid. This acid dissolves some minerals in rocks. When minerals dissolve, the remaining rock is weakened, and it can break into smaller pieces. The next time you see moss or other plants growing on rock, as shown in **Figure 5,** peel back the plant. You'll likely see discoloration of the rock where plant acids are reacting chemically with some of the minerals in the rock.

Figure 5
Moss growing on rocks can cause chemical weathering.

Effect of Oxygen When you see rusty cars, reddish soil, or reddish stains on rock, you witness oxidation, the effects of chemical changes caused by oxygen. When iron-containing materials such as steel are oxidized a chemical reaction causes the material to rust. Rocks chemically weather in a similar way. When some iron-containing minerals are exposed to oxygen, they can weather to minerals that are like rust. This leaves the rock weakened, and it can break apart. As shown in **Figure 6,** some rocks also can be colored red or orange when iron-bearing minerals in them react with oxygen.

Figure 6
Oxidation occurs in rocks and cars.

A Even a tiny amount of iron in rock can combine with oxygen and form a reddish iron oxide.

B The iron contained in metal objects such as this truck also can combine with oxygen and form a reddish iron oxide called rust.

Mini LAB

Rock Dissolving Acids

Procedure

WARNING: *Do not remove goggles until activity, clean up, and handwashing is completed.*

1. Use an **eyedropper** to put several drops of **vinegar** on pieces of **chalk** and **limestone.** Observe the results with a **magnifying glass.**
2. Put several drops of 5% **hydrochloric acid** on the chalk and limestone. Observe the results.

Analysis

1. Describe the effect of the hydrochloric acid and vinegar on chalk and limestone.
2. Research what type of acid vinegar contains.

SECTION 1 Weathering and Soil Formation **229**

Table 2 Factors that Affect Soil Formation

Parent Rock	Slope of Land	Climate	Time	Organisms

Soil

Is soil merely dirt under your feet, or is it something more important? **Soil** is a mixture of weathered rock, organic matter, water, and air that supports the growth of plant life. Organic matter includes decomposed leaves, twigs, roots, and other material. Many factors affect soil formation.

Parent Rock As listed in **Table 2,** one factor affecting soil formation is the kind of parent rock that is being weathered. For example, where limestone is chemically weathered, clayey soil is common because clay is left behind when the limestone dissolves. In areas where sandstone is weathered, sandy soil forms.

The Slope of the Land The **topography,** or surface features, of an area also influences the types of soils that develop. You've probably noticed that on steep hillsides, soil has little chance of developing. This is because rock fragments move downhill constantly. However, in lowlands where the land is flat, wind and water deposit fine sediments that help form thick soils.

Climate Climate affects soil evolution, too. If rock weathers quickly, deep soils can develop rapidly. This is more likely to happen in tropical regions where the climate is warm and moist. Climate also affects the amount of organic material in soil. Soils in desert climates contain little organic material. However, in mild, humid climates, vegetation is lush and much organic material is present. When plants and animals die, decomposition by fungi and bacteria begins. The result is the formation of a dark-colored material called humus, as shown in the soil profile in **Figure 7.** Most of the organic matter in soil is humus. Humus helps soil hold water and provides nutrients that plants need to grow.

TRY AT HOME Mini LAB

Analyzing Soils

Procedure
1. Obtain a sample of **soil** from near your home.
2. Spread the soil out over a piece of **newspaper.**
3. Carefully sort through the soil. Separate out organic matter from weathered rock.
4. Wash hands thoroughly after working with soils.

Analysis
1. Besides the organic materials and the remains of weathered rock, what else is present in the soil?
2. Is some of the soil too fine-grained to tell if it is organic or weathered rock?

Time It takes time for rocks to weather. It can take thousands of years for some soils to form. As soils develop, they become less like the rock from which they formed. In young soils, the parent rock determines the soil characteristics. As weathering continues, however, the soil resembles the parent rock less and less. Thicker, well-developed soils often are found in areas where weathering has gone on undisturbed for a long period of time. For this to happen, soil materials must not be eroded away and new sediment must not be deposited over the land's surface too quickly.

Organisms Organisms influence soil development. Lichens are small organisms that consist of an alga and a fungus that live together for mutual benefit. You may have seen lichens in the form of multicolored patches growing on tree branches or cliff faces. Interestingly, lichens can grow directly on rock. As they grow, they take nutrients from the rock that they are starting to break down, forming a thin soil. After a soil has formed, many types of plants such as grasses and trees can grow.

The roots of these plants further break down the parent rock. Dead plant material such as leaves accumulates and adds organic matter to the soil. Some plants contribute more organic matter to soil than others. For example, soil under grassy areas often is richer in organic matter than soil developing under forests. This is why some of the best farmland in the midwestern United States is where grasslands used to be.

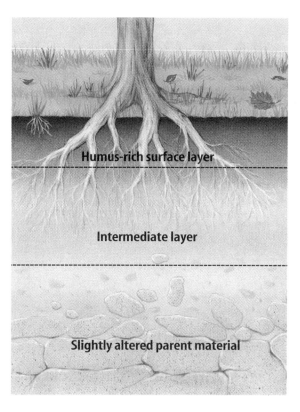

Figure 7
Soils contain layers that are created by weathering, the flow of water and chemicals, and the activities of organisms. *What part do microorganisms play in soil development?*

Section 1 Assessment

1. What are two ways that rocks are mechanically weathered?
2. Name two agents of chemical weathering.
3. How does carbonic acid weather rocks?
4. How does soil form? What factors are important?
5. **Think Critically** How could climate affect rates of mechanical weathering? What about chemical weathering? How are the two kinds of weathering related?

Skill Builder Activities

6. **Comparing and Contrasting** Compare and contrast mechanical weathering caused by ice wedging with mechanical weathering caused by growing roots. **For more help, refer to the Science Skill Handbook.**

7. **Communicating** Write a descriptive poem in your Science Journal that explains different ways rocks are weathered. **For more help, refer to the Science Skill Handbook.**

SECTION 1 Weathering and Soil Formation

Activity
Classifying Soils

Not all soils are the same. Geologists and soil scientists classify soils based on the amounts and kinds of particles they contain.

What You'll Investigate
How is soil texture determined?

Materials
soil sample
stereomicroscope
*hand lens
*Alternate materials

Safety Precautions

Goals
- **Classify** a soil using an identification key.
- **Observe** soil with a stereomicroscope.

Procedure
1. Place a small sample of moistened soil between your fingers. Then follow the directions in the classification key below.
   a. Slide your fingers back and forth past each other. If your sample feels gritty, go to **b**. If it doesn't feel gritty, go to **c**.
   b. If you can mold the soil into a firm ball, it's sandy loam soil. If you cannot mold it into a firm ball, it's sandy soil.
   c. If your sample is sticky, go to **d**. If your sample isn't sticky, go to **e**.
   d. If your sample can be molded into a long, thin ribbon, it's clay soil. If your soil can't be molded into a long, thin ribbon it's clay loam soil.
   e. If your sample is smooth, it's silty loam soil. If it isn't smooth, it's loam soil.
2. After classifying your soil sample, examine it under a microscope. Draw the particles and any other materials that you see.
3. Wash your hands thoroughly after you are finished working with soils.

Conclude and Apply
1. **Determine** the texture of your soil sample.
2. **Describe** two characteristics of loam soil.
3. **Describe** two features of sandy loam soil.
4. Based on your observations with the stereomicroscope, what types of particles and other materials did you see? Did you observe any evidence of the activities of organisms?

Communicating Your Data
Compare your conclusions with those of other students in your class. **For more help, refer to the Science Skill Handbook.**

SECTION 2
Erosion of Earth's Surface

Agents of Erosion

Imagine looking over the rim of the Grand Canyon at the winding Colorado River below or watching the sunset over Utah's famous arches. Features such as these are spectacular examples of Earth's natural beauty, but how can canyons and arches form in solid rock? These features and many other natural landforms are a result of erosion of Earth's surface. **Erosion** is the wearing away and removal of rock or sediment. Erosion occurs because gravity, ice, wind, and water sculpt Earth's surface.

Gravity

Gravity is a force that pulls every object toward every other object. Gravity pulls everything on Earth toward its center. As a result, water flows downhill and rocks tumble down slopes. When gravity alone causes rock or sediment to move down a slope, the erosion is called **mass movement**. Mass movements can occur anywhere hills or mountains are found. One place where they often occur is near volcanoes, as shown in **Figure 8**. Creep, slump, rock slides, and mudflows are four types of mass movements, as seen in **Figure 9**.

As You Read

What You'll Learn
- **Identify** agents of erosion.
- **Describe** the effects of erosion.

Vocabulary
erosion
mass movement
creep
slump
deflation
abrasion
runoff

Why It's Important
Erosion shapes Earth's surface.

Figure 8
The town of Weed, California, was built on top of a landslide that moved down the volcano known as Mount Shasta.

233

NATIONAL GEOGRAPHIC VISUALIZING MASS MOVEMENTS

Figure 9

When the relentless tug of gravity causes a large chunk of soil or rock to move downhill—either gradually or with sudden speed—the result is what geologists call a mass movement. Weathering and water often contribute to mass movements. Several kinds are shown here.

A CREEP When soil on a slope moves very slowly downhill, a mass movement called creep occurs. Some of the trees at right have been gradually bent because of creep's pressure on their trunks.

B SLUMP This cliff in North Dakota shows the effects of the mass movement known as slump. Slumping often occurs after earthquakes or heavy and prolonged rains.

C ROCK SLIDES When rocks break free from the side of a cliff or a mountain, they crash down in what is called a rock slide. Rock slides, like the one at the left in Yosemite National Park, can occur with little warning.

D MUDFLOWS A Japanese town shows the devastation that a fourth type of mass movement—a mudflow—can bring. When heavy moisture saturates sediments, mudflows can develop, sending a pasty mix of water and sediment downhill over the ground's surface.

CHAPTER 8 Weathering and Erosion

Creep Creep is the name for a process in which sediments move slowly downhill, as shown in **Figure 9A.** Creep is common where freezing and thawing occur. As ice expands in soil, it pushes sediments up. Then as soil thaws, the sediments move farther downslope. **Figure 10** shows how small particles of sediment can creep downslope. Over time, creep can move large amounts of sediment, possibly causing damage to some structures. Do you live in an area where you can see the results of creep?

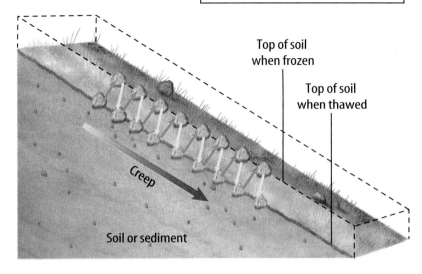

Figure 10
When soil freezes, particles are lifted. When it thaws, the particles are pulled downhill by gravity. Eventually, large amounts of sediment are moved by this process.

Slump A **slump** occurs when a mass of rock or sediment moves downhill along a curved surface, as shown in **Figure 9B.** Slumps are most common in thick layers of loose sediment, but they also form in sedimentary rock. Slumps frequently occur on slopes that have been undercut by erosion, such as those above the bases of cliffs that have been eroded by waves. Slumping of this kind is common along the coast of Southern California where it threatens to destroy houses and other buildings.

Rock Slides Can you imagine millions of cubic meters of rock roaring down a mountain at speeds greater than 100 km/h? This can happen when a rock slide occurs. During a rock slide layers of rock break loose from slopes and slide to the bottom. The rock layers often bounce and break apart during movement. This produces a huge, jumbled pile of rocks at the bottom of the slope, as you can see in **Figure 9C.** Rock slides can be destructive, sometimes destroying entire villages or causing hazards on roads in mountainous areas.

Slumps and rock slides often occur when sediment becomes saturated by rain. Water between sediment grains helps lift up overlying rock and sediment. This makes it easier for the sediment to overcome the forces holding it in place. Can you think of a way some slopes might be protected from slumps and rock slides? Explain.

Mudflows Where heavy rains or melting snow and ice saturate sediments, mudflows, as shown in **Figure 9D,** can develop. A mudflow is a mass of wet sediment that flows downhill over the ground surface. Some mudflows can be thick and flow slowly downhill at rates of a few meters per day. Other mudflows can be much more fluid and move down slope at speeds approaching 70 km/h. This type of mudflow is common on some volcanoes.

Reading Check *What is the slowest of the four kinds of mass movement?*

Figure 11
Glaciers form in cold regions.

A Continental glaciers are located near the poles in Antarctica and Greenland.

B Valley glaciers are found at high elevations on many continents.

☐ Continental Glacier
▨ Valley Glacier

Ice

In some parts of the world, ice is an agent of erosion. In cold regions, more snow might fall than melts. Over many years, the snow can accumulate to form large, deep masses of ice called glaciers. When the ice in a glacier becomes thick enough, its own weight causes it to flow downhill under the influence of gravity. As glaciers move over Earth's surface, they erode materials from some areas and deposit sediment in other areas. **Figure 11** shows the two kinds of glaciers—continental glaciers and valley glaciers.

Today, continental glaciers in polar regions cover about ten percent of Earth. These glaciers are so large and thick that they can bury mountain ranges. Valley glaciers are much smaller and are located in high mountains where the average temperature isn't warm enough to melt the ice sheets. Continental and valley glaciers move and cause erosion.

Glacial Erosion Glaciers can erode rock in two different ways. If the rock that the glacier is sliding over has cracks in it, the ice can pull out pieces of rock. This causes the rock to erode slowly. The loose pieces of rock freeze into the bottom of the glacier and are dragged along as the glacier moves. As these different-sized fragments of rock are dragged over Earth's surface, they scratch the rock below like giant sheets of sandpaper. This scratching is the second way that glaciers can erode rock. Scratching produces large grooves or smaller striations in the rock underneath. The scratching also can wear rock into a fine powder called rock flour.

Research Visit the Glencoe Science Web site at **tx.science.glencoe.com** for more information about glacial erosion and deposition. Communicate to your class what you learned.

Figure 12
Many high-altitude areas owe their distinctive appearance to glacial erosion.

A Mountain glaciers can carve bowl-shaped depressions called cirques.

B Glaciers can widen valleys giving them a U-shaped profile.

Effects of Glacial Erosion Glacial erosion of rock can be a powerful force shaping Earth's surface. In mountains, valley glaciers can remove rock from the mountaintops to form large bowls, called cirques (SURKS), and steep peaks. When a glacier moves into a stream valley, it erodes rock along the valley sides, producing a wider, U-shaped valley. These features are shown in **Figure 12.** Continental glaciers also shape Earth's surface. These glaciers can scour large lakes and completely remove rock layers from the land's surface.

Glacial Deposition Glaciers also can deposit sediments. When stagnant glacier ice melts or when ice melts at the bottom of a flowing glacier or along its edges, the sediment the ice was carrying gets left behind on Earth's surface. This sediment, deposited directly from glacier ice, is called till. Till is a mixture of different-sized particles, ranging from clay to large boulders.

As you can imagine, a lot of melting occurs around glaciers, especially during summer. So much water can be produced that rivers often flow away from the glacier. These rivers carry and deposit sediment. Sand and gravel deposits laid down by these rivers, shown in **Figure 13,** are called outwash. Unlike till, outwash usually consists of particles that are all about the same size.

Figure 13
This valley in New Zealand has been filled with outwash. *How could you distinguish outwash from till?*

Figure 14
In a desert, where small particles have been carried away by wind, larger sediments called desert pavement remain behind.

Wind

If you've had sand blow into your eyes, you've experienced wind as an agent of erosion. When wind blows across loose sediments like silt and sand, it lifts and carries it. As shown in **Figure 14,** wind often leaves behind particles too heavy to move. This erosion of the land by wind is called **deflation.** Deflation can lower the land's surface by several meters.

Wind that is carrying sediment can wear down, or abrade, other rocks just as a sandblasting machine would do. **Abrasion** is a form of erosion that can make pits in rocks and produce smooth, polished surfaces. Abrasion is common in some deserts and in some cold regions with strong winds.

Reading Check *How does abrasion occur?*

When wind blows around some irregular feature on Earth's surface, such as a rock or clump of vegetation, it slows down. This causes sand carried by the wind to be deposited. If this sand deposit continues to grow, a sand dune like that shown in **Figure 15A** might form. Sand dunes move when wind carries sand up one side of the dune and it avalanches down the other, as shown in **Figure 15B.**

Sometimes, wind carries only fine sediment called silt. When this sediment is deposited, an accumulation of silt called loess (LOOS) can blanket Earth's surface. Loess is as fine as talcum powder. Loess often is deposited downwind of some large deserts and near glacial streams.

Figure 15
Wind transportation of sand creates sand dunes.

A Sand dunes do not remain in one location—they migrate.

B As wind blows over a sand dune, sand blows up the windward side and tumbles down the other side. In this way, a sand dune migrates across the land.

Dune movement

Water

You probably have seen muddy water streaming down a street after a heavy rain. You might even have taken off your shoes and waded through the water. Water that flows over Earth's surface is called **runoff.** Runoff is an important agent of erosion. This is especially true if the water is moving fast. The more speed water has, the more material it can carry with it. Water can flow over Earth's surface in several different ways, as you will soon discover.

Figure 16
Water flows over the hood of a car as a thin sheet. *How is this similar to sheet flow on Earth's surface?*

Sheet Flow As raindrops fall to Earth, they break up clumps of soil and loosen small grains of sediment. If these raindrops are falling on a sloped land surface, a thin sheet of water might begin to move downhill. You have observed something similar if you've ever washed a car and seen sheets of water flowing over the hood, as shown in **Figure 16.** When water flows downhill as a thin sheet, it is called sheet flow. This thin sheet of water can carry loose sediment grains with it, and cause erosion of the land. This erosion is called sheet erosion.

Problem-Solving Activity

Can evidence of sheet erosion be seen in a farm field?

If you've ever traveled through parts of your state where there are farms, you might have seen bare, recently cultivated fields. Perhaps the soil was prepared for planting a crop of corn, oats, or soybeans. Do you think sheet erosion can visibly affect the soil in farm fields?

Identifying the Problem

The top layer of most soils is much darker than layers beneath it because it contains more organic matter. This layer is the first to be removed from a slope by sheet flow. How does the photo show evidence of sheet erosion?

Solving the Problem
1. Observe the photo and write a description of it in your Science Journal.

2. Infer why some areas of the field are darker colored than others are. Where do you think the highest point(s) are in this field?
3. Make a generalization about the darker areas of the field.

SECTION 2 Erosion of Earth's Surface

Figure 17
Gullies often form on vegetation-free slopes.

Research Visit the Glencoe Science Web site at **tx.science.glencoe.com** for more information about how running water shapes Earth's surface. Communicate to your class what you learn.

Figure 18
Streams that flow down steep slopes such as this one in Yosemite National Park often have whitewater rapids and waterfalls.

Rills and Gullies Where a sheet of water flows around obstacles and becomes deeper, rills can form. Rills are small channels cut into the sediment at Earth's surface. These channels carry more sediment than can be moved by sheet flow. In some cases, a network of rills can form on a slope after just one heavy rain. Large amounts of sediment can be picked up and carried away by rills.

As runoff continues to flow through the rills, more sediment erodes and the channel widens and deepens. When the channels get to be about 0.5 m across, they are called gullies, as shown in **Figure 17.**

Streams Gullies often connect to stream channels. Streams can be so small that you could jump to the other side or large enough for huge river barges to transport products along their course. Most streams have water flowing through them continually, but some have water only during part of the year.

In mountainous and hilly regions, as in **Figure 18,** streams flow down steep slopes. These streams have a lot of energy and often cut into the rock beneath their valleys. This type of stream typically has white-water rapids and may have waterfalls. As streams move out of the mountains and onto flatter land, they begin to flow more smoothly. The streams might snake back and forth across their valley, eroding and depositing sediments along their sides. All streams eventually must flow into the ocean or a large lake. The level of water in the ocean or lake determines how deeply a river can erode.

Shaping Earth's Surface If you did the Explore activity at the beginning of the chapter, you saw a small model of erosion by a stream. You might not think about them much, but streams are the most important agent of erosion on Earth. They shape more of Earth's surface than ice, wind, or gravity. Over long periods of time, water moving in a stream can have enough power to cut large canyons into solid rock. Many streams together can sculpt the land over a wide region, forming valleys and leaving some rock as hills. Streams also shape the land by depositing sediment. Rivers can deposit sand bars along their course, and can build up sheets of sand across their valleys. When rivers enter oceans or lakes, the water slows and sediment is deposited. This can form large accumulations of sediment called deltas, as in **Figure 19.** The city of New Orleans is built on the delta formed by the Mississippi River.

Figure 19
A triangular area of sediment near the mouth of a river is called a delta. Ancient deltas that are now dry land are often excellent places to grow crops.

Effects of Erosion

As you've learned, all agents of erosion change Earth's surface. Rock and sediment are removed from some areas only to be deposited somewhere else. Where material is removed, canyons, valleys, and mountain bowls can form. Where sediment accumulates, deltas, sand bars, sand dunes, and other features make up the land.

Section Assessment

1. List four agents of erosion. Which of these is the fastest agent of erosion? The slowest? Explain your answers.
2. How does deflation differ from abrasion?
3. How does a cirque form?
4. When do streams deposit sediments? When do they erode them?
5. **Think Critically** Why might a river that was eroding and depositing sediment along its sides start to cut into Earth to form a canyon?

Skill Builder Activities

6. **Recognizing Cause and Effect** Why might a river start filling its valley with sediment? **For more help, refer to the** Science Skill Handbook.

7. **Solving One-Step Equations** If wind is eroding an area at a rate of 2 mm per year and depositing it in a smaller area at a rate of 7 mm per year, how much lower will the first area be in meters after 2 thousand years? How much higher will the second area be? **For more help, refer to the** Math Skill Handbook.

SECTION 2 Erosion of Earth's Surface **241**

Activity: Design Your Own Experiment

Measuring Soil Erosion

During urban highway construction, surface mining, forest harvesting, or agricultural cultivation, surface vegetation can be removed from soil. These practices expose soil to water and wind. Does vegetation significantly reduce soil erosion?

Recognize the Problem
How much does vegetation reduce soil erosion?

Form a Hypothesis
Based on what you've read and observed, hypothesize about how much less soil will be eroded from a sodded field than from bare soil.

Safety Precautions

Wash your hands thoroughly when you are through working with soils.

Possible Materials
blocks of wood   pails (2)
*books   1,000 mL beaker
paint trays (2)   triple-beam balance
soil   calculator
grass sod   watch
water

*Alternate materials

Goals
- **Design** an experiment to measure soil loss from grass-covered soil and from soil without grass cover.
- **Calculate** the percent of soil loss with and without grass cover.

242 CHAPTER 8 Weathering and Erosion

Using Scientific Methods

Test Your Hypothesis

Plan

1. As a group, agree upon the hypothesis and decide how you will test it. Identify which results will falsify or confirm the hypothesis.
2. **List** the steps you will need to take to test your hypothesis. Describe exactly what you will do in each step.
3. **Prepare** a data table in your Science Journal to record your observations.
4. Read over the entire experiment to make sure all steps are in logical order, and that you have all necessary materials.
5. **Identify** all constants and variables and the control of the experiment. A control is a standard for comparing the results of an experiment. One possible control for this experiment would be the results of the treatment for the uncovered soil sample.

Do

1. Make sure your teacher approves your plan before you start.
2. Carry out the experiment step by step as planned.
3. While doing the experiment, record your observations and complete the data table in your Science Journal.

Vegetation and Erosion			
	(A) Mass of Soil at Start	(B) Mass of Eroded Soil	% of Soil Loss (B/A) × 100
Covered Soil Sample			
Uncovered Soil Sample			

Analyze Your Data

1. **Compare** the percent of soil loss from each soil sample.
2. **Compare** your results with those of other groups.
3. What was your control in this experiment? Why?
4. Which were the variables you kept constant? Which did you vary?

Draw Conclusions

1. Did the results support your hypothesis? Explain.
2. **Infer** what effect other types of plants would have in reducing soil erosion. Do you think that grass is better or worse than most other plants at reducing erosion?

Write a letter to the editor of a newspaper. In your letter, **summarize** what you learned in your experiment about the effect of plant cover on soil erosion.

TIME SCIENCE AND HISTORY

SCIENCE CAN CHANGE THE COURSE OF HISTORY!

Acid rain is destroying some of the world's most famous monuments

CRUMBLING

The Taj Mahal in India, the Acropolis in Greece, and the Colosseum in Italy, have stood for centuries. They've survived wars, souvenir-hunters, and natural weathering from wind and rain. But now, something far worse threatens their existence—acid rain. Over the last few decades, this form of pollution has eaten away at some of history's greatest monuments.

Acid rain leads to health and environmental risks. It also harms human-made structures. Most of these structures are made of sandstone, limestone, and marble. Acid rain causes the calcium in these stones to form calcium sulfate, or gypsum. Gypsum's powdery little blotches are sometimes called "marble cancer." When it rains, the gypsum washes away, along with some of the surface of the monument. In many cases, acidic soot falls into the cracks of monuments. When rainwater seeps into the cracks, acidic water is formed, which further damages the structure.

Acid rain has not been kind to this Mayan figure.

Parts of India's Taj Mahal are turning yellow from pollutants.

Greece's Parthenon is slowly being eaten away by acid rain.

MONUMENTS

In Agra, India, the smooth, white marble mausoleum called the Taj Mahal has stood since the seventeenth century. But acid rain is making the surface of the building yellow and flaky. The pollution is caused by hundreds of factories surrounding Agra that emit damaging chemicals.

What moisture, molds, and the roots of vegetation couldn't do in 1,500 years, acid rain is doing in decades. It is destroying the Mayan ruins of Mexico. Pollution is causing statues to crumble and paintings on walls to flake off. The culprits are oil burning refineries and exhaust from tour buses.

Acid rain is a problem affecting national monuments and treasures in just about every urban location in the world. These include the Capitol building in Washington, D.C., churches in Germany, and stained-glass windows in Sweden. Because of pollution, many corroding statues displayed outdoors have been brought inside museums. In London, acid rain has forced workers to repair and replace so much of Westminster Abbey that the structure is becoming a mere copy of the original.

Throughout the world, acid rain has weathered many structures more in the last 20 years than in the 2,000 years before. This is one reason some steps have been taken in Europe and the United States to reduce emissions from the burning of fossil fuels. If these laws don't work, many irreplaceable art treasures may be gone forever.

CONNECTIONS Identify Which monuments and buildings represent the United States? Brainstorm a list with your class. Then choose a monument and, using your school's media center or the Glencoe Science Web site, learn more about it. Is acid rain affecting it in any way?

Science Online
For more information, visit tx.science.glencoe.com

Chapter 8 Study Guide

Reviewing Main Ideas

Section 1 Weathering and Soil Formation

1. Weathering includes processes that break down rock.

2. During mechanical weathering, physical processes break rock into smaller pieces.

3. During chemical weathering, the chemical composition of rocks is changed. *What causes the reddish color of these rocks?*

4. Soil evolves over time from weathered rock. Parent rock, topography, climate, and organisms affect soil formation. *Do you think a thick soil layer could form on this surface? Why or why not?*

Section 2 Erosion of Earth's Surface

1. Erosion is the wearing away and removal of rock. *In the photo below, what evidence do you see that erosion has occurred?*

2. Agents of erosion include gravity, ice, wind, and water. *Which agent of erosion is responsible for this unusual structure?*

3. All agents of erosion move rock and sediment. When energy of motion decreases, sediment is deposited.

4. Erosion and deposition determine the shape of the land.

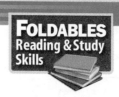

FOLDABLES Reading & Study Skills

After You Read

Identify common characteristics of weathering and erosion and write them on the middle section of your Foldable.

Chapter 8 Study Guide

Visualizing Main Ideas

Fill in the following table, which compares erosion and deposition by different agents.

Erosion and Deposition

Erosional Agent	Evidence of Erosion	Evidence of Deposition
Gravity		material piled at bottom of slopes
Ice	cirques, striations, U-shaped valleys	
Wind		sand dunes, loess
Surface Water	rills, gullies, stream valleys	

Vocabulary Review

Vocabulary Words

a. abrasion
b. chemical weathering
c. creep
d. deflation
e. erosion
f. mass movement
g. mechanical weathering
h. runoff
i. slump
j. soil
k. topography
l. weathering

Using Vocabulary

Use each of the following pairs of terms in a sentence.

1. chemical weathering, mechanical weathering
2. erosion, weathering
3. deflation, runoff
4. mass movement, weathering
5. soil, abrasion
6. soil, erosion
7. mass movement, mechanical weathering
8. weathering, chemical weathering
9. creep, slump
10. topography, runoff

 **Study Tip**

Read the chapters before you go over them in class. Being familiar with the material before your teacher explains it gives you a better understanding and provides you with a good opportunity to ask questions.

Chapter 8 Assessment & TEKS Review

Checking Concepts

Choose the word or phrase that best answers the question.

1. Which of the following agents of erosion forms U-shaped valleys?
   A) gravity C) ice
   B) surface water D) wind

2. In which of these places is chemical weathering most rapid?
   A) deserts C) polar regions
   B) mountains D) tropical regions

3. Which of the following forms when carbon dioxide combines with water?
   A) calcium carbonate C) tannic acid
   B) carbonic acid D) dripstone

4. Which process causes rocks to weather to a reddish color?
   A) oxidation C) carbon dioxide
   B) deflation D) frost action

5. Which type of mass movement occurs when sediments slowly move downhill because of freezing and thawing?
   A) creep C) slump
   B) rock slide D) mudflow

6. Which of the following helps form cirques and U-shaped valleys?
   A) rill erosion C) deflation
   B) ice wedging D) till

7. What is windblown, fine sediment called?
   A) till C) loess
   B) outwash D) delta

8. Which of the following refers to water that flows over Earth's surface?
   A) runoff
   B) slump
   C) chemical weathering
   D) till

9. Which of the following is an example of chemical weathering?
   A) Plant roots grow in cracks in rock and break the rock apart.
   B) Freezing and thawing of water widens cracks in rocks.
   C) Wind blows sand into rock, scratching the rock.
   D) Oxygen causes iron-bearing minerals in rock to break down.

10. Which one of the following erosional agents creates desert pavement?
    A) wind C) water
    B) gravity D) ice

Thinking Critically

11. Explain why mass movement is more common after a heavy rainfall.

12. How does climate affect the development of soils?

13. How could some mass movement be prevented?

14. Would chemical weathering be rapid in Antarctica?

15. Why do caves form only in certain types of rock?

Developing Skills

16. **Recognizing Cause and Effect** Explain how water creates stream valleys.

17. **Forming Hypotheses** Form hypotheses about how deeply water could erode and about how deeply glaciers could erode.

Chapter 8 Assessment

18. **Recognizing Cause and Effect** Explain how valley glaciers create U-shaped valleys.

19. **Classifying** Classify the following by the agent that deposits each: sand dune, delta, till, and loess.

20. **Concept Mapping** Complete the concept map showing the different types of mass movements.

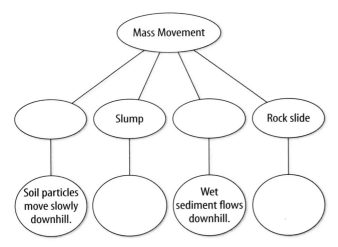

Performance Assessment

21. **Poster** Use photographs from old magazines to make a poster that illustrates different kinds of weathering and erosion. Display your poster in your classroom.

22. **Model** Use polystyrene, cardboard and clay to make a model of a glacier. Include a river of meltwater leading away from the glacier. Use markers to label the areas of erosion and deposition. Show and label areas where till and outwash sediments could be found. Display your model in your classroom.

Technology

Go to the Glencoe Science Web site at **tx.science.glencoe.com** or use the **Glencoe Science CD-ROM** for additional chapter assessment.

TAKS Practice

Geologists measured the amount of cumulative precipitation and the amount of land movement along highway 50 in California to see if there was a relationship between them. *TEKS 6.2 C; 6.4 B*

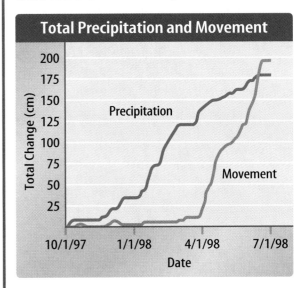

Study the graph and answer the following questions.

1. According to this information, at which precipitation level did soil movement begin?
   A) about 125 cm
   B) about 50 cm
   C) about 75 cm
   D) about 100 cm

2. Based on this information, which of the following is a reasonable conclusion to make about the relationship between precipitation and movement?
   F) As precipitation decreases, movement increases.
   G) As movement decreases, precipitation increases.
   H) There is almost no movement until precipitation reaches a certain level.
   J) There is no relationship.

CHAPTER 9

Science TEKS 6.5 A; 6.6 C; 6.14 B

Groundwater Resources

Why do some springs reach Earth's surface as hot water, but others are clear and cool? In this chapter you'll study how springs and geysers form and the differences between groundwater and surface water. You'll learn how caves and sinkholes form from the action of groundwater. You'll also learn how groundwater becomes polluted, and how the pollutants can be removed.

What do you think?

Science Journal Look at the picture below with a classmate. Discuss what this might be. Here's a hint: *These jewels formed underground.* Write your answer or best guess in your Science Journal.

When it rains, water sinks into the ground at some locations and runs off the surface at others. What causes these differences? Water soaks into the ground if there are spaces in the soil and rock. Do the following activity to investigate these spaces.

Measuring pore space

1. Immerse a dry paper towel in a large beaker of water.
2. Remove the towel. Let it drip into the beaker for 10 s.
3. Squeeze the towel into another beaker. Use a graduated cylinder to measure the volume of water held by the towel.
4. Repeat the experiment using the damp towel.
5. Repeat the experiment with a different brand of paper towel.

Observe

In your Science Journal, record the amount of water held by each towel. Infer why the damp towels and the dry towels don't hold the same amounts of water. Infer which brand of paper towel has more pore space.

Before You Read

Making a Main Ideas Study Fold Make the following Foldable to help you identify the major topics about groundwater resources.

1. Stack two sheets of notebook paper in front of you so the long sides are at the top. Fold both in half from the left side to the right side. Unfold and separate.
2. Cut one sheet along the fold line, from one margin line to the other.
3. Place the second sheet in front of you so the long side is at the top. Cut along the fold line from the bottom of the paper to the margin line. Then cut along the fold line, from the top of the paper to the margin line.
4. Insert the second sheet of paper into the cut of the first paper. Unfold the inserted sheet and align the cuts along the fold of the other sheet. Then fold pages to make a booklet.
5. Title your book *Groundwater Journal*. Title the pages and write what you learn about each topic.

SECTION 1

Groundwater

As You Read

What You'll Learn

- **Explain** the importance of groundwater.
- **Describe** how groundwater moves.
- **Compare and contrast** the formation of springs, artesian wells, and geysers.

Vocabulary

groundwater
porosity
permeable
aquifer
zone of saturation
water table
artesian well
geyser

Why It's Important

Groundwater is one of Earth's most important resources for drinking, washing, and irrigation.

Importance of Groundwater

What do washing clothes, brushing teeth, and taking a bath have in common? These activities wouldn't be possible without freshwater from Earth. Look at the man in **Figure 1.** He is pumping water from deep underground. Every day, people depend on groundwater and other sources of freshwater for drinking, cooking, washing dishes, flushing toilets, and watering lawns. It's used in swimming pools, in industrial plants, and to irrigate crops.

One of the most important sources of freshwater is groundwater. **Groundwater** is water contained in the open spaces, or pores, of soil and rock. There is about 30 times more groundwater than all the surface freshwater on Earth. Water in streams and lakes are examples of surface freshwater. In the United States, 23 percent of the water used by people comes from groundwater. And, more than half of all drinking water in the United States comes from underground sources. Almost everyone who lives in rural areas uses groundwater for drinking and other household uses. If the human population continues to increase, more groundwater will be used.

Figure 1
Every day each person in the United States directly or indirectly uses more than 5,600 L of water.

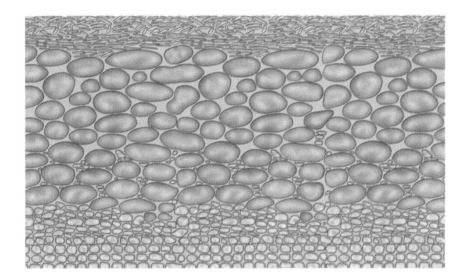

Figure 2
Pore sizes and shapes vary in different kinds of rocks.

Water Beneath Earth's Surface

Unlike Earth's surface water, groundwater normally is not found in pools or channels. Instead, it flows through a series of interconnected pores in soil and some rock. As you learned if you did the Explore Activity, the amount of pore space determines how much water can be held. The pores in soils are the spaces between the fragments and pieces of rock. The pores in rocks are the spaces between the grains and crystals, and also cracks in the rock. Pore size varies, as shown in **Figure 2.** Rocks such as sandstone and loose deposits of sand and gravel can hold large amounts of groundwater.

Porosity Not all soil and rock has the same amount of pore space. The volume of pore space divided by the volume of a rock or soil sample is its **porosity.** Soil or rock with many large pores has high porosity. Even soil or rock with small openings between grains can be porous. But soil or rock with few, small pores has low porosity.

☑ **Reading Check** *What is porosity?*

Permeability When soil or rock has high porosity and the pores are interconnected, water passes through it easily. The ability of rock and soil to transmit water and other fluids is called permeability. Rock that is **permeable** contains many well-connected pores or cracks and allows groundwater to flow through it. Although limestone might contain very little space between grains, it usually has many interconnected cracks. Therefore, limestone is permeable. Impermeable rock, such as shale and granite, has few pores or pores that are not well connected. Groundwater cannot pass through impermeable rock.

Mini LAB

Measuring Porosity

Procedure
1. Put 100 mL of dry **gravel** into a 250-mL **beaker** and 100 mL of dry **sand** into another 250-mL beaker.
2. Fill a **graduated cylinder** with 100 mL of **water.**
3. Pour the water slowly into the beaker with the gravel. Stop pouring the water when it just covers the top of the gravel.
4. Record the volume of the water that is used. This amount is equal to the volume of the pore spaces.
5. Repeat steps 2 through 4 with the beaker containing the sand.

Analysis
1. Which substance has more pore space—gravel or sand?
2. Using the formula below, calculate the porosity of each sediment sample.

$$\frac{\text{Volume of pore spaces}}{\text{100 mL of sediment}} \times 100 = \text{Percent porosity}$$

SECTION 1 Groundwater

Figure 3
This map shows the major aquifers in the United States. The large Ogallala Aquifer stretches from the Texas Panhandle to South Dakota. Other large aquifers are found throughout the eastern two thirds of the country and smaller ones are found in the west.

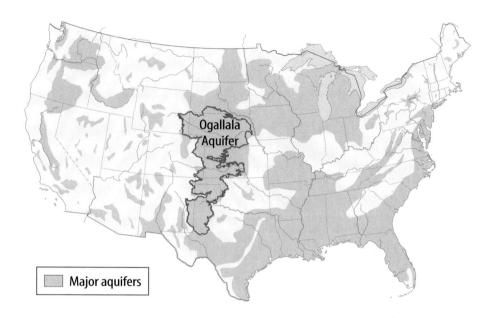

Research Visit the Glencoe Science Web site at **tx.science.glencoe.com** for information about the relationship between groundwater and surface water in a watershed.

Aquifers When it rains or when snow melts, water seeps into permeable soil and rock. You might wonder how deep it can go. Groundwater will keep moving down until it reaches an impermeable layer of rock or sediment. When this happens, the impermeable layer acts like a dam. The water stops seeping downward and begins filling up the pores in the rock and soil above the impermeable layer. An **aquifer** is a layer of permeable rock through which water flows freely. Aquifers act as reservoirs, or storage areas, for groundwater. **Figure 3** shows some of the larger aquifers in the United States. Sand, gravel, sandstone, porous limestone, and highly fractured bedrock of any type can make good aquifers. Shale, mudstone, clay, or other impermeable rocks or sediments that don't contain fractures are not good aquifers. Layers made up of these materials are called aquitards. Aquifers and surface water occur in watersheds, which are large regions drained by a particular river system.

The Water Table The upper part of an aquifer is called the zone of aeration. The pores in this area are partially filled with water, mainly concentrated around the surfaces of grains and fractures. The rest of the pore space in the zone of aeration is filled with air. Below this zone is the **zone of saturation,** where the pores of an aquifer are full of water. The top surface of the zone of saturation is the **water table. Figure 4** shows that the water table meets Earth's surface at lakes, streams, and swamps. Much of the water in some streams is groundwater that has flowed into the stream channels. How deep below ground the water table is depends on rainfall. During heavy rainfall, the water table rises toward Earth's surface. During a drought, the water table falls, and some streams and ponds might dry up.

254 CHAPTER 9 Groundwater Resources

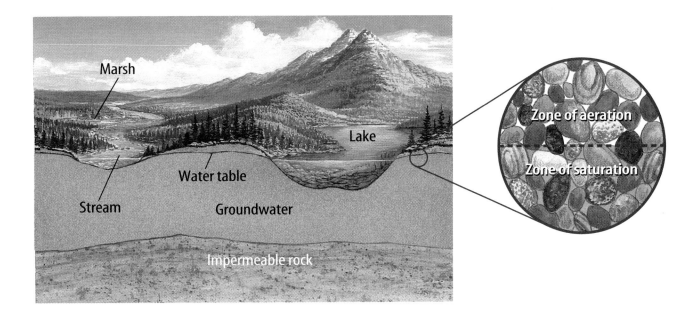

Groundwater Flow

Water below the surface doesn't just sit there—it's on the move. How fast groundwater moves depends on the permeability of the rock and soil that it flows through and the groundwater's pressure. Where permeability is high, groundwater speed is faster. Where permeability is low, its speed is slower. Groundwater pressure is caused by gravity. Groundwater flows from higher-pressure regions toward lower-pressure regions.

You can imagine the shape of the water table as being much like the surface of Earth, with hills and valleys in some areas. In fact, the water table may mimic the shape of the land surface, with higher levels located under hills and lower levels located under valleys. Just as water in a river flows from higher elevations to lower elevations, groundwater flows from where the water table is relatively high to where it is low. The greater the slope of the water table is, the faster groundwater flows. Groundwater can flow toward or away from streams or lakes, depending on the slope of the water table.

Speed of Groundwater Flow Water in rivers or streams flows at speeds ranging from about 3 km to 18 km per hour. Groundwater moves much more slowly. What would happen if you could move only a few centimeters per day? It might take you a year to get from your bedroom to the breakfast table. At that rate, it would take you the rest of your life to walk to school. It's hard to picture something moving that slowly, but that's about the average speed of groundwater. The speed of groundwater moving through an aquifer ranges from about 1.5 m per year to 1.5 m per day. Its speed averages about 2 cm per day.

Figure 4
The water table of an area intersects the surface at lakes, streams, and other wetlands. *What are some relationships between groundwater and surface water in a watershed?*

Chemistry INTEGRATION

Water in some aquifers contains dissolved calcium and magnesium. It is called hard water because it is hard to form suds or lather with soap. Find out if you live in an area that has hard water and how water is treated to make it soft.

SECTION 1 Groundwater **255**

Figure 5
Springs form where the water table meets Earth's surface. *Why are springs often seen on hillsides?*

Springs and Wells

How do people obtain groundwater to use? In some regions, groundwater flows freely out of the ground onto the surface. Where the water table meets Earth's surface, water seeps out to form a spring, shown in **Figure 5.** Some communities get their water from flowing springs.

Reading Check *What is a spring?*

Digging Wells Where springs do not occur, wells must be dug or drilled to obtain water. Most modern wells are drilled using drill rigs mounted on large trucks like the one in **Figure 6.** A drill head, or bit, is placed at the end of a metal pipe. As the bit drills deeper, more pipe is added at the top. Wells more than 300 m deep can be drilled in this way. The hole, or shaft, for the well must be drilled into the zone of saturation. The lower end of the shaft is lined with screens. Then a well casing—a pipe that has slots on the lower part—is set in place inside the hole. The slots allow water in, but keep sand and gravel out. The well casing extends about 0.3 m above ground. The upper part of the shaft around the casing and the surface area around the shaft are sealed with concrete. This prevents surface water from flowing into the well. After such a well has been dug, water can be pumped to the surface. At one time, windmills or hand-operated pumps were used to pump water. Today, most wells have electric-powered pumps.

Figure 6
This truck has drilling equipment that can drill deep water wells in solid rock.

Artesian Springs You learned earlier that pressure on groundwater is caused by gravity. Pressure on groundwater can be high when water is trapped between formations of rock or sediment. Where aquifers are trapped between two layers of impermeable material, water in the sandwiched aquifer is under pressure. The impermeable layers act like the wall of a pipe. Water enters the sandwiched aquifer at the highest part of the aquifer, which puts pressure on water at lower elevations. If a crack or fault cuts through the impermeable layers, groundwater from the aquifer rises toward the surface along the crack. This groundwater can form artesian springs. Artesian springs are common in the Great Plains of the United States, as well as in Australia and northwestern Africa.

Artesian Wells Some communities get water from wells without using pumps. These wells, called **artesian wells,** are drilled into pressurized aquifers, as shown in **Figure 7.** The amount of water pressure depends on the difference in elevation between the highest part of the aquifer and the well. The greater this difference in elevation is, the greater the pressure is. But, some artesian wells do require pumping. Sometimes the top of the well is above the surface to which water under pressure will rise. Then water will not flow freely from the well.

Artesian springs and wells helped shape the history of the United States. The largest aquifer in the United States covers more than 450,000 km^2. Look back at **Figure 3.** Notice that the Ogallala Aquifer stretches from the Texas Panhandle northward to South Dakota. Using artesian springs and wells from this aquifer, pioneers had easy access to water. Early settlements developed around these water supplies.

Research Visit the Glencoe Science Web site at **tx.science.glencoe.com** for information about hot springs. Prepare a travel brochure about one of the sites that would encourage people to visit the springs.

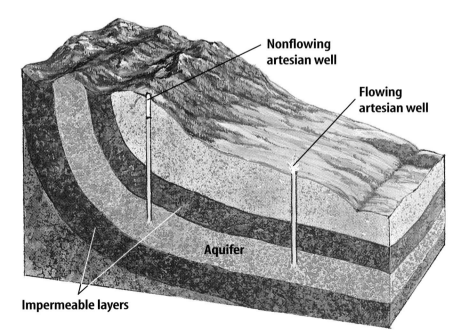

Figure 7
In an artesian well, if the pressure is great enough, water will be forced up to the surface and into the air.

Figure 8
Old Faithful is the most famous geyser in the United States. The time between its eruptions can vary from 35 min to 120 min. As it erupts, it shoots out about 40,000 L of water and steam high into the air.

Geysers

Molten material in Earth can reach temperatures above 650°C. Rock in contact with molten material can heat groundwater to high temperatures. Water heated naturally in this way can form geysers, such as the one shown in **Figure 8.** A geyser is a hot fountain that erupts periodically, shooting water and steam into the air.

Reading Check *What is a geyser?*

Geyser Eruptions Below the surface, a network of fractures and cavities fills with groundwater. Groundwater is heated to temperatures higher than the boiling temperature at the surface. The water expands forcing its way out at the top of a cavity. As some of the water escapes, the pressure inside the cavity drops. This allows the remaining water to boil quickly because of the drop in pressure. Much of the water turns to steam. The steam shoots out the opening, much like steam out of a teakettle, forcing the remaining water out with it. Some of the geyser water reenters the channels to begin the process again. Although geysers erupt repeatedly, many are not entirely predictable. Old Faithful in Wyoming's Yellowstone National Park is more predictable than most geysers.

Groundwater is an important water resource, but in many locations, it is polluted. In the next section, you'll learn about the sources of pollution and how they can be cleaned up.

Section Assessment

1. Why is groundwater important?
2. Why are some rocks more permeable than others are?
3. Why do geysers erupt?
4. Explain what is necessary for a material to be an aquifer.
5. **Think Critically** What happens to the water table if the amount of water pumped from an aquifer is greater than the water seeping in?

Skill Builder Activities

6. **Recognizing Cause and Effect** What causes pressure in an artesian well? **For more help, refer to the** Science Skill Handbook.
7. **Using Percentages** Of the 289.2 billion L of groundwater used in the United States each day, 185.5 billion L are used for irrigation. Calculate the percentage of groundwater that's used for irrigation. **For more help, refer to the** Math Skill Handbook.

Activity

Artesian Wells

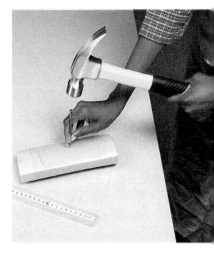

How much water pressure exists at an artesian well? This depends on where it is drilled. The difference between the elevation of the top of an aquifer and the top of the artesian well determines the water pressure. In this activity, you'll observe some variables that affect the difference in elevation and the resulting water pressure.

What You'll Investigate
How do the amount of water in an aquifer and the position of an artesian well affect water pressure at an artesian well?

Materials
tall, soft-plastic bottle with cap
nail or awl
hammer
protractor
sink with water
plastic straw
scissors
metric ruler

Goals
- **Make** a model of an aquifer with artesian wells.
- **Observe** and infer how variables affect the water pressure at artesian wells.

Safety Precautions
Be careful not to pierce yourself with the nail or awl when making the holes in the plastic bottle.

Procedure
1. With the hammer and nail, or an awl, carefully poke two holes in the same side of a plastic bottle. One hole should be 3 cm from the bottom of the bottle, and the other should be 10 cm from the bottom.
2. Cut two pieces of plastic straw, each about 4 cm long. Insert one piece of straw in each of the holes.
3. Hold the bottle over the sink. While your partner covers both straws, fill the bottle with water.
4. Keep the bottle in a vertical position and uncover both holes. Observe and record in your Science Journal what happens at both straws as the water level drops.
5. Repeat the experiment, but this time tilt the bottle at a 30-degree angle from vertical, keeping the straws on the upper side of the bottle. Observe and record what happens.

Conclude and Apply
1. Which part of your bottle apparatus represents the impermeable rock layers? The aquifer? Artesian wells?
2. How does the amount of water in an aquifer affect the water pressure at an artesian well?
3. How does tilting the aquifer affect the water pressure at an artesian well?
4. How does the position of an artesian well in an aquifer affect the pressure of the water at the well?

Communicating Your Data
Make a diagram of your bottle apparatus. Compare your diagram with those of other students. **For more help, refer to the Science Skill Handbook.**

SECTION 2

Groundwater Pollution and Overuse

As You Read

What You'll Learn
- **Identify** the sources of groundwater pollution.
- **Explain** the ways groundwater is cleaned.
- **Explain** what happens when too much water is pumped from an aquifer.

Vocabulary
pollution
sanitary landfill
bioremediation
subsidence

Why It's Important
Good health depends on having clean water.

Sources of Pollution

It's easy to spot some kinds of water pollution. Polluted rivers and lakes can smell bad or have oil slicks. However, it's harder to imagine the many chemicals that find their way into groundwater from polluted surface water and soil. One reason so many chemicals get into groundwater is that many substances dissolve in water that enters the groundwater system.

Pollution comes in many forms and from many different sources. **Pollution** is the contamination of soil, water, air, or other parts of the environment by something harmful. Water pollution comes from two types of sources—point sources and nonpoint sources. An example of point-source pollution is waste dumped directly into water from a pipe or a channel. Nonpoint-source pollution originates over larger areas, such as roadways, shown in **Figure 9,** agricultural areas, and industrial sites. Most groundwater pollution comes from nonpoint sources. These pollutants can seep into soil and eventually travel down into aquifers. **Figure 10** shows some examples of where groundwater pollution comes from.

Figure 9
Salt spread on this road might eventually seep down into groundwater.

260 CHAPTER 9 Groundwater Resources

NATIONAL GEOGRAPHIC VISUALIZING SOURCES OF GROUNDWATER POLLUTION

Figure 10

Pollutants can enter the groundwater of an aquifer in a number of different ways. Pollutants can come from factory, farm, or home. The three-dimensional cutaway below shows some of the major sources of groundwater pollution.

Salt used to melt ice and snow on roads can seep into groundwater.

Toxic substances from mines can dissolve in rainwater or melted snow and seep into groundwater.

In cities, sewer lines can break or leak. Pipelines that leak oil or other toxic substances also contribute to groundwater pollution.

Factories

Aquifer

Impermeable rock

Fertilizers and other chemicals used on cropland wash into the soil when it rains and eventually enter groundwater.

Chemicals and bacteria from leaking landfills can pollute groundwater.

Some septic systems contaminate groundwater with harmful bacteria.

Animal wastes from feed lots and farms can pollute groundwater.

Gasoline and other toxic chemicals oozing from underground storage tanks can poison groundwater supplies.

SECTION 2 Groundwater Pollution and Overuse **261**

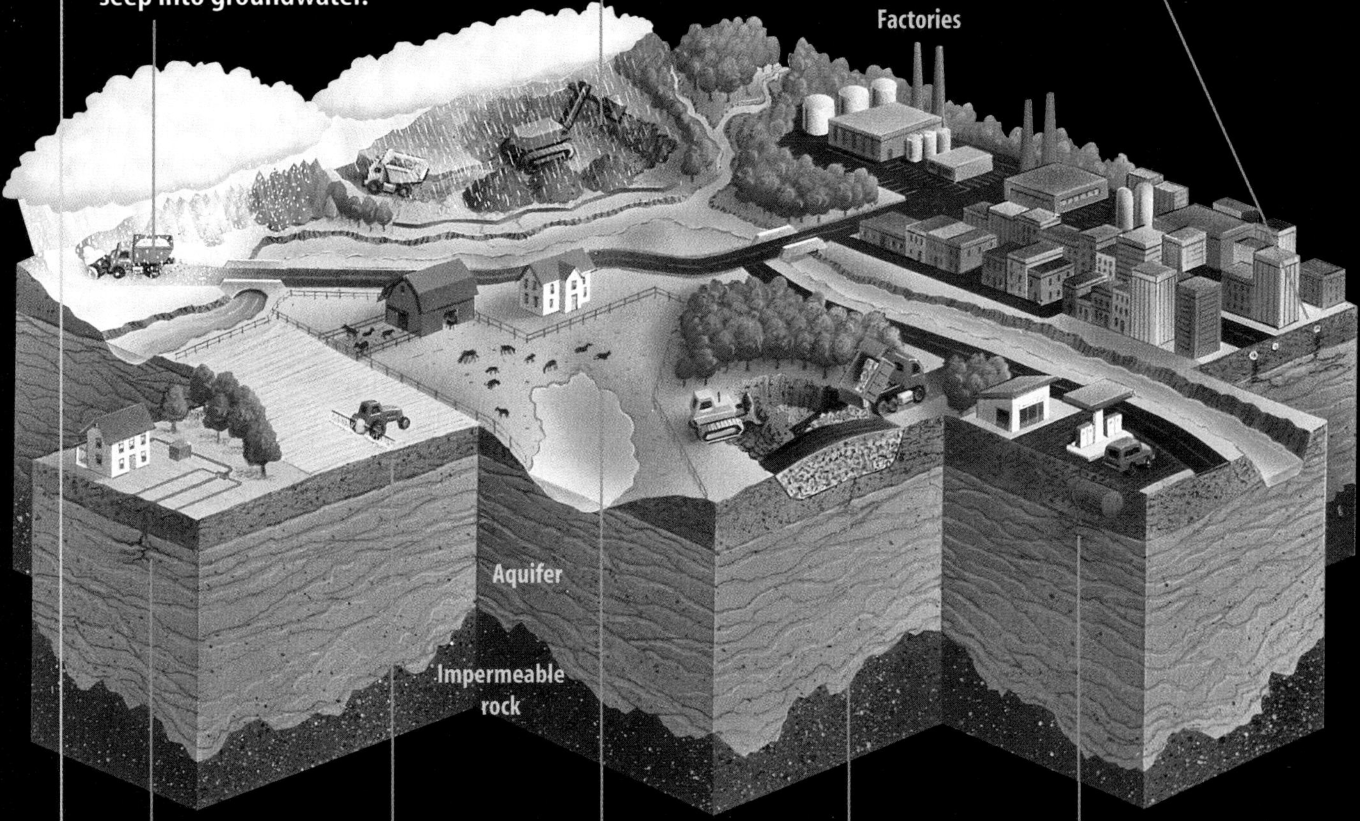

Figure 11
About 55 percent of the solid wastes collected in cities and towns is put in sanitary landfills. The landfills must be checked constantly for leaks so that the soil and the groundwater below it are not polluted.

Landfills Most of the materials that people throw away ends up in landfills. These wastes include food wastes, paper, metals, oils, sprays, batteries, baby diapers, cleaners, cloth, and many other items. Some of these substances are poisonous and cause cancer. Because landfills are dug into the ground when they are constructed, there is a possibility that landfill waste can seep into the ground. To help reduce groundwater pollution, communities construct sanitary landfills, shown in **Figure 11.** In a **sanitary landfill,** there is a lining of plastic or concrete, or the landfill is located in clay-rich soils that trap the liquid waste. Although sanitary landfills greatly reduce the chance that hazardous substances will leak into the surrounding soil and groundwater, some still gets into the environment. You can help reduce this kind of pollution. Contact your local waste management office about where to safely dispose of your family's hazardous wastes. Examples of your hazardous wastes are insect sprays, weed killers, batteries, drain cleaners, bleaches, medicines, and paints.

Road Runoff and Waste Spills Another source of groundwater pollution is runoff from roadways and parking lots. When cars leak oil or gasoline is spilled, groundwater pollution can result. Another risk to groundwater involves the transportation of toxic materials along roadways and railways. Many communities require roadway carriers of such materials to use outer-belt freeways. Outer-belts often are located far enough from towns and cities that a spill would not pose as much of a risk to city wells.

Road Salt Many states use salt to melt ice and snow on roads and highways. Often, salty water runs off roads, where it seeps into the soil and eventually reaches the groundwater. Plowing also pushes salt-containing snow to the sides of roads where it melts and seeps into the ground.

Storage Tanks Gasoline and other substances often are stored in large tanks underground. Storing hazardous materials underground can cause groundwater contamination if the tanks leak. This is being reduced by designing double-walled containers for hazardous substances that resist weathering underground. If left to rust or corrode, leaking pipelines and underground storage tanks pose a risk to aquifers. Federal laws now require that all underground storage tanks be monitored continuously for leaks. If a leak is found, action must be taken for cleanup. Sometimes, this involves removing the tank from the ground.

Septic Systems About one third of all homes in the United States dispose of their wastewater through septic systems. Septic systems consist of a septic tank and an absorption area, shown in **Figure 12.** Solids are removed from the wastewater in the tank. Pipes carry wastewater from the tank to the absorption area. The absorption area then uses the soil and crushed rock to filter and treat the water before it reaches the water table.

If septic systems are not properly designed or maintained, sewage can get into surrounding soil and rock. As it migrates, sewage can carry harmful bacteria. If the bacteria get into drinking water, they can make people ill. Proper installation of septic systems and pumping out tank wastes on a regular basis greatly reduce the chances that groundwater will be polluted.

Health INTEGRATION

Water can carry harmful bacteria, typhoid, dysentery, and hepatitis. Contaminated water can also cause birth defects. If you have a well, why should you have your well water tested on a regular basis?

Figure 12
Although newer septic systems work efficiently, many older systems can allow pollutants to enter the groundwater.

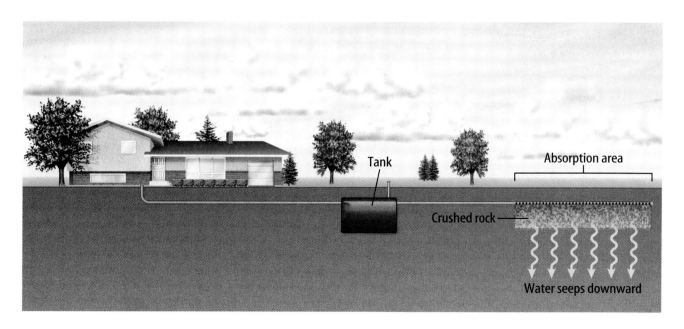

Figure 13
Properly lined holding ponds reduce the chances of groundwater pollution.

Modeling Groundwater Pollution

Procedure
1. Use a **rubber band** to secure a **coffee filter or a sheet of porous paper** to the top of a clear **drinking glass.**
2. Place one spoonful of powdered **coffee, hot chocolate, or** other **powdered drink** onto the filter.
3. Slowly pour **water** over the powdered liquid and examine the water at the bottom of the glass. *Do NOT drink the water.*

Analysis
1. Compare the spoonful of powdered drink to a stockpile or waste pile at an industrial plant. Compare the coffee filter to soil.
2. Infer why the water at the bottom of the glass is discolored. Explain how groundwater below a stockpile or waste pile can be polluted after a heavy rain.

Industrial Wastes One of the major sources of groundwater pollution is from holding ponds like the one in **Figure 13.** These are shallow depressions in the ground that are used for holding industrial chemicals. These ponds sometimes contain concentrated toxic solutions that leak through the soil and into groundwater. If holding ponds are lined with plastic, concrete, clay or other fine-grained sediments, the liquids are less likely to seep into the groundwater supply. However, even lined ponds can overflow during heavy rains, spilling liquid wastes directly into the soil.

✓ **Reading Check** *How can chemical holding ponds cause groundwater pollution?*

Another source of industrial groundwater pollution is from stockpiles and waste piles. A stockpile is a pile of materials that will be used at a later time. A waste pile is a mound of debris that is to be disposed of in the future. When rain falls and dissolves materials in stockpiles and waste piles, toxic materials can get into groundwater.

Mining Wastes When water flows across soil or rock in a mine, toxic metals and other substances can dissolve in the water. Some of these substances combine with water to form acids, such as sulfuric acid. These toxic substances can, in turn, migrate into groundwater.

Today laws require that mining companies make every effort to prevent the flow of pollutants into groundwater. Some of these efforts include changing the pathway of water entering a mine after it rains or snows. Channels and pipes can divert water so that it does not flow across mining wastes. If some water does migrate across mining wastes, it can be captured and treated before entering an aquifer.

Agricultural Runoff When rain falls on fields and lawns that have been treated with pesticides and fertilizers, these substances can move through soil and pollute groundwater. Fertilizers are especially harmful. When ammonia fertilizers decompose, they create nitrates. Water that is high in nitrates creates a serious health problem for infants and elderly people. In some rural areas, people must use bottled water for drinking because of nitrates in their well water.

Feedlots Another source of groundwater pollution is from animal feedlots. A feedlot is where a large number of animals like chickens, cows, sheep, or pigs are raised for food. In these areas, large amounts of animal waste, or manure, collect. Manure contains large concentrations of nitrates and bacteria that can pollute aquifers if the manure is not stored and handled properly. Monitoring wells can be installed around manure storage ponds or tanks. These wells allow groundwater to be constantly monitored for pollution.

Research Visit the Glencoe Science Web site at **tx.science.glencoe.com** to learn more about what farmers are doing to reduce groundwater pollution. Make a poster that explains what you learn.

Problem-Solving Activity

Can stormwater be cleaned and reused for irrigation?

The council of Parfitt Square in Bowden has called a meeting of all residents to discuss implementing a new stormwater management system. Plans will be presented that call for the storage of all stormwater runoff in the aquifer. The water will be cleaned to remove pollutants before storage and then reused for irrigation.

Identifying the Problem

All of the stormwater will be collected in a basin. It will pass through a sediment trap where most of the sediment will settle out. The water then will be treated and sent to the aquifer where it will be stored until it is needed for irrigation. The aquifer is large and contains coarse river gravel. Water in the aquifer has a high concentration of salt, which currently makes it unusable for irrigation. Residents can examine the diagram of the stormwater system and ask questions. What questions would you ask the council members?

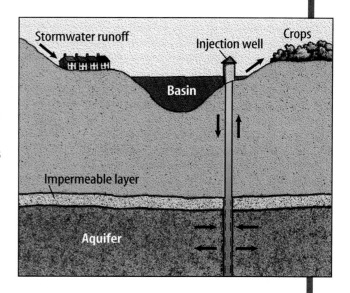

Solving the Problem

1. If the objective is to collect all of the stormwater runoff, how will this affect the aquifer?
2. What would be the advantages of this stormwater mangement system to the neighborhood? Are there any disadvantages?

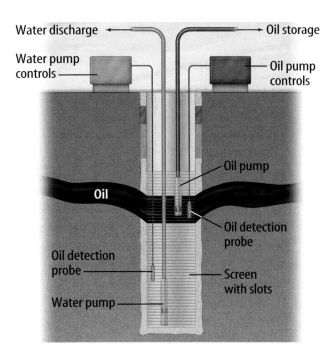

Figure 14
Pumping causes water and oil to move into a well. One pump removes groundwater, while the other pump removes the oil. A control system shuts off the water pump if oil is detected by the probe just above the water pump.

Groundwater Cleanup

Imagine trying to clean paint or oil out of a sponge. You would have to soak the sponge in chemicals, squeeze it, rinse it, and repeat this process several times. Even after this, you may never get all the paint or oil out because some will cling to the sides of the pores. Now, imagine trying to clean the polluted pores in rocks. You aren't allowed to use dangerous chemicals. You can't squeeze rocks. In addition, water moves through the pores slowly. Cleaning groundwater isn't easy.

In many ways, it's easier to prevent groundwater pollution than it is to clean it up. Because water is exchanged among lakes, streams, and aquifers, surface water and groundwater should be cleaned if pollution occurs. Also, the slow speed of groundwater movement sometimes makes cleaning a slow or difficult process.

Removing Pollution Sources The first step in cleaning groundwater is to remove the source of the pollution. In 1980, Congress established a program called Superfund to eliminate the worst hazardous waste sites in the United States. The Environmental Protection Agency (EPA) works in cooperation with states to locate, investigate, and clean up these sites. These sites include abandoned warehouses, manufacturing facilities, processing plants, and landfills. The cleanup operation for Superfund sites is expensive and takes a long time, but it often is possible to clean these and many other point-source pollution sites. Nonpoint-source pollution is much more difficult to eliminate. If the pollution source can be removed, the next task is to clean the groundwater itself.

Cleanup Methods One way of cleaning groundwater is to pump the water out, treat it to remove the pollutants, and then return the water to the aquifer. Another way is to treat water while it remains in an aquifer. The method used for treatment of water left in an aquifer depends on the type of pollutant. In one method, gas is injected under pressure into wells to push the pollutants to the surface where they can be removed. Another method uses steam, heated water, or electricity to increase the flow of pollutants to a well where they can be removed. **Figure 14** shows one way that oil polluting groundwater can be removed.

266 CHAPTER 9 Groundwater Resources

Bioremediation Another way to clean groundwater is by bioremediation. **Bioremediation** is a process that uses living organisms to remove pollutants. Bacteria and fungi often are used in bioremediation because they use organic materials for food. In water that is polluted with sewage and other organic materials, these organisms break down harmful substances into harmless substances.

Some kinds of plants also can be used for bioremediation. The roots of some plants are able to absorb certain dangerous metals. Where groundwater is within 3 m of the ground surface and soil contamination is within 1 m of the ground surface, plants have been used to successfully remove pollutants. Poplar trees like those shown in **Figure 15** have been successfully used to remove petroleum products from groundwater in Ogden, Utah.

 *What is bioremediation?*

Groundwater Shortages

Besides pollution, another groundwater problem exists. It's caused when large populations overpump aquifers. Overpumping means that the amount of groundwater removed from an aquifer is greater than the amount of water flowing into the aquifer. In some parts of the southwestern United States, increased demand for water has caused groundwater to be removed faster than rain and snow can replenish it. When this happens, the water table drops and some wells go dry, as shown in **Figure 16.** This creates water shortages in some regions. You learned earlier about the Ogallala Aquifer. Its water has supported life and agriculture on the High Plains for many years. However, over the years more water has been pumped out of this aquifer than has been replaced by rainfall. It's one of many aquifers that are drying up.

Figure 15
In addition to sunflowers, alfalfa, and clover, poplar trees also have been used for bioremediation. Because tree roots grow more deeply into the ground than the roots of smaller plants, trees can be used to remove deeper pollution.

Figure 16
When the amount of water pumped out of an aquifer is greater than the amount flowing in, some wells go dry.

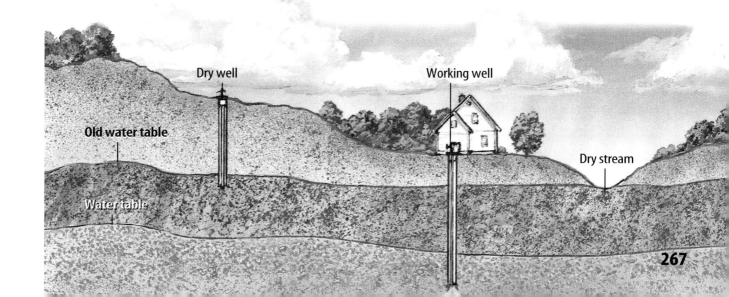

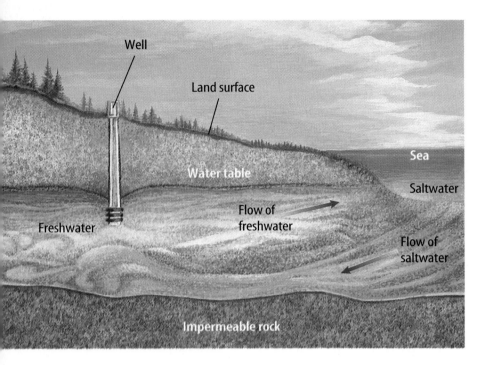

Sinking Land

When water no longer fills the pores in an aquifer, the land above the aquifer sinks. This is called **subsidence** (sub SI dnts). So much groundwater has been removed in parts of California, Florida, and Texas that land surfaces have dropped 5 m or more. In coastal regions, the result of sinking land is a rise in sea level. As a result, flooding along Galveston Bay in Texas has caused some people to abandon their homes.

Figure 17
Pumping of groundwater can reduce freshwater flow toward coastal areas and cause salt water to be drawn toward the freshwater zones of the aquifer.
What will happen when the saltwater layer reaches the bottom of the well?

Seawater Pollution Another problem can arise when too much water is removed from an aquifer along a coast. Seawater can seep into the aquifer and contaminate wells as shown in **Figure 17**. Seawater is more dense than fresh groundwater because it contains dissolved salts. As a result, salt water tends to flow slowly into the lower levels of a coastal aquifer. Notice how the layer of salt water forms an upward cone near the pumping well. This process, called saltwater intrusion, has happened in Florida and Texas, as well as in other places.

Section Assessment

1. What are three sources of groundwater pollution? How do these pollutants get into groundwater?
2. Why is it difficult to clean groundwater?
3. What are two different ways to clean groundwater?
4. What can happen when too much water is pumped from an aquifer?
5. **Think Critically** Benjamin Franklin once said, "When the well's dry, we'll know the worth of water." Explain what you think he meant.

Skill Builder Activities

6. **Recognizing Cause and Effect** Explain how water conservation helps prevent water shortages, subsidence, and seawater pollution of groundwater from oceans. **For more help, refer to the** Science Skill Handbook.
7. **Communicating** You can do many things to help reduce groundwater pollution. List as many as you can think of. When you complete your list, compare it to the lists your classmates have made. **For more help, refer to the** Science Skill Handbook.

SECTION 3
Caves and Other Groundwater Features

Formation of Caves

Some of the most beautiful natural features on Earth are caves like the one shown in **Figure 18.** A **cave,** or cavern, is an underground chamber that opens to the surface. A cave system is a place where many caves are connected by passages. Carlsbad Caverns and Mammoth Cave are two very famous cave systems in the United States. Carlsbad Caverns in New Mexico has at least 48 km of passages. One of the caves in the system is so big it could hold 11 football fields. This cave is 25 stories high. The Mammoth Cave system in Kentucky has about 540 km of mapped passages. It is the longest recorded cave system in the world. Scientists estimate that there could be as many as 900 km of passages that have yet to be discovered and explored in this system.

About 17,000 caves have been found in the United States. Probably many more remain to be discovered and explored. Did you ever wonder what happens to make caves form?

As You Read

What You'll Learn
- **Recognize** how caves change through time.
- **Explain** how cave features form.
- **Explain** what causes sinkholes and disappearing streams.

Vocabulary
cave
dripstone
sinkhole

Why It's Important
As caves evolve, surface features sometimes change.

Figure 18
This cave near Austin, Texas, is one of more than 100 caves in the United States that people can visit to experience the work of groundwater.

Figure 19
Caves form in regions where groundwater dissolves rock.

A Acidic water moves through cracks and pores in the rock.

B Minerals in the rock dissolve in the water and are carried away, forming chambers and passages.

C As the water table drops, the process is repeated on a lower level, leaving the upper cave dry.

Effects of Groundwater Caves form because groundwater is slightly acidic. As you learned earlier, groundwater originates as surface water. Surface water forms when rain falls or snow melts. As water from precipitation falls through the atmosphere and then seeps through soil, it chemically combines with carbon dioxide to form carbonic acid.

Carbonic acid makes groundwater naturally acidic. As this weak acid moves slowly through fractures and pores in certain kinds of rock, the acid reacts with the rock. For example, when carbonic acid passes through limestone, marble, and dolostone, the rocks dissolve easily. Some of the compounds forming these rocks are carried away as ions in solution. **Figure 19** shows how caves change over time.

Formation of Cave Features Groundwater doesn't only dissolve rocks to create caves. It sometimes creates spectacular ceiling, wall, and floor deposits inside the caves. As the water drips and then evaporates, it leaves behind deposits of calcium carbonate called **dripstone. Figure 20** shows different kinds of dripstone formations.

When water drips from the ceiling, it can create long, wavy shapes, much like draperies. It also can form tubes like soda straws that fill with deposits. Some material also is added to the outside of the tubes. Soda-straw formations are the first stage of stalactites. When soda straws are plugged, water trickling down their outside turns them into larger stalactites. Stalactites are dripstone deposits that hang down from the ceilings of caves, much like icicles. As water drips to the floor, towers of stalagmites can build up toward the ceiling. Where stalactites join stalagmites, columns are created. Round pearl-like objects, called cave pearls, form when water drips into shallow cave ponds.

 What is dripstone?

> **Research** Visit the Glencoe Science Web site at **tx.science.glencoe.com** for more information about Carlsbad Caverns and Mammoth Cave. Communicate to your class what you learn.

Figure 20
Dripping water creates beautiful formations in some caves.

A Find the stalactites, stalagmites, and columns.

B At times, water contains minerals that result in draperies with dark orange, reddish, or brown bands. These draperies are called cave bacon.

C Cave pearls come in a variety of sizes and colors.

Figure 21
In one day, land collapsed to create this sinkhole at Winter Park, Florida.

Sinkholes In some caves, so much rock dissolves that the cave's roof can no longer support the land above it and the land collapses into the cave. The depression it leaves is called a **sinkhole.** In regions that overlie thin, delicate roofs of caves, dangerous situations can arise. Roadways, housing developments, and agricultural areas are vulnerable when rock and soil collapse rapidly into an underground cavern. Sinkholes often develop in zones of fractured limestone. When groundwater enters the limestone along the fractures, rock dissolves near the cracks and caves form. **Figure 21** shows a sinkhole that formed at Winter Park, Florida. Sinkholes also occur when so much groundwater is pumped out that the water table drops and flooded caves become dry. When water is no longer in the cave to help support the roof, the roof of the cave can collapse. An area where there are many sinkholes is called a karst area.

Disappearing Streams Walking along a karst area, such as central Kentucky, you might see sinkholes and lakes. Some of these lakes formed when sinkholes filled with water. You might also see some strange streams. They flow along the surface and then suddenly disappear. Where do they go? These disappearing streams fall into caves and continue flowing as underground streams. **Figure 22** shows one way that scientists track these disappearing streams.

Figure 22
Fluorescent dye is poured into streams where they sink into caves. The dye can be tracked to help determine how and where underground water moves.

Figure 23
Most karst regions are in central and southern states where limestone is found and where rainfall is abundant.

Location of Karst Areas In the United States, karst areas with caves, sinkholes, and disappearing streams are common in Missouri, Tennessee, Kentucky, Indiana, Alabama, Florida, and Texas, among other places, as shown in **Figure 23**. Worldwide, karst areas are found in southern Europe, southern Australia, southeastern Asia, Puerto Rico, and Cuba.

As you've learned in this chapter, groundwater is one of Earth's most important resources. It provides many communities with water, and creates caves. Unfortunately, groundwater also can transport pollutants. Understanding how groundwater moves through soil and rock is a first step in learning how to clean, conserve, and protect groundwater resources that are vital for life.

Section 3 Assessment

1. How does carbonic acid help form caves?
2. How does dripstone form? What cave features form from dripstone?
3. What causes a sinkhole?
4. How do disappearing streams differ from regular surface streams?
5. **Think Critically** In western states, some areas with limestone don't contain karst features. Explain.

Skill Builder Activities

6. **Comparing and Contrasting** Compare and contrast stalactites and stalagmites. **For more help, refer to the** Science Skill Handbook.
7. **Using a Word Processor** Use a word processing program to write a creative short story or poem titled, *My Trip Down a Disappearing Stream.* **For more help, refer to the** Technology Skill Handbook.

SECTION 3 Caves and Other Groundwater Features

Activity

Pollution in Motion

Groundwater pollution is a serious problem in the United States. It becomes more serious as the use of this resource grows. When water seeps into the ground, it can carry with it many pollutants. When the polluted water reaches an aquifer, a pollution plume forms as the contaminated groundwater flows through the aquifer. The movement of a pollutant through an aquifer depends on the permeability of the aquifer, the water pressure, whether the pollutant dissolves in water, and other factors. In this activity, you'll investigate and observe the movement of pollution through a model aquifer.

What You'll Investigate

How does a pollutant move through an aquifer?

Goals

- **Observe** the flow of groundwater pollution in a model of an aquifer.
- **Observe and infer** the direction pollution flows when water is pumped from an aquifer.

Materials

clear-plastic box
aquarium gravel
water
pump (from a bottle of hair spray or liquid soap)
30 mL of salt water
food coloring
5-cm × 5-cm piece of cloth
1,000-mL beaker
small rubber band
50-mL graduated cylinder

Safety Precautions

This technician is testing well water for pollutants.

274 CHAPTER 9 Groundwater Resources

Using Scientific Methods

Procedure

1. Add clean gravel to a clear-plastic box until it is three-fourths full.
2. Pour water into the box until it's just below the top of the sediment.
3. Cover the bottom end of the pump with cloth and secure with a rubber band. Push the pump into the sediment near one end of the box.
4. Use a few drops of food coloring to dye the salt water.
5. Pour salt water on top of the sediment at the end of the box away from the pump.
6. Observe the salt water's movement.
7. Sketch the box, the sediment, and the direction that the salt water moves inside the box.
8. **Predict** what will happen to the movement of the salt water when you pump water out of the box.

9. Begin pumping water into the empty beaker and observe the movement of the salt water.
10. Complete your drawing of the salt water movement.

Conclude and Apply

1. **Infer** why the salt water moved in the direction it did before you began pumping water out of the box.
2. **Infer** what caused the movement of the salt water after you began pumping the water.
3. **Infer** what would happen to the direction of the saltwater flow if you used two pumps located next to each other.
4. **Infer** what would happen to the direction of the saltwater flow if you used two pumps that were located at different ends of the box.
5. **Infer** how pumping water from an aquifer affects the movement of pollutants.
6. **Infer** how an increasing population in an area could affect the movement of pollutants in an aquifer and the water quality in people's wells.

Compare your drawings with drawings made by other students. Compare your conclusions with those of other students.

ACTIVITY **275**

Science Stats

Caves

Did you know...

...Wind Cave is one of the world's oldest caves. It was formed more than 320 million years ago. Wind cave doesn't have many stalactites or stalagmites. It's known for its boxwork, an unusual cave formation comprised of thin calcite fins resembling honeycombs.

Boxwork

Wind Cave, South Dakota

...Mammoth Cave is the longest cave in the world. Beneath central Kentucky, 540 km of twisting passages snake through Mammoth Cave. Straightened out, these passageways would stretch the distance from Chicago, Illinois, to Cleveland, Ohio.

...The tallest cave passage in the world is in Deer Cave in Malaysia. A rock arch curves 120 m above the surface of the cave's lake. That's about as tall as a 30-story building.

Deer Cave

Connecting To Math

Lechuguilla Cave

...The deepest cave in the U.S. is in New Mexico. The Lechuguilla (lay chew GEE uh) Cave, in Carlsbad Caverns National Park, is more than 475 m deep. Cave explorers, called spelunkers, are still discovering deeper parts of this cave.

Chauvet Cave, France

...The oldest known cave paintings are in the Chauvet (shoh VAY) Cave in France. The paintings of mammoth, reindeer, bison, and other animals are thought to be about 31,000 years old.

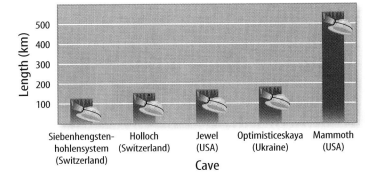

Longest Measured Caves in the World

Do the Math

1. Deer Cave has a volume of 21 million m^3. If about 8.4 Great Pyramids would fit within Deer Cave, what is the approximate volume of the Great Pyramid?
2. According to the above graph, Mammoth Cave is about how many times longer than Jewel Cave?
3. The famous Lascaux (la SKOH) cave paintings in France date back 13,000 years. How many centuries separate the prehistoric artists of Chauvet from the artists of Lascaux?

Go Further

Go to **tx.science.glencoe.com** or a library to find out about the size and history of a cave that was not discussed in this feature.

SCIENCE STATS

Chapter 9 Study Guide

Reviewing Main Ideas

Section 1 Groundwater

1. Groundwater is an important resource.
2. Groundwater moves slowly through pores in soil and rock. An aquifer is a formation of rock or sediment in which water flows freely through pore spaces.
3. In the zone of saturation, pores in an aquifer are full of water. The water table is the top surface of this zone. The water table is also the surface of lakes and rivers.
4. Springs form where the water table is exposed on hillsides. In confined aquifers where water pressure is high, water rises toward the surface when artesian wells are drilled. When groundwater is heated, geysers can form. *Why do geysers like this one erupt repeatedly?*

Section 2 Groundwater Pollution and Overuse

1. Sources of groundwater pollution include landfills, street runoff, accidental hazardous waste spills, holding ponds and septic systems, industrial wastes, mining wastes, and agricultural and feedlot runoff. *How could this landfill cause groundwater pollution?*

2. After the source of pollution is removed, polluted groundwater often can be cleaned. Sometimes it's removed from the aquifer, cleaned, and put back. Other times, it's treated while still in the aquifer. Bioremediation uses organisms to clean water.
3. Too much pumping of water from an aquifer can result in water shortages, subsidence, and saltwater intrusion.

Section 3 Caves and Other Features

1. Water can combine with carbon dioxide in air or soil to form a weak acid. Slightly acidic groundwater dissolves some types of rock to form caves.
2. As water drips within a cave, it leaves behind calcium carbonate. This forms dripstone structures. *How did these columns form?*

3. When the ceiling in a cave can no longer support the ground above it, it collapses. The hole that forms is a sinkhole. Surface streams can disappear into caves.

FOLDABLES Reading & Study Skills

After You Read

On the last page of your Foldable, explain what conclusions you have reached about the importance of groundwater and how it can be protected.

Chapter 9 Study Guide

Visualizing Main Ideas

Fill in the following concept map on groundwater shortages.

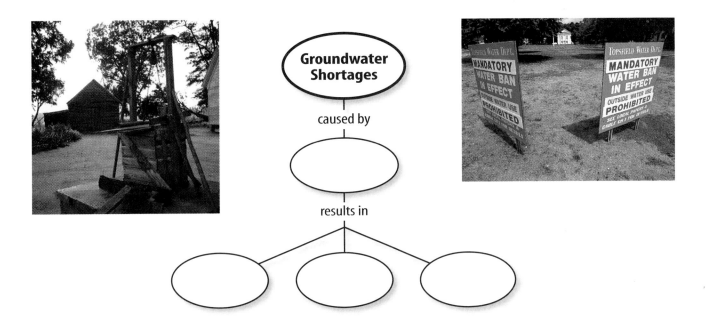

Vocabulary Review

Vocabulary Words

a. aquifer
b. artesian well
c. bioremediation
d. cave
e. dripstone
f. geyser
g. groundwater
h. permeable
i. pollution
j. porosity
k. sanitary landfill
l. sinkhole
m. subsidence
n. water table
o. zone of saturation

 **Study Tip**

With a study partner, read aloud to each other from a chapter. Then discuss what you've been reading.

Using Vocabulary

Explain the differences between the vocabulary words in each of the following pairs.

1. water table, zone of saturation
2. artesian well, geyser
3. subsidence, sinkhole
4. pollution, groundwater
5. sinkhole, cave
6. aquifer, groundwater
7. permeable, zone of saturation
8. bioremediation, subsidence
9. cave, water table
10. dripstone, sinkhole

Chapter 9 Assessment & TEKS Review

Checking Concepts

Choose the word or phrase that best answers the question.

1. Which of the following describes rock with unconnected pores?
   A) impermeable   C) aquifers
   B) permeable   D) saturated

2. Where do most solid wastes end up?
   A) holding ponds C) septic systems
   B) landfills   D) feedlots

3. Which of the following is an example of an impermeable material?
   A) limestone   C) clay
   B) sandstone   D) gravel

4. Which of the following is an example of nonpoint-source pollution?
   A) runoff from a farmer's field
   B) a leaking underground storage tank
   C) a broken pipeline
   D) a leaking landfill

5. Which of the following was set up by Congress to clean up hazardous waste sites?
   A) Safe Drinking Water Act
   B) Superfund
   C) The Clean Water Act
   D) Bioremediation

6. What happens when water no longer fills the pores of an aquifer?
   A) An artesian well forms.
   B) A spring forms.
   C) A geyser erupts.
   D) Subsidence occurs.

7. Which layer of an aquifer has its pores filled with both air and water?
   A) water table
   B) impermeable rock
   C) zone of saturation
   D) zone of aeration

8. To which of the following states does the Ogallala Aquifer provide water?
   A) Florida   C) California
   B) Ohio   D) Texas

9. If 100 mL of sediment holds 35 mL of water, what is its porosity?
   A) 65 percent   C) 50 percent
   B) 35 percent   D) 100 percent

10. Which cave formation forms when structures from the ceiling join with structures on the floor?
    A) stalactite   C) cave pearl
    B) stalagmite   D) column

Thinking Critically

11. Discuss what affects the speed of groundwater flowing through rock.

12. Under what conditions would the water table in an aquifer be higher than normal? What could cause the water table to be lower than normal?

13. A homeowner decides not to have the tank on his septic system pumped out regularly. How could this create health problems?

14. Why might some towns choose to spread sand, rather than salt, on icy roadways?

15. A property owner's well runs dry when a nearby housing development drills several new wells as shown below. Explain what might have caused this to occur.

Chapter 9 Assessment

Developing Skills

16. Interpreting Scientific Illustrations This picture shows stalagmites and stalactites in an underwater cave. Was the water there when these cave features were forming? Explain your answer.

17. Classifying Which sources of groundwater pollution involve sewage?

18. Drawing Conclusions Look back at **Figure 19.** If the water table continues to drop, infer what might happen to the stream and the development of caves.

19. Testing Hypotheses Water tests at a city well indicate high levels of an organic pollutant. Some people have suggested that the pollution is coming from an industrial site east of town. Describe how the city could determine if this is true.

20. Comparing and Contrasting Compare and contrast porosity and permeability.

Performance Assessment

21. Make a Diorama Use clay or a mixture of flour, salt, and water to make a diorama of a cave. Include at least three different kinds of cave formations in your diorama. Display your diorama in the classroom.

Technology

Go to the Glencoe Science Web site at **tx.science.glencoe.com** or use the **Glencoe Science CD-ROM** for additional chapter assessment.

THE PRINCETON REVIEW — TAKS Practice

The Environmental Protection Agency presented a report to congress in 1998 detailing the use of groundwater in the United States. They placed their results in the following circle graph. *TEKS 6.2 C*

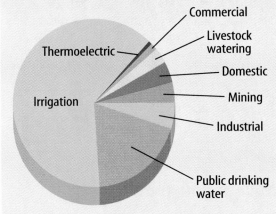

National Groundwater Use

Study the circle graph and answer the following questions.

1. According to this information, which of the following uses the least amount of groundwater?
   A) irrigation   C) mining
   B) industrial   D) thermoelectric

2. According to the circle graph, the percentage of total groundwater used for irrigation is approximately _____.
   F) 30%   H) 65%
   G) 50%   J) 75%

3. The circle graph reflects groundwater used in the _____.
   A) nation   C) state
   B) world   D) county

CHAPTER 10

Science TEKS 6.2 C; 6.5 A; 6.8 B; 6.14 C

The Atmosphere in Motion

Suddenly, it's windier and there's a drop in the temperature. You look up and see a wall of clouds quickly approaching from the northwest. Perhaps this storm will bring rain. The air that surrounds Earth, called the atmosphere, is constantly moving. In this chapter you'll learn about the atmosphere, its structure, clouds, weather, and the water cycle.

What do you think?

Science Journal Look at the picture below with a classmate. Can you guess what it is? Here's a hint: *They can be as small as peas or as large as grapefruit.* Write your answer in your Science Journal.

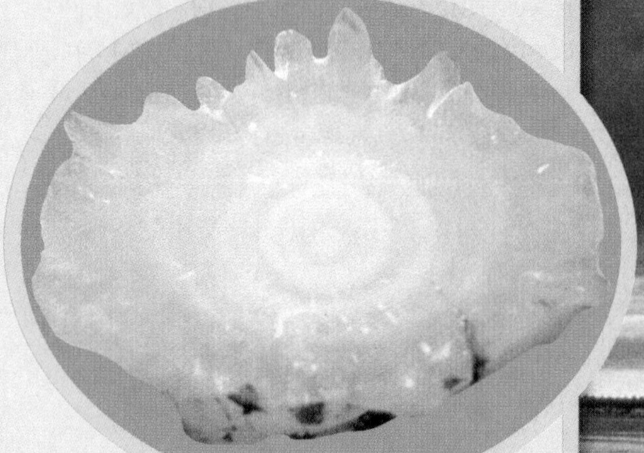

The temperature of air affects the movement of gas molecules. In the activity below, you will increase and then decrease the temperature of air and observe the changes that occur as a result of the movement of air molecules.

Model the effects of temperature on molecules in air

1. With your finger, rub a mixture of water and dishwashing liquid across the top of a narrow-necked plastic bottle until a thin film forms over the opening.
2. Place the bottle in a beaker that is half-filled with hot water and observe what happens to the soap film.
3. Without breaking the film, remove the bottle from the hot water and place it in a beaker that is half-filled with ice water. Observe what happens to the film.

Observe
In your Science Journal, describe what you observed. Then infer what happened to change the shape of the film on top of the bottle.

Before You Read

Making a Know-Want-Learn Study Fold Make the following Foldable to help identify what you already know and what you want to know about the atmosphere.

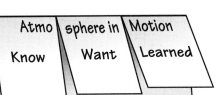

1. Place a sheet of paper in front of you so the long side is at the top. Fold the paper in half from top to bottom.
2. Fold both sides in. Unfold the paper so three sections show.
3. Through the top thickness of paper, cut along each of the fold lines to the top fold, forming three tabs. Label *Atmosphere in Motion* across the front of the paper. Label the tabs *Know, Want,* and *Learned* as shown.
4. Before you read the chapter, write what you already know about the atmosphere under the left tab. Under the middle tab, write what you want to know.
5. As you read the chapter, add to or correct what you have written under the tabs.

283

SECTION 1

The Atmosphere

As You Read

What You'll Learn
- **Explain** why air has pressure.
- **Describe** the composition of the atmosphere.
- **Describe** how energy causes water on Earth to cycle.

Vocabulary
atmosphere
aerosol
troposphere
water cycle

Why It's Important
Movements within the atmosphere create weather changes.

Investigating Air

Air, air . . . everywhere. It's always there. You take it for granted, but without it, Earth would be unfit for life. The **atmosphere**—the layer of gases surrounding Earth—provides Earth with all the gases necessary to support life. It protects living things against harmful doses of ultraviolet and X-ray radiation. At the same time it absorbs and distributes warmth.

Galileo Galilei (1564–1642), an Italian astronomer and physicist, suspected that air was more than just empty space. He weighed a flask, then injected air into it and weighed it again. As shown in **Figure 1,** Galileo observed that the flask weighed more after injecting the air. He concluded that air must have weight and therefore must contain matter. Today scientists know that the atmosphere has other properties, as well. Air stores and releases heat and holds moisture. Because it has weight, air can exert pressure. All of these properties, when combined with energy from the Sun, create Earth's daily weather.

Composition of the Atmosphere

What else do scientists know about the atmosphere? Because it is composed of matter and has mass, it is subject to the pull of gravity. This is what keeps the atmosphere around Earth and prevents it from moving into space. Because it exerts pressure in all directions, you barely notice the atmosphere. Yet its weight is equal to a layer of water more than 10 m deep covering Earth. Scientists also know that the atmosphere is composed of a mixture of gases, liquid water, and microscopic particles of solids and other liquids.

✓ **Reading Check** *What is Earth's atmosphere composed of?*

Figure 1
The flask with air injected weighs more than the flask with no air injected.

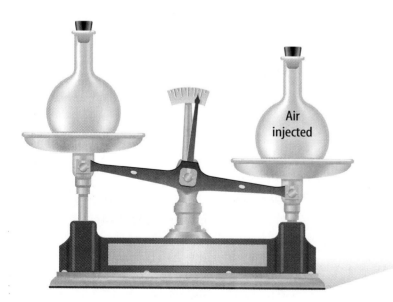

284 CHAPTER 10 The Atmosphere in Motion

Gases Although the atmosphere contains many gases, two of them make up approximately 99 percent of the total. **Figure 2** shows a graph of the gases found in the atmosphere. Nitrogen (N_2) is the most abundant gas—it makes up 78 percent of the atmosphere. Oxygen (O_2), the gas necessary for human life, makes up 21 percent. A variety of trace gases makes up the remaining one percent.

Of the trace gases, two have important roles within the atmosphere. Water vapor (H_2O) makes up from 0.0 to 4.0 percent of the atmosphere and is critical to weather. In its liquid state, water is responsible for clouds and precipitation, and therefore the water that sustains life on Earth. The other important trace gas, carbon dioxide (CO_2), is present in small amounts. Carbon dioxide is needed for plants to make food. Also, carbon dioxide in the atmosphere absorbs heat and emits it back toward Earth's surface, helping keep Earth warm.

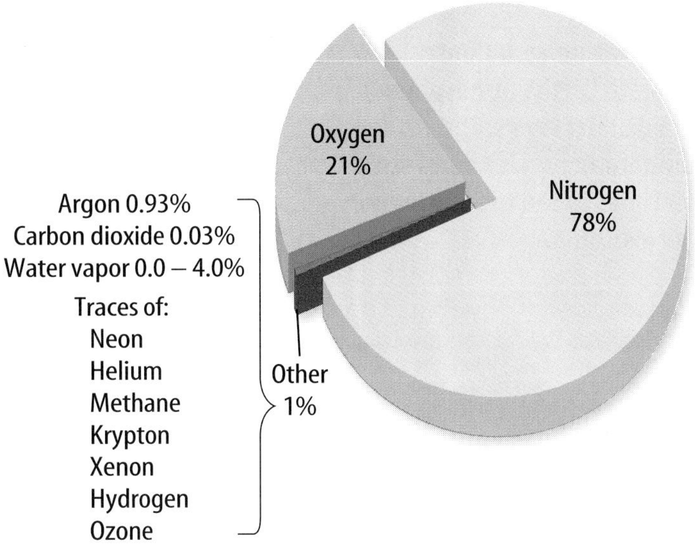

Aerosols Solids such as dust, salt, and pollen and tiny liquid droplets such as acids in the atmosphere are called **aerosols** (AR uh sahlz). Dust enters the atmosphere when wind picks tiny soil particles off the ground or when ash is emitted from volcanoes. Salt enters the atmosphere when wind blows across the oceans. Pollen enters the atmosphere when it is released by plants. Such human activities as burning coal in power plants also release aerosols into the air. Some aerosols, such as those given off by the volcano in **Figure 3**, reflect incoming solar energy, which can affect weather and climate.

Figure 2
The percentages of gases in the atmosphere vary slightly. For example, water vapor makes up from 0.0 to 4.0 percent of the atmosphere. *When the percentage of water vapor is higher, what happens to the percentages of other gases?*

Figure 3
Volcanoes add many aerosols to the atmosphere. Volcanic aerosols can remain suspended in the atmosphere for months or even years. *What do you think happens if many aerosols are in the atmosphere?*

SECTION 1 The Atmosphere

Figure 4
Temperature variations separate Earth's atmosphere into distinct layers. The white temperature scale shows temperatures in the thermosphere and exosphere.

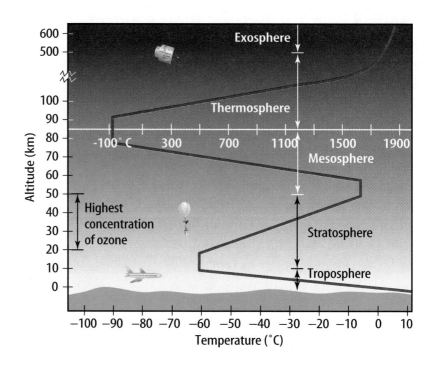

Health INTEGRATION

Ozone in the stratosphere shields Earth's surface from the Sun's ultraviolet (UV) radiation. However, scientists have discovered that the ozone layer has been damaged, allowing more UV radiation to reach Earth. This radiation can cause skin cancers and cataracts, which damage vision. What should you do to protect your skin and eyes when you are outdoors?

Layers of the Atmosphere

The atmosphere is divided into the layers that you see in **Figure 4**. These layers are based on temperature changes that occur with altitude. Each atmospheric layer has unique properties. Find each layer as you read about it. The lower layers are the troposphere and stratosphere. The upper layers are the mesosphere, the thermosphere, and the exosphere.

Troposphere The **troposphere** (TROH puh sfihr) is the atmospheric layer closest to Earth's surface. Notice that it extends upward to about 10 km. The troposphere contains about three fourths of the matter in Earth's entire atmosphere and nearly all of its clouds and weather. The atmosphere absorbs some of the Sun's energy and reflects part of it back to space. However, about 50 percent of the Sun's energy passes through the troposphere and reaches Earth's surface. This energy heats Earth. The atmosphere near Earth's surface is heated by the process of conduction. This means that the source of most of the troposphere's heat is Earth's surface. Therefore, temperatures in the troposphere are usually warmest near the surface and tend to cool as altitude increases. Temperatures cool at a rate of about 6.5 Celsius degrees per kilometer of altitude. If you ever climb a mountain, you will notice that it gets colder as you go higher.

 **Reading Check** *What is the troposphere?*

Stratosphere Above the troposphere is the stratosphere (STRAH tuh sfihr). The stratosphere extends from about 10 km to about 50 km above Earth's surface. As shown in **Figure 4,** most atmospheric ozone is contained in the stratosphere. This ozone absorbs much of the Sun's ultraviolet radiation. As a result, the stratosphere warms as you go upward through it, which is just the opposite of the troposphere. Without the ozone in this layer, too much radiation would reach Earth's surface, causing health problems for plants and animals.

Upper Layers Above the stratosphere is the mesosphere (ME zuh sfihr). This layer extends from approximately 50 km to 85 km above Earth's surface. This layer contains little ozone, so much less heat is absorbed. Notice in **Figure 4** how the temperature in this layer drops to the lowest temperatures in the atmosphere.

The thermosphere (THUR muh sfihr) is above the mesosphere. The thermosphere extends from about 85 km to approximately 500 km above Earth's surface. Temperatures increase rapidly in this layer to more than 1,700°C. The thermosphere layer filters out harmful X rays and gamma rays from the Sun.

Because of intense interaction with the Sun's radiation, atoms can become electrically charged particles called ions. For this reason a part of the thermosphere and mesosphere is called the *ionosphere* (i AN nuh sfihr). This layer of ions is useful because it can reflect AM radio waves, as shown in **Figure 5,** making long-distance communication possible. If the interaction between the Sun's radiation and this layer is too active, however, the quality of radio reception is reduced. Radio signals break up and a lot of static can be heard.

The outermost layer of the atmosphere is the exosphere. It extends outward to where space begins and contains few atoms. No clear boundary separates the exosphere from space.

Earth's Water

Earth often is referred to as the water planet. This is because Earth's surface is about 70 percent water. Because water can exist in three separate states it can be stored throughout the entire land-ocean-atmosphere system. As **Table 1** shows, water exists as solid snow or ice in glaciers. In oceans, lakes, and rivers water exists as a liquid and in the atmosphere it exists as gaseous water vapor.

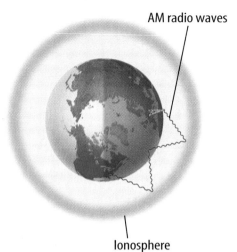

Figure 5
Radio waves are reflected by the ionosphere.

Table 1 Distribution of Earth's Water

Location	Amount of Water (%)
Oceans	97.2
Ice caps and glaciers	2.05
Groundwater	0.62
Rivers and lakes	0.009
Atmosphere	0.001
Total (rounded)	100.00

NATIONAL GEOGRAPHIC VISUALIZING THE WATER CYCLE

FIGURE 6

As the diagram below shows, energy for the water cycle is provided by the Sun. Water continuously cycles between oceans, land, and the atmosphere through the processes of evaporation, transpiration, condensation, and precipitation.

▲ Droplets inside clouds join to form bigger drops. When they become heavy enough, they fall as rain, snow, or some other form of precipitation.

▲ As it rises into the air, water vapor cools and condenses into water again. Millions of tiny water droplets form a cloud.

▲ Rain runs off the land into streams and rivers. Water flows into lakes and oceans. Some water is taken up by plants.

▲ Water evaporates from oceans, lakes, and rivers. Plants release water vapor through transpiration.

The Water Cycle Earth's water is in constant motion in a never-ending process called the **water cycle,** shown in **Figure 6.** The Sun's radiant energy powers the cycle. Water on Earth's surface—in oceans, lakes, rivers, and streams—absorbs energy and stores it as heat. When water has enough heat energy, it changes from liquid water into water vapor in a process called evaporation. Water vapor then enters the atmosphere.

Evaporation occurs from all bodies of water, no matter how large or small. Have you ever noticed that a puddle of water left on the sidewalk from a rainstorm disappears after a while? The water evaporates into the atmosphere. Water also is transferred into the atmosphere from plant leaves in a process called transpiration. As water vapor moves up through the atmosphere, it becomes cooler. The molecules begin to slow down. Eventually, the water molecules change back into droplets of liquid water. This process is called condensation.

Reading Check *How do evaporation and condensation differ?*

Water droplets grow in size when two or more droplets run into each other and combine to form a larger droplet. Eventually, these droplets become large enough to be visible, forming a cloud. If the water droplets continue to grow, they become too large to remain suspended in the atmosphere and fall to Earth as precipitation. You will learn about the different forms of precipitation in the next section. After it is on the ground, some water evaporates. Most water enters streams or soaks into the soil to become groundwater. Much of this water eventually makes its way back to lakes or to the oceans, where more evaporation occurs and the water cycle continues.

TRY AT HOME Mini LAB

Observing Condensation and Evaporation

Procedure
1. Fill a **glass** with **ice water.** Make sure that the outside of the glass is dry.
2. Let the glass stand for 10 min and observe what happens on the outside of the glass.
3. Pour 250 mL of **water** into a small **saucepan**.
4. Boil the water on a **stove** for 10 min.
5. After the water has cooled, pour it into a **measuring cup** and measure its volume.

Analysis
1. Infer why water droplets formed on the outside of the glass.
2. Infer where some of the water in the saucepan went when it was boiled.
3. Compare condensation and evaporation.

Section Assessment

1. Why does air have pressure?
2. Name three solid particles that can be found in the atmosphere.
3. Starting at Earth's surface, what are the five layers of the atmosphere?
4. What are four processes that are part of the water cycle?
5. **Think Critically** Why is it possible for a high mountain at the equator to be covered by snow?

Skill Builder Activities

6. **Recognizing Cause and Effect** Use a computer to create a diagram illustrating the interactions between the Sun's energy and the matter in the water cycle. **For more help, refer to the** Technology Skill Handbook.

7. **Using an Electronic Spreadsheet** Make a spreadsheet and a circle graph of the information listed in **Table 1. For more help, refer to the** Technology Skill Handbook.

SECTION
2 Earth's Weather

As You Read

What You'll Learn
- **List** the factors of weather.
- **Compare** ways that heat is transferred on Earth.
- **Describe** the formation of different kinds of clouds and precipitation.
- **Explain** what causes wind.

Vocabulary
weather
humidity
dew point
relative humidity
precipitation

Why It's Important
Weather affects your life every day.

Weather

Your favorite television show is interrupted by a special weather bulletin. Heavy snow is expected in your area during the night. Will the schools be closed? Will people be able to get to work? How might this weather affect your family? **Weather** describes the current condition of the atmosphere. Factors of weather include temperature, cloud cover, wind speed, wind direction, humidity, and air pressure. It is the task of meteorologists (mee tee uh RAH luh jists) to monitor all weather data continuously in an attempt to forecast weather.

Temperature You learned earlier that the Sun's radiant energy powers the water cycle. In fact, the Sun is the source of almost all of the energy on Earth. When the Sun's rays reach Earth, energy is absorbed. All molecules are constantly in motion, but when they absorb more energy, they move faster and farther apart, as you can see in **Figure 7.** Temperature measures how fast air molecules are moving. When air molecules are moving rapidly, temperature is high. Temperature is measured with a thermometer that uses a particular scale. In science, thermometers with the Celsius scale are used.

Figure 7

A All molecules move. **B** When heated, molecules move faster. Temperature is a measure of the average movement of molecules. The faster they're moving, the higher the temperature is.

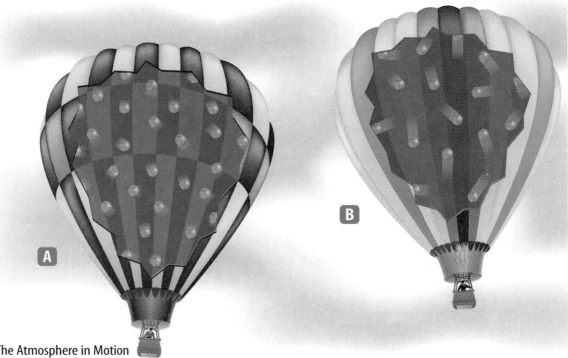

Figure 8
Energy from the Sun warms Earth's surface. Conduction and convection transfer heat on Earth.

Energy Transfer Fast-moving molecules transfer energy to slower-moving molecules when they bump into each other. The transfer of energy that results when molecules collide is called conduction. It is conduction that transfers heat from Earth's surface to those molecules in the air that are in contact with it. After it is in the atmosphere, heated air will move upward as long as it is warmer than the surrounding air. The rising air cools as it gets higher. If it becomes cooler than the surrounding air, it will sink. The process of warm air rising and cool air sinking is called convection. It is the main way heat is transferred throughout the atmosphere. Both processes are shown in **Figure 8.**

Crickets chirp more often and rattlesnakes rattle faster when they're warm. How could these animals be used as natural thermometers?

Atmospheric Pressure As you have learned, because of the attraction of gravity, air has weight. Therefore, the weight of air exerts pressure. Air pressure decreases with altitude in the atmosphere. This is because as you go higher, the weight of the atmosphere above you is less.

Temperature and pressure are related. When air is heated, its molecules move faster, and the air expands. This makes the air less dense, which is why heated air gets moved upward. Less dense air also exerts less pressure on anything below it, creating lower pressure. Cooled, sinking air becomes more dense as the molecules slow down and move closer together, creating more pressure. Therefore, rising air generally means lower pressure and sinking air means higher pressure. Air pressure varies over Earth's surface.

SECTION 2 Earth's Weather **291**

Humidity As air warms up, it can cause water that is in contact with it to evaporate to form water vapor. The amount of water vapor in the atmosphere is called **humidity.** The graph in **Figure 9** shows how temperature affects how much moisture can be present in the air. When air is warmer, evaporation occurs more quickly, and more water vapor can be added to the air. More water vapor can be present in warm air than in cool air. When air is holding as much water vapor as it can, it is said to be saturated and condensation can occur. The temperature at which this takes place is called the **dew point.**

Relative Humidity Suppose a mass of air is chilled. The actual amount of water vapor in the air doesn't change unless condensation occurs, but the amount of moisture that can be evaporated into it decreases. **Relative humidity** is a measure of the amount of water vapor that is present compared to the amount that could be held at a specific temperature. As air cools, relative humidity increases if the amount of water vapor present doesn't change. When air is holding all of the water vapor it can at a particular temperature, it has 100 percent relative humidity.

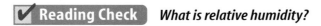 **Reading Check** *What is relative humidity?*

Sometimes local TV weather reports give the dew point on summer days. If the dew point is close to the air temperature, the relative humidity is high. If the dew point is much lower than the air temperature, relative humidity is low.

Figure 9
This graph shows how temperature affects the amount of water vapor that air can hold. *If air temperature is 50°C, how much water vapor can it hold? If the temperature drops to 10°C, how much can it hold?*

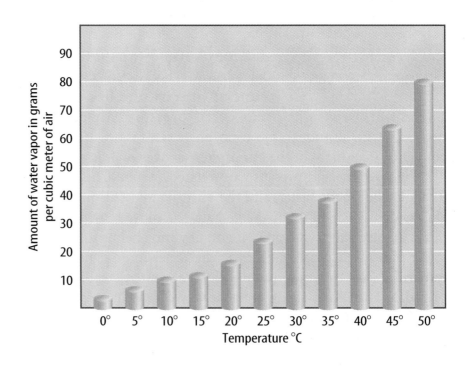

Clouds

One of the best indications that Earth has an atmosphere in motion is the presence of clouds. Clouds form when air rises, cools to its dew point, and becomes saturated. Water vapor in the air then condenses onto small particles in the atmosphere. If the temperature is not too cold, the clouds will be made of small drops of water. If the temperature is cold enough, clouds can consist of small ice crystals.

Clouds commonly are classified according to the altitude at which they begin to form. The most common classification method is one that separates clouds into low, middle, or high groups. Some cloud types are shown in **Figure 10**.

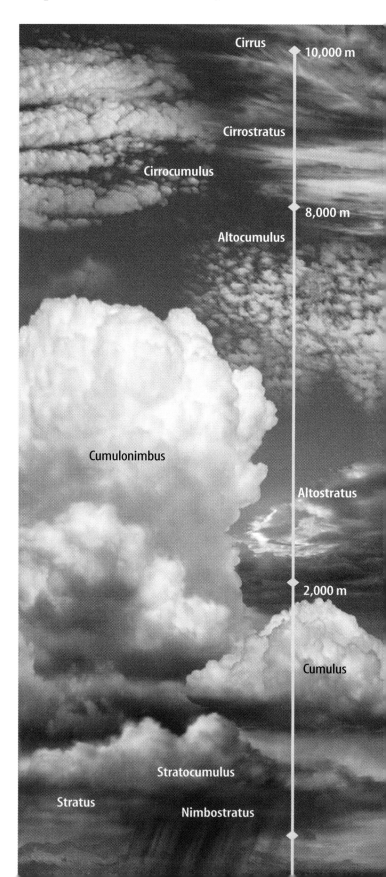

Figure 10
Clouds are grouped according to how high they are above the ground. The types of clouds can be used to predict weather.

Low Clouds The low-cloud group consists of clouds that form at about 2,000 m or less in altitude. These clouds include the cumulus (KYEW myuh lus) type, which are puffy clouds that form when air currents rise, carrying moisture with them. Sometimes cumulus clouds are fair weather clouds. However, when they have high vertical development, they can produce thunder, lightning, and heavy rain. Another type of low cloud includes layered stratus (STRA tus) clouds. Stratus clouds form dull, gray sheets that can cover the entire sky. Nimbostratus (nihm boh STRA tus) clouds form low, dark, thick layers that blot out the Sun. If you see either of these types of clouds, you can expect some kind of precipitation. Fog is a type of stratus cloud that is in contact with the ground.

Middle Clouds Clouds that form between about 2,000 m and 8,000 m are known as the middle-cloud group. Most of these clouds are of the layered variety. Their names often have the prefix *alto-* in front of them, such as altocumulus and altostratus. Sometimes they contain enough moisture to produce light precipitation. Middle clouds can be made up of a mixture of liquid water and ice crystals.

High and Vertical Clouds Some clouds occur in air that is so cold they are made up entirely of ice crystals. Because this usually happens high in the atmosphere, these are known as the high-cloud group. They include cirrus (SIHR us) clouds, which are wispy, high-level clouds. Another type is cirrostratus clouds, which are high, layered clouds that sometimes cover the entire sky.

Some clouds can extend vertically throughout all the levels of the atmosphere. These are clouds of vertical development, and the most common type is cumulonimbus (kyew myuh loh NIHM bus). When you see the term *nimbus* attached to a cloud name, it usually means the cloud is creating precipitation. Cumulonimbus clouds can create the heaviest precipitation of all. Better known as thunderstorm clouds, they start to form at heights of less than 1,000 m but can build to more than 16,000 m high.

Precipitation

When drops of water or crystals of ice become too large to be suspended in a cloud, they fall as **precipitation.** Precipitation can be in the form of rain, freezing rain, sleet, snow, or hail. The type of precipitation that falls depends on the temperature of the atmosphere and the temperature of Earth's surface. **Figure 11** shows the conditions necessary for the formation of rain and snow. Hail consists of balls of ice that form within cumulonimbus clouds. Within the storm cloud, strong winds toss ice crystals up and down, as shown in **Figure 12.** As the ice crystals move up and down, droplets of water freeze around them. Hailstones keep growing until they are too heavy for the winds to keep up. Then they fall to the ground.

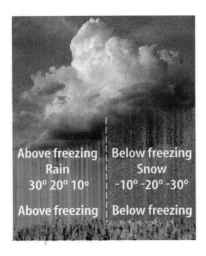

Figure 11
If the temperature of the air and ground are above freezing, precipitation will fall as rain. If the temperature of the air is below freezing, snow will fall.

Figure 12
Hailstones develop in cumulonimbus clouds. Most hailstones are the size of peas, but some can reach the size of softballs. *What does this tell you about the strength of the winds in the cloud?*

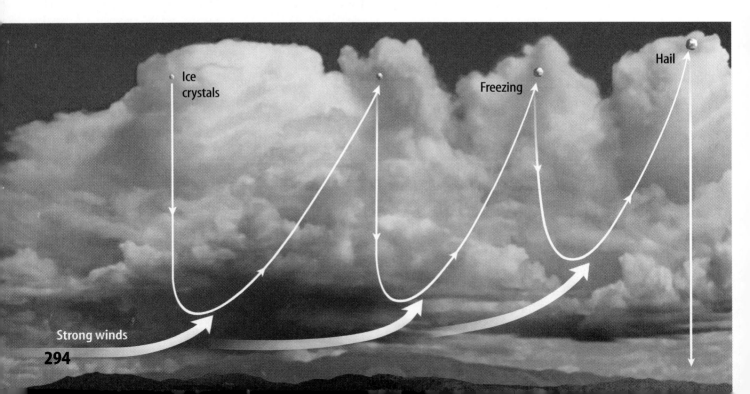

Wind

As you learned earlier, air pressure is related to temperature. When molecules in the atmosphere are heated, they move more rapidly and spread apart. The air becomes less dense and is moved upward. This causes regions of low air pressure. When cooled, those molecules move more slowly and move closer together. The air becomes more dense and sinks, forming regions of high pressure. Typically, air moves from high-pressure areas toward low-pressure areas. Because pressure and temperature are directly related, wind can be thought of simply as air moving from one temperature or pressure area to another. The greater the difference in temperature or pressure between two areas, the stronger the winds that blow between them will be. Wind speed is measured by an instrument called an anemometer (an uh MAHM ut ur), which indicates wind speed by how fast an array of cups that catch the wind rotate. The fastest wind speed ever measured was 371 km/h measured on Mount Washington, New Hampshire, in 1934.

Math Skills Activity

MATH TEKS 6.2 C; 6.4 A; 6.5; 6.11 A, D; 6.12 A

Calculating Speed

Example Problem

Air moves from an area of high air pressure to an area of low air pressure. The wind that is created travels a distance of 14 km in 2 h. What is the wind speed?

Solution

1 *This is what you know:*   distance: $d = 14$ km
                                               time: $t = 2$ h

2 *This is what you need to find:*   speed (rate): r

3 *This is the equation you need to use:* $r = d/t$

4 *Substitute the known values:* $r = 14$ km$/2$ h $= 7$ km/h

Check your work by multiplying your answer by the time. Do you calculate the same distance that was given?

Practice Problems

1. Air moves from a cool area to a warmer area. The wind that is created moves 20 km in 2 h. What is the wind speed?
2. Air moves from an area of high air pressure to an area of low air pressure. The wind that is created travels a distance of 69 km in 3 h. What is the wind speed?

For more help, refer to the **Math Skill Handbook.**

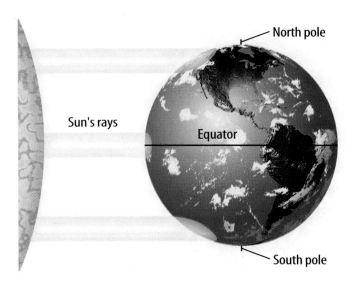

Figure 13
Near the equator, the angle of the Sun's rays is more direct than at higher latitudes.

General Air Circulation Look at **Figure 13**. In any given year, the Sun's rays strike Earth more directly near the equator than near the poles. As a result, Earth's tropical areas heat up more than the polar regions do. Because of this imbalance of heat, warm air flows toward the poles from the tropics and cold air flows toward the equator from the poles. Because Earth rotates, this moving air is deflected to the right in the northern hemisphere and to the left south of the equator. This is known as the Coriolis (kor ee OH lus) effect.

 *What is the Coriolis effect?*

Surface Winds **Figure 14** shows Earth's major surface winds. Air at the equator is heated by the direct rays of the Sun. This air expands, becomes less dense, and gets pushed upward. Farther from the equator, at about 30° latitude, the air is somewhat cooler. This air sinks and flows toward the equator. As this air flows, it is turned by the Coriolis effect, creating steady winds called the trade winds. Trade winds also are called tropical easterlies because they blow in a general east-to-west direction. Find the trade winds in **Figure 14**.

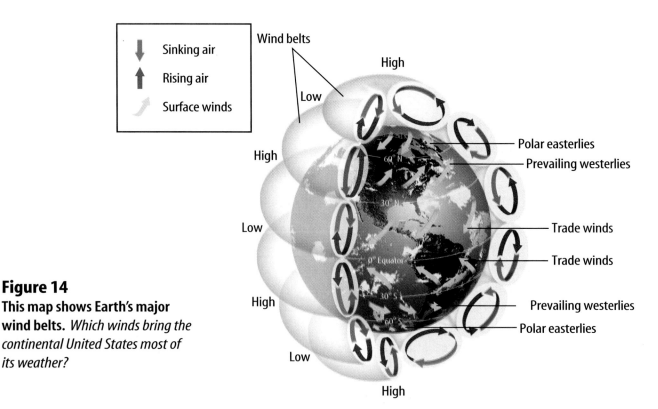

Figure 14
This map shows Earth's major wind belts. *Which winds bring the continental United States most of its weather?*

Westerlies and Easterlies Major wind cells also are located between 30° and 60° latitude north and south of the equator. They blow from the west and are called the prevailing westerlies. These winds are important because they form the boundary between cold air from the poles and milder air closer to the equator. Many of Earth's major weather systems form along these boundaries, so these regions are known for frequent storms.

Near the poles, cold, dense air sinks and flows away from the poles. It is replaced by warmer air flowing in from above. As the cold air flows away from the poles, it is turned by the Coriolis effect. These winds, the polar easterlies, blow from the east.

Figure 15
Weather forecasters often show the position of a jet stream to help explain the movements of weather systems.

Jet Streams Within the zone of prevailing westerlies are bands of strong winds that develop at higher altitudes. Called jet streams, they are like giant rivers of air, as shown in **Figure 15**. They blow near the top of the troposphere from west to east at the northern and southern boundaries of the prevailing westerlies. Their positions in latitude and altitude change from day to day and from season to season. Jet streams are important because weather systems move along their paths.

Other Winds Besides the major winds, other winds constantly are forming around the world because of heating and cooling differences. Slight differences in pressure create gentle breezes. Great differences create strong winds. The most destructive winds occur when air rushes into the center of low pressure. This can cause severe weather like tornadoes and hurricanes, which you'll learn about in the next section.

Research Visit the Glencoe Science Web site at **tx.science.glencoe.com** for more information about jet streams. Communicate to your class what you learn.

Section Assessment

1. How is Earth's surface heated?
2. What happens when water vapor rises and cools?
3. How is temperature related to air pressure and humidity?
4. What causes wind?
5. **Think Critically** Why doesn't precipitation fall from every cloud?

Skill Builder Activities

6. **Comparing and Contrasting** Compare and contrast conduction and convection. **For more help, refer to the Science Skill Handbook.**
7. **Communicating** Use a dictionary to explain why the term *doldrums* is a good description for air near the equator. **For more help, refer to the Science Skill Handbook.**

SECTION 3: Air Masses and Fronts

As You Read

What You'll Learn
- Explain the ways that air masses and fronts form.
- Discuss the causes of severe weather.

Vocabulary
air mass
front
tornado
hurricane

Why It's Important
By understanding how weather changes, you can better plan your outdoor activities.

Air Masses

Weather can change quickly. It can be sunny with calm winds in the morning and turn stormy by noon. Weather changes quickly when a different air mass enters an area. An **air mass** is a large body of air that develops over a particular region of Earth's surface.

Types of Air Masses A mass of air that remains over a region for a few days acquires the characteristics of the area over which it occurs. For example, an air mass over tropical oceans becomes warm and moist. **Figure 16** shows the location of the major air masses that affect weather in North America.

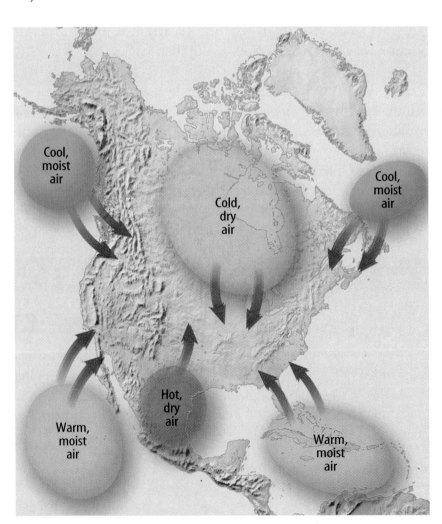

Figure 16
Six major air masses develop in North America. *What are the characteristics of an air mass that originates over the North Pacific Ocean?*

298 CHAPTER 10 The Atmosphere in Motion

Fronts

Where air masses of different temperatures meet, a boundary between them, called a **front,** is created. Along a front, the air doesn't mix. Because cold air is more dense, it sinks beneath warm air. The warm air is forced upward and winds develop. Fronts usually bring a change in temperature as they pass, and they always bring a change in wind direction. The four kinds of fronts are shown in **Figures 17, 18,** and **19.**

Reading Check *What is a front?*

Cold Fronts When a cold air mass advances and pushes under a warm air mass, the warm air is forced to rise. The boundary is known as a cold front, shown in **Figure 17A.** As water condenses, clouds and precipitation develop. If the air is pushed upward quickly enough, a narrow band of violent storms can result. Cumulus and cumulonimbus clouds can develop. As the name implies, a drop in temperature occurs with a cold front.

Warm Fronts If warm air is advancing into a region of colder air, a warm front is formed. Notice in **Figure 17B** that warm, less dense air slides up and over the colder, denser air mass. As the warm air mass moves upward, it cools. Water vapor condenses and precipitation occurs over a wide area. As a warm front approaches, high cirrus clouds are seen where condensation begins. The clouds become progressively lower as you get nearer the front.

Research Visit the Glencoe Science Web site at **tx.science.glencoe.com** to learn more about air masses and fronts. Communicate to your class what you learn.

Figure 17
Cold and warm fronts always bring changes in the weather.

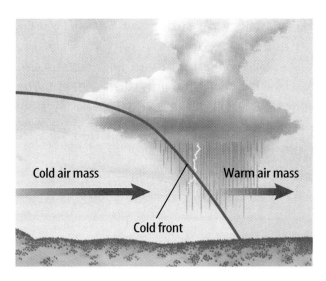

A A cold front often produces short periods of storms with heavy precipitation. After the front passes, wind changes direction, skies begin to clear, and the temperature usually drops.

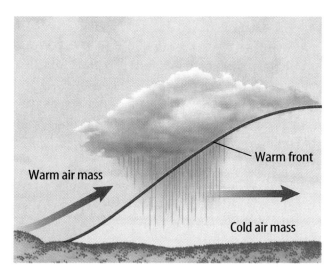

B A warm front usually produces a long period of steady precipitation over a wide area. After the front passes, the sky clears, wind direction changes, and the temperature rises.

Figure 18
A stationary front can result in days of steady precipitation over the same area.

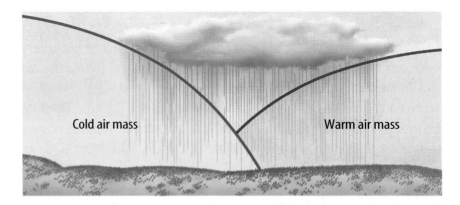

Stationary Fronts A stationary front, shown in **Figure 18,** is a front where a warm air mass and a cold air mass meet but neither advances. This kind of front can remain in the same location for several days. Cloudiness and precipitation occur along the front. Some precipitation can be heavy because the front moves so little.

Occluded Fronts **Figure 19** illustrates how an occluded front forms when a fast-moving cold front overtakes a slower warm front. Another less common occluded front occurs when a warm front overtakes a cold front. Both types can produce cloudy weather with precipitation.

High- and Low-Pressure Centers

In areas where pressure is high, air sinks. As it reaches the ground, it spreads outward away from the high-pressure center. As it spreads, the Coriolis effect turns the air in a clockwise direction in the northern hemisphere. Because the air is sinking, moisture cannot rise and condense, so air near a high-pressure center is usually dry with few clouds.

As air flows into a low-pressure center, it rises and cools. Eventually, the air reaches its dew point and the water vapor condenses, forming clouds and precipitation. Because of the Coriolis effect, air circulates in a counterclockwise direction in the northern hemisphere in a low-pressure center.

Figure 19
An occluded front produces weather similar to, but less severe than, the weather along a cold front.

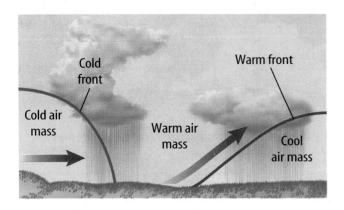
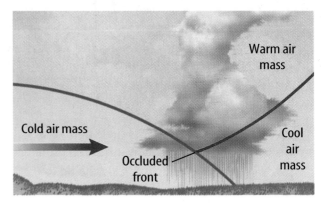

Severe Weather

Severe weather causes strong winds and heavy precipitation. People can be injured and property can be damaged. How can you prepare for severe weather? To prepare, you must first understand it.

Thunderstorms Thunderstorms develop from cumulonimbus clouds. Recall that cumulonimbus clouds often form along cold fronts where air is forced rapidly upward, causing water droplets to form. Falling droplets collide with other droplets and grow bigger. As these large droplets fall, they cool the surrounding air, creating downdrafts that spread out at the surface. These are the strong winds associated with thunderstorms. Dangerous hail can develop in these storms.

Lightning and thunder also are created in cumulonimbus clouds. Where air uplifts rapidly, electric charges form, as shown in **Figure 20**. Lightning is the energy flow that occurs between areas of opposite electrical charge. A bolt of lightning can be five times hotter than the Sun's surface. Its extreme temperature heats the air nearby, causing it to expand rapidly. Then the air cools quickly and contracts, producing a rapid rush of air molecules. The sound produced due to the rapid expansion and contraction of this heated air is the thunder you hear.

Reading Check *What causes thunder?*

Mini LAB

Creating a Low-Pressure Center

Procedure
1. Fasten a **birthday candle** firmly to the bottom of a **pie pan or plate** with **clay**.
2. Fill a **tall, narrow jar** halfway with **water**, and pour the water into the pan or plate.
3. Light the candle and invert the jar over the candle. Set the jar mouth down into the water and rest it on a **penny**.
4. In your **Science Journal**, write a brief description of what happens to the water level inside the jar when the candle goes out.

Analysis
1. Infer what happens to the air inside the jar when the candle is lit.
2. Infer what happens to air inside the jar when the candle goes out, and why water rises in the jar when this happens.

Figure 20
During a thunderstorm, the bottom of the storm cloud has a negative charge. The ground has a positive charge. The negative charge rushes toward the ground. At the same time, the positive charge rushes toward the cloud.

Tornadoes Along some frontal boundaries, cumulonimbus clouds create severe weather. If conditions are just right, updrafts of rising air can start to spin into a rotating vortex. This creates a funnel cloud. **Figure 21** shows the steps in the formation of a funnel cloud. If the funnel cloud reaches Earth's surface, it becomes a tornado like the one shown. A **tornado** is a violent, whirling wind that moves in a narrow path over land. Although tornadoes are usually less than 200 m in diameter, seldom travel on the ground for more than 10 km, and generally last less than 15 min, they are extremely destructive. The powerful updrafts into the low pressure in the center of a tornado act like a giant vacuum cleaner, sucking up anything in its path.

A Strong updrafts and downdrafts develop within cumulonimbus clouds when warm, moist air meets cool, dry air.

B Winds within the clouds cause air to spin faster and faster.

C A funnel of spinning air drops downward through the base of the cloud toward the ground.

Figure 21
A tornado's winds can reach nearly 500 km/h, and it can move across the ground at speeds of up to 100 km/h.

Figure 22
Hurricanes begin as low-pressure areas over warm oceans.

A Air circulation in a hurricane produces updrafts and downdrafts. The downdrafts prevent cloud formation, creating the calm eye of the storm.

B As seen from a satellite, the swirling storm clouds of a hurricane are easy to spot.

Hurricanes Unlike tornadoes, hurricanes can last for weeks and travel thousands of kilometers. The diameter of a hurricane can be up to 1,000 km. A **hurricane** is a large storm that begins as an area of low pressure over tropical oceans. The hurricanes that affect the East Coast and Gulf Coast of the United States often begin over the Atlantic Ocean west of Africa. Look at **Figure 22A** as you read how a hurricane forms. The Coriolis effect causes winds to rotate counterclockwise around the center of the storm. As the storm moves, carried along by upper wind currents, it pulls in moisture. The heat energy from the moist air is converted to wind. When the winds reach 120 km/h, the low-pressure area is called a hurricane. The sustained winds in a hurricane can reach 250 km/h with gusts up to 300 km/h. **Figure 22B** shows a satellite photo of a hurricane over the ocean.

Sometimes a hurricane spends its entire existence at sea and is a danger only to ships. However, when a hurricane passes over land, high winds, tornadoes, heavy rains, and storm surge pound the affected region. Crops can be destroyed, land flooded, and people and animals killed or injured. After the storm begins traveling over land, however, it no longer has the warm, moist air to provide it with energy, and it begins losing power. Gradually, its winds decrease and the storm disappears.

Research Visit the Glencoe Science Web site at **tx.science.glencoe.com** for information about severe weather watches and warnings and what you should do to keep safe. Make a poster about what you learn and share it with your class.

SECTION 3 Air Masses and Fronts

Weather Safety In the United States, the National Weather Service carefully monitors weather. Using weather instruments like Doppler radar, shown in **Figure 23,** as well as weather balloons, satellites, and computers, the position and strength of storms are watched constantly. Predicting the movement of storms is sometimes difficult because the conditions that affect them are always changing. If the National Weather Service believes conditions are right for severe weather to develop in a particular area, it issues a severe weather watch. If the severe weather already is occurring or has been indicated by radar, a warning is issued.

Watches and warnings Watches and warnings are issued for severe thunderstorms, tornadoes, tropical storms, hurricanes, blizzards, and floods. Local radio and television stations announce watches and warnings, along with the National Weather Service's own radio network, called NOAA (NOH ah) Weather Radio.

In dealing with severe weather, the best preparation is to understand how storms develop and to know what to do during watch and warning advisories. During a watch, you should stay tuned to a radio or television station and have a plan of action in case a warning is issued. If the National Weather Service does issue a warning, you should take immediate action to protect yourself.

Figure 23
These scientists are placing weather instruments in the path of a tornado. Their research helps forecasters better understand and predict tornadoes.

Section 3 Assessment

1. Describe how an air mass forms and how its characteristics are determined.
2. Describe the formation of a cold front.
3. What kind of weather occurs along a stationary front?
4. What causes lightning?
5. **Think Critically** Why do low-pressure centers usually have clouds and high-pressure centers have clear skies?

Skill Builder Activities

6. **Comparing and Contrasting** Compare and contrast a hurricane and a tornado. **For more help, refer to the Science Skill Handbook.**
7. **Solving One-Step Equations** Calculate the average speed of a hurricane if it travels 3,500 km in nine days. What is the average speed of a tornado that travels 8 km in 10 min? **For more help, refer to the Math Skill Handbook.**

Activity

Modeling Air Masses and Fronts

Fronts form where air masses of different temperatures meet. In this activity, you'll make two air masses and observe the movement of a front.

What You'll Investigate
How does a moving air mass create a front?

Materials
plastic box with lid
water
smoke paper and match
lamp with reflector
ice

Goals
- **Model** air masses.
- **Observe** the location of a front, and infer the type of front that forms.

Safety Precautions
Wear your safety glasses and apron at all times when the lamp is on. Keep the cord on the lamp away from the water.

Procedure
1. Cover the bottom of the box with 5 cm of water. Be sure that the inside walls of the box are dry.
2. Light the smoke paper. Blow it out, directing the smoke into the box so that the box is filled with smoke.
3. Place the lid upside down on the box.
4. Put five ice cubes inside one end of the lid.
5. Turn on the lamp so that it shines directly above the other end of the box.
6. **Observe** what happens inside the box and on the bottom of the lid. Record your observations.
7. Move the lamp so that it shines over the midsection of the box. Observe and record what happens inside the box and on the bottom of the lid.

Conclude and Apply
1. **Infer** where clouds form.
2. **Describe** where condensation and precipitation form.
3. **Identify** what section of your box represents a cold air mass. Identify what section represents a warm air mass.
4. **Infer** where the weather front is in your box.
5. **Describe** where the front moved.
6. **Identify** which kind of front you made.

Communicating Your Data
Compare your results with other students' results. **For more help, refer to the** Science Skill Handbook.

Activity: Design Your Own Experiment

Creating Your Own Weather Station

The weather across Texas can be unpredictable at times. Being able to forecast severe weather such as thunderstorms, tornadoes, and flash floods can save property or lives. Weather stations, such as the National Weather Service Station in New Braunfels, Texas, use sophisticated instruments such as satellites to help predict weather patterns. More simple instruments that can be found in a weather station include thermometers for measuring temperature, barometers for observing changes in air pressure, anemometers for measuring wind speed, and rain gauges for measuring precipitation. In this activity, you will design and assemble your own weather station to monitor and predict the weather.

Recognize the Problem

How could you use weather instruments and design your own weather station to monitor and predict weather conditions?

Form a Hypothesis

Based on your reading in the text and your own experiences with the weather, form a hypothesis about how accurately you could predict future weather conditions using the weather instruments in your weather station.

Goals

- Use weather instruments for measuring air pressure, wind data, temperature, and precipitation.
- **Design** a weather station using your weather instruments.
- **Evaluate** current weather conditions and predict future conditions using your weather station.

Possible Materials

peanut butter jar
olive jar
permanent marker
metric ruler
meterstick
confetti
*shredded tissue paper
wind vane
anemometer
compass
coffee can
barometer
thermometer
*alternate materials

Safety Precautions

306 CHAPTER 10 The Atmosphere in Motion

Using Scientific Methods

Test Your Hypothesis

Plan

1. **Decide** on the materials you will need to construct a rain gauge. A wide mouth jar is best for rain, and a small, tall jar is best for accurately measuring the rain collected in the larger jar. Decide how you will mark your jars to measure centimeters of rainfall.
2. To measure wind speed you can use an anemometer or you can make a wind-speed scale. Lightweight materials can be dropped from a specific height, and the distance the wind carries them can be measured with a meterstick. A compass can be used to determine wind direction. A wind vane also can be used to determine wind direction.
3. **Decide** where you will place your thermometer. Avoid placing it in direct sunlight.
4. **Decide** where you will place your barometer.
5. Prepare a data table in your Science Journal or on a computer to record your observations.
6. **Describe** how you will use your weather instruments to evaluate current weather conditions and predict future conditions.

Do

1. Ask your teacher to examine your plans and your data table before you start.
2. **Assemble** your weather instruments.
3. Use the weather instruments to monitor weather conditions for several days and to predict future weather conditions.
4. **Record** your weather data.

Analyze Your Data

Compare your weather data with those given on the nightly news or in the newspaper.

1. How well did your weather equipment measure current weather conditions?
2. How accurate were your weather predictions?
3. **Compare** your barometer readings with the dates it rained in your area.

Draw Conclusions

1. Did the results of your experiment support your hypothesis?
2. **Identify** ways your weather instruments could be improved for greater accuracy.
3. **Predict** how accurate your weather predictions would be if you used your instruments for a year.

ACTIVITY 307

TIME SCIENCE AND Society

SCIENCE ISSUES THAT AFFECT YOU!

How Zoos Prepare for Hurricanes

Humans aren't the only ones to take cover when hurricanes strike

Flamingos are herded into the zoo's rest room for safety.

As you step into the rest room, you notice a crunch under your feet. When you look up expecting to see sinks and stalls, you see a flock of pink flamingos standing on a bed of straw. What's going on here?

The Miami Metrozoo is preparing for a hurricane, which means herding all of the flamingos into the shelter of the rest room. Why so much fuss?

In 1992, the Metrozoo was devastated by Hurricane Andrew, which killed five mammals and 50 to 75 birds. The zoo, along with many other Florida zoos, has since been forced to rethink how it gets ready for a hurricane and how to deal with its residents after the storm blows over.

First, zoos must get their facilities ready for hurricane season—the period from roughly early August through early November. Zoos keep plenty of food and water stored in refrigerated trucks during hurricane season. They keep cell phones handy to help with communication in case of a power outage. They line up places for the animals to stay if they must move them after the storm. When a hurricane warning is given, workers make sure that all of the zoo's vehicles are filled with fuel to help with rescues after the storm.

Shovels, rakes, buckets, mops, and other objects that possibly could be tossed around are tied down or taken inside in case of strong winds so they won't hit animals.

Before the Storm

Where do the animals go before a storm at Metrozoo? The lions, tigers, bears, and monkeys are kept in their solid, strong, concrete overnight pens. Poisonous snakes must be bagged because it could be disastrous if they escaped. Other small animals are put into whatever containers can be found, including dog carriers and shipping crates. Some animals are shipped to warehouses or to other zoos that can care for them and are out of the hurricane's path.

Some animals can trust their instincts to tell them what to do. Larger animals may be given the option of coming under shelter or staying out and braving the storm. According to a spokesperson from Seaworld, "The killer whales stay under water longer," which is what they would do in the wild.

Even after the animals are locked up tight, zookeepers worry that the animals could be hurt psychologically by the storm. After Hurricane Andrew, some frightened animals were running around after the storm or just sitting alone. For many zookeepers, the most frustrating thing is being unable to go to an animal and hold it and say, "It's going to be okay."

Rhinos make their way around fallen trees after Hurricane Andrew damaged the zoo in Miami, Florida.

CONNECTIONS Make a List List animal safety tips in case of weather disaster in your area. What should you have on hand to keep your pets safe and as unafraid as possible? What should you do with your pets in case of a weather disaster? If you live on a farm, how can you keep the livestock safe?

SCIENCE Online For more information, visit tx.science.glencoe.com

Chapter 10 Study Guide

Reviewing Main Ideas

Section 1 The Atmosphere

1. The atmosphere is made of gases, liquids, and solids.

2. The troposphere is warmest near the surface and grows cooler with height. Above the troposphere are four additional layers of the atmosphere, each with different characteristics.

3. Atmospheric pressure develops because gases, liquids, and solids in the air contain matter, so they have mass and therefore are pulled by gravity toward Earth's surface.

4. Water circulates between Earth's surface and the atmosphere in the water cycle. *What is shown here?*

Section 2 Earth's Weather

1. Conduction and convection are two ways that heat is distributed on Earth.

2. Temperature affects the amount of moisture that the air can hold. Warm air can hold more moisture than cold air can. When relative humidity is 100 percent, condensation takes place and clouds form.

3. Precipitation occurs when droplets become too heavy to be supported by the air. The type of precipitation depends on the temperatures of the air and the ground. *How were the hailstones shown at the bottom of column one formed?*

4. Wind is the force created by air molecules moving from high-pressure centers to low-pressure centers.

Section 3 Air Masses and Fronts

1. Air masses are dry or moist and warm or cool, depending on where they originate.

2. Fronts develop where air masses of different temperatures collide, forming a boundary. The four kinds of fronts are cold, warm, stationary, and occluded.

3. Severe weather develops from low-pressure centers. Thunderstorms and tornadoes often form near fronts. Hurricanes develop from lows over tropical waters.

4. Knowing what to do when weather watch and warning advisories are made can save your life. *What does the warning on the television screen mean?*

After You Read

FOLDABLES
Reading & Study Skills

Without looking at the chapter, record all you've learned about the atmosphere in motion under the right tab of your Foldable.

Chapter 10 Study Guide

Visualizing Main Ideas

Fill in the following concept map on air masses and fronts.

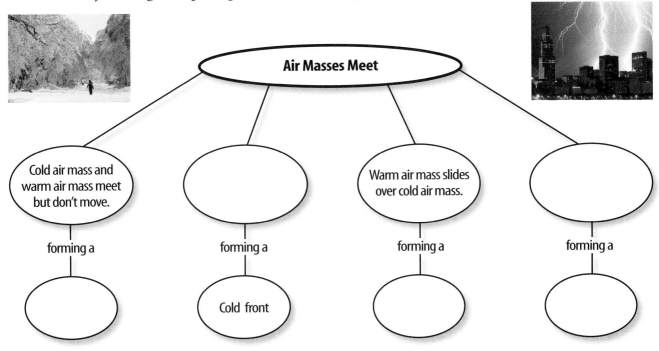

Vocabulary Review

Vocabulary Words

a. aerosol
b. air mass
c. atmosphere
d. dew point
e. front
f. humidity
g. hurricane
h. precipitation
i. relative humidity
j. tornado
k. troposphere
l. water cycle
m. weather

Using Vocabulary

Replace the underlined words with the correct vocabulary words.

1. The <u>water cycle</u> describes the current condition of the atmosphere.

2. The boundary between different air masses is called a <u>dew point</u>.

3. A <u>hurricane</u> is a violent, whirling wind that forms over land.

4. Dust, salt, pollen, and acid droplets in the atmosphere are called <u>precipitation</u>.

5. A large body of air that develops over a particular region of Earth's surface is called an <u>atmosphere</u>.

6. The amount of water vapor in the atmosphere is called <u>weather</u>.

THE PRINCETON REVIEW — Study Tip

Outline the chapters to make sure that you're understanding the key ideas. Writing down the main points of the chapter will help you review facts, remember important details, and understand larger themes.

Chapter 10 Assessment & TEKS Review

Checking Concepts

Choose the word or phrase that best answers the question.

1. Which layer of Earth's atmosphere contains ozone that protects living things from too much ultraviolet radiation?
   A) thermosphere   C) stratosphere
   B) ionosphere   D) troposphere

2. Which front forms when a cold air mass overtakes a warm air mass?
   A) warm   C) cold
   B) stationary   D) occluded

3. Which of the following causes low-pressure centers to rotate counterclockwise in the northern hemisphere?
   A) trade winds   C) Coriolis effect
   B) prevailing westerlies D) jet stream

4. Air at 30°C can hold 30 g of water vapor per cubic meter of air. If the air is holding 15 g of water vapor, what is its relative humidity?
   A) 15 percent   C) 50 percent
   B) 30 percent   D) 100 percent

5. Which of the following scientists first proved that air has weight and must contain matter?
   A) Robert Hooke   C) Robert Boyle
   B) Evangelista Torricelli D) Galileo Galilei

6. Which atmospheric layer reflects radio waves back to Earth?
   A) troposphere   C) stratosphere
   B) ionosphere   D) exosphere

7. Which step of the water cycle occurs when water is heated and changes to water vapor?
   A) evaporation   C) precipitation
   B) condensation D) transpiration

8. Which kind of cloud touches the ground?
   A) altostratus   C) stratocumulus
   B) stratocirrus   D) fog

9. Which occurs when energy is transferred by colliding molecules?
   A) precipitation   C) radiation
   B) conduction   D) convection

10. Which of the following kind of precipitation occurs when strong winds toss ice crystals up and down within a cloud?
    A) rain   C) snow
    B) freezing rain D) hail

Thinking Critically

11. Explain two reasons why hurricanes are so dangerous to people.

12. How do human activities add gases and aerosols to the atmosphere?

13. Why is air pressure greater at sea level than on top of a mountain?

14. What happens to the movement of air molecules when they are heated?

15. Why does air temperature usually change as a front passes by?

Developing Skills

16. **Making and Using Tables** Complete the table using the words *tornado, polar easterlies, westerlies, trade winds,* and *jet stream.*

Characteristics	Wind Type
Blow near the poles	
Move storms across United States	
Often called tropical easterlies	
Narrow stream of air	
Swirling funnels	

Chapter 10 Assessment

17. Comparing and Contrasting Compare and contrast condensation and precipitation.

18. Interpreting Scientific Illustrations Use **Figure 9** to determine how much water vapor air can hold if the temperature is 40°C.

19. Recognizing Cause and Effect How can a cloud produce both rain and hail?

20. Classifying You observe a tall puffy cloud. Rain is falling from its lower surface. How would you classify this cloud?

21. Concept Mapping Complete the concept map of the water cycle below.

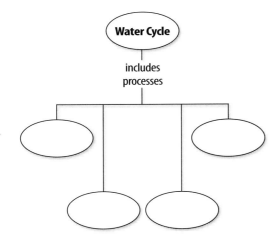

Performance Assessment

22. Pamphlet Research three destructive hurricanes and make a pamphlet of the information you collect. Discuss the paths the hurricanes took, how fast they moved, and the damage they caused.

TECHNOLOGY

Go to the Glencoe Science Web site at **tx.science.glencoe.com** or use the **Glencoe Science CD-ROM** for additional chapter assessment.

TAKS Practice

A student was preparing for an exam on the atmosphere. The student created the following table to study. *TEKS 9.2 D; 9.9 D*

Layers of the Atmosphere	
Thermosphere	Filters out gamma rays
Ionosphere	Reflects AM radio waves
Mesosphere	Contains little ozone
Stratosphere	UV radiation-absorbing ozone
Troposphere	Contains most clouds

Use this table to find the best answer to the following questions.

1. According to the table, in which layer does Earth's atmosphere absorb most of the Sun's UV radiation?
   A) troposphere
   B) stratosphere
   C) mesosphere
   D) thermosphere

2. The ions in the ionosphere are electrically charged from solar radiation. Which has been one of the greatest benefits of the ions in the ionosphere?
   F) They reflect radio waves.
   G) They absorb ozone.
   H) They filter out gamma rays.
   J) They form rain clouds.

Reading Comprehension

Read the passage. Then read each question that follows the passage. Decide which is the best answer to each question.

Obsidian Uses

From the top of the highest mountain to the bottom of the deepest ocean, Earth is made mostly of rock. Geologists classify rocks according to three different categories depending upon how the rocks were formed. These are igneous, sedimentary, and metamorphic.

Igneous rocks are formed from rock that melted and later cooled and solidified. When temperature and pressure conditions are just right, often deep within Earth, rocks will melt. This molten rock, called magma, moves up toward the surface of Earth over time or might even reach the surface quickly in a volcanic eruption. When magma cools rapidly, few or no crystals form in the rock. This gives the rock a glasslike look.

Obsidian is a type of volcanic rock. It is often referred to as volcanic glass because it has a smooth surface. It was a prized material among prehistoric cultures because it fractures with sharp edges and can be used as a weapon or tool. Knives, arrowheads, and spear points were made from obsidian. Prehistoric people also used obsidian as mirrors. In modern times, obsidian has been used for surgical scalpel blades.

The beauty and mystery of igneous rocks have inspired many artists. Some artists carefully sculpt already cooled volcanic rock into beautiful, one-of-a-kind pieces of art.

> **Test-Taking Tip** As you read the passage, write a one-sentence summary for each paragraph.

This is obsidian, a type of volcanic glass.

1. In this passage, the word <u>molten</u> means
   Reading TEKS 6.9 B
   A) deep
   B) melted
   C) igneous
   D) cooled

2. According to the passage, the three categories, or groups, of rocks are _____.
   Reading TEKS 6.9 B
   F) igneous, metamorphic, and sedimentary
   G) volcanic, glassy, and irregular
   H) chemical, organic, and detrital
   J) residual, original, and primitive

3. Which conclusion is best supported by information given in the passage?
   Reading TEKS 6.10 F
   A) When magma cools rapidly, it produces rocks with many large crystals.
   B) Igneous rocks can be made into tools and pieces of art.
   C) Obsidian was used by prehistoric cultures to build stone houses.
   D) Igneous rock forms when rocks weather and erode.

TAKS Practice

Reasoning and Skills

Read each question and choose the best answer.

Igneous Rocks

Formed	Light-colored	Dark-colored
Below Earth's Surface	Granite	Gabbro
At Earth's Surface	Rhyolite	Basalt

1. According to the chart, lava that flows onto the surface from a volcano should cool to form the dark-colored rock _____.
   Science TEKS 6.6 C
   - A) granite
   - B) gabbro
   - C) rhyolite
   - D) basalt

   Test-Taking Tip Reread the question and think about the color of the rock and where the rock was formed.

2. Earth's crust is estimated to be composed of 46% oxygen, 28% silicon, 8% aluminum, and 18% other elements. Which area of the graph represents aluminum?
   Science TEKS 6.4 B
   - F) Q
   - G) R
   - H) S
   - J) T

   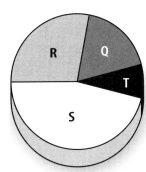

   Test-Taking Tip Think about the quantities of the element the question refers to compared to the sizes of each portion of the graph.

Selected Rock Data

Volume of Sample (cm³)	Mass of Sample (g)
1	3
2	6
3	9
4	12
5	?

3. These data were collected by determining the mass and volume of four different samples of the same rock. If everything remains the same, what will be the mass of a fifth sample whose volume is 5 cm³?
   Science TEKS 6.2 C
   - A) 9 g
   - B) 10 g
   - C) 15 g
   - D) 20 g

   Test-Taking Tip If you know the mass of a 1 cm³ sample of the rock, you can use math to calculate the mass of a 5-cm³ sample.

Consider this question carefully before writing your answer on a separate sheet of paper.

4. Many igneous rocks contain crystals. Geologists have observed that igneous rocks containing large crystals formed slowly, while those containing small crystals formed more quickly. Design an experiment, using sugar and water, to show that the rate at which crystals form affects their size.
   Science TEKS 6.2 A

   Test-Taking Tip You can use a supersaturated solution and seed crystals to speed up the rate at which the sugar crystals form.

STANDARDIZED TEST PRACTICE 315

UNIT 4
Solar System

How Are Inuit & Astronauts Connected?

For thousands of years, people known as the Inuit have lived in Arctic regions. In the early 1900s, an American naturalist spent time among the Inuit in Canada. The naturalist watched the Inuit preserve fish and meat by freezing them in the cold northern air. Months later, when the people thawed and cooked the food, it was tender and tasted fresh. Eventually, the naturalist returned to the United States, perfected a quick-freezing process, and began marketing frozen foods. Later, inventors found a way to remove most of the water from frozen foods. This process, called freeze-drying, produces a lightweight food that can be stored at room temperature and doesn't spoil. Freeze-dried foods are carried by all sorts of adventurers—including astronauts.

SCIENCE CONNECTION

SPACE TRAVEL Many discoveries and technological advances came together to make it possible for people to travel into space. Working in small groups, investigate the history of space travel, from the invention of rockets to the development of liquid fuels, radio communications, and the special suits that astronauts wear. As a class, compile the information to create a space-travel timeline that traces these inventions and breakthroughs through time.

CHAPTER 11

Science TEKS 6.6 A; 6.13 A, B

Space Technology

Stars and planets have always fascinated humans. We admire their beauty, and our nearest star—the Sun—provides energy that enables life to exist on Earth. For centuries, people have studied space from the ground. But, in the last few decades, space travel has allowed us to get a closer look. In this chapter, you'll learn how space is explored with telescopes, rockets, probes, satellites, and space shuttles. You'll see how astronauts like Shannon Lucid, shown here, now can spend months living and working aboard space stations.

What do you think?

Science Journal Look at the picture below with a classmate. Discuss what you think this might be or what might be happening. Here's a hint: *It's part of a dusty trail that's far, far away.* Write your answer or best guess in your Science Journal.

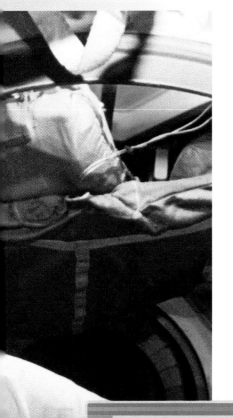

Explore Activity

You might think exploring space with a telescope is easy because the visible light coming from stars is so bright and space is dark. But space contains massive clouds of gases, dust, and other debris called nebulae that block part of the starlight traveling to Earth making it more difficult for astronomers to observe deep space. What does visible light look like when viewed through clouds of dust or gas?

Model visible light seen through nebulae

1. Turn on a lightbulb and darken the room.
2. View the lightbulb through a sheet of dark plastic.
3. View the lightbulb through different-colored plastic sheets.
4. View the light through a variety of different-colored balloons such as yellow, blue, red, and purple. Observe how the light changes when you slowly let the air out of each balloon.

Observe
Write a paragraph in your Science Journal describing how this activity modeled the difficulty astronomers have when viewing stars through thick nebulae.

Before You Read

Making a Sequence Study Fold Identifying a sequence helps you understand what you are experiencing and predict what might occur next. Before you read this chapter, make the following Foldable to prepare you to learn about the sequence of space exploration.

1. Place a sheet of paper in front of you so the short side is at the top. Fold the paper in half from the left side to the right side.
2. Fold the top and bottom in to divide the paper into thirds. Unfold the paper so three sections show.
3. Through the top thickness of paper, cut along each of the fold lines to the left fold, forming three tabs. Label the tabs "Past", "Present", and "Future", as shown.
4. As you read the chapter, write what you learn under the tabs.

SECTION 1

Radiation from Space

As You Read

What You'll Learn
- **Explain** the electromagnetic spectrum.
- **Identify** the differences between refracting and reflecting telescopes.
- **Recognize** the differences between optical and radio telescopes.

Vocabulary
electromagnetic spectrum
refracting telescope
reflecting telescope
observatory
radio telescope

Why It's Important
You can learn much about space without traveling there.

Electromagnetic Waves

As you just read, living in space now is possible. The same can't be said, though, for space travel to distant galaxies. If you've dreamed about racing toward distant parts of the universe—think again. Even at the speed of light, it would take years and years to reach even the nearest stars.

Light from the Past When you look at a star, the light that you see left the star many years ago. Although light travels fast, distances between objects in space are so great that it sometimes takes millions of years for the light to reach Earth.

The light and other energy leaving a star are forms of radiation. Radiation is energy that is transmitted from one place to another by electromagnetic waves. Because of the electric and magnetic properties of this radiation, it's called electromagnetic radiation. Electromagnetic waves carry energy through empty space and through matter.

Electromagnetic radiation is everywhere around you. When you turn on the radio, peer down a microscope, or have an X ray taken—you're using various forms of electromagnetic radiation.

Figure 1
The electromagnetic spectrum ranges from gamma rays with wavelengths of less than 0.000 000 000 01 m to radio waves more than 100,000 m long. *How does frequency change as wavelength shortens?*

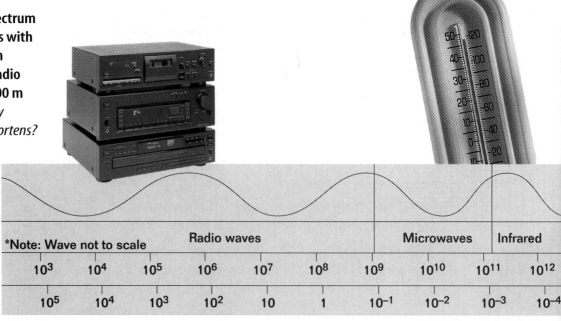

320 CHAPTER 11 Space Technology

Electromagnetic Radiation Sound waves, which are a type of mechanical wave, can't travel through empty space. How, then, do we hear the voices of the astronauts while they're in space? When astronauts speak into a microphone, the sound waves are converted into electromagnetic waves called radio waves. The radio waves travel through space and through Earth's atmosphere. They're then converted back into sound waves by electronic equipment and audio speakers.

Radio waves and visible light from the Sun are just two types of electromagnetic radiation. Other types include gamma rays, X rays, ultraviolet waves, infrared waves, and microwaves. **Figure 1** shows these forms of electromagnetic radiation arranged according to their wavelengths. This arrangement of electromagnetic radiation is called the **electromagnetic spectrum.** Forms of electromagnetic radiation also differ in their frequencies. Frequency is the number of times a wave vibrates per unit of time. The shorter the wavelength is, the more vibrations will occur, as shown in **Figure 1.**

Speed of Light Although the various electromagnetic waves differ in their wavelengths, they all travel at 300,000 km/s in a vacuum. This is called the speed of light. Visible light and other forms of electromagnetic radiation travel at this incredible speed, but the universe is so large that it takes millions of years for the light from most stars to reach Earth.

When electromagnetic radiation from stars and other objects reaches Earth, scientists use it to learn about its source. One tool for studying electromagnetic radiation from distant sources is a telescope.

Health INTEGRATION

Many newspapers include an ultraviolet (UV) index to urge people to minimize their exposure to the Sun. Compare the wavelengths and frequencies of red and violet light, shown below in **Figure 1.** Infer what properties of UV light cause damage to tissues of organisms.

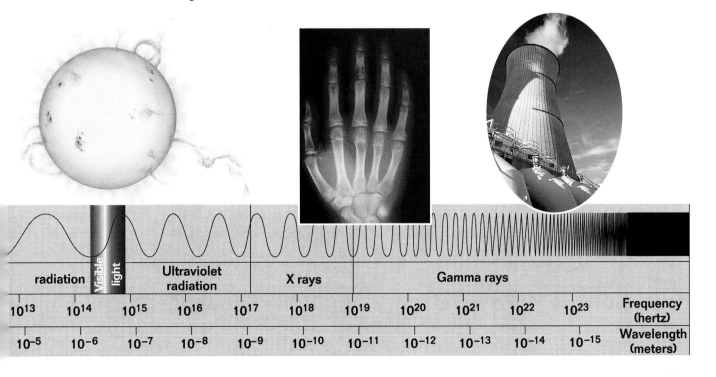

Optical Telescopes

Optical telescopes use light, which is a form of electromagnetic radiation, to produce magnified images of objects. Light is collected by an objective lens or mirror, which then forms an image at the focal point of the telescope. The focal point is where light that is bent by the lens or reflected by the mirror comes together to form a point. The eyepiece lens then magnifies the image. The two types of optical telescopes are shown in **Figure 2.**

A **refracting telescope** uses convex lenses, which are curved outward like the surface of a ball. Light from an object passes through a convex objective lens and is bent to form an image at the focal point. The eyepiece magnifies the image.

A **reflecting telescope** uses a curved mirror to direct light. Light from the object being viewed passes through the open end of a reflecting telescope. This light strikes a concave mirror, which is curved inward like a bowl and located at the base of the telescope. The light is reflected off the interior surface of the bowl to the focal point where it forms an image. Sometimes, a smaller mirror is used to reflect light into the eyepiece lens, where it is magnified for viewing.

Using Optical Telescopes Most optical telescopes used by professional astronomers are housed in buildings called **observatories.** Observatories often have dome-shaped roofs that can be opened up for viewing. However, not all telescopes are located in observatories. The *Hubble Space Telescope* is an example.

Figure 2
These diagrams show how each type of optical telescope collects light and forms an image.

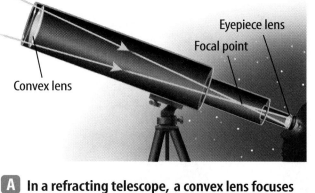

A In a refracting telescope, a convex lens focuses light to form an image at the focal point.

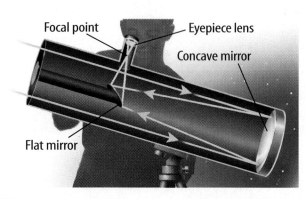

B In a reflecting telescope, a concave mirror focuses light to form an image at the focal point.

C Optical telescopes are widely available for use by individuals.

Hubble Space Telescope The *Hubble Space Telescope* was launched in 1990 by the space shuttle *Discovery*. Because *Hubble* is located outside Earth's atmosphere, which absorbs and distorts some of the energy received from space, it should have produced clear images. However, when the largest mirror of this reflecting telescope was shaped, a mistake was made. As a result, images obtained by the telescope were not as clear as expected. In December 1993, a team of astronauts repaired the *Hubble Space Telescope* by installing a set of small mirrors designed to correct images obtained by the faulty mirror. Two more missions to service *Hubble* were carried out in 1997 and 1999, shown in **Figure 3**. Among the objects viewed by *Hubble* after it was repaired in 1999 was a large cluster of galaxies known as Abell 2218.

Reading Check *Why is* **Hubble** *located outside Earth's atmosphere?*

Figure 3
The *Hubble Space Telescope* was serviced at the end of 1999. Astronauts replaced devices on *Hubble* that are used to stabilize the telescope.

SECTION 1 Radiation from Space

TRY AT HOME Mini LAB

Observing Effects of Light Pollution

Procedure
1. Obtain a **cardboard tube** from an empty roll of paper towels.
2. Go outside on a clear night about two hours after sunset. Look through the cardboard tube at a specific constellation decided upon ahead of time.
3. Count the number of stars you can see without moving the observing tube. Repeat this three times.
4. Calculate the average number of observable stars at your location.

Analysis
1. Compare and contrast the number of stars visible from other students' homes.
2. Explain the causes and effects of your observations.

Large Reflecting Telescopes Since the early 1600s, when the Italian scientist Galileo Galilei first turned a telescope toward the stars, people have been searching for better ways to study what lies beyond Earth's atmosphere. For example, the twin Keck reflecting telescopes, shown in **Figure 4**, have segmented mirrors 10 m wide. Until 2000, these mirrors were the largest reflectors ever used. To cope with the difficulty of building such huge mirrors, the Keck telescope mirrors are built out of many small mirrors that are pieced together. In 2000, the European Southern Observatory's telescope, in Chile, consisted of four 8.2-m reflectors, making it the largest optical telescope in use.

 Reading Check *About how long have people been using telescopes?*

Active and Adaptive Optics The most recent innovations in optical telescopes involve active and adaptive optics. With active optics, a computer corrects for changes in temperature, mirror distortions, and bad viewing conditions. Adaptive optics is even more ambitious. Adaptive optics uses a laser to probe the atmosphere and relay information to a computer about air turbulence. The computer then adjusts the telescope's mirror thousands of times per second, which lessens the effects of atmospheric turbulence. Telescope images are clearer when corrections for air turbulence, temperature changes, and mirror-shape changes are made.

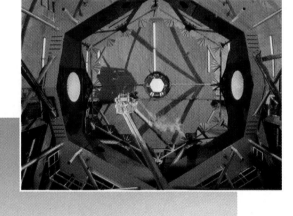

Figure 4
The twin Keck telescopes on Mauna Kea in Hawaii can be used together, more than doubling their ability to distinguish objects. A Keck reflector is shown in the inset photo. Currently, plans include using these telescopes, along with four others to obtain images that will help answer questions about the origin of planetary systems.

Radio Telescopes

As shown in the spectrum illustrated in **Figure 1,** stars and other objects radiate electromagnetic energy of various types. Radio waves are an example of long-wavelength energy in the electromagnetic spectrum. A **radio telescope,** such as the one shown in **Figure 5,** is used to study radio waves traveling through space. Unlike visible light, radio waves pass freely through Earth's atmosphere. Because of this, radio telescopes are useful 24 hours per day under most weather conditions.

Radio waves reaching Earth's surface strike the large, concave dish of a radio telescope. This dish reflects the waves to a focal point where a receiver is located. The information allows scientists to detect objects in space, to map the universe, and to search for signs of intelligent life on other planets.

Later in this chapter, you'll learn about the instruments that travel into space and send back information that telescopes on Earth's surface cannot obtain.

Figure 5
This radio telescope is used to study radio waves traveling through space.

Section 1 Assessment

1. What is the difference between radio telescopes and optical telescopes?
2. If red light has a longer wavelength than blue light, which has a greater frequency?
3. Compare and contrast refracting and reflecting telescopes.
4. How does adaptive optics in a telescope help solve problems caused by atmospheric turbulence?
5. **Think Critically** It takes light from the closest star to Earth (other than the Sun) about four years to reach Earth. If intelligent life were on a planet circling that star, how long would it take for scientists on Earth to send them a radio transmission and for the scientists to receive their reply?

Skill Builder Activities

6. **Sequencing** Sequence these electromagnetic waves from longest wavelength to shortest wavelength: *gamma rays, visible light, X rays, radio waves, infrared waves, ultraviolet waves,* and *microwaves.* **For more help, refer to the Science Skill Handbook.**

7. **Solving One-Step Equations** The magnifying power (*Mp*) of a telescope is determined by dividing the focal length of the objective lens (FL_{obj}) by the focal length of the eyepiece lens (FL_{eye}) using the following equation:

$$Mp = FL_{obj}/FL_{eye}$$

If FL_{obj} = 1,200 mm and FL_{eye} = 6 mm, what is the telescope's magnifying power? **For more help, refer to the Math Skill Handbook.**

Activity

Building a Reflecting Telescope

Nearly four hundred years ago, scientist Galileo Galilei saw what no human had ever seen before. Using the telescope he built, Galileo discovered moons revolving around Jupiter, observed craters on the Moon in detail, and saw sunspots on the surface of the Sun. What was it like to make these discoveries? You will find out as you make your own reflecting telescope.

What You'll Investigate
How do you construct a reflecting telescope?

Materials
flat mirror
shaving or cosmetic mirror (a curved, concave mirror)
magnifying lenses of different magnifications (3–4)

Goals
- **Construct** a reflecting telescope.
- **Observe** magnified images using the telescope and different magnifying lenses.

Safety Precautions
WARNING: *Never observe the Sun directly or with mirrors.*

Procedure
1. Position the cosmetic mirror so that you can see the reflection of the object you want to look at. Choose an object such as the Moon, a planet, or an artificial light source.
2. Place the flat mirror so that it is facing the cosmetic mirror.
3. Adjust the position of the flat mirror until you can see the reflection of the object in it.
4. View the image of the object in the flat mirror with one of your magnifying glasses. Observe how the lens magnifies the image.
5. Use your other magnifying lenses to view the image of the object in the flat mirror. Observe how the different lenses change the image of the object.

Conclude and Apply
1. **Describe** how the image of the object changed when you used different magnifying lenses.
2. **Identify** the part or parts of your telescope that reflected the light of the image.
3. **Identify** the part or parts of your telescope that magnified the image of the object.
4. **Explain** how the three parts of your telescope worked to reflect and magnify the light of the object.
5. **Infer** how the materials you used would have differed if you had constructed a refracting telescope instead of a reflecting telescope.
6. **Determine** whether you should discard the materials used in this activity or save them for future uses.

SECTION 2

Early Space Missions

The First Missions into Space

You're offered a choice—front-row-center seats for this weekend's rock concert, or a copy of the video when it's released. Wouldn't you rather be right next to the action? Astronomers feel the same way about space. Even though telescopes have taught them a great deal about the Moon and planets, they want to learn more by going to those places or by sending spacecraft where humans can't go.

Rockets The space program would not have gotten far off the ground using ordinary airplane engines. To break free of gravity and enter Earth's orbit, spacecraft must travel at speeds greater than 11 km/s. The space shuttle and several other spacecrafts are equipped with special engines that carry their own fuel. **Rockets,** like the one in **Figure 6,** are engines that have everything they need for the burning of fuel. They don't even require air to carry out the process. Therefore, they can work in space, which has no air. The simplest rocket engine is made of a burning chamber and a nozzle. More complex rockets have more than one burning chamber.

Rocket Types Two types of rockets are distinguished by the type of fuel they use. One type is the liquid-propellant rocket and the other is the solid-propellant rocket. Solid-propellant rockets are generally simpler but they can't be shut down after they are ignited. Liquid-propellant rockets can be shut down after they are ignited and can be restarted. The space shuttle uses liquid-propellant rockets and solid-propellant rockets.

As You Read

What You'll Learn
- **Compare and contrast** natural and artificial satellites.
- **Identify** the differences between artificial satellites and space probes.
- **Explain** the history of the race to the Moon.

Vocabulary
rocket
satellite
orbit
space probe
Project Mercury
Project Gemini
Project Apollo

Why It's Important
Early missions that sent objects and people into space began a new era of human exploration.

Figure 6
Rockets differ according to the types of fuel used to launch them. This rocket uses liquid oxygen for fuel.

SECTION 2 Early Space Missions **327**

Figure 7
In this view of the shuttle, a red-colored external liquid fuel tank is behind a white, solid rocket booster.

Rocket Launching Solid-propellant rockets use a powdery or rubberlike fuel and a liquid such as liquid oxygen. The burning chamber of a rocket is a tube that has a nozzle at one end. As the solid propellant burns, hot gases exert pressure on all inner surfaces of the tube. The tube pushes back on the gas except at the nozzle where hot gases escape. Thrust builds up and pushes the rocket forward.

Liquid-propellant rockets use a liquid fuel and, commonly, liquid oxygen, stored in separate tanks. To ignite the rocket, the liquid oxygen is mixed with the liquid fuel in the burning chamber. As the mixture burns, forces are exerted and the rocket is propelled forward. **Figure 7** shows the space shuttle, with both types of rockets, being launched.

Math Skills Activity

Using a Grid to Draw

Points are defined by two coordinates, called an ordered pair. To plot an ordered pair, find the first number on the horizontal x-axis and the second on the vertical y-axis. The point is placed where these two coordinates intersect. Line segments are drawn to connect points.

Example Problem
Using an x-y grid and point coordinates, draw a symmetrical house.

Solution

1. On a piece of graph paper, label and number the x-axis 0 to 6 and the y-axis 0 to 6, as shown here.

2. Plot the following points and connect them with straight line segments, as shown here. (1,1), (5,1), (5,4), (3,6), (1,4).

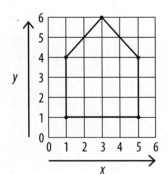

Section	Points
1	(1,−8) (3,−13) (6,−21) (9,−21) (9,−17) (8,−15) (8,−12) (6,−8) (5,−4) (4,−3) (4,−1) (5,1) (6,3) (8,3) (9,4) (9,7) (7,11) (4,14) (4,22) (−9,22) (−9,10) (−10,5) (−11,−1) (−11,−7) (−9,−8) (−8,−7) (−8,−1) (−6,3) (−6,−3) (−6,−9) (−7,−20) (−8,−21) (−4,−21) (−4,−18) (−3,−14) (−1,−8)
2	(0,11) (2,13) (2,17) (0,19) (−4,19) (−6,17) (−6,13) (−4,11)
3	(−4,9) (1,9) (1,5) (−1,5) (−2,6) (−4,6)

Practice Problem

Label and number the x-axis −12 to 10 and the y-axis −22 to 23. Draw an astronaut by plotting and connecting the points in each section. Do not draw segments to connect points in different sections.

For more help, refer to the **Math Skill Handbook.**

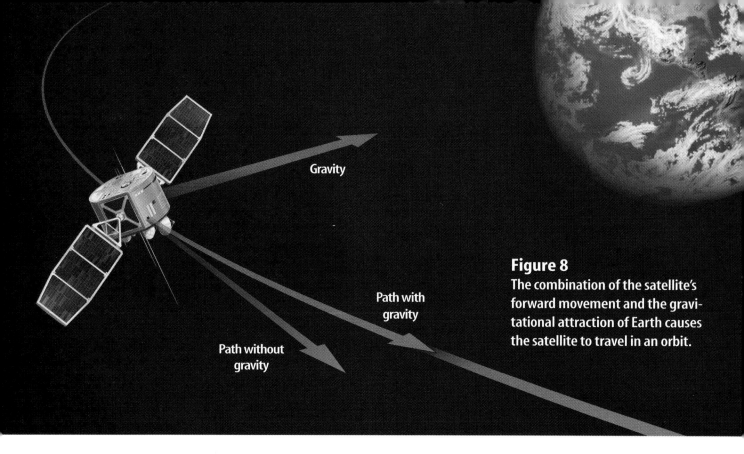

Figure 8
The combination of the satellite's forward movement and the gravitational attraction of Earth causes the satellite to travel in an orbit.

Satellites The space age began in 1957 when the former Soviet Union used a rocket to send *Sputnik I* into space. *Sputnik I* was the first artificial satellite. A **satellite** is any object that revolves around another object. When an object enters space, it travels in a straight line unless a force, such as gravity, makes it turn. Earth's gravity pulls a satellite toward Earth. The result of the satellite traveling forward while at the same time being pulled toward Earth is a curved path, called an **orbit,** around Earth. This is shown in **Figure 8.** *Sputnik I* orbited Earth for 57 days before gravity pulled it back into the atmosphere, where it burned up.

Figure 9
Data obtained from the satellite *Terra,* launched in 1999, illustrates the use of space technology to study Earth. This false-color image includes data on spring growth, sea-surface temperature, carbon monoxide concentrations, and reflected sunlight, among others.

Satellite Uses *Sputnik I* was an experiment to show that artificial satellites could be made and placed into orbit around Earth.

Today, thousands of artificial satellites orbit Earth. Communication satellites transmit radio and television programs to locations around the world. Other satellites gather scientific data, like those shown in **Figure 9,** which can't be obtained from Earth, and weather satellites constantly monitor Earth's global weather patterns.

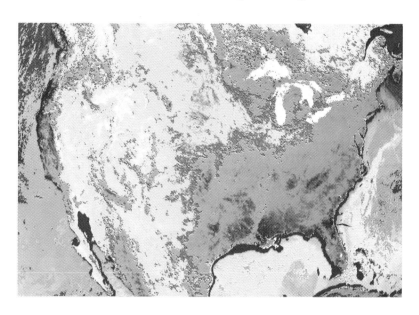

SECTION 2 Early Space Missions

Chemistry INTEGRATION

The *Viking* landers collected samples of the soil on Mars. Tests concluded that the soil on Mars contains iron. Based on this information, write a paragraph in your Science Journal about why Mars is called the red planet.

Space Probes

Not all objects carried into space by rockets become satellites. Rockets also can be used to send instruments into space to collect data. A **space probe** is an instrument that gathers information and sends it back to Earth. Unlike satellites that orbit Earth, space probes travel far into the solar system as illustrated in **Figure 10.** Some even have traveled out of the solar system. Space probes, like many satellites, carry cameras and other data-gathering equipment, as well as radio transmitters and receivers that allow them to communicate with scientists on Earth. **Table 1** shows some of the early space probes launched by the National Aeronautics and Space Administration (NASA).

Table 1 Some Early Space Missions

Mission Name		Date Launched	Destination	Data Obtained
Mariner 2		August 1962	Venus	verified high temperatures in Venus's atmosphere
Pioneer 10		March 1972	Jupiter	sent back photos of Jupiter—first probe to encounter an outer planet
Viking 1		August 1975	Mars	orbiter mapped the surface of Mars; lander searched for life on Mars
Magellan		May 1989	Venus	mapped Venus's surface and returned data on the composition of Venus's atmosphere

CHAPTER 11 Space Technology

NATIONAL GEOGRAPHIC VISUALIZING SPACE PROBES

Figure 10

Probes have taught us much about the solar system. As they travel through space, these car-size craft gather data with their onboard instruments and send results back to Earth via radio waves. Some data collected during these missions are made into pictures, a selection of which is shown here.

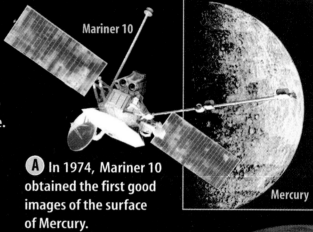

A In 1974, Mariner 10 obtained the first good images of the surface of Mercury.

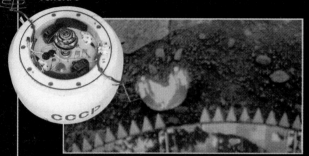

B A Soviet Venera probe took this picture of the surface of Venus on March 1, 1982. Parts of the spacecraft's landing gear are visible at the bottom of the photograph.

D In 1990, Magellan imaged craters, lava domes, and great rifts, or cracks, on the surface of Venus.

C The Voyager 2 mission included flybys of the outer planets Jupiter, Saturn, Uranus, and Neptune. Voyager took this photograph of Neptune in 1989 as the craft sped toward the edge of the solar system.

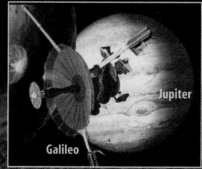

E NASA's veteran space traveler Galileo nears Jupiter in this artist's drawing. The craft arrived at Jupiter in 1995 and sent back data, including images of Europa, one of Jupiter's 16 moons, seen below in a color-enhanced view.

Science Online

Collect Data Visit the Glencoe Science Web site at **tx.science.glencoe.com** to get the latest information on Galileo's discoveries. Record your information in your Science Journal.

Figure 11
Future missions will be needed to determine whether life exists on Europa.

Voyager and Pioneer Probes Space probes *Voyager 1* and *Voyager 2* were launched in 1977 and now are heading toward deep space. *Voyager 1* flew past Jupiter and Saturn. *Voyager 2* flew past Jupiter, Saturn, Uranus, and Neptune. These probes now are exploring beyond the solar system as part of the Voyager Interstellar Mission. Scientists expect these probes to continue to transmit data to Earth for at least 20 more years.

Pioneer 10, launched in 1972, was the first probe to survive a trip through the asteroid belt and encounter an outer planet, Jupiter. As of 2000, *Pioneer 10* is more than 11 billion km from Earth, and will continue beyond the solar system. The probe carries a gold medallion with an engraving of a man, a woman, and Earth's position in the galaxy.

Galileo Launched in 1989, *Galileo* reached Jupiter in 1995. In July 1995, *Galileo* released a smaller probe that began a five-month approach to Jupiter. The small probe took a parachute ride through Jupiter's violent atmosphere in December 1995.

Before being crushed by the atmospheric pressure, it transmitted information about Jupiter's composition, temperature, and pressure to the satellite orbiting above. *Galileo* studied Jupiter's moons, rings, and magnetic fields and then relayed this information to scientists who were waiting eagerly for it on Earth.

Studies of Jupiter's moon Europa by *Galileo* indicate that an ocean of water may exist under the surface of Europa. A cracked outer layer of ice makes up Europa's surface, shown in **Figure 11**. The cracks in the surface may be caused by geologic activity that heats the ocean underneath the surface. Sunlight penetrates these cracks, further heating the ocean and setting the stage for the possible existence of life on Europa. *Galileo* ended its study of Europa in 2000. More advanced probes will be needed to determine whether life exists on this icy moon.

Reading Check *What features on Europa suggest the possibility of life existing on this moon?*

In October and November of 1999, *Galileo* approached Io, another one of Jupiter's moons. It came within 300 km and took photographs of a volcanic vent named Loki, which emits more energy than all of Earth's volcanoes combined. *Galileo* also discovered eruption plumes that shoot gas made of sulfur and oxygen.

Moon Quest

Throughout the world, people were shocked when they turned on their radios and television sets in 1957 and heard the radio transmissions from *Sputnik I* as it orbited Earth. All that *Sputnik I* transmitted was a sort of beeping sound, but people quickly realized that launching a human into space wasn't far off.

In 1961, Soviet cosmonaut Yuri A. Gagarin became the first human in space. He orbited Earth and returned safely. Soon, President John F. Kennedy called for the United States to send humans to the Moon and return them safely to Earth. His goal was to achieve this by the end of the 1960s. The race for space was underway.

The U.S. program to reach the Moon began with **Project Mercury.** The goals of Project Mercury were to orbit a piloted spacecraft around Earth and to bring it back safely. The program provided data and experience in the basics of space flight. On May 5, 1961, Alan B. Shepard became the first U.S. citizen in space. In 1962, *Mercury* astronaut John Glenn became the first U.S. citizen to orbit Earth. **Figure 12** shows Glenn preparing for liftoff.

Figure 12
An important step in the attempt to reach the Moon was John Glenn's first orbit around Earth.

✔ **Reading Check** *What were the goals of Project Mercury?*

Project Gemini The next step in reaching the Moon was called **Project Gemini.** Teams of two astronauts in the same *Gemini* spacecraft orbited Earth. One *Gemini* team met and connected with another spacecraft in orbit—a skill that would be needed on a voyage to the Moon.

The *Gemini* spacecraft was much like the *Mercury* spacecraft, except it was larger and easier for the astronauts to maintain. It was launched by a rocket known as a *Titan II*, which was a liquid fuel rocket.

In addition to connecting spacecraft in orbit, another goal of Project *Gemini* was to investigate the effects of space travel on the human body.

Along with the Mercury and Gemini programs, a series of robotic probes was sent to the Moon. *Ranger* proved that a spacecraft could be sent to the Moon. In 1966, *Surveyor* landed gently on the Moon's surface, indicating that the Moon's surface could support spacecraft and humans. The mission of *Lunar Orbiter* was to take pictures of the Moon's surface that would help determine the best future lunar landing sites.

Modeling a Satellite

WARNING: *Stand a safe distance away from classmates. Use heavy string.*

Procedure
1. Tie one end of a 2-m-long **string** to a **rubber stopper.**
2. Thread the string through a 15-cm piece of **hose.**
3. Tie the other end of the string securely to several large **steel nuts.**
4. Swing the rubber stopper in a circle above your head. Swing the stopper at different speeds.

Analysis
Based upon your observations, explain how a satellite stays in orbit above Earth.

Figure 13
The Lunar Rover vehicle was first used during the *Apollo 15* mission. Riding in the moon buggy, *Apollo 15, 16,* and *17* astronauts explored large areas of the lunar surface.

Project Apollo The final stage of the U.S. program to reach the Moon was **Project Apollo.** On July 20, 1969, *Apollo 11* landed on the Moon's surface. Neil Armstrong was the first human to set foot on the Moon. His first words as he stepped onto its surface were, "That's one small step for man, one giant leap for mankind." Edwin Aldrin, the second of the three *Apollo 11* astronauts, joined Armstrong on the Moon, and they explored its surface for two hours. While they were exploring, Michael Collins remained in the Command Module; Armstrong and Aldrin then returned to the Command Module before beginning the journey home. A total of six lunar landings brought back more than 2,000 samples of moon rock and soil for study before the program ended in 1972. **Figure 13** shows an astronaut exploring the Moon's surface from the Lunar Rover vehicle.

Sharing Knowledge During the past three decades, most missions in space have been carried out by individual countries, often competing to be the first or the best. Today, countries of the world cooperate more and work together, sharing what each has learned. Projects are being planned for cooperative missions to Mars and elsewhere. As you read the next section, you'll see how the U.S. program has progressed since the days of Project Apollo and what may be planned for the future.

Section 2 Assessment

1. Explain why Neptune has eight satellites even though it is not orbited by human-made objects.
2. *Galileo* was considered a space probe as it traveled to Jupiter. Once there, however, it became an artificial satellite. Explain.
3. List several discoveries made by the *Voyager 1* and *Voyager 2* space probes.
4. Draw a time line beginning with *Sputnik* and ending with Project Apollo. Include descriptions of important missions.
5. **Think Critically** Is Earth a satellite of any other body in space? Explain.

Skill Builder Activities

6. **Using a Word Processor** Use a computer (home, library, computer lab) to describe different types of transportation needed for space travel. Include space probes, rockets, crewed capsules, and rovers in your descriptions. **For more help, refer to the Technology Skill Handbook.**
7. **Solving One-Step Equations** Suppose a spacecraft were launched at a speed of 40,200 km/h. Express this speed in kilometers per second. **For more help, refer to the Math Skill Handbook.**

SECTION 3
Current and Future Space Missions

The Space Shuttle

Imagine spending millions of dollars to build a machine, sending it off into space, and watching its 3,000 metric tons of metal and other materials burn up after only a few minutes of work. That's exactly what NASA did with the rocket portions of spacecraft for many years. The early rockets were used only to launch a small capsule holding astronauts into orbit. Then sections of the rocket separated from the rest and burned when reentering the atmosphere.

A Reusable Spacecraft NASA administrators, like many others, realized that it would be less expensive and less wasteful to reuse resources. The reusable spacecraft that transports astronauts, satellites, and other materials to and from space is called the **space shuttle,** shown in **Figure 14,** as it is landing.

At launch, the space shuttle stands on end and is connected to an external liquid-fuel tank and two solid-fuel booster rockets. When the shuttle reaches an altitude of about 45 km, the emptied, solid-fuel booster rockets drop off and parachute back to Earth. These are recovered and used again. The external liquid-fuel tank separates and falls back to Earth, but it isn't recovered.

Work on the Shuttle After the space shuttle reaches space, it begins to orbit Earth. There, astronauts perform many different tasks. In the cargo bay, astronauts can conduct scientific experiments and determine the effects of spaceflight on the human body. When the cargo bay isn't used as a laboratory, the shuttle can launch, repair, and retrieve satellites. Then the satellites can be returned to Earth or repaired onboard and returned to space. After a mission, the shuttle glides back to Earth and lands like an airplane. A large landing field is needed as the gliding speed of the shuttle is 335 km/h.

As You Read

What You'll Learn
- **Explain** the benefits of the space shuttle.
- **Identify** the usefulness of orbital space stations.
- **Explore** future space missions.

Vocabulary
space shuttle
space station

Why It's Important
Many future space missions have planned experiments that may benefit you.

Figure 14
The space shuttle is designed to make many trips into space.

Figure 15
Astronauts performed a variety of tasks while living and working in space onboard *Skylab*.

Figure 16
Russian and American scientists have worked together to further space exploration.

Space Stations

Astronauts can spend only a short time living in the space shuttle. Its living area is small, and the crew needs more room to live, exercise, and work. A **space station** has living quarters, work and exercise areas, and all the equipment and support systems needed for humans to live and work in space.

In 1973, the United States launched the space station *Skylab,* shown in **Figure 15.** Crews of astronauts spent up to 84 days there, performing experiments and collecting data on the effects on humans of living in space. In 1979, the abandoned *Skylab* fell out of orbit and burned up as it entered Earth's atmosphere.

Crews from the former Soviet Union have spent more time in space, onboard the space station *Mir,* than crews from any other country. Cosmonaut Dr. Valery Polyakov returned to Earth after 438 days in space studying the long-term effects of weightlessness.

Cooperation in Space

In 1995, the United States and Russia began an era of cooperation and trust in exploring space. Early in the year, American Dr. Norman Thagard was launched into orbit aboard the Russian *Soyuz* spacecraft, along with two Russian cosmonaut crewmates. Dr. Thagard was the first U.S. astronaut launched into space by a Russian booster and the first American resident of the Russian space station *Mir.*

In June 1995, Russian cosmonauts rode into orbit onboard the space shuttle *Atlantis,* America's 100th crewed launch. The mission of *Atlantis* involved, among other studies, a rendezvous and docking with the space station *Mir.* The cooperation that existed on this mission, as shown in **Figure 16,** continued through eight more space shuttle-*Mir* docking missions. Each of the eight missions was an important step toward building and operating the *International Space Station.* In 2001, the abandoned *Mir* space station fell out of orbit and burned up upon reentering the atmosphere. Cooperation continued as the *International Space Station* began to take form.

The International Space Station The *International Space Station (ISS)* will be a permanent laboratory designed for long-term research projects. Diverse topics will be studied, including research on the growth of protein crystals. This particular project will help scientists determine protein structure and function, which is expected to enhance work on drug design and the treatment of many diseases.

The *ISS* will draw on the resources of 16 nations. These nations will build units for the space station, which then will be transported into space onboard the space shuttle and Russian launch rockets. The station will be constructed in space. **Figure 17** shows what the completed station will look like.

Figure 17
This is a picture of what the proposed *International Space Station* will look like when it is completed in 2005.

 *What is the purpose of the* **International Space Station?**

Phases of *ISS* NASA is planning the *ISS* program in phases. Phase One, now concluded, involved the space shuttle-*Mir* docking missions. Phase Two began in 1998 with the launch of the Russian-built *Zarya Module,* also known as the Functional Cargo Block, and will end with the delivery of a U.S. laboratory onboard the space shuttle. The first assembly of *ISS* occurred in December of 1998 when a space shuttle mission attached the Unity module to *Zarya*. During Phase Two, crews of three people were delivered to the space station.

Living in Space The project will continue with Phase Three when the Japanese Experiment Module, the European Columbus Orbiting Facility, and another Russian lab will be delivered.

It is hoped that the *International Space Station* will be completed in 2005. A three- or four-person crew then should be able to work comfortably onboard the station. A total of 47 separate launches will be required to take all the components of the *ISS* into space and prepare it for permanent habitation. NASA plans for crews of astronauts to stay onboard the station for several months at a time. NASA already has conducted numerous tests to prepare crews of astronauts for extended space missions. One day, the station could be a construction site for ships that will travel to the Moon and Mars.

Research Visit the Glencoe Science Web site at **tx.science.glencoe.com** for more information on the *International Space Station.* Share your information with the class.

SECTION 3 Current and Future Space Missions **337**

Figure 18
Gulleys, channels, and aprons of sediment imaged by the *Mars Global Surveyor* are similar to features on Earth known to be caused by flowing water. This water is thought to seep out from beneath the surface of Mars.

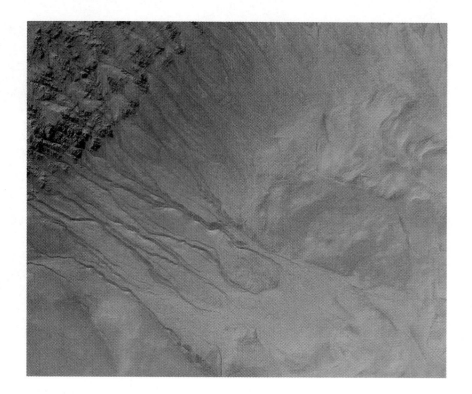

Exploring Mars

Two of the most successful missions in recent years were the 1996 launchings of the *Mars Global Surveyor* and the *Mars Pathfinder*. *Surveyor* orbited Mars, taking high-quality photos of the planet's surface as shown in **Figure 18**. *Pathfinder* descended to the Martian surface, using rockets and a parachute system to slow its descent. Large balloons absorbed the shock of landing. *Pathfinder* carried technology to study the surface of the planet, including a remote-controlled robot rover called Sojourner. Using information gathered by studying photographs taken by *Surveyor*, scientists determined that water recently had seeped to the surface of Mars in some areas.

✔ **Reading Check** *What type of data were obtained by the* Mars Global Surveyor?

Although the *Mars Global Surveyor* and the *Mars Pathfinder* missions were successful, not all the missions to Mars have met with the same success. The *Mars Climate Orbiter*, launched in 1998, was lost in September of 1999. An incorrect calculation of the force that the thrusters were to exert caused the spacecraft to be lost. Engineers had used English units instead of metric units. Then, in December of 1999, the *Mars Polar Lander* was lost just as it was making its descent to the planet. This time, it is believed that the spacecraft thought it had landed and shut off its thrusters too soon. NASA tried to make contact with the lander but never had any success.

New Millennium Program

To continue space missions into the future, NASA has created the New Millennium Program (NMP). The goal of the NMP is to develop advanced technology that will let NASA send smart spacecraft into the solar system. This will reduce the amount of ground control needed. They also hope to reduce the size of future spacecraft to keep the cost of launching them under control. NASA's challenge is to prove that certain cutting-edge technologies, as well as mission concepts, work in space.

Exploring the Moon

Does water exist in the craters of the Moon's poles? This is one question NASA intends to explore with data gathered from the *Lunar Prospector* spacecraft shown in **Figure 19.** Launched in 1998, the *Lunar Prospector's* one-year mission was to orbit the Moon, mapping its structure and composition. Early data obtained from the spacecraft indicate that hydrogen might be present in the rocks of the Moon's poles. Hydrogen is one of the elements found in water. Scientists now hypothesize that ice on the floors of the Moon's polar craters may be the source of this hydrogen. Ice might survive indefinitely at the bottom of these craters because it would always be shaded from the Sun.

Other things could account for the presence of hydrogen. It could be from solar wind or certain minerals. The *Lunar Prospector* was directed to crash into a crater at the Moon's south pole when its mission ended in July 1999. Scientists hoped that any water vapor thrown up by the collision could be detected using special telescopes. However, it didn't work. Further studies are needed to determine if water exists on the Moon.

Data Update For an online update on the New Millenium Program, visit the Glencoe Science Web site at **tx.science.glencoe.com**

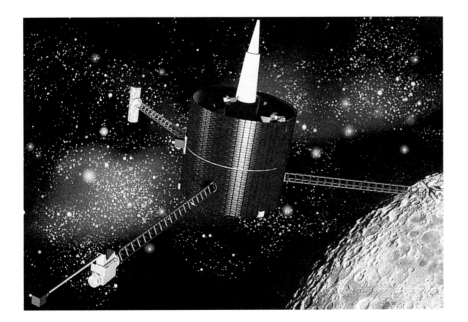

Figure 19
The *Lunar Prospector* analyzed the Moon's composition during its one-year mission.

Cassini

In October 1997, NASA launched the space probe *Cassini*. This probe's destination is Saturn. *Cassini*, shown in **Figure 20,** will not reach its goal until 2004. At that time, the space probe will explore Saturn and surrounding areas for four years. One part of its mission is to deliver the European Space Agency's *Huygens* probe to Saturn's largest moon, Titan. Some scientists theorize that Titan's atmosphere may be similar to the atmosphere of early Earth.

Figure 20
Cassini is currently on its way to Saturn. After it arrives, it will spend four years studying Saturn and its surrounding area.

The Next Generation Space Telescope Not all space missions involve sending astronauts or probes into space. Plans are being made to launch a new space telescope that is capable of observing the first stars and galaxies in the universe. The *Next Generation Space Telescope*, shown in **Figure 21,** will be the successor to the *Hubble Space Telescope*. As part of the Origins project, it will provide scientists with the opportunity to study the evolution of galaxies, the production of elements by stars, and the process of star and planet formation. To accomplish these tasks, the telescope will have to be able to see objects 400 times fainter than those currently studied with ground-based telescopes such as the twin Keck telescopes. NASA hopes to launch the *Next Generation Space Telescope* as early as 2009.

Figure 21
The *Next Generation Space Telescope* will attempt to observe stars and galaxies that formed early in the history of the universe.

Everyday Space Technology Items developed for space exploration don't always stay in space. In fact, many of today's cutting-edge technologies are modifications of research or technology used in the space program. For example, NASA space suit technology, shown in **Figure 22,** was used to give a child with a skin disorder the opportunity to play outside. Without the suit, the child could have been seriously hurt by the Sun's rays.

Space technology also has been used to understand, diagnose, and treat heart disease. Programmable pacemakers, developed through space technology, have given doctors more programming possibilities and more detailed information about their patients' health.

Other advances include ribbed swimsuits that reduce water resistance. Also, badges have been designed that warn workers of toxic chemicals in the air by turning color when the wearer is exposed to a particular chemical. Jet engines capable of much higher speeds than current jet engines are being developed as well.

A new technology that may prevent many accidents also has been developed. Equipment on emergency vehicles causes traffic lights to turn yellow and then red for other vehicles approaching the same intersections. The equipment activates the traffic lights when fast-moving emergency vehicles come close to such an intersection, preventing crashes.

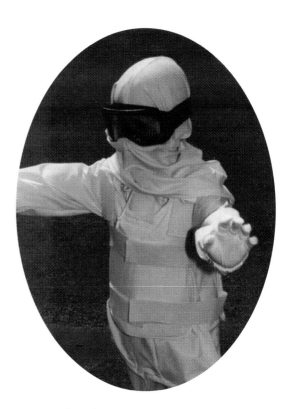

Figure 22
Space technology has helped children go places and do things that they otherwise wouldn't be able to do.

Section 3 Assessment

1. What is the main advantage of the space shuttle?
2. Why were the space shuttle-*Mir* docking missions so important?
3. What is the *International Space Station* used for? Describe how the *ISS* could help future space missions.
4. Describe Phase Three of the *International Space Station* program.
5. **Think Critically** What makes the space shuttle more versatile than earlier spacecraft?

Skill Builder Activities

6. **Making and Using Tables** Make a table of the discoveries from missions to the Moon and Mars. **For more help, refer to the Science Skill Handbook.**
7. **Communicating** Suppose you're in charge of assembling a crew for a new space station. Select 50 people to do a variety of jobs, such as farming, maintenance, scientific experimentation, and so on. In your Science Journal, list and explain your choices. **For more help, refer to the Science Skill Handbook.**

ACTIVITY: Use the Internet

Star Sightings

For thousands of years, humans have used the stars to learn about Earth. From star sightings, you can map the change of seasons, navigate the oceans, and even determine the size of Earth.

Polaris, or the North Star, has occupied an important place in human history. The location of Polaris is not affected by Earth's rotation. At any given observation point, it always appears at the same angle above the horizon. At Earth's north pole, Polaris appears directly overhead. At the equator, it is just above the northern horizon. Polaris provides a standard from which other locations can be measured. Such star sightings can be made using the astrolabe, an instrument used to measure the height of a star above the horizon.

Recognize the Problem

How can you determine the size of Earth?

Form a Hypothesis

Think about what you have learned about sightings of Polaris. How does this tell you that Earth is round? Knowing that Earth is round, form a hypothesis about how you can estimate the circumference of Earth based on star sightings.

Goals
- **Record** your sightings of Polaris.
- **Share** the data with other students to calculate the circumference of Earth.

Safety Precautions
WARNING: *Do not use the astrolabe during the daytime to observe the Sun.*

Data Sources
SCIENCE*Online* Go to the Glencoe Science Web site at **tx.science.glencoe.com** to obtain instructions on how to make an astrolabe. Also visit the Web site for more information about the location of Polaris, and for data from other students.

342 CHAPTER 11 Space Technology

Using Scientific Methods

Test Your Hypothesis

Plan

1. Obtain an astrolabe or construct one using the instructions posted on the Glencoe Science Web site.
2. **Design** a data table in your Science Journal similar to the one below.

Polaris Observations		
Your Location:		
Date	Time	Astrolabe Reading

3. Decide as a group how you will make your observations. Does it take more than one person to make each observation? When will it be easiest to see Polaris?

Do

1. Make sure your teacher approves your plan before you start.
2. Carry out your observations.
3. **Record** your observations in your data table.
4. Average your readings and post them in the table provided on the Glencoe Science Web site.

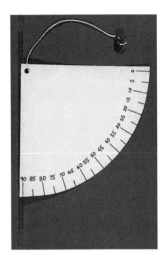

Analyze Your Data

1. **Research** the names of cities that are at approximately the same longitude as your hometown. Gather astrolabe readings at the Glencoe Science Web site from students in one of those cities.
2. **Compare** your astrolabe readings. Subtract the smaller reading from the larger one.
3. Determine the distance between your star sighting location and the other city.
4. **Calculate** the circumference of Earth using the following relationship.
   Circumference = (360°) × (distance between locations)/ difference between readings

Draw Conclusions

1. How does the circumference of Earth that you calculated compare with the accepted value of 40,079 km?
2. What are some possible sources of error in this method of determining the size of Earth? What improvements would you suggest?

Communicating Your Data

 Find this *Use the Internet* activity on the Glencoe Science Web site at **tx.science.glencoe.com** Create a poster that includes a table of your data and data from students in other cities. Perform a sample circumference calculation for your class.

TIME SCIENCE AND Society
SCIENCE ISSUES THAT AFFECT YOU!

Cities in Space

Should the U.S. spend money to colonize space?

Humans have landed on the Moon, and spacecrafts have landed on Mars. But these space missions are just small steps that may lead to a giant new space program. As technology improves, humans may be able to visit and even live on other planets. The twenty-first century may turn science fiction into science fact. But is it worth the time and money involved?

Those in favor of living in space point to the International Space Station that already is orbiting Earth. It's an early step toward establishing floating cities where astronauts can live and work. The 94 billion dollar station may pave the way for "ordinary people" to live in space, too. As Earth's population continues to increase and there is less room on this planet, why not create ideal cities on another planet or a floating city in space? That reason, combined with the fact that there is little pollution in space, makes the idea appealing to many.

Critics of colonizing space think we should spend the hundreds of billions of dollars that it would cost to colonize space on projects to help improve people's lives here on Earth. Building better housing, developing ways to feed the hungry, finding cures for diseases, and increasing funds for education should come first, these people say. And, critics continue, if people want to explore, why not explore right here on Earth? "The ocean floor is Earth's last frontier," says one person. "Why not explore that?"

If humans were to move permanently to space, the two most likely destinations would be Mars and the Moon, both bleak places.

But those in favor of moving to these places say humans could find a way to make them livable. They argue that humans have made homes in harsh climates and in rugged areas, and people can meet the challenges of living in space.

Choosing Mars

Mars may be the best place to live. Photos suggest that the planet once had liquid water on its surface. If that water is now frozen underground, humans may someday be able to tap into it.

NASA is studying whether it makes sense to send astronauts and scientists to explore Mars. An international team would live there for about 500 days, collecting and studying soil and rock samples for clues as to whether Mars is a planet that could be settled. NASA says this journey could begin as early as 2009.

But a longer-range dream to transform Mars into an Earthlike place with breathable air and usable water is just that—a dream. Some small steps are being taken to make that dream more realistic. Experimental plants are being developed that could absorb Mars' excess carbon dioxide and release oxygen. Solar mirrors, already available, could warm Mars' surface.

Those for and against colonizing space agree on one thing—setting up colonies on Mars or the Moon will take large amounts of money, research, and planning. It also will take the same spirit of adventure that has led history's pioneers into so many bold frontiers—deserts, the Poles, and the sky.

Is the International Space Station a small step toward colonizing space?

An early Mars colony might look something like this. Settlers would live in air-filled domes and even grow crops.

CONNECTIONS Debate Research further information about colonizing space. Make a list of the pros and cons for colonizing space. Do you think the United States should spend money to create space cities or use it now to improve lives of people on Earth? Debate with your class.

science *Online*
For more information, visit tx.science.glencoe.com

Chapter 11 Study Guide

Reviewing Main Ideas

Section 1   Radiation from Space

1. The arrangement of electromagnetic waves according to their wavelengths is the electromagnetic spectrum.

2. Optical telescopes produce magnified images of objects. *What does this reflecting telescope use to focus light that produces an image?*

3. Radio telescopes collect and record radio waves given off by some space objects.

Section 2   Early Space Missions

1. A satellite is an object that revolves around another object. The moons of planets are natural satellites. Artificial satellites are those made by people.

2. A space probe travels into the solar system, gathers data, and sends them back to Earth. *How far can space probes, like the one pictured here, travel?*

3. Early American piloted space programs included the Gemini and Apollo Projects.

Section 3   Recent and Future Space Missions

1. Space stations provide the opportunity to conduct research not possible on Earth. The *International Space Station* is being constructed in space with the cooperation of more than a dozen nations.

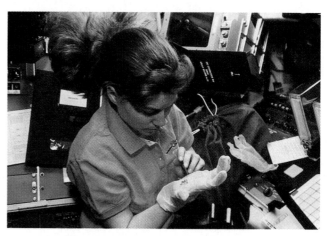

2. The space shuttle is a reusable spacecraft that carries astronauts, satellites, and other cargo to and from space. *What special obstacles must astronauts overcome when they conduct research in space?*

3. Space technology is used to solve problems on Earth not related to space travel. Advances in engineering related to space travel have led to problem solving in medicine and environmental sciences, among other fields.

After You Read

Use what you've learned to predict the future of space exploration. Record your predictions under the Future tab of your Foldable.

346 CHAPTER STUDY GUIDE

Chapter 11 Study Guide

Visualizing Main Ideas

Complete the following concept map about the race to the Moon. Use the following phrases: first satellite, Project Gemini, Project Mercury, team of two astronauts orbits Earth, Project Apollo.

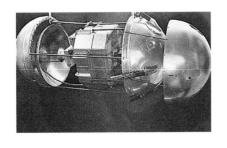

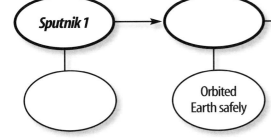
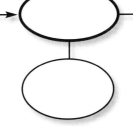

Sputnik 1 → ◯ → Orbited Earth safely → ◯ → First human on the Moon

Vocabulary Review

Vocabulary Words

a. electromagnetic spectrum
b. observatory
c. orbit
d. Project Apollo
e. Project Gemini
f. Project Mercury
g. radio telescope
h. reflecting telescope
i. refracting telescope
j. rocket
k. satellite
l. space probe
m. space shuttle
n. space station

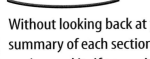 **Study Tip**

Without looking back at your textbook, write a summary of each section of a chapter after you've read it. If you write it in your own words, you will remember it better.

Using Vocabulary

Each of the following sentences is false. Make each sentence true by replacing the underlined word(s) with the correct vocabulary word(s).

1. A <u>radio</u> telescope uses lenses to bend light.

2. A <u>space probe</u> is an object that revolves around another object in space.

3. <u>Project Apollo</u> was the first piloted U.S. space program.

4. A <u>satellite</u> carries people and tools to and from space.

5. In the <u>space station</u>, electromagnetic waves are arranged according to their wavelengths.

Chapter 11 Assessment & TEKS Review

Checking Concepts

Choose the word or phrase that best answers the question.

1. Which spacecraft has sent images of Venus to scientists on Earth?
   A) *Voyager* C) *Apollo 11*
   B) *Viking*   D) *Magellan*

2. Which kind of telescope uses mirrors to collect light?
   A) radio   C) refracting
   B) electromagnetic D) reflecting

3. What was *Sputnik I*?
   A) the first telescope
   B) the first artificial satellite
   C) the first observatory
   D) the first U.S. space probe

4. Which kind of telescope can be used during the day or night and during bad weather?
   A) radio   C) refracting
   B) electromagnetic D) reflecting

5. When fully operational, what is the maximum number of people who will crew the *International Space Station*?
   A) 3   C) 15
   B) 7   D) 50

6. Which space mission's goal was to put a spacecraft into orbit and bring it back safely?
   A) Project Mercury C) Project Gemini
   B) Project Apollo   D) *Viking I*

7. Which of the following is a natural satellite of Earth?
   A) *Skylab*   C) the Sun
   B) the space shuttle   D) the Moon

8. What does the space shuttle use to place a satellite into space?
   A) liquid-fuel tank C) mechanical arm
   B) booster rocket   D) cargo bay

9. What was *Skylab*?
   A) a space probe   C) a space shuttle
   B) a space station D) an optical telescope

10. What part of the space shuttle is reused?
    A) liquid-fuel tanks C) booster engines
    B) *Gemini* rockets   D) Saturn rockets

Thinking Critically

11. Describe any advantages that a Moon-based telescope would have over an Earth-based telescope.

12. Would a space probe to the Sun's surface be useful? Explain.

13. Which do you think is a wiser method of exploration—space missions with people onboard or robotic space probes? Why?

14. Suppose two astronauts are outside the space shuttle orbiting Earth. The audio speaker in the helmet of one astronaut quits working. The other astronaut is 1 m away and shouts a message. Can the first astronaut hear the message? Explain.

15. Space probes have crossed Pluto's orbit, but never have visited the planet. Explain.

Developing Skills

16. **Making and Using Tables** Copy and complete the table below. Use information from several resources.

United States Space Probes		
Probe	Launch Date(s)	Planets or Objects Visited
Vikings 1 and *2*		
Galileo		
Lunar Prospector		
Pathfinder		

Chapter 11 Assessment

17. Concept Mapping Use the following phrases to complete the concept map about rocket launching: *thrust pushes rocket forward, rocket breaks free of Earth's gravity, propellant is ignited,* and *hot gases exert pressure on walls of burning chamber.*

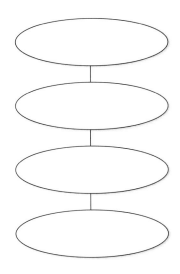

18. Classifying Classify the following as a satellite or a space probe: *Cassini, Sputnik I, Hubble Space Telescope,* space shuttle, and *Voyager 2.*

19. Comparing and Contrasting Compare and contrast space probes and satellites.

Performance Assessment

20. Poem Research a space probe launched within the last five years. Write a poem that includes its destination, the goals for its mission and something about the individuals who crewed the flight.

Technology

Go to the Glencoe Science Web site at **tx.science.glencoe.com** or use the **Glencoe Science CD-ROM** for additional chapter assessment.

TAKS Practice

Scientists use several different kinds of telescopes to make observations about space. Information about some types of telescopes is listed in the table below.
Science TEKS 6.2c; 6.4a; 6.13b

1. According to the table, a refracting telescope focuses light using a _____.
   A) convex lens
   B) mirror
   C) dish
   D) receiver

Types of Telescopes

Telescope	Use	How it Works
Optical Refracting Telescope	Produces magnified images of distant objects	Uses a convex lens to bend and focus light
Optical Reflecting Telescope	Produces magnified images of distant objects	Uses mirrors to reflect and focus light
Radio Telescope	Collects radio waves from space	Large dish reflects and focuses waves to receiver.

2. While in space, the *Hubble Space Telescope* needed its largest mirror repaired in 1993. While costly, the repair mission was a huge success. According to the chart, the *Hubble Space Telescope* is _____.
   F) an optical refracting telescope
   G) an optical reflecting telescope
   H) a radio telescope
   J) an optical radio receiver

CHAPTER 12

Science TEKS 6.5 A, B; 6.13 A, B

The Solar System and Beyond

It doesn't feel as if Earth is moving. Does Earth move through space? Does the Moon? What's out there besides Earth, the Moon, the Sun, and stars? In this chapter you will find the answers to these questions. In addition, you will learn why the Moon changes its appearance, how comets appear, and where meteorites come from. You also will read about constellations, galaxies, and the life cycles of stars.

What do you think?

Science Journal Look at the picture below with a classmate. Discuss what you think this might be or what is happening. Here's a hint: *It's a time exposure.* Write your answer or best guess in your Science Journal.

350

When you gaze at the night sky, what do you see? On a clear night, the sky is full of sparkling points of light. With the unaided eye, you can see dozens—no, hundreds—of these sparkles. How many stars are there?

Estimate grains of rice

1. Using white crayon or chalk and a ruler, draw grid lines on a sheet of black construction paper, dividing it into 5-cm squares.
2. Spill 4 g of rice grains onto the black paper.
3. Count the number of grains of rice in one square. Repeat this step with a different square. Add the number of grains of rice in the two squares, then divide this number by two to calculate the average number of grains of rice in the two squares.
4. Multiply this number by the number of squares on the paper. This will give you an estimate of the number of grains of rice on the paper.

Observe
How could scientists use this same method to estimate the number of stars in the sky? In your Science Journal, describe the process scientists might use.

Before You Read

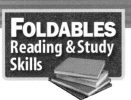

Making an Organizational Study Fold Make the following Foldable to help you organize your thoughts into clear categories about the solar system and beyond.

1. Stack six sheets of paper in front of you so the short sides are at the top.
2. Slide the top sheet up so that about four centimeters of the next sheet show. Move each sheet up so about four centimeters of the next sheet show.
3. Fold the sheets top to bottom to form 12 tabs. Staple along the top fold.
4. Label the tabs *Sun, Mercury, Venus, Earth, Mars, Jupiter, Saturn, Uranus, Neptune, Pluto, Beyond the Solar System: Stars,* and *Beyond the Solar System: Galaxies.*
5. Before you read the chapter, write what you know about each under the tabs. As you read the chapter, correct and add to what you've written.

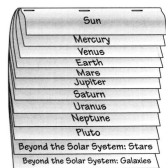

SECTION 1

Earth's Place in Space

As You Read

What You'll Learn
- **Explain** why Earth has seasons.
- **Describe** the motions that cause Moon phases.

Vocabulary
rotation
orbit
revolution
eclipse

Why It's Important
When you understand Earth's movements you'll understand night and day as well as the seasons.

Earth Moves

You wake up, stretch and yawn, then glance out your window to see the first rays of dawn. By lunchtime, the Sun is high in the sky. As you sit down to dinner in the evening, the Sun appears to sink below the horizon. Although it seems like the Sun moves across the sky, it is Earth that is moving.

Earth's Rotation Earth spins in space like a twirling figure skater. Your planet spins around an imaginary line running through its center called an axis. **Figure 1** shows how Earth spins on its axis.

The spinning of Earth on its axis is called Earth's **rotation** (roh TAY shun). Earth rotates once every 24 h. The Sun appears each morning due to Earth's rotation. Throughout the day, Earth continues to rotate and the Sun appears to move across the sky. In the evening, the Sun seems to go down because the place where you are on Earth is rotating away from the Sun.

You can see how this works by standing and facing a lamp. Pretend you are Earth and the lamp is the Sun. Now, without pivoting your head, turn around slowly in a counterclockwise direction. The lamp seems to move across your vision, then disappear. You rotate until you finally see the lamp again. The lamp didn't move—you did. When you rotated, you were like Earth rotating in space, causing different parts of the planet to face the Sun at different times. The rotation of Earth—not movement of the Sun—causes night and day.

Reading Check *Why does the Sun appear to move across the sky?*

Because the Sun only appears to move across the sky, this movement is called apparent motion. Can you think of any other objects you encounter that might display apparent motion?

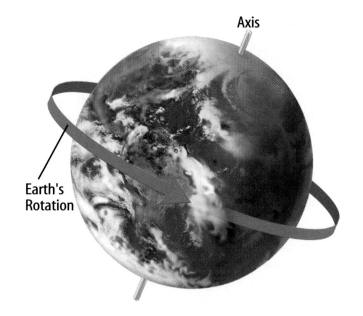

Figure 1
The rotation of Earth on its axis causes night and day.

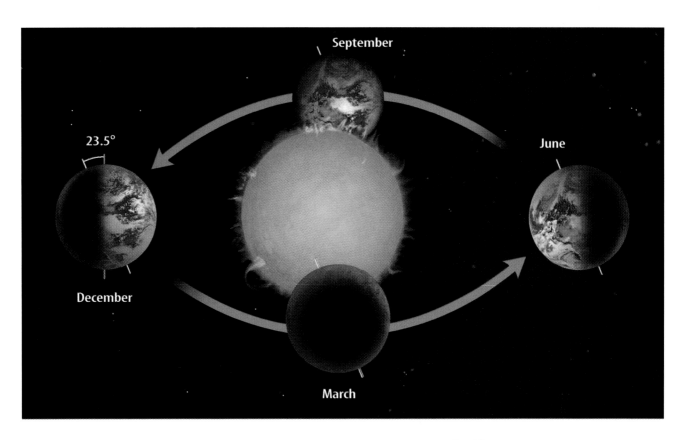

Earth's Revolution Earth rotates in space, but it also moves in other ways. Like an athlete running around a track, Earth moves around the Sun in a regular, curved path called an **orbit**. The movement of Earth around the Sun is known as Earth's **revolution** (reh vuh LEW shun). A year on Earth is the time it takes for Earth to complete one revolution, as seen in **Figure 2**.

Seasons Who doesn't love summer? The long, warm days are great for swimming, biking, and relaxing. Why can't summer last all year? Blame it on Earth's axis and revolution around the Sun. The axis is not straight up and down like a skyscraper—it is slightly tilted. It's due to this tilt and Earth's revolution that you experience seasons.

Look at **Figure 2**. Summer occurs when your part of Earth is tilted toward the Sun. Then it receives more direct sunlight and thus more energy from the Sun than the part of Earth that is tilted away from the Sun. The Sun appears high in the sky. The days are long and the nights are short. Six months later, when the part of Earth that you live on is tilted away from the Sun, you have winter. During this time, the slanted rays of the Sun are weak. The Sun appears low in the sky. The days are short and the nights are long. Autumn and spring occur when Earth is not tilted toward or away from the Sun.

 *What causes seasons?*

Figure 2
Earth takes one year to revolve around the Sun. The Sun's rays strike more directly in the summer, so they are more powerful than the weak, spread-out rays that strike in winter.

Collect Data Visit the Glencoe Science Web site at **tx.science.glencoe.com** for data on Earth's distance from the Sun at various times of the year. Why does the distance change? Communicate to your class what you learn.

SECTION 1 Earth's Place in Space

Movements of the Moon

Imagine a fly buzzing around the head of a jogger on a track. That's how the Moon moves around Earth. Just like that relentless fly around the jogger's head, the Moon constantly circles Earth as Earth revolves around the Sun. The Moon revolves around Earth once every 27.3 days. As you probably have noticed, the Moon does not always look the same from Earth. Sometimes it looks like a big, glowing disk. Other times, it appears to be a thin sliver.

Moon Phases How many different Moon shapes have you seen? Have you seen the Moon look round or maybe like a half circle? Although the Moon looks different at different times of the month, it doesn't change. What does change is the way the Moon appears from Earth. These changes are called phases of the Moon. **Figure 3** shows the various phases of the Moon.

Light from the Sun The phase of the Moon that you see on any given night depends on the positions of the Moon, the Sun, and Earth in space. The Sun lights up the Moon, just as it lights up Earth. Also, just as half of Earth experiences day while the other half experiences night, one half of the Moon is lit by the Sun while the other half is dark. It takes the Moon about one month to go through its phases. During that time, also called a lunar cycle, you see different portions of the daylight side of the Moon. Once each cycle, when the Moon and Sun are on opposite sides of Earth, you can see all of the lit portion of the Moon. This is called a full moon. Nearly two weeks later, the Moon is on the same side as the Sun and it is a new moon. Half the Moon is still lit by the Sun, but none of that half is visible to you. The Moon appears in a slightly different shape each night throughout the lunar cycle as it circles Earth and goes from full to new to full again. The Moon is waning during that portion of the month when it changes from full to new. A waxing moon grows bigger each night on the way from new to full.

Figure 3
The phase of the Moon is determined by the relative positions of the Sun, Earth, and the Moon.
Which photo below shows a full moon?

Waning crescent
New moon
Third quarter
Waxing crescent
Waning gibbous
First quarter
Full moon
Waxing gibbous

✓ **Reading Check** *Describe the lunar cycle.*

Figure 4
During a solar eclipse, the Moon moves between the Sun and Earth. The Sun's corona is visible during a total solar eclipse. *What phase must the Moon be in for a solar eclipse to occur?*

Solar Eclipse Have you ever tried to watch TV with someone standing between you and the screen? You can't see a thing. The picture from the screen can't reach your eyes because someone is blocking it. Sometimes the Moon is like that person standing in front of the TV. It moves between the Sun and Earth in a position that blocks sunlight from reaching Earth. The Moon's shadow travels across parts of Earth. This event, shown in **Figure 4,** is an example of an **eclipse** (ih KLIHPS). Because it is an eclipse of the Sun, it is known as a solar eclipse. The Moon is much smaller than the Sun, so it casts a tiny shadow on Earth. Sunlight is blocked completely only in the small area of Earth where the Moon's darker shadow falls. In that area, the eclipse is said to be a total solar eclipse.

Reading Check *What causes solar eclipses?*

Due to the small size of the shadow—about 269 km wide—only a lucky few get to experience each solar eclipse. For the few minutes the total eclipse lasts, the sky darkens, flowers close, and some planets and brighter stars appear. The Sun's spectacular corona, its pearly white, outermost layer, appears. Far more people will be in the lighter part of the Moon's shadow and will experience a partial solar eclipse.

Mini LAB

Observing Distance and Size

Procedure
1. Place a **basketball** on a **table** at the front of the classroom. Then stand at the back of the room.
2. Holding a **penny,** extend your arm, close one eye, and try to block the ball from sight with the penny.
3. Slowly move the penny closer to you until it completely blocks your view of the basketball.

Analysis
1. In your **Science Journal,** describe what you observed. When did the penny block your view of the basketball?
2. A small object can sometimes block a larger object from view. Explain how this relates to a solar eclipse.

SECTION 1 Earth's Place in Space **355**

Figure 5
During a lunar eclipse, Earth moves between the Sun and the Moon. The Moon often appears red during a lunar eclipse. *Why is a lunar eclipse more common than a solar eclipse?*

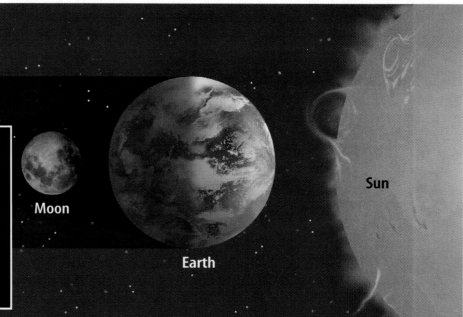

Lunar Eclipse Sometimes Earth gets between the Sun and the Moon, blocking sunlight from reaching the Moon. When Earth's shadow falls on the Moon, an eclipse of the Moon occurs, which is called a lunar eclipse. Earth's shadow is big compared to the Moon, so everyone on the nighttime side of Earth, weather permitting, gets to see a lunar eclipse. When eclipsed, the full moon grows faint and sometimes turns deep red, as shown in **Figure 5.**

Section 1 Assessment

1. Explain the difference between Earth's revolution and rotation.
2. Describe how Earth's revolution and the tilt of its axis contribute to the seasons.
3. Explain why Earth's shadow often covers the entire Moon during a lunar eclipse, but only a small part of Earth is covered by the Moon's shadow during a solar eclipse.
4. Which phase of the Moon would occur during a lunar eclipse?
5. **Think Critically** The tilt of Earth's axis contributes to the seasons. What would seasons be like if Earth's axis were not tilted?

Skill Builder Activities

6. **Concept Mapping** Draw a Venn diagram in your Science Journal. In one circle, write what you know about solar eclipses. In the second circle, write what you know about lunar eclipses. Where the circles overlap, write the facts that apply to lunar and solar eclipses. **For more help, refer to the** Science Skill Handbook.
7. **Solving One-Step Equations** Light travels 300,000 km/s. There are 60 s in 1 min. If it takes 8 min for the Sun's light to reach Earth, how far is the Sun from Earth? **For more help, refer to the** Math Skill Handbook.

356 CHAPTER 12 The Solar System and Beyond

Activity

Moon Phases

The Moon is Earth's nearest neighbor in space. However, the Sun, which is much farther away, affects how you see the Moon from Earth. In this activity, you'll observe how the positions of the Sun, the Moon, and Earth cause the different phases of the Moon.

What You'll Investigate
How do the positions of the Sun, the Moon, and Earth affect the phases of the Moon?

Materials
drawing paper (several sheets)
softball
flashlight
scissors

Goals
- **Model and observe** Moon phases.
- **Record and label** phases of the Moon.
- **Infer** how the positions of the Sun, the Moon, and Earth affect phases of the Moon.

Safety Precautions

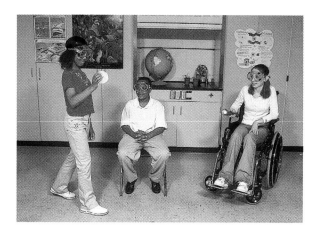

Procedure

1. Turn on the flashlight and darken other lights in the room. Select a member of your group to hold the flashlight. This person will be the Sun. Select another member of your group to hold up the softball so that the light shines directly on the ball. The softball will be the Moon in your experiment.

2. Everyone else represents Earth and should sit between the Sun and the Moon.

3. **Observe** how light shines on the Moon. Draw the Moon, being careful to add shading to represent its dark portion.

4. The student who is holding the Moon should begin to walk in a slow circle around the group, stopping at least seven times at different spots. Each time the Moon stops, observe it, draw it, and shade in its dark portion.

Conclude and Apply

1. **Compare and contrast** your drawings with those of other students. Discuss similarities and differences in the drawings.

2. In your own words, explain how the positions of the Sun, the Moon, and Earth affect the phase of the Moon that is visible from Earth.

3. **Compare** your drawings with **Figure 3.** Which phase is the Moon in for each drawing? Label each drawing with the correct Moon phase.

Communicating Your Data

Use your drawings to make a poster explaining phases of the moon. **For more help, refer to the** Science Skill Handbook.

SECTION 2 The Solar System

As You Read

What You'll Learn
- **Explain** how to measure distance in the solar system.
- **List** the various objects in the solar system.
- **Describe** important characteristics of each planet.

Vocabulary
solar system
astronomical unit
comet
meteorite

Why It's Important
Much can be learned about Earth by studying the other planets.

Distances in Space

Imagine that you are an astronaut living in the future, doing research on a space station in orbit around Earth. You've been working hard for a long time and need a vacation. Where will you go? How about a tour of the solar system? The **solar system,** shown in **Figure 6,** is made up of the nine planets and numerous other objects that orbit the Sun, all held in place by the Sun's immense gravity. How long will your tour take?

Reading Check *What holds the solar system together?*

The *Voyagers 1* and *2* spacecraft left Earth in 1977 to explore the solar system. It took *Voyager 2* two years to reach the planet Jupiter, four years to pass Saturn, and more than eight years to reach Uranus. *Voyager 2* passed Neptune, the farthest planet on its itinerary, 12 years after it left the launchpad on Earth.

Figure 6
The Sun is the center of the solar system, which is made up of the nine planets and other objects that orbit the Sun.

358 CHAPTER 12 The Solar System and Beyond

Measuring Space Distances in space are hard to imagine because space is so vast. Suppose you had to measure your pencil, the hallway outside your classroom, and the distance from your home to school. Would you use the same units for each measurement? No. You probably would measure your pencil in centimeters. You would use something bigger to measure the length of the hallway, such as meters. You might measure the trip from your home to school in kilometers. Larger units are used to measure longer distances. Imagine trying to measure the trip from your home to school in centimeters. If you didn't lose count, you'd end up with a huge number.

Research Visit the Glencoe Science Web site at **tx.science.glencoe.com** for more information about measuring distances in space. Communicate to your class what you learn.

Astronomical Unit Kilometers are fine for measuring long distances on Earth, such as the distance from New York to Chicago (about 1,200 km). Even bigger units are needed to measure vast distances in space. One such measure is the **astronomical** (as truh NAHM ih kul) **unit.** An astronomical unit equals 150 million km, which is the mean distance from Earth to the Sun. Astronomical unit is abbreviated *AU*. If something is 3 *AU* away from the Sun, then the object is three times farther from the Sun than Earth is. The *AU* is a convenient unit for measuring distances in the solar system.

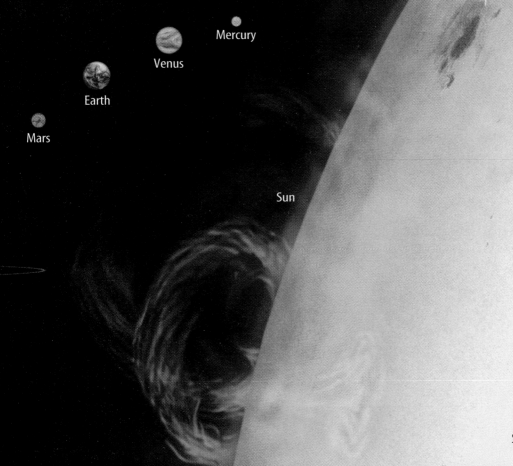

SECTION 2 The Solar System

Life Science INTEGRATION

Recent discoveries indicating that water might occasionally seep from the surface of Mars have caused some people to hypothesize that microscopic life might exist on Mars. Research this hypothesis and critique its strengths and weaknesses using current scientific evidence and information.

Touring the Solar System

Now you know a little more about how to measure distances in the solar system. Next, you can travel outward from the Sun and take a look at the objects in the solar system. Maybe you can find a nice destination for your next vacation. Strap yourself into your spacecraft and get ready to travel. It's time to begin your journey. What will you see first?

Inner Planets

The first group of planets you pass are the inner planets. These planets are mostly solid, with minerals similar to those on Earth. As with all the planets, much of what is known comes from spacecraft that send data back to Earth. Various spacecraft took the photographs shown in **Figure 7** and the rest of this section. Some were taken while in orbit and others upon landing.

Mercury The first planet that you will visit is the one that is closest to the Sun. Mercury, shown in **Figure 7A,** is the second-smallest planet. Its surface has many craters. Craters form when meteorites, which are chunks of rock or metal that fall from the sky, strike a planet's surface. You will read about meteorites later in this section. Because of Mercury's small size and low gravity, most gases that could form an atmosphere escape into space. The nearly absent atmosphere and the closeness of this planet to the Sun cause great extremes in temperature on Mercury. Its surface temperature can reach 430°C during the day and drop to −180°C at night, making the planet unfit for life.

Reading Check *Why does Mercury have almost no atmosphere?*

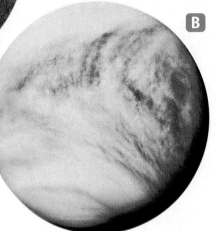

Figure 7
A Mercury is the closest planet to the Sun. Like the Moon, its surface is scarred by craters. **B** Earth's closest neighbor, Venus, is covered in clouds.

Venus You won't be able to see much at your next stop, shown in **Figure 7B.** Venus, the second-closest planet to the Sun, is hard to see because its surface is surrounded by thick clouds. These clouds trap the solar energy that reaches the surface of Venus. That energy causes surface temperatures to hover around 470°C—hot enough to bake a clay pot.

Earth Home sweet home. You've reached Earth, the third planet from the Sun. You didn't realize how unusual your home planet was until you saw other planets. Earth's surface temperatures allow water to exist as a solid, a liquid, and a gas. Also, ozone in Earth's atmosphere works like a screen to limit the number of ultraviolet (ul truh VI uh lut) rays that reach the planet's surface. Ultraviolet rays are harmful rays from the Sun. Because of Earth's atmosphere, life can thrive on the planet. You would like to linger on Earth, shown in **Figure 8,** but you have six more planets to explore.

Mars Has someone else been here? You see signs of earlier visits to Mars, the fourth of the inner planets. Tiny robotic explorers have been left behind. However, it wasn't a person who left them here. Spacecraft that were sent from Earth to explore Mars's surface left the robots. If you stay long enough and look around, you might notice that Mars, shown in **Figure 9,** has seasons and polar ice caps. Signs indicate that the planet once had liquid water. Water might even be shaping the surface of Mars today. You'll also notice that the planet looks red. That's because the rocks on its surface contain iron oxide, which is rust. Two small moons, Phobos and Deimos, orbit Mars.

Figure 8
As far as scientists know, Earth is the only planet that supports life. It is one of the four inner planets.

Figure 9
Mars often is called the Red Planet. *What causes Mars's surface to appear red?*

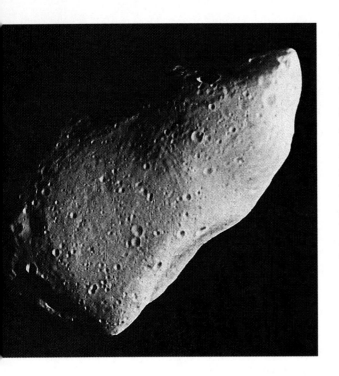

Asteroid Belt Look out for asteroids. On the next part of your trip, you must make your way through the asteroid belt that lies between Mars and the next planet, Jupiter. As you can see in **Figure 10,** asteroids are pieces of rock made of minerals similar to those that formed the rocky planets and moons. In fact, these asteroids might have become a planet if it weren't for the giant planet, Jupiter. Jupiter's huge gravitational force might have prevented a small planet from forming in the area of the asteroid belt. The asteroids also might be the remains of larger bodies that broke up in collisions. The asteroid belt separates the solar system's planets into two groups—the inner planets, which you've already visited, and the outer planets, which are coming next.

✓ Reading Check   *What are asteroids?*

Figure 10
This close-up of the asteroid Gaspra was taken by the *Galileo* spacecraft in 1991.

Figure 11
Jupiter is the largest planet in the solar system. This gas giant has 28 moons.

Outer Planets

Moving past the asteroids, you come to the outer planets. The outer planets are Jupiter, Saturn, Uranus, Neptune, and Pluto. Let's hope you aren't looking for places to stop and rest. Trying to stand on most of these planets would be like trying to stand on a cloud. That's because all of the outer planets, except Pluto, are huge balls of gas called gas giants. Each might have a solid core, but none of them has a solid surface. The gas giants have lots of moons, also called satellites, which orbit the planets just like Earth's Moon orbits Earth. They have rings surrounding them that are made of dust and ice. The only outer planet that doesn't have rings is Pluto. Pluto also differs from the other outer planets because it is composed of ice and rock.

Jupiter If you're looking for excitement, you'll find it on Jupiter, which is the largest planet in the solar system and the fifth from the Sun. It also has the shortest day—less than 10 h long—which means this giant planet is spinning faster than any other planet. Watch out for a huge, red whirlpool near the middle of the planet. That's the Great Red Spot, a giant storm on Jupiter's surface. Jupiter, shown in **Figure 11,** almost looks like a miniature solar system. It has 28 moons. One called Ganymede (GA nih meed) is larger than the planet Mercury. Ganymede, along with two other moons, Europa and Callisto, might have liquid water under its icy crust. Another of Jupiter's moons, Io, has more active volcanoes than any other object in the solar system.

Saturn You might have thought that Jupiter was unusual. Wait until you see Saturn, the sixth planet from the Sun. You'll be dazzled by its rings, shown in **Figure 12.** Saturn's several broad rings are made up of hundreds of smaller rings, which are made up of pieces of ice and rock. Some of these pieces are like specks of dust. Others are many meters across. Saturn is orbited by at least 30 moons, the largest of which is Titan. Titan has an atmosphere that resembles the atmosphere on Earth in primitive times. Some scientists hypothesize that Titan's atmosphere might provide clues about how life formed on Earth.

Uranus After Saturn, you come to Uranus, the seventh planet from the Sun. Uranus warrants a careful look because of the interesting way it spins on its axis. The axis of most planets is tilted just a little, somewhat like the handle of a broom that is leaning against a wall. Uranus, also shown in **Figure 12,** is nearly lying on its side. Its axis is tilted almost even with the plane of its orbit like a broomstick lying on the floor. Uranus's atmosphere is made mostly of hydrogen with smaller amounts of helium and methane. The methane gives Uranus its distinctive bluish-green color. Uranus has rings and is thought to have at least 21 moons.

Figure 12
Saturn and Uranus are two of the four gas giant planets.

Problem-Solving Activity

How can you model distances in the solar system?

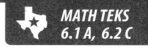

MATH TEKS
6.1 A, 6.2 C

The distances between the planets and the Sun are unimaginably large but definitely measurable. Astronomers have developed a system of measurement to describe these distances in space. Could you represent these vast distances in a simple classroom model? Use your knowledge of SI and your ability to read a data table to find out.

Identifying the Problem

The table to the right shows the distances in astronomical units between the planets and the Sun. Notice that the inner planets are fairly close together, and the outer planets are far apart. Study the distances carefully, then answer the questions.

Solving the Problem

1. Based on the distances shown in the table, how would you make a scale model of the solar system that would fit in your classroom? What unit would you use to show the distances between the planets?

2. Show the conversion between astronomical units and the SI unit you would use for your model.

Solar System Data	
Planet	Distance from the Sun (AU)
Mercury	0.39
Venus	0.72
Earth	1.00
Mars	1.52
Jupiter	5.20
Saturn	9.54
Uranus	19.19
Neptune	30.07
Pluto	39.48

SECTION 2 The Solar System

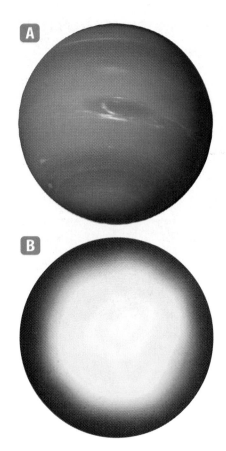

Figure 13
The outermost planets are **A** Neptune and **B** Pluto. This is the best image available of Pluto, which was not visited by *Voyager* spacecraft.

Neptune Neptune is the next stop in your space travel. Neptune, shown in **Figure 13A,** is the eighth planet from the Sun. Between 1979 and 1999, Pluto was closer to the Sun than Neptune was because their orbits overlap. However, even then Neptune was considered the eighth planet. Neptune's atmosphere is composed of hydrogen, helium, and methane. Methane and helium give the planet a blue color. Neptune is the last of the big, gas planets with rings around it. It also has eight moons. Triton, the largest of these, has geysers that shoot gaseous nitrogen into space. The low number of craters on Triton indicates that lava still flows onto its surface.

Pluto The last planet that you come to on your tour is Pluto, a small, rocky planet with a frozen crust. Pluto was discovered in 1930 and is farthest from the Sun. It is the smallest planet in the solar system—smaller even than Earth's Moon—and the one scientists know the least about. It is the only planet in the solar system that has never been visited by a spacecraft. Pluto, shown in **Figure 13B,** has one moon, Charon, which is nearly half the size of the planet itself.

Comets

A **comet** is a large body of ice and rock that travels around the Sun in an elliptical orbit. These objects are like dirty snowballs that measure a few kilometers across. Comets might originate in a cloud of objects far beyond the orbit of Pluto known as the Oort Cloud. This belt is 50,000 AU from the Sun. Some comets also originate in the Kuiper Belt, which lies just beyond the orbit of Pluto. As a comet approaches the Sun, radiation vaporizes some of the material. Solar winds blow vaporized gas and dust away from the comet, forming what appears from Earth as a bright, glowing tail, shown in **Figure 14**.

Reading Check *Where do comets come from?*

Figure 14
The tails of comets point away from the Sun, pushed by solar wind. Solar wind is a stream of charged particles heading outward from the Sun. *Why do comets appear to glow?*

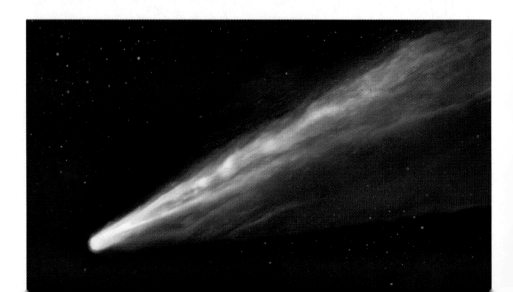

Meteorites Occasionally, chunks of extraterrestrial rock and metal fall to Earth. **Meteorites** are any fragments from space that survive their plunge through the atmosphere and land on Earth's surface. Small ones are no bigger than pebbles. The one in **Figure 15** has a mass of 14.5 metric tons. Hundreds of meteorites fall to Earth each year. Luckily, strikes on buildings or other human-made objects are rare. In fact, only a tiny fraction of the meteorites that fall are ever found. Scientists are extremely interested in those that are, as they yield important clues from space. For example, many seem to be about 4.5 billion years old, which provides a rough estimate of the age of the solar system. Several thousand meteorites have been collected in Antarctica, where moving ice sheets concentrate them in certain areas. Any rock seen on an ice sheet in Antarctica is probably a meteorite, because few other rocks are exposed. Meteorites can be one of three types—irons, stones, and stoney-irons. Irons are almost all iron, with some nickel mixed in. Stones are rocky. The rarest, stoney irons, are a mixture of metal and rock.

Figure 15
This meteorite on display at the American Museum of Natural History in New York has a mass of 14.5 metric tons. *Why are meteorites rare?*

Section 2 Assessment

1. Explain how the astronomical unit is useful for measuring distances in space.
2. In general, how are the outer planets different from the inner planets? How are they alike?
3. Describe the objects other than planets that are located within Earth's solar system.
4. How is Saturn's largest satellite different from satellites of other planets?
5. **Think Critically** Larger units of measure are used to express increasingly larger distances. How do scientists express tiny distances, such as the distances between molecules or atoms?

Skill Builder Activities

6. **Developing Multimedia Presentations** Use your knowledge of the solar system to develop a multimedia presentation. You might begin by drawing a labeled poster that includes the Sun, the planets with their moons, the asteroid belt, and comets. **For more help, refer to the** Technology Skill Handbook.

7. **Using an Electronic Spreadsheet** Using the table in the Problem Solving Activity, make a spreadsheet showing the distances of the planets from the Sun. Add columns for additional data such as day and year lengths and the diameters of each planet. **For more help, refer to the** Technology Skill Handbook.

SECTION 3

Stars and Galaxies

As You Read

What You'll Learn
- **Explain** how a star is born.
- **Describe** the galaxies that make up the universe.
- **Explain** how to measure distances in space beyond Earth's solar system.

Vocabulary
constellation
supernova
galaxy
light-year

Why It's Important
Understanding the vastness of the universe will help you appreciate Earth's place in space.

Figure 16
Find the Big Dipper in the constellation Ursa Major. *Why do you think people call it the Big Dipper?*

Stars

Every night, a whole new world opens to you as the stars come out. The fact is, stars are always in the sky. You can't see them during the day because the Sun's light makes Earth's atmosphere so bright that it hides them. The Sun is a star, too. In fact, it is the closest star to Earth. You can't see the Sun at night because as Earth rotates, your part of Earth is facing away from it.

Constellations Ursa Major, Orion, Taurus—do these names sound familiar? They are **constellations** (kahn stuh LAY shunz), or groups of stars that form patterns in the sky. **Figure 16** shows some constellations.

Constellations are named after animals, objects, and people—real or imaginary. Many of the names that early Greek astronomers gave to the constellations are still in use. However, throughout history, different groups of people have seen different things in the constellations. In early England, people thought the Big Dipper, found in the constellation Ursa Major, looked like a plow. Native Americans saw a horse and rider. To the Chinese, it looked like a governmental official and his helpers moving on a cloud. What image does the Big Dipper bring to your mind?

Starry Colors When you glance at the sky on a clear night, the stars look like tiny pinpoints of light. From a distance, they look alike, but stars are different sizes and colors.

Most stars in the universe are cool and small. However, some smaller and medium-sized stars can be hot, and many larger stars are fairly cool. How is a star's temperature measured? The color of a star is a clue. Just as the red flames in a campfire are cooler, red stars are the coolest visible stars. Yellow stars are of medium temperature. Bluish-white stars, like the blue flames on a gas stove, are the hottest. The Sun is a yellow, medium-sized star. The giant, red star called Betelgeuse (BEE tul jews) is much bigger than the Sun. If this huge star were in the same place as Earth's Sun, it would swallow Mercury, Venus, Earth, and Mars.

Reading Check *How is star color related to temperature?*

Apparent Magnitude Look at the sky on a clear night and you can easily notice that some stars are brighter than others. A system called apparent magnitude is used for classifying how bright a star appears from Earth. The dimmest stars that are visible to the unaided eye measure 6 on the apparent magnitude scale. A star with an apparent magnitude of 5 is 2.5 times brighter. The smaller the number is, the brighter the star is. The brightest star in the sky, Sirius, has an apparent magnitude of −1.5, and the Sun's apparent magnitude is −26.7.

Compared to other stars, the Sun is medium in size and temperature. It looks so bright because it is so close to Earth. Apparent magnitude is a measure of how bright a star looks from Earth but not a measure of its actual brightness, known as absolute magnitude. As **Figure 17** shows, a small, close star would look brighter than a giant star that is far away.

TRY AT HOME Mini LAB

Modeling Constellations

Procedure
1. Draw a dot pattern of a constellation on a piece of **black construction paper**. Choose a known constellation or make up your own.
2. With an adult's help, cut off the end of a **cardboard cylinder** such as an oatmeal box. You now have a cylinder with both ends open.
3. Place the cylinder over the constellation. Trace around the rim. Cut the paper along the traced line.
4. **Tape** the paper to the end of the cylinder. Using a **pencil**, carefully poke holes through the dots on the paper.
5. Place a **flashlight** inside the open end of the cylinder. Darken the room and observe your constellation on the ceiling.

Analysis
1. Turn on the overhead light and view your constellation again. Can you still see it? Why or why not?
2. The stars are always in the sky, even during the day. How is the overhead light similar to the Sun? Explain.

Figure 17
This flashlight looks brighter than the car headlights because it is closer. In a similar way, a small but close star will appear brighter than a more distant, giant star.

SECTION 3 Stars and Galaxies

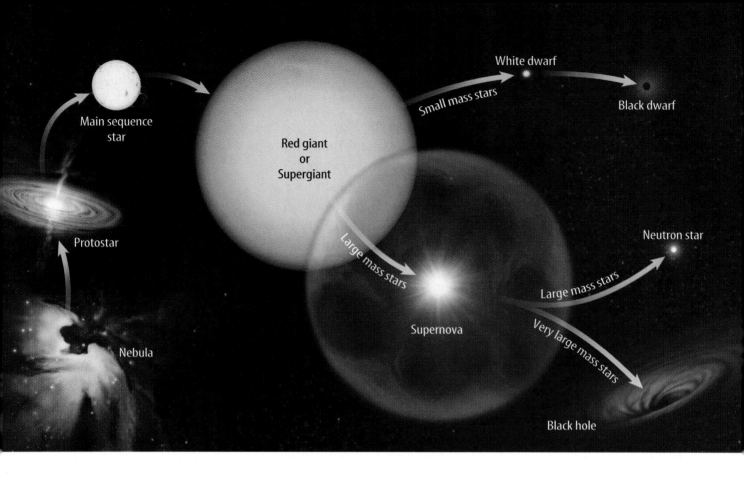

Figure 18
The events in the lifetime of a star depend greatly on the star's mass. *What happens to supergiants when their cores collapse?*

Collect Data Visit the Glencoe Science Web site at **tx.science.glencoe.com** for data on supernovas that astronomers have observed in distant galaxies over the past year. How many were found? How far away did these events occur? Communicate to your class what you learn.

The Lives of Stars

You've grown up and changed a lot since you were born. You've gone through several stages in your life, and you'll go through many more. Stars go through stages in their lives, too.

The stages a star goes through in its life depend on the star's size. When a medium-sized star like the Sun uses up some of the gases in its center, it expands to become a giant star. The Sun will become a giant in about 5 billion years. At that time it will expand to cover the orbits of Mercury, Venus, and possibly Earth. It will remain that way for about a billion years. The Sun then will lose its outer shell and shrink to a hot white dwarf. Eventually, it will cool and become a black dwarf. Stars more massive than the Sun complete their life cycles in shorter periods of time. The smallest stars shine the longest. **Figure 18** illustrates how the course of a star's life is determined by its mass.

Reading Check *What stages does a star go through in its life?*

Scientists hypothesize that stars begin their lives as huge clouds of gas and dust. The force of gravity, which causes attraction between objects, causes the dust and gases to move closer together. When this happens, temperatures within the cloud begin to rise. A star is formed when this cloud gets so dense and hot that the atoms within it merge. This process is known as fusion, and it changes matter to the energy that powers the star.

Supergiants When a large star begins to use up the fuel in its core, it becomes a supergiant. Over time, the core of a supergiant collapses. Then a **supernova** occurs, in which the outer part of the star explodes and becomes bright. For a few brief days, the supernova might shine more brightly than a whole galaxy. The dust and gases released by this explosion, shown in **Figure 19,** eventually might form other stars.

Meanwhile, the core of the supergiant is still around. It now is called a neutron star. If the neutron star is massive enough, it could become a black hole rapidly. Black holes, shown in **Figure 20,** are so dense that even light cannot escape their gravity. Light shone into them disappears, and no light can escape from them.

Galaxies

What do you see when you look at the night sky? If you live in a city, you might not see much. The glare from city lights makes it hard to see the stars. If you go to a dark place, far from the lights of towns and cities, you can see much more. In a dark area, you might be able to use a powerful telescope to see dim clumps of stars grouped together. These groups of stars are galaxies (GA luk seez). A **galaxy** is a group of stars, gas, and dust held together by gravity.

Types of Galaxies You now know how planets and stars differ from one another. Galaxies come in different shapes and sizes, too. The three major types of galaxies are elliptical, spiral, and irregular. Elliptical galaxies are very common. They're shaped like huge footballs or spheres. Spiral galaxies have arms radiating outward from the center, somewhat like a giant pinwheel. As shown in **Figure 21,** some spiral galaxies have bar-shaped centers. Irregular galaxies are just that—irregular. They come in all sorts of different shapes and can't be classified easily. Irregular galaxies are usually smaller than other galaxies. They are also common.

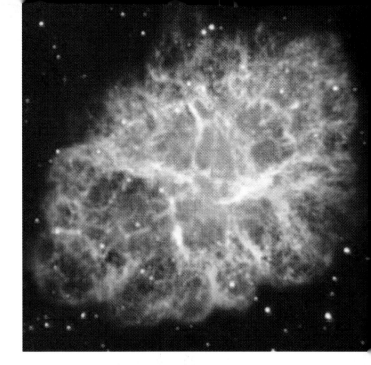

Figure 19
This photo shows the remains of a supernova located trillions of kilometers from Earth.

Figure 20
A black hole has so much gravity that not even light can escape. This drawing shows a black hole stripping gas from a nearby star.
How do black holes form?

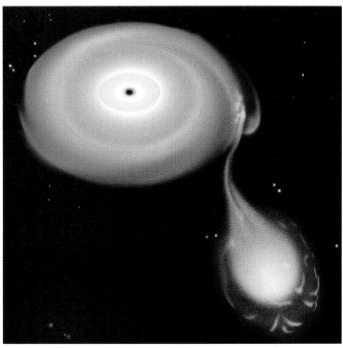

NATIONAL GEOGRAPHIC VISUALIZING GALAXIES

Figure 21

Most stars visible in the night sky are part of the Milky Way Galaxy. Other galaxies, near and far, vary greatly in size and mass. The smallest galaxies are just a few thousand light-years in diameter and a million times more massive than the Sun. Large galaxies—which might be more than 100,000 light-years across—have a mass several trillion times greater than the Sun. Astronomers group galaxies into four general categories, as shown here.

▲ **ELLIPTICAL GALAXIES** They are nearly spherical to oval in shape and consist of a tightly packed group of relatively old stars.

▶ **SPIRAL GALAXIES** Spiral galaxies consist of a large, flat disk of interstellar gas and dust with star clusters extending from the disk in a spiral pattern. The Andromeda Galaxy, one of the Milky Way Galaxy's closest neighbors, is a spiral galaxy.

◀ **BARRED SPIRAL GALAXIES** Sometimes the flat disk that forms the center of a spiral galaxy is elongated into a bar shape. Two arms containing clusters of stars swirl out from either end of the bar, forming what is known as a barred spiral galaxy.

▲ **IRREGULAR GALAXIES** A few galaxies are neither spiral nor elliptical. Their shape seems to follow no set pattern, so astronomers have given them the general classification of irregular.

The Milky Way Galaxy Which type of galaxy do you live in? Look at **Figure 22.** You live in the Milky Way, which is a giant spiral galaxy. Hundreds of billions of stars are in the Milky Way, including the Sun. Just as Earth revolves around the Sun, stars revolve around the centers of galaxies. The Sun revolves around the center of the Milky Way about once every 240 million years.

A View from Within You can see part of the Milky Way as a band of light across the night sky. However, you can't see the whole Milky Way. To understand why, think about boarding a Ferris wheel and looking straight up. Can you really tell what the ride looks like? Because you are at the bottom looking up, you get a limited view. Your view of the Milky Way from Earth is like the view of the Ferris wheel from the bottom. As you can see in **Figure 23,** you can view only parts of this galaxy because you are within it.

Reading Check *Why can't you see the entire Milky Way from Earth?*

The faint band of light across the sky that gives the Milky Way its name is the combined glow of stars in the galaxy's disk. In 1609, when the Italian astronomer Galileo looked at the Milky Way with a telescope, he showed that the band was actually made of countless individual stars. The galaxy is vast—bigger and brighter than most of the galaxies in the universe. Every star you see in the sky with your naked eye is a member of the Milky Way Galaxy.

Figure 22
The Sun, one of billions of stars in the galaxy, is located toward the edge of the Milky Way.

Figure 23
This is the view of the Milky Way from inside the galaxy. *Why is it called the Milky Way?*

SECTION 3 Stars and Galaxies **371**

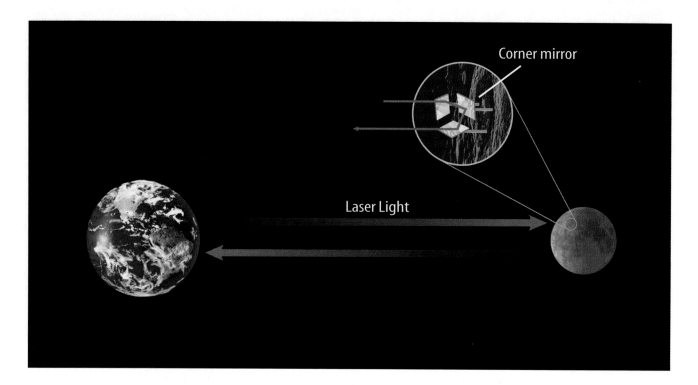

Figure 24
The constant speed of light through space helps astronomers in many ways. For example, the distance to the Moon has been determined by bouncing a laser beam off mirrors left by *Apollo 11* astronauts.

The Milky Way belongs to a cluster of galaxies called the Local Group. Scientists have determined that galaxies outside of the Local Group are moving away from Earth. Based on this, what can you infer about the size of the universe? Research the phenomenon known as red shift and describe to the class how it has helped astronomers learn about the universe.

Speed of Light The speed of light is unique. Light travels through space at about 300,000 km/s—so fast it could go around Earth seven times in 1 s. You can skim across ocean waves quickly on a speedboat, but no matter how fast you go, you can't gain on light waves. It's impossible to go faster than light. Most galaxies are moving away from the Milky Way and a few are moving closer, but the light from all galaxies travels toward Earth at the same speed. The constant speed of light is useful to astronomers, as shown in **Figure 24**.

Light-Years Earlier you learned that distances between the planets are measured in astronomical units. However, distances between galaxies are vast. Measuring them requires an even bigger unit of measure. Scientists often use light-years to measure distances between galaxies. A **light-year** is the distance light travels in one year—about 9.5 trillion km.

✓ **Reading Check** *Why is a light-year better than an astronomical unit for measuring distances between galaxies?*

Would you like to travel back in time? In a way, that's what you're doing when you look at a galaxy. The galaxy might be millions of light-years away. The light that you see started on its journey long ago. You are seeing the galaxy as it was millions of years ago. On the other hand, if you could look at Earth from this distant galaxy, you would see events that happened here millions of years ago. That's how long it takes the light to travel the vast distances through space.

The Universe

Each galaxy contains billions of stars. Some might have as many stars as the Milky Way, and a few might have more. As many as 100 billion galaxies might exist. All these galaxies with all of their countless stars make up the universe.

Look at **Figure 25.** The *Hubble Space Telescope* spent ten days in 1995 photographing a tiny sector of the sky to produce this image. More than 1,500 galaxies were discovered. Astronomers think a similar picture would appear if they photographed any other sector of the sky. In this great vastness of exploding stars, black holes, star-filled galaxies, and empty space is one small planet called Earth. If you reduced the Sun to the size of a period on this page, the next-closest star would be more than 16 km away. Earth looks even lonelier when you consider that the universe also seems to be expanding. Most other galaxies are moving away at speeds as fast as 20,000 km/s. In relation to the immensity of the universe, Earth is an insignificant speck of dust. Could it be the only place where life exists?

Figure 25
The Hubble Deep Field Image shows hundreds of galaxies in one tiny sector of the sky. *What does this image tell you about the sky?*

 Reading Check *How do other galaxies move relative to Earth?*

Section Assessment

1. What are the three major types of galaxies? What type of galaxy is the Milky Way?
2. Describe how a star forms.
3. Describe the size and temperature of most stars that exist in the universe.
4. How long has light from a star that is 50 light-years away been traveling when it reaches Earth?
5. **Think Critically** Some stars might no longer be in existence, but you still see them in the night sky. Why?

Skill Builder Activities

6. **Making Models** The Milky Way is 100,000 light-years in diameter. Outline a plan for how you would build a model of the Milky Way. **For more help, refer to the** Science Skill Handbook.

7. **Communicating** Observe the stars in the night sky. In your Science Journal, draw the stars you observed. Now draw your own constellation based on those stars. Give your constellation a name. Why did you choose that name? **For more help, refer to the** Science Skill Handbook.

SECTION 3 Stars and Galaxies **373**

Activity: Design Your Own Experiment

Space Colony

Many fictional movies and books describe astronauts from Earth living in space colonies on other planets. Some of these make-believe societies seem far-fetched. So far, humans haven't built a space colony on another planet. However, if it happens, what would it look like?

Recognize the Problem

How would conditions on a planet affect the type of space colony that might be built there?

Form a Hypothesis

Research a planet. Review conditions on the surface of the planet. Make a hypothesis about the things that would have to be included in a space colony to allow humans to survive on the planet.

Possible Materials
drawing paper
markers
books about the planets

Goals
- **Infer** what a space colony might look like on another planet.
- **Classify** planetary surface conditions.
- **Draw** a space colony for a planet.

Using Scientific Methods

Test Your Hypothesis

Plan

1. Select a planet and study the conditions on its surface.
2. **Classify** the surface conditions in the following ways.
    a. solid, liquid, or gas
    b. hot, cold, or a range of temperatures
    c. heavy atmosphere, thin atmosphere, or no atmosphere
    d. bright or dim sunlight
    e. unique conditions
3. **List** the things that humans need to survive. For example, humans need air to breathe. Does your planet have air that humans can breathe, or would your space colony have to provide the air?
4. Make a table for the planet showing its surface conditions and the features the space colony would have to have so that humans could survive on the planet.
5. **Discuss** your decisions as a group to make sure they make sense.

Do

1. Make sure your teacher approves your plan before you start.
2. **Draw** a picture of the space colony. Draw another picture showing the inside of the space colony. Label the parts of the space colony and explain how they aid in the survival of its human inhabitants.

Analyze Your Data

1. **Compare and contrast** your space colony with those of other students who researched the same planet you did. How are they alike? How are they different?
2. Would you change your space colony after seeing other groups' drawings? If so, what changes would you make? Explain your reasoning.

Draw Conclusions

1. What was the most interesting thing you learned about the planet you studied?
2. Was your planet a good choice for a space colony? Explain.
3. Would humans want to live on your planet? Why or why not?
4. Could your space colony be built using present technology? Explain.

Communicating Your Data

Present your drawing and your table to the class. Make a case for why your planet would make a good home for a space colony. **For more help, refer to the** Science Skill Handbook.

ACTIVITY 375

Science and Language Arts

The Sun and the Moon
A Korean Folktale

Respond to the Reading

1. What is the purpose of this folktale?
2. Describe the personalities of the girl and the king. What clues does the folktale give us about each of their personalities?

The two children lived peacefully in the Heavenly Kingdom, until one day the Heavenly King said to them, "We can not allow anyone to sit here and idle away the time. So I have decided on duties for you. The boy shall be the Sun, to light the world of men, and the girl shall be the Moon, to shine by night." Then the girl answered, "Oh King, I am not familiar with the night. It would be better for me not to be the Moon." So the King made her the Sun instead, and made her brother the Moon.

It is said that when she became the Sun, the people used to gaze up at her in the sky. But she was modest, and greatly embarrassed by this. So she shone brighter and brighter, so that it was impossible to look at her directly. And that is why the Sun is so bright, that her modesty might be forever respected.

Cause and Effect

Understanding Literature

Cause and Effect In the folktale you just read, a story was created to explain why the Sun and the Moon exist, as well as why you should never look directly at the Sun. When one event brings about a second event, you are dealing with cause and effect. In this folktale, the Heavenly King says that no one is allowed to be idle in the Heavenly Kingdom. This is a cause. The effect is that the girl and boy are given the duties of being the Sun and the Moon. Many cultures create their own explanations, like this folktale, of why things happen or exist.

Science Connection In this chapter, you learned that a cause-and-effect relationship between Earth and the Sun is responsible for the changing seasons. According to the scientific explanation, the tilt of Earth's axis and Earth's revolution around the Sun cause the seasons to change. When the part of Earth you live on is tilted towards the Sun, you experience summer. When your part of Earth is tilted away from the Sun, you experience winter. Autumn and spring occur when Earth is not tilted toward or away from the Sun.

Linking Science and Writing

Create a Folktale Many early cultures used stories called folktales or myths to explain things that they didn't understand scientifically. Think of something that happens in nature that you don't understand scientifically. Write a one-page folktale that explains why it happens. You might explain what causes thunder, why the sky is blue, or how a spider knows how to spin its web.

Career Connection

Astronaut and physician

Dr. Mae Jemison finds ways to use space to help humans on Earth. She was a Science Mission Specialist on the space shuttle *Endeavor.* In space, she studied bone cells and biofeedback. On Earth, she directs the Jemison Institute. This institute brings new technology to countries that need it. It is working on a satellite system called Alafiya—which is Yoruba for "good health." Alafiya helps countries share information about health care. Using the satellite in space, people can learn about health education, disease prevention, and health resources.

SCIENCE *Online* To learn more about a career as an astronaut, visit the Glencoe Science Web site at **tx.science.glencoe.com**.

SCIENCE AND LANGUAGE ARTS

Chapter 12 Study Guide

Reviewing Main Ideas

Section 1 Earth's Place in Space

1. Earth spinning on its axis is called rotation. This movement causes night and day.

2. Earth orbits the Sun in a regular, curved path. This movement is known as Earth's revolution. Earth's revolution and the tilt of its axis are responsible for seasons.

3. The Moon moves, too, as it orbits Earth. The different positions of Earth, the Sun, and the Moon in space cause Moon phases and eclipses. *Explain the difference between a lunar eclipse and the solar eclipse shown here.*

Section 2 The Solar System

1. The solar system is made up of the nine planets and numerous other objects that orbit the Sun. Planets are classified as inner planets or outer planets.

2. The inner planets—Mercury, Venus, Earth, and Mars—are closest to the Sun.

3. The outer planets—Jupiter, Saturn, Uranus, Neptune, and Pluto—are much farther away. Most of the outer planets are large, gas giants with rings and moons. *How is Pluto different from the other outer planets?*

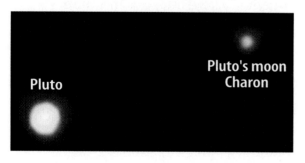

Section 3 Stars and Galaxies

1. Constellations are groups of stars that form patterns in the sky. Although stars might look the same from Earth, they differ greatly in temperature, size, and color. The Sun, for instance, is a medium-sized, yellow star. *What color are the hottest stars? The coolest?*

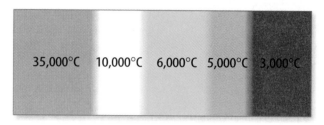

35,000°C   10,000°C   6,000°C   5,000°C   3,000°C

2. Stars begin as gas and dust that are pulled together by gravity. Eventually, a star begins to produce light as hydrogen atoms fuse. The course of the life cycle for each star is determined by its size. Supernovas and black holes are the results of huge stars that have completed their life cycles.

3. Galaxies are groups of stars, gas, and dust held together by gravity. The three main types of galaxies are elliptical, spiral, and irregular. You live in the Milky Way, a spiral galaxy. Distances between galaxies are measured in light-years. A light-year is the distance light travels in one year—about 9.5 trillion km. *Why are special units needed for studying distances in space?*

After You Read

FOLDABLES Reading & Study Skills

To help you review characteristics of the solar system, stars, and galaxies, use the Foldable you made at the beginning of the chapter.

Chapter 12 Study Guide

Visualizing Main Ideas

Use the following terms to fill in the concept map below: asteroid belt, galaxy, universe, inner planets, comets and meteorites, *and* outer planets.

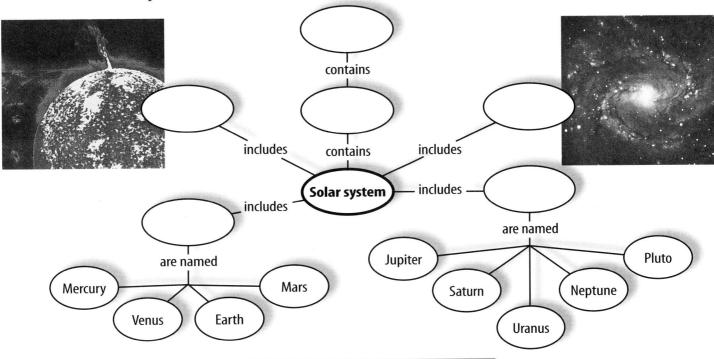

Vocabulary Review

Vocabulary Words

a. astronomical unit
b. comet
c. constellation
d. eclipse
e. galaxy
f. light-year
g. meteorite
h. orbit
i. revolution
j. rotation
k. solar system
l. supernova

 **Study Tip**

Design experiments that you could perform to test some of the principles discussed in this chapter.

Using Vocabulary

Each question below asks about a vocabulary word from the list. Write the word that best answers each question.

1. What event occurs when Earth's shadow falls on the Moon or when the Moon's shadow falls on Earth?

2. Which motion of Earth produces day and night and causes the planets and stars to rise and set?

3. What is a large group of stars, gas, and dust held together by gravity called?

4. What is a group of stars that forms a pattern in the sky called?

5. Which movement of Earth causes it to travel around the Sun?

CHAPTER STUDY GUIDE 379

Chapter 12 Assessment & TEKS Review

Checking Concepts

Choose the word or phrase that best answers the question.

1. What is caused by the tilt of Earth's axis and Earth's revolution?
   A) eclipses   C) day and night
   B) phases   D) seasons

2. What is occurring when the Moon's phases are waning?
   A) Phases are growing larger.
   B) Phases are growing smaller.
   C) a full moon
   D) a new moon

3. An astronomical unit equals the distance from Earth to which of the following?
   A) the Moon   C) Mercury
   B) the Sun   D) Pluto

4. Earth is which planet from the Sun?
   A) first   C) third
   B) second   D) fourth

5. How many galaxies could be in the universe?
   A) 1 billion   C) 50 billion
   B) 10 billion   D) 100 billion

6. Which results from Earth's rotation?
   A) night and day   C) Moon phases
   B) summer and winter   D) solar eclipses

7. What unit often is used to measure large distances in space, such as between stars or galaxies?
   A) kilometer   C) light-year
   B) astronomical unit   D) centimeter

8. How many planets are in the solar system?
   A) six   C) eight
   B) seven   D) nine

9. Which object's shadow travels across part of Earth during a solar eclipse?
   A) the Moon   C) Mars
   B) the Sun   D) a comet

10. If a star is massive enough, what can result after it produces a supernova?
    A) a galaxy   C) a black dwarf
    B) a black hole   D) a superstar

Thinking Critically

11. What conditions on Earth allow life to thrive?

12. Which of the planets in the solar system seems most like Earth? Which seems most different? Explain your answers using facts about the planets.

13. How might a scientist predict the day and time of a solar eclipse?

14. Which of the Moon's motions are real? Which are apparent? Explain why each occurs.

Developing Skills

15. **Making and Using Tables** Research the size, period of rotation, and period of revolution for each planet. Show this information in a table. How do tables help you better understand information?

16. **Comparing and Contrasting** Compare and contrast the inner planets and the outer planets.

17. **Making a Model** Based on what you have learned about the Sun, the Moon, and Earth, model a lunar or a solar eclipse using simple classroom materials.

Chapter 12 Assessment

18. Concept Mapping Complete the following concept map using these terms: *full, red surface, corona, solar,* and *few.*

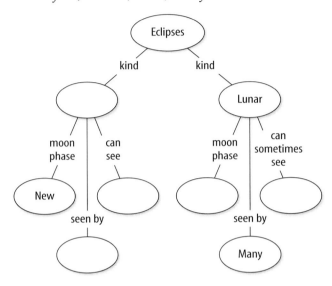

19. Comparing and Contrasting Compare and contrast Earth and other bodies in the solar system in terms of their ability to support life.

Performance Assessment

20. Model Make a three-dimensional model including a light source for the Sun that shows how Earth's tilted axis and its orbit around the Sun combine to cause changes in the lengths of day and night throughout the year.

21. Poster Research the moons of Jupiter, Saturn, Uranus, or Neptune. Make a poster showing the special characteristics of these moons. Display your poster for your class.

Technology

Go to the Glencoe Science Web site at **tx.science.glencoe.com** or use the **Glencoe Science CD-ROM** for additional chapter assessment.

 **TAKS Practice**

The way that Earth moves in space creates night and day and different seasons. The picture below shows Earth and the Sun in space. *TEKS 6.2 C; 6.3 C; 6.13 A*

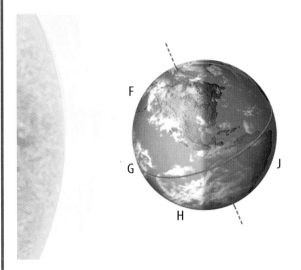

Study the picture and answer the following questions.

1. Which of the following processes contributes to the fact that it is summertime at location F?
   A) The rotation of Earth
   B) The core of Earth
   C) The tilt of Earth on its axis
   D) The phases of the Moon

2. In the picture of Earth, it is nighttime at location _____.
   F) F   H) H
   G) G   J) J

3. Why would location G not experience very much seasonal temperature change?
   A) It is facing the Sun.
   B) It is near Earth's equator.
   C) It is in the southern hemisphere.
   D) It is in the northern hemisphere.

CHAPTER ASSESSMENT 381

Reading Comprehension

Read the passage. Then read each question that follows the passage. Decide which is the best answer to each question.

Teamwork: *The International Space Station*

Sixteen different countries have been working together since 1998 to design and put together a space station. When the *International Space Station* is finished, it will be spacious enough for three astronauts to live and work inside it for months. It will have the same amount of room inside it as a 747 jumbo jet.

The space station will be so big that it needs to be brought into space in parts. It will take more than 40 flights to ship everything that is needed. A Russian rocket already has brought the first piece of the space station into orbit, and a U.S. space shuttle also has brought many of the pieces and supplies into orbit. Astronauts are using robotic arms to connect the different pieces together.

The completed space station will be able to house seven science laboratories inside it. Because the space station is orbiting Earth, there will only be a slight force of gravity on people and objects inside the space station. This weaker gravity is called microgravity. Scientists are excited to try experiments in space because they are interested in seeing how materials and organisms work differently when there is very little gravity to affect them. They hope to study new drugs and new treatments for diseases like cancer. They will study how flames and materials behave differently when they are in microgravity.

The first crew to live and work in the *International Space Station* arrived in November 2000. There were one American astronaut, Bill Shepard, and two Russian cosmonauts, Yuri Gidenko and Sergei Krikalev. Many countries will be watching the space station, curious to see what life is like in this little home so far above Earth.

Test-Taking Tip List important details on a separate page as you read.

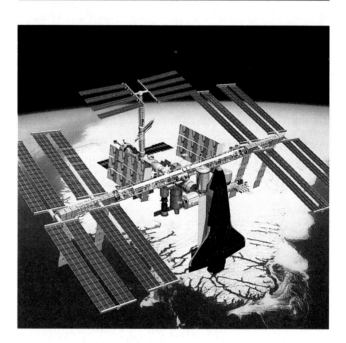

1. In this passage, the word <u>international</u> means ____. *Reading TEKS 6.9 B*
   A) found in orbit in outer space
   B) involving more than one country
   C) belonging to the United States
   D) not connected to anyone else

2. According to the passage, the parts of the space station are being put together ____. *Reading TEKS 6.10 G*
   F) using robotic arms
   G) on Earth before the space station is launched
   H) inside a U.S. space shuttle
   J) inside a laboratory in the space station

TAKS Practice

Reasoning and Skills

Read each question and choose the best answer.

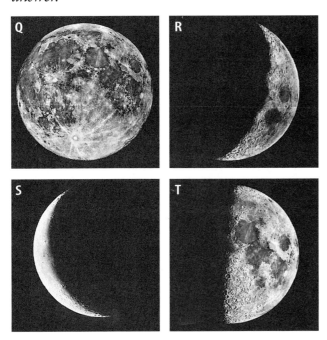

1. Which of the above pictures depicts the moon in its first quarter phase?
   Science TEKS 6.13 A
   A) Q
   B) R
   C) S
   D) T

   Test-Taking Tip Think about the meaning of the word "quarter." Select the image that looks like a quarter of the Moon.

2. What is shown in the lower left photo?
   Science TEKS 6.5 A; 6.13 A
   F) a solar eclipse
   G) a new moon
   H) a lunar eclipse
   J) a comet

   Test-Taking Tip In the picture, there is a shadow moving across the surface of the Moon.

3. Which of these best explains why temperatures on Venus are so much hotter than temperatures on Earth? *Science TEKS 6.13 A*
   A) Winters are longer on Earth than on Venus.
   B) Venus has a thick atmosphere composed mainly of carbon dioxide.
   C) Earth is tilted toward the Sun.
   D) Earth is closer to the Sun than Venus.

   Test-Taking Tip Review what causes the greenhouse effect.

Consider this question carefully before writing your answer on a separate sheet of paper.

4. Scientists often use the unit light-years to measure distances between galaxies and stars. One light-year is the distance light travels in one year, or about 9.5 trillion kilometers. Why would scientists use light years rather than kilometers to measure distance in space? *Science TEKS 6.13 B*

   Test-Taking Tip Think about the numbers scientists would have to use if they measured distances in the universe in kilometers. Then carefully write your answer.

STANDARDIZED TEST PRACTICE 383

UNIT 5
Living Systems

How Are Seaweed & Cell Cultures Connected?

In the 1800s, many biologists were interested in studying one-celled microorganisms. But to study them, the researchers needed to grow, or culture, large numbers of these cells. And to culture them properly, they needed a solid substance on which the cells could grow. One scientist tried using nutrient-enriched gelatin, but the gelatin had drawbacks. It melted at relatively low temperatures—and some microorganisms digested it. Fannie Eilshemius Hesse came up with a better option. She had been solidifying her homemade jellies using a substance called agar, which is derived from red seaweed (such as the one seen in the background here). It turned out that nutrient-enriched agar worked perfectly as a substance on which to culture cells. On the two types of agar in the dishes below, so many cells have grown that, together, they form dots and lines.

SCIENCE CONNECTION

CELL REPRODUCTION Under ideal conditions, some one-celled microorganisms can reproduce very quickly by cell division. Suppose you placed one cell in a dish of nutrient-enriched agar. Twenty minutes later, the cell divided to form two cells. Assuming that the cells continue to divide every twenty minutes, how many would be in the dish an hour after the first division? Two hours after the first division? Make a graph that illustrates the pattern of cell reproduction.

CHAPTER 13

Science TEKS 6.10 A, B, C; 6.11 B; 6.12 A, B

Life's Structure and Function

The world around you is filled with organisms that you could overlook, or even be unable to see. Some of these organisms are one-celled and some are many-celled. The monster in this photograph is a louse crawling across human skin. It can be seen in great detail with a microscope that is found in many classrooms. You can study the cells of smaller organisms with other kinds of microscopes.

What do you think?

Science Journal Look at the picture below with a classmate. Discuss what you think is happening. Here's a hint: *Not every battlefield is found on land or at sea.* Write your answer or best guess in your Science Journal.

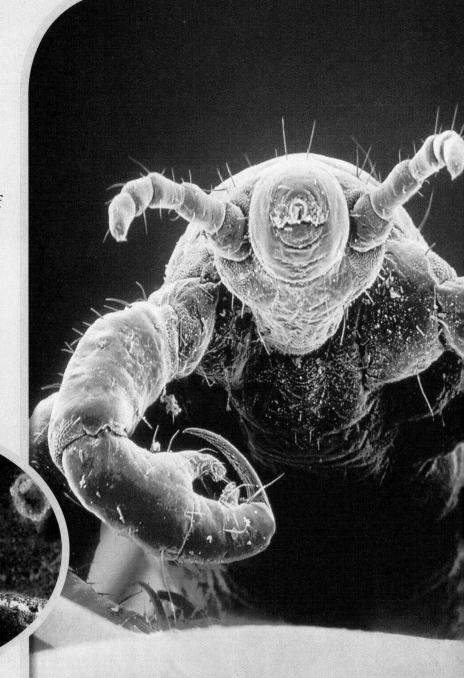

386

If you look around your classroom, you can see many things of all sizes. With the aid of a hand lens, you can see more details. You might examine a speck of dust and discover that it is a living or dead insect. In the following activity, use a hand lens to search for the smallest thing you can find in the classroom.

Measure a small object

1. Obtain a hand lens from your teacher. Note its power (the number followed by ×, shown somewhere on the lens frame or handle).

2. Using the hand lens, look around the room for the smallest object you can find.

3. Measure the size of the image as you see it with the hand lens. To estimate the real size of the object, divide that number by the power. For example, if it looks 2 cm long and the power is 10×, the real length is about 0.2 cm.

Observe

In your Science Journal, describe what you observe. Did the details become clearer? Explain.

Before You Read

Making a Main Ideas Study Fold Make the following Foldable to help you identify the main ideas or major topics on cells.

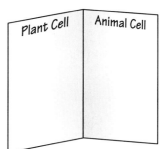

1. Place a sheet of paper in front of you so the long side is at the top. Fold the paper in half from the left side to the right side. Then unfold.

2. Label the left side of the paper *Plant Cell*. Label the right side of the paper *Animal Cell*, as shown.

3. Before you read the chapter, draw a plant cell on the left side of the paper and an animal cell on the right side of the paper.

4. As you read the chapter, change and add to your drawings.

SECTION 1

Cell Structure

As You Read

What You'll Learn
- **Identify** names and functions of each part of a cell.
- **Explain** how important a nucleus is in a cell.
- **Compare** tissues, organs, and organ systems.

Vocabulary
cell membrane
cytoplasm
cell wall
organelle
nucleus
chloroplast
mitochondrion
ribosome
endoplasmic reticulum
Golgi body
tissue
organ

Why It's Important
If you know how organelles function, it's easier to understand how cells survive.

Common Cell Traits

Living cells are dynamic and have several things in common. A cell is the smallest unit that is capable of performing life functions. All cells have an outer covering called a **cell membrane.** Inside every cell is a gelatinlike material called **cytoplasm** (SI toh plaz uhm). In the cytoplasm of every cell is hereditary material that controls the life of the cell.

Comparing Cells Cells come in many sizes. A nerve cell in your leg could be a meter long. A human egg cell is no bigger than the dot on this **i**. A human red blood cell is about one-tenth the size of a human egg cell. A bacterium is even smaller—8,000 of the smallest bacteria can fit inside one of your red blood cells.

A cell's shape might tell you something about its function. The nerve cell in **Figure 1** has many fine extensions that send and receive impulses to and from other cells. Though a nerve cell cannot change shape, muscle cells and some blood cells can. In plant stems, some cells are long and hollow and have openings at their ends. These cells carry food and water throughout the plant.

Figure 1
The shape of the cell can tell you something about its function. These cells are drawn 700 times their actual size.

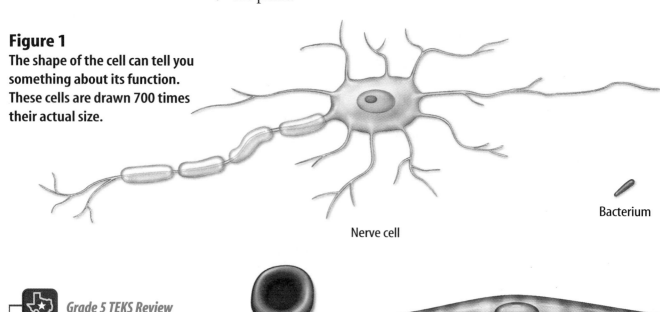

Nerve cell

Bacterium

Red blood cell

Muscle cell

Grade 5 TEKS Review
For a review of the Grade 5 TEKS *Systems* see page 497.

388 CHAPTER 13 Life's Structure and Function

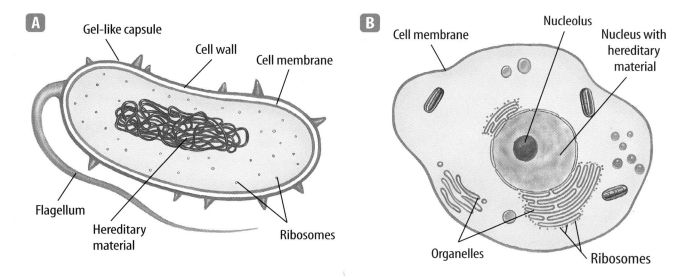

Cell Types
Scientists have found that cells can be separated into two groups. One group has no membrane-bound structures inside the cell and the other group does, as shown in **Figure 2**. Cells without membrane-bound structures are called prokaryotic (proh KAYR ee yah tihk) cells. Cells with membrane-bound structures are called eukaryotic (yew KAYR ee yah tihk) cells.

Reading Check *Into what two groups can cells be separated?*

Cell Organization
Each cell in your body has a specific function. You might compare a cell to a busy delicatessen that is open 24 hours every day. Raw materials for the sandwiches are brought in often. Some food is eaten in the store, and some customers take their food with them. Sometimes food is prepared ahead of time for quick sale. Wastes are put into trash bags for removal or recycling. Similarly, your cells are taking in nutrients, secreting and storing chemicals, and breaking down substances 24 hours every day.

Cell Wall
Just like a deli that is located inside the walls of a building, some cells are enclosed in a cell wall. The cells of plants, algae, fungi, and most bacteria are enclosed in a cell wall. **Cell walls** are tough, rigid outer coverings that protect the cell and give it shape.

A plant cell wall, as shown in **Figure 3**, mostly is made up of a carbohydrate called cellulose. The long, threadlike fibers of cellulose form a thick mesh that allows water and dissolved materials to pass through it. Cell walls also can contain pectin, which is used in jam and jelly, and lignin, which is a compound that makes cell walls rigid. Plant cells responsible for support have a lot of lignin in their walls.

Figure 2
Examine these drawings of cells.
A Prokaryotic cells are only found in one-celled organisms, such as bacteria. **B** Protists, fungi, plants and animals are made of eukaryotic cells. *What differences do you see between them?*

Figure 3
The protective cell wall of a plant cell is outside the cell membrane.

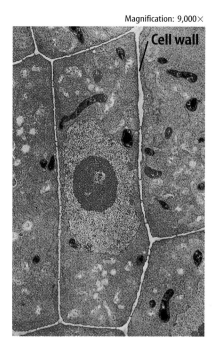

SECTION 1 Cell Structure

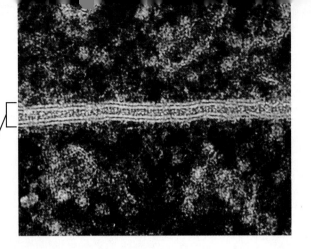

Figure 4
The cell membrane is made up of a double layer of fatlike molecules.

Cell membrane

Cell Membrane The protective layer around all cells is the cell membrane, as shown in **Figure 4.** If cells have cell walls, the cell membrane is inside of it. The cell membrane regulates interactions between the cell and the environment. Water is able to move freely into and out of the cell through the cell membrane. Food particles and some molecules enter and waste products leave through the cell membrane.

Cytoplasm Cells are filled with a gelatinlike substance called cytoplasm that constantly flows inside the cell membrane. Many important chemical reactions occur within the cytoplasm.

Throughout the cytoplasm is a framework called the cytoskeleton, which helps the cell maintain or change its shape. Cytoskeletons enable some cells to move. An amoeba, for example, moves by stretching and contracting its cytoskeleton. The cytoskeleton is made up of thin, hollow tubes of protein and thin, solid protein fibers, as shown in **Figure 5.** Proteins are organic molecules made up of amino acids.

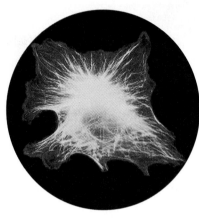

Figure 5
Cytoskeleton, a network of fibers in the cytoplasm, gives cells structure and helps them maintain shape.

Reading Check *What is the function of the cytoskeleton?*

Most of a cell's life processes occur in the cytoplasm. Within the cytoplasm of eukaryotic cells are structures called **organelles.** Some organelles process energy and others manufacture substances needed by the cell or other cells. Certain organelles move materials, while others act as storage sites. Most organelles are surrounded by membranes. The nucleus is usually the largest organelle in a cell.

Nucleus The nucleus is like the deli manager who directs the store's daily operations and passes on information to employees. The **nucleus,** shown in **Figure 6,** directs all cell activities and is separated from the cytoplasm by a membrane. Materials enter and leave the nucleus through openings in the membrane. The nucleus contains the instructions for everything the cell does. These instructions are found on long, threadlike, hereditary material made of DNA. DNA is the chemical that contains the code for the cell's structure and activities. During cell division, the hereditary material coils tightly around proteins to form structures called chromosomes. A structure called a nucleolus also is found in the nucleus.

Modeling Cytoplasm

Procedure
1. Add 100 mL of **water** to a clear container.
2. Add **unflavored gelatin** and stir.
3. Shine a **flashlight** through the solution.

Analysis
1. Describe what you see.
2. How does a model help you understand what cytoplasm might be like?

Figure 6
Refer to these diagrams of a typical animal cell (top) and plant cell (bottom) as you read about cell structures and their functions.

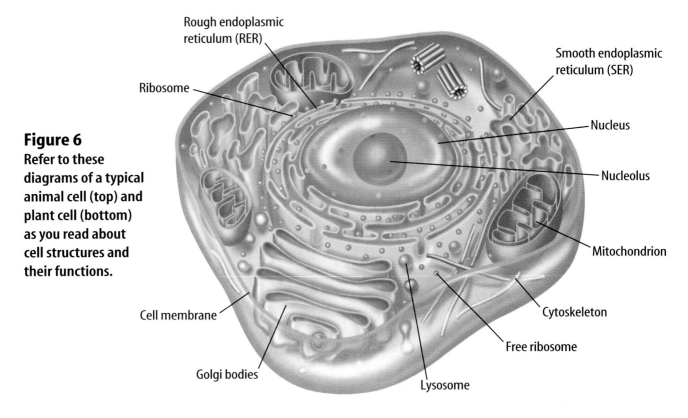

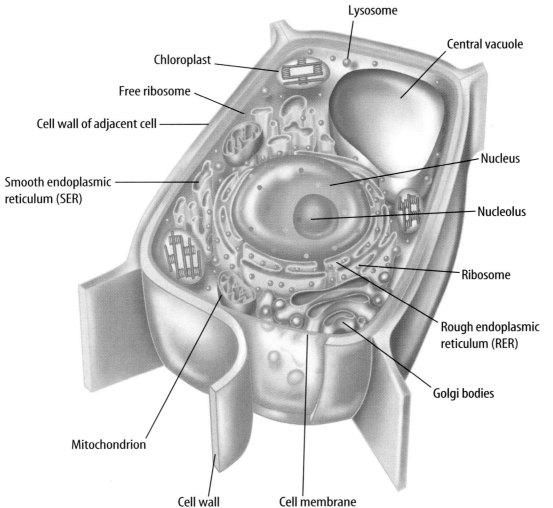

SECTION 1 Cell Structure **391**

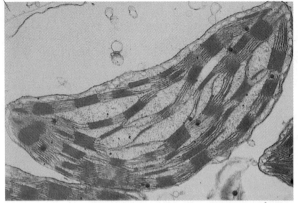
Magnification: 37,000×

Energy-Processing Organelles Cells require a continuous supply of energy to process food, make new substances, eliminate wastes, and communicate with each other. In plant cells, food is made in green organelles in the cytoplasm called **chloroplasts** (KLOR uh plasts), as shown in **Figure 7.** Chloroplasts contain the green pigment chlorophyll, which gives leaves and stems their green color. Chlorophyll captures light energy that is used to make a sugar called glucose. Glucose molecules store the captured light energy as chemical energy. Many cells, including animal cells, do not have chloroplasts for making food. They must get food from their environment.

The energy in food is stored until it is released by the mitochondria. **Mitochondria** (mi tuh KAHN dree uh) (singular, *mitochondrion*), such as the one shown in **Figure 8,** are organelles where energy is released from breaking down food into carbon dioxide and water. Just as the gas or electric company supplies fuel for the deli, a mitochondrion releases energy for use by the cell. Some types of cells, such as muscle cells, are more active than other cells. These cells have large numbers of mitochondria. Why would active cells have more or larger mitochondria?

Figure 7
Chloroplasts are organelles that use sunlight to make sugar from carbon dioxide and water. They contain chlorophyll, which gives most leaves and stems their green color.

Figure 8
Mitochondria are known as the powerhouses of the cell because they release energy that is needed by the cell from food.
What types of cells might contain many mitochondria?

Manufacturing Organelles One substance that takes part in nearly every cell activity is protein. Proteins are part of cell membranes. Other proteins are needed for chemical reactions that take place in the cytoplasm. Cells make their own proteins on small structures called **ribosomes.** Even though ribosomes are considered organelles, they are not membrane bound. Some ribosomes float freely in the cytoplasm; and others are attached to the endoplasmic reticulum. Ribosomes are made in the nucleolus and move out into the cytoplasm. Ribosomes receive directions from the hereditary material in the nucleus on how, when, and in what order to make specific proteins.

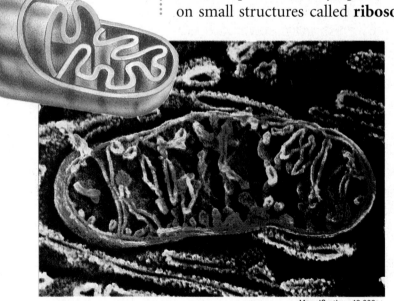
Magnification: 48,000×

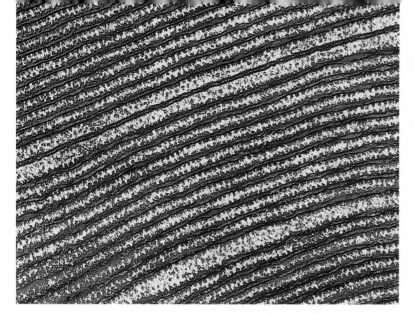

Figure 9
Endoplasmic reticulum (ER) is a complex series of membranes in the cytoplasm of the cell. *What would smooth ER look like?*

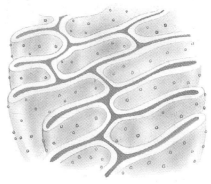

Processing, Transporting, and Storing Organelles

The **endoplasmic reticulum** (en duh PLAZ mihk • rih TIHK yuh lum) or ER, as shown in **Figure 9,** extends from the nucleus to the cell membrane. It is a series of folded membranes in which materials can be processed and moved around inside of the cell. The ER takes up a lot of space in some cells.

The endoplasmic reticulum may be "rough" or "smooth." ER that has no attached ribosomes is called smooth endoplasmic reticulum. This type of ER processes other cellular substances such as lipids that store energy. Ribsomes are attached to areas on the rough ER. There they carry out their job of making proteins that are moved out of the cell or used within the cell.

 *What is the difference between rough ER and smooth ER?*

After proteins are made in a cell, they are transferred to another type of cell organelle called the Golgi (GAWL jee) bodies. The **Golgi bodies**, as shown in **Figure 10,** are stacked, flattened membranes. The Golgi bodies sort proteins and other cellular substances and package them into membrane-bound structures called vesicles. The vesicles deliver cellular substances to areas inside the cell. They also carry cellular substances to the cell membrane where they are released to the outside of the cell.

Just as a deli has refrigerators for temporary storage of some its foods and ingredients, cells have membrane-bound spaces called vacuoles for the temporary storage of materials. A vacuole can store water, waste products, food, and other cellular materials. In plant cells, the vacuole may make up most of the cell's volume.

Figure 10
The Golgi body packages materials and moves them to the outside of the cell. *Why are materials removed from the cell?*

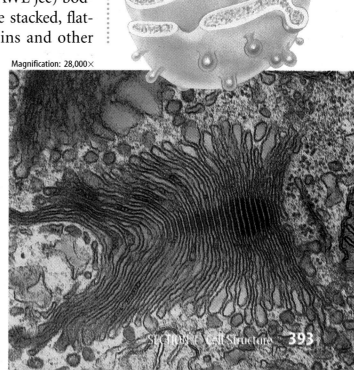

Magnification: 28,000×

Environmental Science INTEGRATION

Just like a cell, you can recycle materials. Paper, plastics, aluminum, and glass are materials that can be recycled into usable items. Make a promotional poster to encourage others to recycle.

Recycling Organelles Active cells break down and recycle substances. Organelles called lysosomes (LI suh sohmz) contain digestive chemicals that help break down food molecules, cell wastes, and worn-out cell parts. In a healthy cell, chemicals are released into vacuoles only when needed. The lysosome's membrane prevents the digestive chemicals inside from leaking into the cytoplasm and destroying the cell. When a cell dies, a lysosome's membrane disintegrates. This releases digestive chemicals that allow the quick breakdown of the cell's contents.

✔ **Reading Check** *What is the function of the lysosome's membrane?*

Math Skills Activity

MATH TEKS 6.2 C; 6.3 A, C; 6.8 B

Calculate the Ratio of Surface Area to Volume of Cells

Example Problem

Assume that a cell is like a cube with six equal sides. Find the ratio of surface area to volume for a cube that is 4 cm high.

Solution

1. *This is what you know:* A cube has 6 equal sides of 4 cm × 4 cm.

2. *This is what you want to find:* the ratio (R) of surface area to volume for each cube

3. *These are the equations you use:* surface area (A) = width × length × 6
   volume (V) = length × width × height
   R = A/V

4. *Solve for surface area and volume, then solve for the ratio:*
   $A = 4 \text{ cm} \times 4 \text{ cm} \times 6 = 96 \text{ cm}^2$
   $V = 4 \text{ cm} \times 4 \text{ cm} \times 4 \text{ cm} = 64 \text{ cm}^3$
   $R = 96 \text{ cm}^2 / 64 \text{ cm}^3 = 1.5 \text{ cm}^2/\text{cm}^3$

Check your answer by multiplying the ratio by the volume. Do you calculate the surface area?

Practice Problems

1. Calculate the ratio of surface area to volume for a cube that is 2 cm high. What happens to this ratio as the size of the cube decreases?

2. If a 4-cm cube doubled just one of its dimensions—length, width, or height—what would happen to the ratio of surface area to volume?

For more help, refer to the **Math Skills Handbook**.

From Cell to Organism

Many one-celled organisms perform all their life functions by themselves. Cells in a many-celled organism, however, do not work alone. Each cell carries on its own life functions while depending in some way on other cells in the organism.

In **Figure 11,** you can see cardiac muscle cells grouped together to form a tissue. A **tissue** is a group of similar cells that work together to do one job. Each cell in a tissue does its part to keep the tissue alive.

Tissues are organized into organs. An **organ** is a structure made up of two or more different types of tissues that work together. Your heart is an organ made up of cardiac muscle tissue, nerve tissue, and blood tissues. The cardiac muscle tissue contracts, making the heart pump. The nerve tissue brings messages that tell the heart how fast to beat. The blood tissue is carried from the heart to other organs of the body.

 *What type of tissues make up your heart?*

A group of organs working together to perform a certain function is an organ system. Your heart, arteries, veins, and capillaries make up your cardiovascular system. In a many-celled organism, several systems work together in order to perform life functions efficiently. Your nervous, circulatory, respiratory, muscular, and other systems work together to keep you alive.

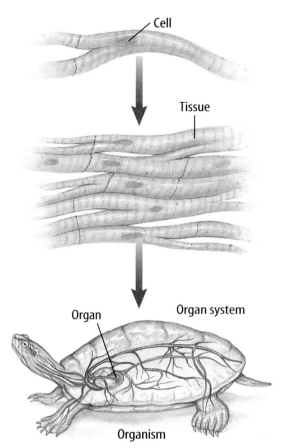

Figure 11
In a many-celled organism, cells are organized into tissues, tissues into organs, organs into systems, and systems into an organism.

Section Assessment

1. Explain the important role of the nucleus in the life of a cell.
2. Compare and contrast the energy processing organelles.
3. Why are digestive enzymes in a cell enclosed in a membrane-bound organelle?
4. How are cells, tissues, organs, and organ systems related?
5. **Think Critically** Identify how the spindle shape of muscle cells relates to what happens when you contract a muscle in your arm.

Skill Builder Activities

6. **Interpreting Scientific Illustrations** Examine the illustrations of the animal cell and the plant cell in **Figure 6** and make a list of differences and similarities between them. **For more help, refer to the** Science Skill Handbook.

7. **Communicating** Your textbook compared some cell functions to that of a deli. In your Science Journal, write an essay that explains how a cell is like your school or town. **For more help, refer to the** Science Skill Handbook.

Activity

Comparing Cells

If you compared a goldfish to a rose, you would find them unlike each other. Are their individual cells different also? Try this activity to compare plant and animal cells.

What You'll Investigate
How do human cheek cells and plant cells compare?

Materials
microscope
microscope slide
coverslip
forceps
tap water
dropper
Elodea plant
prepared slide of human cheek cells

Goal
- **Compare and contrast** an animal cell and a plant cell.

Safety Precautions

Procedure
1. Copy the data table in your Science Journal. Check off the cell parts as you observe them.

Cell Observations		
Cell Part	**Cheek**	***Elodea***
Cytoplasm		
Nucleus		
Chloroplasts		
Cell Wall		
Cell Membrane		

2. Using forceps, make a wet-mount slide of a young leaf from the tip of an *Elodea* plant.

3. **Observe** the leaf on low power. Focus on the top layer of cells.

4. Switch to high power and focus on one cell. In the center of the cell is a membrane-bound organelle called the central vacuole. Observe the chloroplasts—the green, disk-shaped objects moving around the central vacuole. Try to find the cell nucleus. It looks like a clear ball.

5. **Draw** the *Elodea* cell. Label the cell wall, cytoplasm, chloroplasts, central vacuole, and nucleus. Return to low power and remove the slide. Properly dispose of the slide.

6. **Observe** the prepared slide of cheek cells under low power.

7. Switch to high power and observe the cell nucleus. Draw and label the cell membrane, cytoplasm, and nucleus. Return to low power and remove the slide.

Conclude and Apply
1. **Compare and contrast** the shapes of the cheek cell and the *Elodea* cell.
2. What can you conclude about the differences between plant and animal cells?

Communicating Your Data
Draw the two kinds of cells on one sheet of paper. Use a green pencil to label the organelles found only in plants, a red pencil to label the organelles found only in animals, and a blue pencil to label the organelles found in both. **For more help, refer to the Science Skill Handbook.**

SECTION 2 Viewing Cells

Magnifying Cells

The number of living things in your environment that you can't see is much greater than the number that you can see. Many of the things that you cannot see are only one cell in size. To see most cells, you need to use a microscope.

Trying to see separate cells in a leaf, like the ones in **Figure 12,** is like trying to see individual photos in a photo mosaic picture that is on the wall across the room. As you walk toward the wall, it becomes easier to see the individual photos. When you get right up to the wall, you can see details of each small photo. A microscope has one or more lenses that enlarge the image of an object as though you are walking closer to it. Seen through these lenses, the leaf appears much closer to you, and you can see the individual cells that carry on life processes.

Early Microscopes In the late 1500s, the first microscope was made by a Dutch maker of reading glasses. He put two magnifying glasses together in a tube and got an image that was larger than the image that was made by either lens alone.

In the mid 1600s, Antonie van Leeuwenhoek, a Dutch fabric merchant, made a simple microscope with a tiny glass bead for a lens, as shown in **Figure 13.** With it, he reported seeing things in pond water that no one had ever imagined. His microscope could magnify up to 270 times. Another way to say this is that his microscope could make the image of an object 270 times larger than its actual size. Today you would say his lens had a power of 270×. Early compound microscopes were crude by today's standards. The lenses would make an image larger, but it wasn't always sharp or clear.

As You Read

What **You'll Learn**
- **Compare** the differences between the compound light microscope and the electron microscope.
- **Summarize** the discoveries that led to the development of the cell theory.
- **Relate** the cell theory to modern biology.

Vocabulary
cell theory

Why **It's Important**
Humans are like other living things because they are made of cells.

Figure 12
Individual cells become visible when a plant leaf is viewed using a microscope with enough magnifying power.

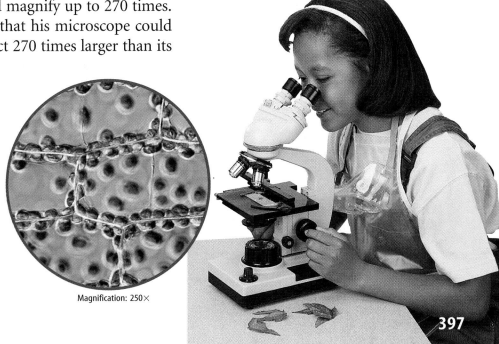

Magnification: 250×

NATIONAL GEOGRAPHIC VISUALIZING MICROSCOPES

Figure 13

Microscopes give us a glimpse into a previously invisible world. Improvements have vastly increased their range of visibility, allowing researchers to study life at the molecular level. A selection of these powerful tools—and their magnification power—is shown here.

▶ **Up to 250×** LEEUWENHOEK MICROSCOPE Held by a modern researcher, this historic microscope allowed Leeuwenhoek to see clear images of tiny freshwater organisms that he called "beasties."

▼ **Up to 2,000×** BRIGHTFIELD / DARKFIELD MICROSCOPE The light microscope is often called the brightfield microscope because the image is viewed against a bright background. A brightfield microscope is the tool most often used in laboratories to study cells. Placing a thin metal disc beneath the stage, between the light source and the objective lenses, converts a brightfield microscope to a darkfield microscope. The image seen using a darkfield microscope is bright against a dark background. This makes details more visible than with a brightfield microscope. Below are images of a *Paramecium* as seen using both processes.

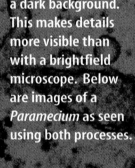

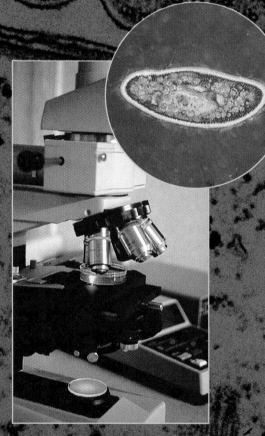

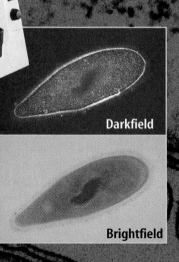

Darkfield

Brightfield

▲ **Up to 1,500×** FLUORESCENCE MICROSCOPE This type of microscope requires that the specimen be treated with special fluorescent stains. When viewed through this microscope, certain cell structures or types of substances glow, as seen in the image of a *Paramecium* above.

▶ **Up to 1,000,000×** **TRANSMISSION ELECTRON MICROSCOPE** A TEM aims a beam of electrons through a specimen. Denser portions of the specimen allow fewer electrons to pass through and appear darker in the image. Organisms, such as the *Paramecium* at right, can only be seen when the image is photographed or shown on a monitor. A TEM can mag-nify hundreds of thousands of times.

◀ **Up to 1,500×** **PHASE-CONTRAST MICROSCOPE** A phase-contrast microscope emphasizes slight differences in a specimen's capacity to bend light waves, thereby enhancing light and dark regions without the use of stains. This type of microscope is especially good for viewing living cells, like the *Paramecium* above left. The images from a phase-contrast microscope can only be seen when the specimen is photographed or shown on a monitor.

▶ **Up to 200,000×** **SCANNING ELECTRON MICROSCOPE** An SEM sweeps a beam of electrons over a specimen's surface, causing other electrons to be emitted from the specimen. SEMs produce realistic, three-dimensional images, which can only be viewed as photographs or on a monitor, as in the image of the *Paramecium* at right. Here a researcher compares an SEM picture to a computer monitor showing an enhanced image.

SECTION 2 Viewing Cells **399**

TRY AT HOME Mini LAB

Observing Magnified Objects

Procedure
1. Look at a **newspaper** through the curved side and through the flat bottom of an **empty, clear glass.**
2. Look at the newspaper through a **clear glass bowl** filled with **water** and then with a **magnifying glass.**

Analysis
In your Science Journal, compare how well you can see the newspaper through each of the objects.

Physics INTEGRATION

A magnifying glass is a convex lens. All microscopes use one or more convex lenses. In your Science Journal, diagram a convex lens and describe its shape.

Modern Microscopes Scientists use a variety of microscopes to study organisms, cells, and cell parts that are too small to be seen with the human eye. Depending on how many lenses a microscope contains, it is called simple or compound. A simple microscope is similar to a magnifying glass. It has only one lens. A microscope's lens makes an enlarged image of an object and directs light toward your eye. The change in apparent size produced by a microscope is called magnification. Microscopes vary in powers of magnification. Some microscopes can make images of individual atoms.

The microscope you probably will use to study life science is a compound light microscope, similar to the one in the Reference Handbook at the back of this book. The compound light microscope has two sets of lenses—eyepiece lenses and objective lenses. The eyepiece lenses are mounted in one or two tubelike structures. Images of objects viewed through two eyepieces, or stereomicroscopes, are three-dimensional. Images of objects viewed through one eyepiece are not. Compound light microscopes usually have two to four movable objective lenses.

Magnification The powers of the eyepiece and objective lenses determine the total magnifications of a microscope. If the eyepiece lens has a power of 10× and the objective lens has a power of 43×, then the total magnification is 430× (10× times 43×). Some compound microscopes, like those in **Figure 13,** have more powerful lenses that can magnify an object up to 2,000 times its original size.

Electron Microscopes Things that are too small to be seen with other microscopes can be viewed with an electron microscope. Instead of using lenses to direct beams of light, an electron microscope uses a magnetic field in a vacuum to direct beams of electrons. Some electron microscopes can magnify images up to one million times. Electron microscope images must be photographed or electronically produced.

Several kinds of electron microscopes have been invented, as shown in **Figure 13.** Scanning electron microscopes (SEM) produce a realistic, three-dimensional image. Only the surface of the specimen can be observed using an SEM. Transmission electron microscopes (TEM) produce a two-dimensional image of a thinly-sliced specimen. Details of cell parts can be examined using a TEM. Scanning tunneling microscopes (STM) are able to show the arrangement of atoms on the surface of a molecule. A metal probe is placed near the surface of the specimen and electrons flow from the tip. The hills and valleys of the specimen's surface are mapped.

400 CHAPTER 13 Life's Structure and Function

Development of the Cell Theory

During the seventeenth century, scientists used their new invention, the microscope, to explore the newly discovered microscopic world. They examined drops of blood, scrapings from their own teeth, and other small things. Cells weren't discovered until the microscope was improved. In 1665, Robert Hooke cut a thin slice of cork and looked at it under his microscope. To Hooke, the cork seemed to be made up of empty little boxes, which he named cells.

Table 1 The Cell Theory

All organisms are made up of one or more cells.	An organism can be one cell or many cells like most plants and animals.
The cell is the basic unit of organization in organisms.	Even in complex organisms, the cell is the basic unit of structure and function.
All cells come from cells.	Most cells can divide to form two new, identical cells.

In the 1830s, Matthias Schleiden used a microscope to study plant parts. He concluded that all plants are made of cells. Theodor Schwann, after observing many different animal cells, concluded that all animals also are made up of cells. Eventually, they combined their ideas and became convinced that all living things are made of cells.

Several years later, Rudolf Virchow hypothesized that cells divide to form new cells. Virchow proposed that every cell came from a cell that already existed. His observations and conclusions and those of others are summarized in the **cell theory,** as described in **Table 1.**

 *Who made the conclusion that all animals are made of cells?*

Section Assessment

1. Explain why the invention of the microscope was important in the study of cells.
2. What is stated in the cell theory?
3. What is the difference between a simple and a compound light microscope?
4. What was Virchow's contribution to the cell theory?
5. **Think Critically** Why would it be better to look at living cells than at dead cells?

Skill Builder Activities

6. **Concept Mapping** Using a network tree concept map, compare a compound light microscope to an electron microscope. **For more help, refer to the** Science Skill Handbook.

7. **Solving One-Step Equations** Calculate the magnifications of a microscope that has an $8\times$ eyepiece, and $10\times$ and $40\times$ objectives. **For more help, refer to the** Math Skill Handbook.

SECTION 3 Viruses

As You Read

What You'll Learn
- **Explain** how a virus makes copies of itself.
- **Identify** the benefits of vaccines.
- **Investigate** some uses of viruses.

Vocabulary
virus
host cell

Why It's Important
Viruses infect nearly all organisms, usually affecting them negatively yet sometimes affecting them positively.

What are viruses?

Cold sores, measles, chicken pox, colds, the flu, and AIDS are diseases caused by nonliving particles called viruses. A **virus** is a strand of hereditary material surrounded by a protein coating. Viruses don't have a nucleus or other organelles. They also lack a cell membrane. Viruses, as shown in **Figure 14,** have a variety of shapes. Because they are too small to be seen with a light microscope, they were discovered only after the electron microscope was invented. Before that time, scientists only hypothesized about viruses.

How do viruses multiply?

All viruses can do is make copies of themselves. However, they can't do that without the help of a living cell called a **host cell.** Crystalized forms of some viruses can be stored for years. Then, if they enter an organism, they can multiply quickly.

Once a virus is inside of a host cell, the virus can act in two ways. It can either be active or it can become latent, which is an inactive stage.

Figure 14
Viruses come in a variety of shapes.

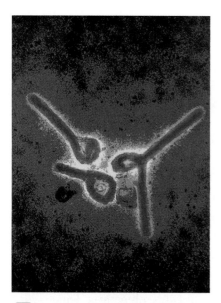

A Filoviruses do not have uniform shapes. Some of these *Ebola* viruses have a loop at one end.

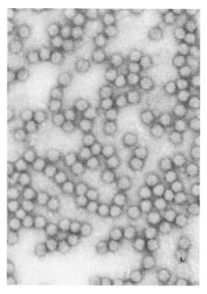

B The potato leafroll virus, *Polervirus*, damages potato crops worldwide.

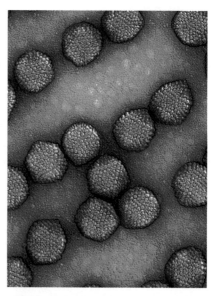

C This is just one of the many adenoviruses that can cause the common cold.

Figure 15
An active virus multiplies and destroys the host cell.
A The virus attaches to a specific host cell. **B** The virus's hereditary material enters the host cell. **C** The hereditary material of the virus causes the cell to make viral hereditary material and proteins. **D** New viruses form inside of the host cell. **E** New viruses are released as the host cell bursts open and is destroyed.

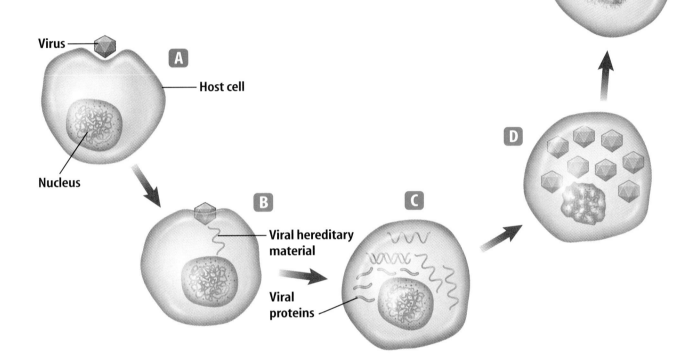

Active Viruses When a virus enters a cell and is active, it causes the host cell to make new viruses. This process destroys the host cell. Follow the steps in **Figure 15** to see one way that an active virus functions inside a cell.

Latent Viruses Some viruses can be latent. That means that after the virus enters a cell, its hereditary material can become part of the cell's hereditary material. It does not immediately make new viruses or destroy the cell. As the host cell reproduces, the viral DNA is copied. A virus can be latent for many years. Then, at any time, certain conditions, either inside or outside your body, can activate the virus.

If you have had a cold sore on your lip, a latent virus in your body has become active. The cold sore is a sign that the virus is active and destroying cells in your lip. When the cold sore disappears, the virus has become latent again. The virus is still in your body's cells, but it is hiding and doing no apparent harm.

Research Visit the Glencoe Science Web site at **tx.science.glencoe.com** for information on viruses. What environmental stimuli might activate a latent virus? Record your answer in your Science Journal.

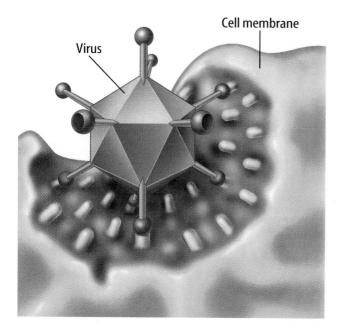

Figure 16
Viruses and the attachment sites of the host cell must match exactly. That's why most viruses infect only one kind of host cell.

How do viruses affect organisms?

Viruses attack animals, plants, fungi, protists, and all prokaryotes. Some viruses can infect only specific kinds of cells. For instance, many viruses, such as the potato leafroll virus, are limited to one host species or to one type of tissue within that species. A few viruses affect a broad range of hosts. An example of this is the rabies virus. Rabies can infect humans and many other animal hosts.

A virus cannot move by itself, but it can reach a host's body in several ways. For example, it can be carried onto a plant's surface by the wind or it can be inhaled by an animal. In a viral infection, the virus first attaches to the surface of the host cell. The virus and the place where it attaches must fit together exactly, as shown in **Figure 16**. Because of this, most viruses attack only one kind of host cell.

Viruses that infect bacteria are called bacteriophages (bak TIHR ee uh fay juhz). They differ from other kinds of viruses in the way that they enter bacteria and release their hereditary material. Bacteriophages attach to a bacterium and inject their hereditary material. The entire cycle takes about 20 min, and each virus-infected cell releases an average of 100 viruses.

Fighting Viruses

Vaccines are used to prevent disease. A vaccine is made from weakened virus particles that can't cause disease anymore. Vaccines have been made to prevent many diseases, including measles, mumps, smallpox, chicken pox, polio, and rabies.

Reading Check *What is a vaccine?*

The First Vaccine Edward Jenner is credited with developing the first vaccine in 1796. He developed a vaccine for smallpox, a disease that was still feared in the early twentieth century. Jenner noticed that people who got a disease called cowpox didn't get smallpox. He prepared a vaccine from the sores of people who had cowpox. When injected into healthy people, the cowpox vaccine protected them from smallpox. Jenner didn't know he was fighting a virus. At that time, no one understood what caused disease or how the body fought disease.

SCIENCE Online

Collect Data Scientists have determined that *Marburg* virus, *Ebola zaire,* and *Ebola reston* belong to the virus family Filoviridae. Visit the Glencoe Science Web site at **tx.science.glencoe.com** for the latest information about these viruses. Share your results with your class.

Treating and Preventing Viral Diseases Antibiotics are used to treat bacterial infections. They are ineffective against any viral disease. One way your body can stop viral infections is by making interferons. Interferons are proteins that protect cells from viruses. These proteins are produced rapidly by infected cells and move to noninfected cells in the host. They cause the noninfected cells to produce protective substances.

Antiviral drugs can be given to infected patients to help fight a virus. A few drugs show some effectiveness against viruses but some have limited use because of their adverse side effects.

Public health measures for preventing viral diseases include vaccinating people, improving sanitary conditions, quarantining patients, and controlling animals that spread the disease. Yellow fever was wiped out completely in the United States through mosquito-control programs. Annual rabies vaccinations protect humans by keeping pets and farm animals free from infection. To control the spread of rabies in wild animals such as coyotes and wolves, wildlife workers place bait containing an oral rabies vaccine, as shown in **Figure 17,** where wild animals will find it.

Research with Viruses

You might think viruses are always harmful. However, through research, scientists are discovering helpful uses for some viruses. One use, called gene therapy, is being tried on cells with defective genes. Normal hereditary material is substituted for a cell's defective hereditary material. The normal material is enclosed in viruses. The viruses then "infect" targeted cells, taking the new hereditary material into the cells to replace the defective hereditary material. Using gene therapy, scientists hope to help people with genetic disorders and find a cure for cancer.

Figure 17
This oral rabies bait is being prepared for an aerial drop by the Texas Department of Health as part of their Oral Rabies Vaccination Program. This five-year program has prevented the expansion of rabies into Texas.

Section 3 Assessment

1. Describe the structure of viruses and explain how viruses multiply.
2. How are vaccines beneficial?
3. How might some viruses be helpful?
4. How might viral diseases be prevented?
5. **Think Critically** Explain why a doctor might not give you any medication if you have a viral disease.

Skill Builder Activities

6. **Concept Mapping** Make an events chain concept map to show what happens when a latent virus becomes active. **For more help, refer to the** Science Skill Handbook.
7. **Using a Word Processor** Make an outline of the cycle of an active virus. **For more help, refer to the** Technology Skill Handbook.

Activity: Design Your Own Experiment

Comparing Light Microscopes

You're a technician in a police forensic laboratory. You use a stereomicroscope and a compound light microscope in the laboratory. A detective just returned from a crime scene with bags of evidence. You must examine each piece of evidence under a microscope. How do you decide which microscope is the best tool to use?

Recognize the Problem

Will all of the evidence that you've collected be viewable through both microscopes?

Form a Hypothesis

Compare the items to be examined under the microscopes. Which microscope will be used for each item?

Possible Materials

compound light microscope
stereomicroscope
items from the classroom—include some living or once-living items (8)
microscope slides and coverslips
plastic petri dishes
distilled water
dropper

Goals

- **Learn** how to correctly use a stereomicroscope and a compound light microscope.
- **Compare** the uses of the stereomicroscope and compound light microscope.

Safety Precautions

Thoroughly wash your hands when you have completed this experiment.

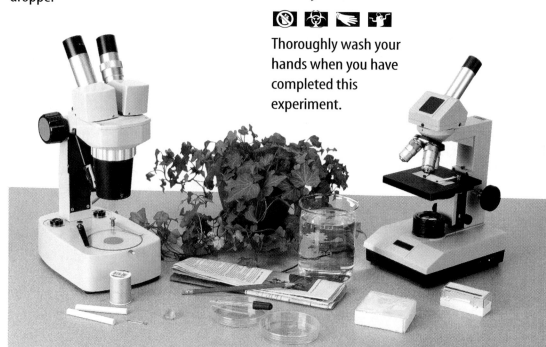

Using Scientific Methods

Test Your Hypothesis

Plan

1. As a group, decide how you will test your hypothesis.
2. **Describe** how you will carry out this experiment using a series of specific steps. Make sure the steps are in a logical order. Remember that you must place an item in the bottom of a plastic petri dish to examine it under the stereomicroscope and you must make a wet mount of any item to be examined under the compound light microscope. For more help, see the Reference Handbook.
3. If you need a data table or an observation table, design one in your Science Journal.

Do

1. Make sure your teacher approves the objects you'll examine, your plan, and your data table before you start.
2. Carry out the experiment.
3. While doing the experiment, record your observations and complete the data table.

Analyze Your Data

1. **Compare** the items you examined with those of your classmates.
2. Based on this experiment, classify the eight items you observed.

Draw Conclusions

1. **Infer** which microscope a scientist might use to examine a blood sample, fibers, and live snails.
2. **List** five careers that require people to use a stereomicroscope. List five careers that require people to use a compound light microscope. Enter the lists in your Science Journal.
3. If you examined an item under a compound light microscope and a stereomicroscope, how would the images differ?
4. Which microscope was better for looking at large, or possibly live items?

In your Science Journal, **write** a short description of an imaginary crime scene and the evidence found there. **Sort** the evidence into two lists—items to be examined under a stereomicroscope and items to be examined under a compound light microscope. **For more help, refer to the Science Skill Handbook.**

ACTIVITY **407**

TIME SCIENCE AND HISTORY

SCIENCE CAN CHANGE THE COURSE OF HISTORY!

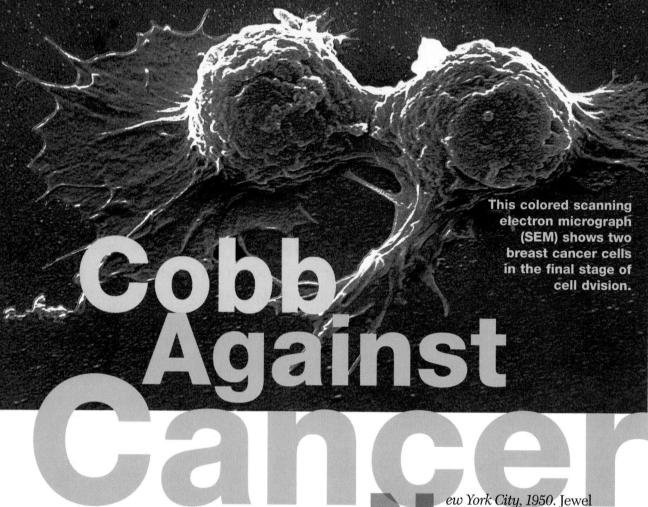

This colored scanning electron micrograph (SEM) shows two breast cancer cells in the final stage of cell dvision.

Magnification: 2,000×

Cobb Against Cancer

New York City, 1950. Jewel Plummer put yet another slide onto the stage of her microscope and clipped it into place. She switched to the high power objective, looked through the eyepiece, and turned the fine adjustment a tiny bit to bring her subject—cells from a cancerous tumor—into focus. She switched back to low power and removed the slide. She had found no change in the tumor cells. The drug that doctors had used wasn't killing or slowing the growth rate of those cancer cells. Sighing, she reached for the next slide. Maybe the slightly different drug they had used on that batch of cells would be the answer....

Jewel Plummer Cobb is a cell biologist who did important background research on the use of drugs against cancer. She removed cells from cancerous tumors and cultured them in the lab. Then, in a controlled study, she tried a series of different drugs against batches of the same cells. Her goal was to find the right drug to cure each patient's particular cancer. Cobb never met that goal, but her research laid the groundwork for modern chemotherapy—the use of chemicals to treat people with cancer.

Role Model

Jewel Cobb also influenced the course of science in a different way. She served as dean or president of several universities, retiring as president of the University of California at Fullerton. In her role as a college official, she was able to promote equal opportunity for students of all backgrounds, especially in the sciences.

Light Up a Cure

Vancouver, British Columbia, 2000. While Cobb herself was only able to infer what was going on inside a cell from its reactions to various drugs, her work has helped others go further. Building on Cobb's work, Professor Julia Levy and her research team at the University of British Columbia actually go inside cells and even inside organelles to work against cancer. One technique they are pioneering is the use of light to guide cancer drugs to the right cells. First, the patient is given a chemotherapy drug that reacts to light. Next, a fiber optic tube is inserted into the tumor. Finally, laser light is passed through the tube. The light activates the light-sensitive drug—but only in the tumor itself. This technique keeps healthy cells healthy but kills sick cells on the spot.

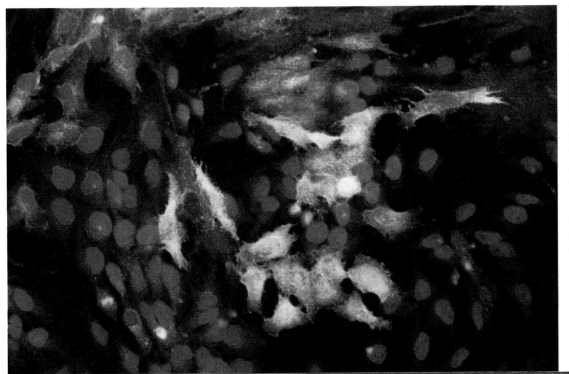

The image to the left shows human cervical cells magnified 125 times that have been attacked by cancer. The light blue areas at the center are keratin, a kind of protein. The cell nuclei are stained blue, and the red areas are fibroblasts, a kind of connective-tissue cell. These are the first human cells used to research cancer. This type of cell grows well in a lab, and is used in research worldwide.

CONNECTIONS Write Report on Cobb's experiments on cancer cells. What were her dependent and independent variables? What would she have used as a control? What sources of error did she have to guard against? Answer the same questions about Levy's work.

SCIENCE Online
For more information, visit tx.science.glencoe.com

Chapter 13 Study Guide

Reviewing Main Ideas

Section 1 Cell Structure

1. There are two basic cell types. Cells without membrane-bound structures are called prokaryotic cells. Cells with membrane-bound structures are called eukaryotic cells.

2. Most of the life processes of a cell occur within the cytoplasm.

3. Cell functions are performed by organelles under the control of DNA in the nucleus.

4. Organelles such as mitochondria and chloroplasts process energy.

5. Proteins take part in nearly every cell activity.

6. Golgi bodies and vacuoles transport substances, rid the cell of wastes, and store cellular materials. *What does this organelle do?*

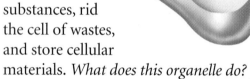

7. Most many-celled organisms are organized into tissues, organs, and organ systems that perform specific functions to keep an organism alive.

Section 2 Viewing Cells

1. A simple microscope has just one lens. A compound light microscope has eyepiece lenses and objective lenses.

2. To calculate the magnification of a microscope, multiply the power of the eyepiece by the power of the objective lens.

3. An electron microscope uses a beam of electrons instead of light to produce an image of an object.

4. Things that are too small to be viewed with a light microscope can be viewed with an electron microscope. This is an SEM of an ant. *How do you know if the ant is alive or dead?*

5. According to the cell theory, the cell is the basic unit of life. Organisms are made of one or more cells, and all cells come from other cells.

Section 3 Viruses

1. A virus is a structure containing hereditary material surrounded by a protein coating.

2. A virus can make copies of itself only when it is inside a living host cell.

3. Viruses cause diseases in animals, plants, fungi, and bacteria. *Why don't scientists consider viruses like these in the photo to be living organisms?*

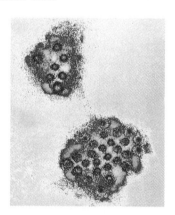

After You Read

On the inside of the Main Ideas Study Fold you made at the beginning of the chapter describe the characteristics of each type of cell.

Chapter 13 Study Guide

Visualizing Main Ideas

Complete the following concept map of the basic units of life.

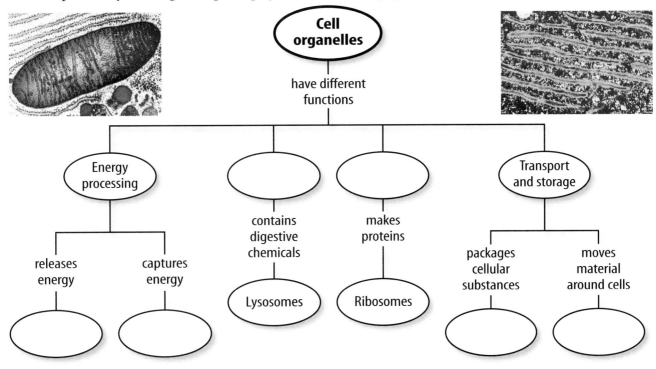

Vocabulary Review

Vocabulary Words

a. cell membrane
b. cell theory
c. cell wall
d. chloroplast
e. cytoplasm
f. endoplasmic reticulum
g. Golgi body
h. host cell
i. mitochondrion
j. nucleus
k. organ
l. organelle
m. ribosome
n. tissue
o. virus

 **Study Tip**

In order to understand the information that a graph is trying to communicate, write out a sentence that talks about the relationship between the *x*-axis and *y*-axis in the graph.

Using Vocabulary

Using the vocabulary words, give an example of each of the following.

1. found in every organ
2. smaller than one cell
3. a plant-cell organelle
4. part of every cell
5. powerhouse of a cell
6. used by biologists
7. contains hereditary material
8. a structure that surrounds the cell
9. can be damaged by a virus
10. made up of cells

Chapter 13 Assessment & TEKS Review

Checking Concepts

Choose the word or phrase that best answers the question.

1. What structure allows only certain things to pass in and out of the cell?
   A) cytoplasm   C) ribosomes
   B) cell membrane   D) Golgi body

2. Which microscope uses lenses to magnify?
   A) compound light microscope
   B) scanning electron microscope
   C) transmission electron microscope
   D) atomic force microscope

3. What is made of folded membranes that move materials around inside the cell?
   A) nucleus
   B) cytoplasm
   C) Golgi body
   D) endoplasmic reticulum

4. Which scientist gave the name *cells* to structures he viewed?
   A) Hooke   C) Schleiden
   B) Schwann   D) Virchow

5. What organelle helps recycle old cell parts?
   A) chloroplast   C) lysosome
   B) centriole   D) cell wall

6. Which of the following is a viral disease?
   A) tuberculosis   C) smallpox
   B) anthrax   D) tetanus

7. What are structures in the cytoplasm of a eukaryotic cell called?
   A) organs   C) organ systems
   B) organelles   D) tissues

8. Which microscope can magnify up to a million times?
   A) compound light microscope
   B) stereomicroscope
   C) transmission electron microscope
   D) atomic force microscope

9. Which of the following is part of a bacterial cell?
   A) a cell wall   C) mitochondria
   B) lysosomes   D) a nucleus

10. Which of the following do groups of different tissues form?
    A) organ   C) organ system
    B) organelle   D) organism

Thinking Critically

11. Why is it difficult to treat a viral disease?

12. What type of microscope would be best to view a piece of moldy bread? Explain.

13. What would happen to a plant cell that suddenly lost its chloroplasts?

14. What would happen to this animal cell if it didn't have ribosomes?

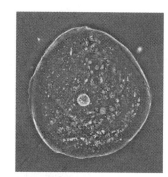

15. How would you decide whether an unknown cell was an animal cell, a plant cell, or a bacterial cell?

Developing Skills

16. **Concept Mapping** Make an events-chain concept map of the following from simple to complex: *small intestine, circular muscle cell, human,* and *digestive system.*

17. **Interpreting Scientific Illustrations** Use the illustrations in **Figure 1** to describe how the shape of a cell is related to its function.

18. **Making and Using Graphs** Use a computer to make a line graph of the following data. At 37°C there are 1.0 million viruses; at, 37.5°C, 0.5 million; at 37.8°C, 0.25 million; at 38.3°C, 0.1 million; and at 38.9°C, 0.05 million.

Chapter 13 Assessment

19. Comparing and Contrasting Complete the following table to compare and contrast the structures of a prokaryotic cell to those of a eukaryotic cell.

Cell Structures		
Structure	Prokaryotic Cell	Eukaryotic Cell
Cell Membrane		Yes
Cytoplasm	Yes	
Nucleus		Yes
Endoplasmic Reticulum		
Golgi Bodies		

20. Making a Model Make and illustrate a time line to show the development of the cell theory. Begin with the development of the microscope and end with Virchow. Include the contributions of Leeuwenhoek, Hooke, Schleiden, and Schwann.

Performance Assessment

21. Model Use materials that resemble cell parts or that represent their functions to make a model of a plant cell or an animal cell. Make a key to the cell parts to explain your model.

22. Poster Research the history of vaccinations. Contact your local Health Department for current information. Display your results on a poster.

Technology

Go to the Glencoe Science Web site at **tx.science.glencoe.com** or use the **Glencoe Science CD-ROM** for additional chapter assessment.

THE PRINCETON REVIEW TAKS Practice

A scientist is studying living cells. Below is an image of one of the cells that is being studied. This image represents what a scientist sees when he or she uses a tool in the laboratory. *TEKS 6.2 A, B*

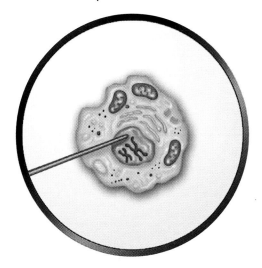

Closely examine the image above then answer the following questions.

1. If the pointer shown above with the cell is 10 micrometers in length, then about how wide is this cell?
   A) 20 micrometers
   B) 10 micrometers
   C) 5 micrometers
   D) 0.1 micrometers

2. Which of the following tools is the scientist probably using to view the living cell?
   F) telescope
   G) endoplasmic reticulum
   H) compound light microscope
   J) kaleidoscope

CHAPTER 14

Science TEKS 6.10A, B; 6.11A, B, C

The Role of Genes in Inheritance

Why don't all horses in a herd of wild mustangs look exactly alike? How does a foal develop inside its mother? In this chapter, you'll find the answers to these questions. You also will learn how plants produce new plants that can be the same as the one parent plant or can be different from the two parent plants. Also, you will learn how characteristics are passed from parents to their offspring.

What do you think?

Science Journal Look at the picture below with a classmate. Discuss what you think this might be or what is happening. Here's a hint: *They are all identical.* Write your answer or best guess in your Science Journal.

When you peel a banana or bite into an apple, you're probably only thinking about the taste and sweet smell of the fruit. You usually don't think about how and why the fruit was formed. Oranges, and most of the fruits you eat, contain seeds. Making seeds is one way that reproduction is carried out by living things. For life to continue, all living things must produce more living things that are similar to themselves.

Compare seeds

WARNING: *Do not eat the orange.*

1. Obtain half of an orange from your teacher. Peel the orange and remove all of the seeds.
2. Examine, count, and measure the length of each seed. Record these data in your Science Journal.
3. When you finish, dispose of your orange half as instructed by your teacher. Wash your hands.

Observe

Describe in your Science Journal why you think the seeds are different from one another.

Before You Read

Making a Know-Want-Learn Study Fold Make the following Foldable to help you identify what you already know and what you want to know about the role of genes in inheritance.

1. Place a sheet of paper in front of you so the short side is at the top. Fold the paper in half from top to bottom.
2. Fold both sides in, forming three equal sections. Unfold the paper so three sections show.
3. Through the top thickness of paper, cut along each of the fold lines to the topfold, forming three tabs. Label each tab *Know, Want,* and *Learned* as shown.
4. Before you read the chapter, write what you know about genes and inheritance under the left tab. Under the middle tab, write what you want to know.
5. As you read the chapter, add to or correct what you have written under the tabs.

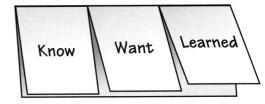

415

SECTION 1
Continuing Life

As You Read

What You'll Learn
- **Describe** how cells divide.
- **Identify** the importance of reproduction for living things.
- **Compare and contrast** sexual and asexual reproduction.
- **Describe** the structure and function of DNA.

Vocabulary
DNA
mitosis
asexual reproduction
cloning
sexual reproduction
sex cell
meiosis
fertilization

Why It's Important
All living things, including you, inherit characteristics from their parents.

Reproduction

If you look carefully in a pond in the spring, you may see frog or toad eggs. Frogs reproduce by laying hundreds of eggs in gooey clumps. Tadpoles can hatch from these eggs and mature into adult frogs, as shown in **Figure 1.** Some other kinds of organisms, including humans, usually produce only one offspring at a time. How do frogs and all of the other living things on Earth produce offspring that are similar to themselves?

The Importance of Reproduction Organisms produce offspring through the process of reproduction. Reproduction is important to all living things. Without reproduction, species could not continue. Hereditary material is passed from parent to offspring during reproduction. This material is found inside cells. It is made up of the chemical deoxyribonucleic (dee AHK sih ri boh noo klay ihk) acid, called DNA. **DNA** controls how offspring will look and how they will function by controlling what proteins each cell will produce. The DNA that all living things pass on determines many of their offspring's characteristics. Although organisms reproduce in different ways, reproduction always involves the transfer of hereditary information.

Figure 1
When frogs reproduce, they continue their species.

A Adult frogs reproduce by laying and then fertilizing eggs.

B These frog eggs can hatch into tadpoles.

C These tadpoles can develop into adult frogs.

Chemistry INTEGRATION

Life's Code You've probably seen or heard about science fiction movies in which DNA is used to grow prehistoric animals. What makes up DNA? How does it work?

DNA is found in all cells. If a cell has a nucleus, DNA is in structures called chromosomes. All of the information that is in your DNA is called your genetic information. You can think of DNA as a genetic blueprint that contains all of the instructions for making an organism what it is. Your DNA controls the texture of your hair, the shape of your ears, and even how you digest the food you had for lunch.

If you could look at DNA in detail, you would see that it is shaped like a twisted ladder. This structure, shown in **Figure 2,** is the key to how DNA works. The two sides of the ladder form the backbone of the DNA molecule. The sides support the rungs of the ladder. It is the rungs that hold all the genetic information. Each rung of the ladder is made up of a pair of chemicals called bases. There are only four bases, and they pair up very specifically. A DNA ladder has billions of rungs, and the bases are arranged in thousands of different orders. The secret of DNA has to do with the order or sequence of bases along the DNA ladder. The sequence forms a code. From this DNA code the cell gets instructions about what substances to make, how to make them, and when to make them.

SCIENCE *Online*

Research Visit the Glencoe Science Web site at **tx.science.glencoe.com** for more information about the Human Genome Project. Communicate with your class what you have learned.

Bases

Bases

Figure 2
The sequence of bases that are the rungs of the DNA molecule forms a code. This code contains the instructions for all of your body's characteristics and processes.

SECTION 1 Continuing Life

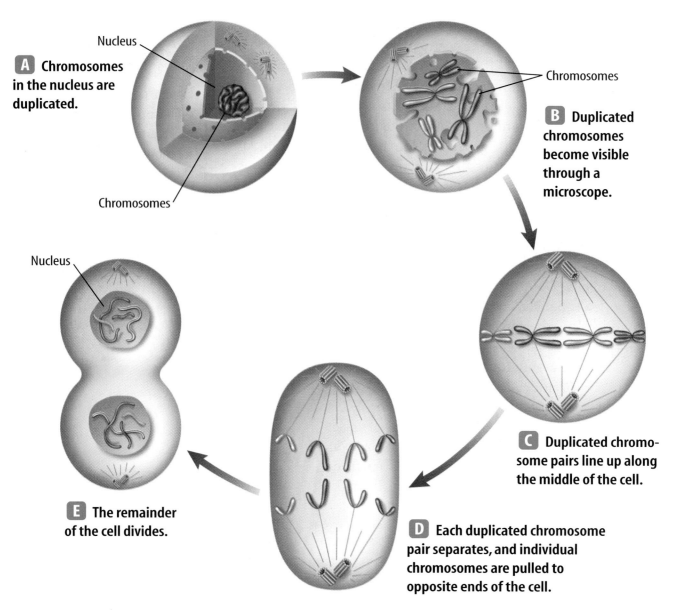

A Chromosomes in the nucleus are duplicated.

B Duplicated chromosomes become visible through a microscope.

C Duplicated chromosome pairs line up along the middle of the cell.

D Each duplicated chromosome pair separates, and individual chromosomes are pulled to opposite ends of the cell.

E The remainder of the cell divides.

Figure 3
During cell division, cells go through several steps to produce two cells with identical nuclei.

Cell Division

How did you become the size you are now? The cells of your body are formed by cell division. Cell division has two big steps. First, DNA in the nucleus is copied. Then the nucleus divides into two identical nuclei. Each new nucleus receives a copy of the DNA. Division of the nucleus is called **mitosis** (mi TOH sus). Mitosis is the process that results in two nuclei, each with the same genetic information. You can follow the process of mitosis in **Figure 3**. After mitosis has taken place, the rest of the cell divides into two cells of about equal size. Almost all the cells in any plant or animal undergo mitosis. Whether it occurs in a plant or an animal, cell division results in growth and replaces aging, missing, or injured cells.

Grade 5 TEKS Review

For a review of the Grade 5 TEKS strand *Characteristics of Offspring*, see page 511.

Reading Check *During cell division, why must the DNA be duplicated before the nucleus divides?*

Reproduction by One Organism

Shoots growing from the eyes of a potato are a form of reproduction. Reproduction in which a new organism is produced from a part of another organism by cell division is called **asexual** (ay SEK shul) **reproduction.** In asexual reproduction, all the DNA in the new organism comes from one other organism. The DNA of the growing potato eye is the same as the DNA in the rest of the potato.

Some one-celled organisms such as bacteria, divide in half, forming two cells. Before the one-celled organism divides, its DNA copies itself. After it has divided, each new organism has an exact copy of the first organism's DNA. The two new cells are alike. The first organism no longer exists.

Budding and Regeneration Many plants and species of mushroom, and even a few animals reproduce asexually. **Figure 4A** shows asexual reproduction in hydra, a relative of jellyfish and corals. When a hydra reproduces asexually, a new individual grows on it by a process called budding. As you can see, the hydra bud has the same shape and characteristics as the parent organism. The bud matures and eventually breaks away to live on its own.

In a process called regeneration (ree jen uh RAY shun), some organisms are able to replace body parts that have been lost because of an injury. Sea stars can grow a new arm if one is broken off. Lizards, such as chameleons, can grow a new tail if theirs is broken off, as shown in **Figure 4B.**

Mini LAB

Observing Yeast Budding

Procedure
1. Use a **dropper** to place a drop of a **prepared yeast** and **sugar mixture** onto a **microscope slide.** Place a **coverslip** on the slide.
2. Examine the slide with a **microscope** under low power, then high power.
3. Record your observations in your **Science Journal.**
4. Make a new slide after 5 min. Examine the slide under low power, then under high power.
5. Record your observations in your Science Journal.

Analysis
1. What did you observe on the first slide?
2. What might account for any differences between what you observed on the first slide and the second slide?

Figure 4
Cell division can result in asexual reproduction or replacement of body parts.

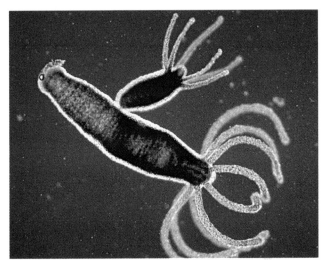

A Hydra reproduce asexually by budding.

B A chameleon can regenerate, or regrow, a tail if it is broken off.

Figure 5 These African violets are clones.

Cloning What would it be like if humans or other animals were exact copies of each other? Making copies of organisms is called **cloning.** The new organism produced is called a clone. The clone receives DNA from just one parent cell. It has the same DNA as the parent cell. In many ways, cloning is not a new technology. In the past, most cloning was done with plants. Gardeners clone plants when they take cuttings of a plant's stems, leaves, or roots. They can grow many identical plants from one, as shown in **Figure 5.**

Only since the 1990s has cloning large animals become possible. In 1997, it was announced that an adult Finn Dorset sheep had been cloned. The new sheep, named Dolly, was the first successfully cloned mammal. The real value of Dolly is that scientists now have a better understanding of how cells reproduce.

Sex Cells and Reproduction

Does a human baby look exactly like its father or its mother? Usually, the baby has features of both of its parents. The baby might have her dad's hair color and her mom's eye color. However, the baby probably doesn't look exactly like either of her parents. That's because humans, as well as many other organisms, are the products of sexual (SEK shul) reproduction. In **sexual reproduction** a new organism is produced from the DNA of two cells. **Sex cells,** as shown in **Figure 6,** are the specialized cells that carry DNA and join in sexual reproduction. During this process, DNA from each sex cell contributes to the formation of a new individual and that individual's traits.

Reading Check *What results from sexual reproduction?*

Figure 6 Specialized cells called sex cells are involved in reproduction. A female sex cell usually is called an egg and a male sex cell is usually called a sperm. **A** A human egg cell and **B** a human sperm cell contains 23 chromosomes.

A

Magnification: 700×

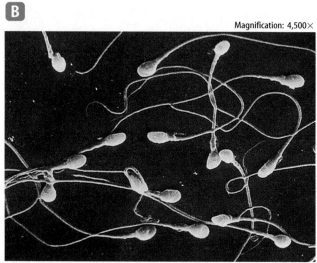

B

Magnification: 4,500×

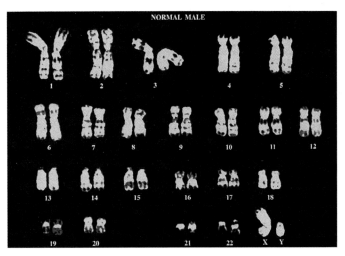

Figure 7 Each chromosome in a human body is made of DNA. All 23 pairs of chromosomes of one person are shown in this photograph.

Production of Sex Cells

Recall that your body is made up of different types of cells most of which were formed by mitosis. When a skin cell, a bone cell, or another body cell divides, it produces two new cells by cell division. Each cell has DNA that is identical to the original cell. Recall that DNA can be found in structures called chromosomes. A human body cell has 46 chromosomes arranged in 23 pairs, as shown in **Figure 7**. Each chromosome of a pair has genetic information about the same things. For example, if one chromosome has information about hair color, its mate also will have information about hair color.

Sex cells are different. Instead of being formed by cell division like body cells are, sex cells are formed by **meiosis** (mi OH sus). **Table 1** compares cell division and sex cell formation. Only certain cells in reproductive organs undergo the process of meiosis. Before meiosis begins, DNA is duplicated. During meiosis, the nucleus divides twice. Four sex cells form, each with half the number of chromosomes of the original cell. Human eggs and sperm contain only 23 chromosomes each—one chromosome from each pair of chromosomes. That way, when a human egg and sperm join in a process called **fertilization,** the result is a new individual with a full set of 46 chromosomes. **Figure 8** shows how sex cells pair up to form a cell with a full set of chromosomes that develops into a new human being.

Health INTEGRATION

In humans, sex cell production and fertilization can be affected by cigarette smoking. Cigarette smoking can decrease the number of sperm produced in the male body. Also, some of the sperm produced by a male that smokes may be deformed and unable to fertilize an egg.

Table 1 Cell Division and Sex Cell Formation in Humans

	Cell Division	Sex Cell Formation
Process used	Mitosis	Meiosis
DNA duplicated?	Yes	Yes
Nucleus divides	Once	Twice
Number of cells formed	2	4
Chromosome number of beginning cell	46	46
Chromosomes in each new cell	46	23

NATIONAL GEOGRAPHIC VISUALIZING HUMAN REPRODUCTION

Figure 8

Humans, like most animals and plants, reproduce sexually. In sexual reproduction, a new and genetically unique individual is produced when a female sex cell and a male sex cell join in a process called fertilization.

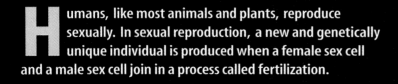

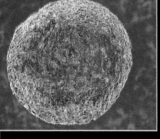

▲ A woman's sex cell, an egg, has only 23 chromosomes—half the amount contained in human body cells.

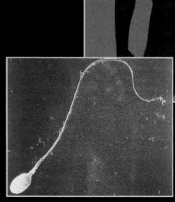

◀ A male's sex cell, a sperm, also contains only 23 chromosomes.

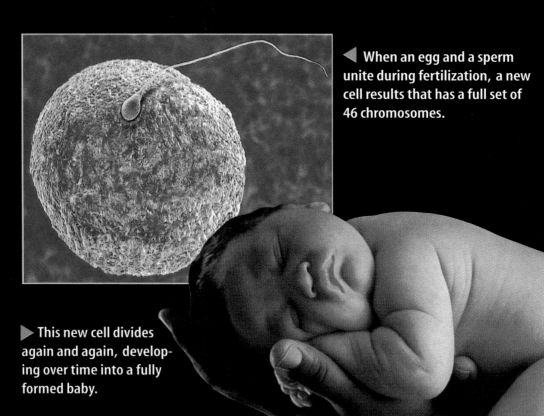

◀ When an egg and a sperm unite during fertilization, a new cell results that has a full set of 46 chromosomes.

▶ This new cell divides again and again, developing over time into a fully formed baby.

Sex Cells in Plants

Plants can reproduce sexually. How this occurs is different for each plant group. But in every case, a sperm and an egg join to create a new cell that eventually becomes a plant.

It may seem that flowers are just a decoration for many plants, but flowers contain structures for reproducing. Male flower parts produce pollen, which contains sperm cells. Female flower parts produce eggs. When a sperm and an egg join, a new cell forms. In most flowers, rapid changes begin soon after fertilization. The cell divides many times and becomes enclosed in a protective seed. The petals and most other flower parts fall off. A fruit that contains seeds soon develops, as shown in **Figure 9**.

Figure 9
An apple flower will develop into an apple containing seeds if the eggs in the female reproductive structure are fertilized.

Section Assessment

1. How does the outcome of meiosis differ from the outcome of mitosis?
2. Why is reproduction an important process for all living things?
3. Explain why offspring produced by asexual reproduction are identical to the parent that produced them.
4. How does DNA control how an organism looks and functions?
5. **Think Critically** For a species, what are some advantages of reproducing asexually? Of reproducing sexually? Of having the ability to do either?

Skill Builder Activities

6. **Comparing and Contrasting** List similarities and differences between the three types of asexual reproduction. **For more help, refer to the** Science Skill Handbook.
7. **Solving One-Step Equations** A female bullfrog produces 350 eggs. All of the eggs are fertilized and hatch in one season. Assume that half of the tadpoles are male and half are female. If all the female tadpoles survive and, one year later, produce 350 eggs each, how many eggs would be produced? **For more help, refer to the** Math Skill Handbook.

SECTION 1 Continuing Life

Activity

Getting DNA from Onion Cells

DNA contains the instructions for the many processes that occur in a cell. DNA is found in all cells. In this activity, you will be able to see the actual DNA of one living thing—an onion.

What You'll Investigate
How is DNA taken out of cells?

Materials
prepared onion mixture (125 mL)
toothpicks
small beaker
*measuring cup
rubbing alcohol (125 mL)
large beaker
*other glass container
magnifying glass
*microscope
*Alternate materials

Goals
- **Separate** DNA from onion cells.
- **Practice** laboratory skills.

Safety Precautions

Be sure to wear an apron and goggles throughout this activity. Keep hands away from face.

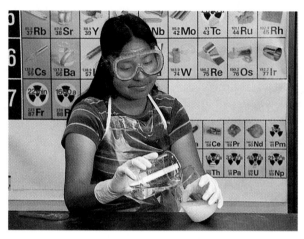

Procedure
1. Obtain 125 mL of prepared onion mixture from your teacher. Empty it into the large glass beaker or container.
2. Slowly pour 125 mL of rubbing alcohol down the side of the container onto the mixture. The alcohol should form a layer on top of the onion mixture.
3. **Observe** the gooey strings floating to the top. These strings are DNA.
4. Use a toothpick to gently stir the alcohol layer. Use another toothpick to remove the gooey DNA.
5. **Observe** DNA with a magnifying glass or a microscope. Record your observations in your Science Journal.
6. When you're finished, pour all liquids into containers provided by your teacher.

Conclude and Apply
1. Based on what you know about DNA, predict whether the DNA that you removed would look different from the DNA of other types of plants.
2. **Infer** whether this method of taking DNA out of cells could be used to compare the amount of DNA between different organisms. Explain your answer.

Communicating Your Data

Compare and contrast your findings with those of other students in your class. Explain in your Science Journal why your findings were the same or different from those of other students. **For more help, refer to the Science Skill Handbook.**

SECTION 2 Genetics—The Study of Inheritance

Heredity

When you go to a family reunion or browse through family pictures, like the one in **Figure 10,** you can't help but notice similarities and differences among your relatives. You notice that your mother's eyes look just like your grandmother's, and one uncle is tall while his brothers are short. These similarities and differences are the result of the way traits are passed from one generation to the next. **Heredity** (huh RE duh tee) is the passing of traits from parents to offspring. Solving the mystery of heredity has been one of the great success stories of biology.

Look around at the students in your classroom. What makes each person an individual? Is it hair or eye color? Is it the shape of a nose or the arch in a person's eyebrows? Eye color, hair color, skin color, nose shape, and many other features, including those inside an individual that can't be seen, are traits that are inherited from a person's parents. A trait is a physical characteristic of an organism. Every organism, including yourself, is made up of many traits. The study of how traits are passed from parents to offspring is called **genetics** (juh NET ihks).

Reading Check *What traits could you pass to your offspring?*

As You Read

What You'll Learn
- **Explain** how traits are inherited.
- **Relate** chromosomes, genes, and DNA to one another.
- **Discuss** how mutations add variation to a population.

Vocabulary
heredity
genetics
gene
variation
mutation

Why It's Important
You will understand why you have certain traits.

Figure 10
Family members often share similar physical features. These traits can be something obvious, like curly hair, or less obvious, such as color blindness.

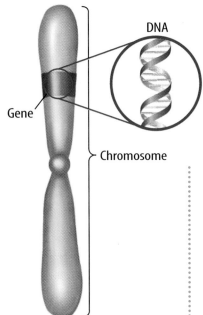

Figure 11
Hundreds of genes are located on each chromosome.

Genes All traits are inherited. Half of your genetic information came from your father, and half came from your mother. This information was contained in the chromosomes of the sperm and egg that joined and formed the cell that eventually became you.

As **Figure 11** shows, all chromosomes contain genes (JEENZ). A **gene** is a small section of DNA on a chromosome that has information about a trait. Humans have thousands of different genes arranged on 23 pairs of chromosomes. Genes control all of the traits of organisms—even traits that can't be seen, such as the size and shape of your stomach and your blood type. Genes provide all of the information needed for growth and life.

What determines traits?

Recall that in body cells, such as skin cells or muscle cells, chromosomes are in pairs. One pair of chromosomes can contain genes that control many different traits. Each gene on one chromosome of the pair has a similar gene on the other chromosome of the pair. Each gene of a gene pair is called an allele (uh LEEL), as shown in **Figure 12.** The genes that make up a gene pair might or might not contain the same information about a trait. For example, the genes for the flower color trait in pea plants might be purple or white. If a pair of chromosomes contains different alleles for a trait, that trait is called a hybrid (HI brud). When a trait has two identical alleles, it's called pure.

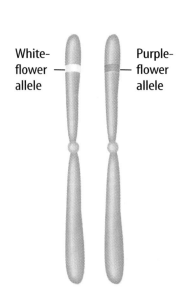

Figure 12
Pea flowers can be purple or white. The chromosome pair from a pea plant shows that both chromosomes have an allele for the flower color trait. A pea plant with this chromosome pair would have purple flowers.

Figure 13
Widow's peak is a dominant allele in humans. *How many alleles for widow's peak might this person have?*

Dominant and Recessive Alleles The combination of alleles in a gene pair determines how a trait will be shown, or expressed, in an organism. In pea plants and other organisms, that depends on something called dominance (DAW muh nunts). Dominance means that one allele covers over or masks another allele of the trait. For instance, if a pea plant has one purple-flower allele and one white-flower allele or two purple-flower alleles, its flowers will be purple. Purple is the dominant flower color in pea plants. The dominant allele is seen when the trait is hybrid or dominant pure. White flowers, the masked allele, are said to be recessive. Recessive alleles are seen only when a trait is recessive pure.

Humans also have traits that are controlled by dominant and recessive alleles. These traits are controlled in the same way that dominant and recessive alleles are controlled in plants. To show a recessive allele, a person needs to inherit two copies of the recessive allele for that trait—one from their mother and one from their father. To show a dominant allele, a person can have either one or two alleles for the trait. One dominant allele in humans is the presence of a widow's peak, as shown in **Figure 13.**

Expression of Traits The traits of an organism are coded in the organism's DNA. However, the environment can play an important role in the way that a trait is shown, or expressed. You may know a person whose dark hair lightens when exposed to sunlight, or a person whose light skin darkens in sunlight. Human hair color and skin color are traits that are coded for by genes, but the environment can change the way that the traits appear. The environment can affect the expression of traits in every kind of organism, including bacteria, fungi, plants, and animals.

Modeling Probability

Procedure
1. Flip a **coin** ten times. Count the number of heads and the number of tails.
2. Record these data in a data table in your **Science Journal.**
3. Now flip the coin twenty times. Count the number of heads and tails.
4. Record these data in a data table in your Science Journal.

Analysis
1. What results did you expect when you flipped the coin ten times? Twenty times?
2. Were your observed results closer to your expected results when you flipped the coin more times?
3. How is the flipping of a coin similar to the joining of egg and sperm at fertilization?

Passing Traits to Offspring

How are traits passed from parents to offspring during fertilization? The flower color trait in pea plants can be used as an example. Suppose a hybrid purple-flowered pea plant (one with two different alleles for flower color) is mated with a white-flowered pea plant. What color flowers will the offspring have?

The traits that a new pea plant will inherit depend upon which genes are carried in each plant's sex cells. Remember that sex cells are produced during meiosis. In sex cell formation, pairs of chromosomes duplicate, then separate as the four sex cells form. Therefore, gene pairs also separate. As a result, each sex cell contains one allele for each trait. Because the purple-flowered plant in **Figure 14** is a hybrid, half of its sex cells contain the purple-flower allele and half contain the white-flower allele. On the other hand, the white-flowered plant is recessive pure. The gene pair for flower color has two white alleles. All of the sex cells that it makes contain only the white-flower allele.

In fertilization, one sperm will join with one egg. Many events, such as flipping a coin and getting either heads or tails, are a matter of chance. In the same way, chance is involved in heredity. In the case of the pea plants, the chance was equal that the new pea plant would receive either the purple-flower allele or the white-flower allele from the hybrid plant.

Figure 14
The traits an organism has depends upon which genes were carried in the parents' sex cells. This diagram shows how the flower color trait is passed in pea plants.

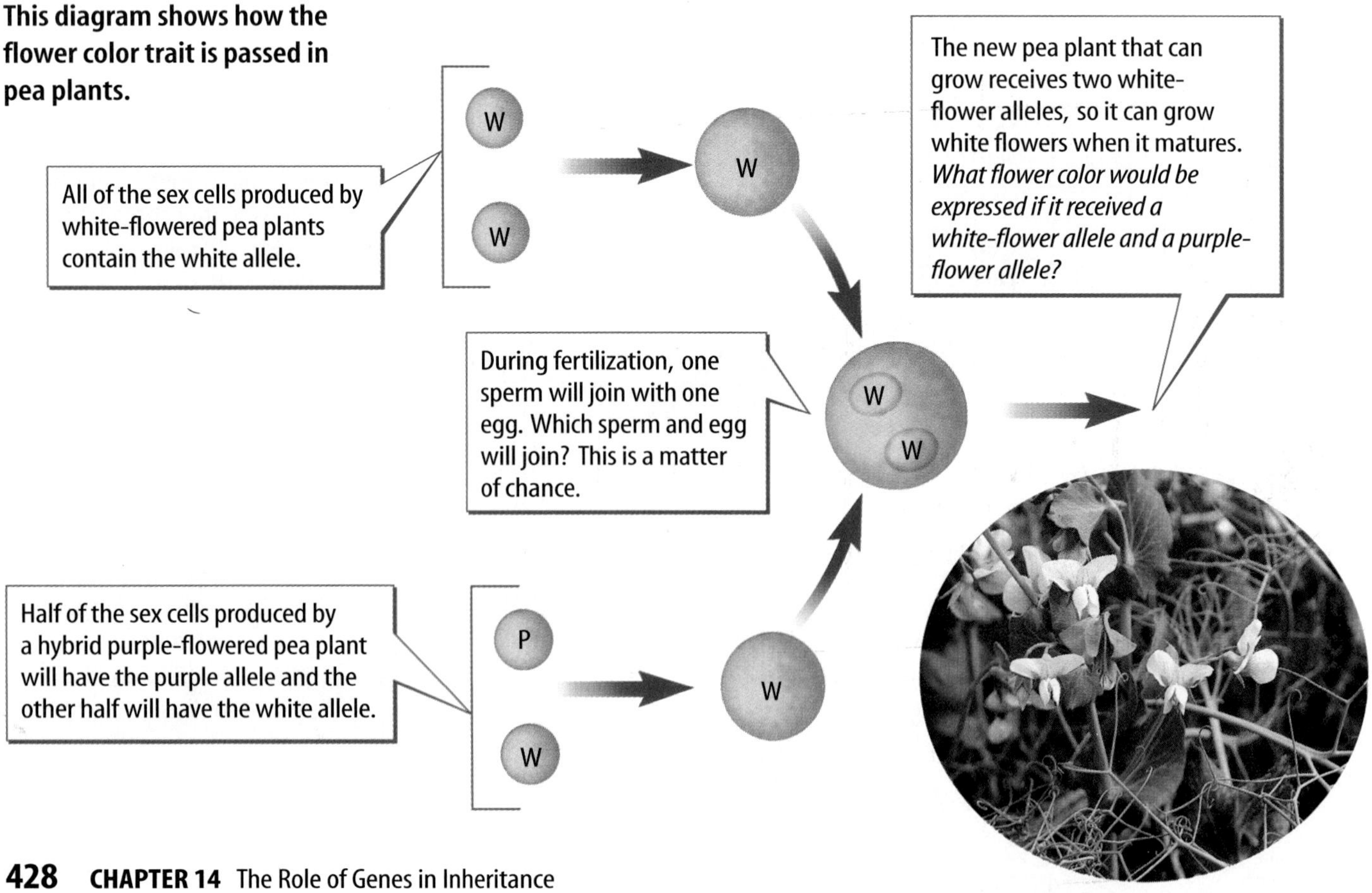

All of the sex cells produced by white-flowered pea plants contain the white allele.

During fertilization, one sperm will join with one egg. Which sperm and egg will join? This is a matter of chance.

The new pea plant that can grow receives two white-flower alleles, so it can grow white flowers when it matures. *What flower color would be expressed if it received a white-flower allele and a purple-flower allele?*

Half of the sex cells produced by a hybrid purple-flowered pea plant will have the purple allele and the other half will have the white allele.

428 CHAPTER 14 The Role of Genes in Inheritance

Differences in Organisms

Now you know why a baby can have characteristics of either of its parents. The inherited genes from his or her parents determine hair color, skin color, eye color, and other traits. But what accounts for the differences, or variations (vayr ee AY shuns), in a family? **Variations** are the different ways that a certain trait appears.

Math Skills Activity

MATH TEKS 6.2 C, 6.3 B

Calculating Possible Alleles in Sex Cells

When sex cells form, each allele separates from its partner. Each sex cell will contain only one allele for each trait. What percent of sex cells formed will contain a certain allele?

Example Problem

A parent is a hybrid for a certain trait. That means that the parent has a dominant and a recessive allele for that trait. What percent of the parent's sex cells will contain the dominant allele?

Solution

1. *This is what you know:*
   the number of possible alleles for the trait from the parent, 2

2. *This is what you need to find:*
   percent of sex cells with the dominant allele

3. *This is the equation you need to use:*

$$\frac{100\%}{\text{number of possible alleles from the parent}} = x$$

4. *Solve the equation for x:*

$$\frac{100\%}{2} = x \qquad 50\% = x$$

Check your answer by multiplying the number of possible alleles from the parent by the percent of sex cell with the dominant allele. Do you get 100%?

Practice Problem

The attached-earlobes trait in humans is a recessive trait. What percent of the sex cells produced by a parent with attached earlobes would have an allele for this trait?

For more help, refer to the **Math Skill Handbook.**

Figure 15
Traits in humans that show great variation usually are controlled by more than one gene pair.

 The members of this family have different hair color.

B Height is a trait that has many variations.

Multiple Alleles and Multiple Genes Earlier, you learned how the flower color trait in pea plants is passed from parent to offspring. Flower color in pea plants shows a simple pattern of inheritance. Sometimes, though, the pattern of inheritance of a trait is not so simple. Many traits in organisms are controlled by more than two alleles. For example, in humans, multiple alleles A, B, and O control blood types.

Traits also can be controlled by more than one gene pair. For humans, hair color, as shown in **Figure 15A,** height, as shown in **Figure 15B,** weight, eye color, and skin color, are traits that are controlled by several gene pairs. This type of inheritance is the reason for the differences, or variations, in a species.

Earth Science INTEGRATION

The ultraviolet rays of the Sun can cause mutations that lead to skin cancer. Limiting your exposure to the Sun and using sunscreen are two ways to avoid ultraviolet rays. In your Science Journal write down one other suggestion for limiting your exposure to the Sun's ultraviolet rays.

Mutations—The Source of New Variation If you've searched successfully through a patch of clover for one with four leaves instead of three, you've come face-to-face with a mutation (myew TAY shun). A four-leaf clover is the result of a mutation. The word *mutate* simply means "to change." In genetics, a **mutation** is a change in a gene or chromosome. This can happen because of an error during meiosis or mitosis or because of something in the environment. Many mutations happen by chance.

Reading Check *What is a mutation?*

What are the effects of mutations? Sometimes mutations affect the way cells grow, repair, and maintain themselves. This type of mutation is usually harmful to the organism. Many mutations, such as a four-leaf clover, have a neutral effect. Whether a mutation is beneficial, harmful, or neutral, all mutations add variation to the genes of a species.

Figure 16
Dairy cattle are bred selectively for the amount of milk that they can produce.

Selective Breeding Sometimes, a mutation produces a different version of a trait that many people find attractive. To continue this trait, selective breeding is practiced.

Nearly all breeding of animals is based on their observable traits and is controlled, instead of being random. For many years cattle, like the one in **Figure 16,** have been bred on the basis of how much milk they produced. Racehorses are bred according to how fast they run. It eventually was learned that in a few generations, breeding closely related animals produced an increased percentage of offspring with the desired traits.

Research Visit the Glencoe Science Web site at **tx.science.glencoe.com** for more information about selective breeding. Communicate with your class information about one type of animal that is bred selectively.

Section Assessment

1. What is heredity?
2. Describe which alleles for a trait must be present for a recessive allele to be expressed.
3. How are the chromosomes in human body cells arranged?
4. Explain how mutations add variation to the genes of a species.
5. **Think Critically** What might happen if two hybrid purple-flowered pea plants are mated? What possible flower colors could the offspring have? Explain.

Skill Builder Activities

6. **Concept Mapping** Make a concept map that shows the relationships between the following concepts: *genetics, genes, chromosomes, DNA, variation,* and *mutation.* **For more help, refer to the** Science Skill Handbook.

7. **Communicating** Research to find what a transgenic organism is, then find books or articles about these organisms. In your Science Journal, write a paragraph summary of your findings. **For more help, refer to the** Science Skill Handbook.

SECTION 2 Genetics—The Study of Inheritance

Activity
Use the Internet

Genetic Traits: The Unique You

What makes you unique? Unless you have an identical twin, no other person has the same combination of genes as you do. To learn more about three human genetic traits, you will collect data about your classmates. When you compare the data you collected with data from other students, you'll see that patterns develop in the frequency of types of traits that are present within a group of people.

Recognize the Problem

How are three genetic traits expressed among your classmates?

Form a Hypothesis

Genetic traits can be dominant or recessive. When you survey your classmates about the three characteristics being studied, do you think more people will have the dominant trait or the recessive trait?

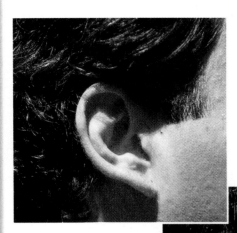

Attached earlobe

Detached earlobe

Goals
- **Identify** genetic traits.
- **Collect** data about three specific human genetic traits.
- **Investigate** what are dominant and recessive alleles.
- **Graph** your results and then communicate them to other students.

Data Source

SCIENCE *Online* Go to the Glencoe Science Web site at **tx.science.glencoe.com** to get more information about human genetic traits and for data collected by other students.

432 CHAPTER 14 The Role of Genes in Inheritance

Using Scientific Methods

Test your Hypothesis

Plan

1. **Research** general information about human genetic traits.
2. **Search** reference sources to find out which form of each characteristic being studied is dominant and which form is recessive.
3. **Survey** the students in your class to collect data about the three genetic traits being studied.

Do

1. Make sure your teacher approves your plan before you start.
2. **Record** your data in your Science Journal. Use frequency data tables to organize your data.

Analyze Your Data

1. **Record** the total number of people included in your survey.
2. **Calculate** the number of people who show each form of each of the three traits that are being studied. Record each of these numbers in your Science Journal.
3. **Graph** the data you collected on a bar graph. Bars should represent the numbers of students exhibiting each of the different genetic traits you investigated.
4. **Compare** the data among each of the three genetic traits you explored.

Draw Conclusions

1. Think about the genetic traits you investigated. Which traits were most common in the people you surveyed?
2. Might surveying a larger group of people give different results?
3. Which genetic traits are least commonly found?
4. In the people you surveyed, were dominant alleles present more often than recessive alleles were?

Communicating Your Data

SCIENCE Online Find this *Use the Internet* activity on the Glencoe Science Web site at **tx.science.glencoe.com**. **Post** your data in the table provided. **Compare** your data to that of other students. **Combine** your data with that of other students and **graph** the combined data on a bar graph.

ACTIVITY 433

TIME SCIENCE AND Society

SCIENCE ISSUES THAT AFFECT YOU!

These twins, separated at birth and reunited as adults, had the same kind of job, drove the same kind of car, and had the same hobbies.

SEPARATED

When Barbara Herbert was about 40, she met her long lost twin sister, Daphne Goodship. She had not seen her since infancy. The two grew up in separate homes with separate families. Because they are identical twins, it makes sense that they look alike. After all, they share the same genes. What was shocking, however, was the number of coincidences in their lives. Although they were not in contact while growing up, they shared identical experiences.

Both women...
- dropped out of school at age 14
- got jobs working for the local government
- met their future husbands at age 16
- gave birth to two boys and one girl
- are squeamish about blood and heights
- drink their coffee cold.

As if this isn't strange enough, the first time Barbara and Daphne met, they both wore cream-colored dresses and brown velvet jackets. Both have a habit of pushing up their nose with the palm of their hand. Both independently nicknamed this behavior "squidging."

AT BIRTH

Are genes or the people who raised you more more important in determining personality?

In the Genes?

Barbara and Daphne are part of an ongoing scientific study at Minnesota's Center for Twin and Adoption Research, which examines twins who were separated at birth. Thomas Bouchard, the psychologist who founded the research center, studies these "separated" twins to note their similarities and differences. It is helping him, and others, better understand what is stronger in a person's development—genetic makeup or how and by whom these twins were raised.

There are a couple of tests that help scientists analyze this nature versus nurture question. Identical twins make ideal subjects for these tests because their genetic makeup is exactly the same—even their brains! First, a psychological assessment is made, using personality tests, job interest questions, mental ability, and I.Q. tests. Then scientists analyze the twins' background environments, including where they were raised, what their parents were like, and what schools they attended. These tests help determine whether a person's habits and personality are based on genetic makeup or social interactions.

Each year, about 90,000 people gather at the International Twins Day Festival in Twinsburg, Ohio.

Bouchard recently observed a pair of twins as they reunited at the Minneapolis airport. He was stunned at their similar behavior. Although the twins had not seen each other in more than 30 years, they finished each other's sentences without interrupting each other. Both twins said they felt like they have known each other all their lives. And, perhaps, thanks to their genes, they have!

CONNECTIONS Interview Find a pair of identical twins that go to your school or live in your community. Make a list of 10 questions and interview each of the twins separately. Write down their answers, or tape-record them. Compare the responses, then share your findings with the class.

Online For more information, visit tx.science.glencoe.com

Chapter 14 Study Guide

Reviewing Main Ideas

Section 1 Continuing Life

1. Reproduction is an important process for all living things.

2. During reproduction, information stored in DNA is passed from parent to offspring.

3. Mitosis is the process that results in two nuclei with the same genetic information.

4. Organisms can reproduce sexually or asexually. *Why does sexual reproduction provide more variety in a species, such as these dogs, than asexual reproduction does?*

5. DNA is shaped liked a twisted ladder. An organism's DNA contains all of the information about how it will look and function.

Section 2 Genetics—The Study of Inheritance

1. Genetics is the study of how traits are passed from parent to offspring. *What traits might have been passed from this bird to her offspring?*

2. Genes are small sections of DNA on chromosomes. Each gene has information about a specific trait.

3. Chromosomes are found in pairs. For each gene on a particular chromosome, a gene with information about the same trait can be found on the other chromosome of the chromosome pair. Each gene of a gene pair is called an allele.

4. The way a trait is shown depends on the combination of dominant and recessive alleles carried on the chromosome pair.

5. Differences in organisms can result from mutations. Mutations are changes in a gene or chromosome. Mutations are a source of variation in populations.

6. Selective breeding allows favorable traits of organisms to be passed from one generation to the next. *What trait might these horses have been bred selectively for?*

After You Read

Record what you learned about genes and inheritance under the right tab of the Foldable you made at the beginning of the chapter.

Chapter 14 Study Guide

Visualizing Main Ideas

Complete the following concept map on reproduction.

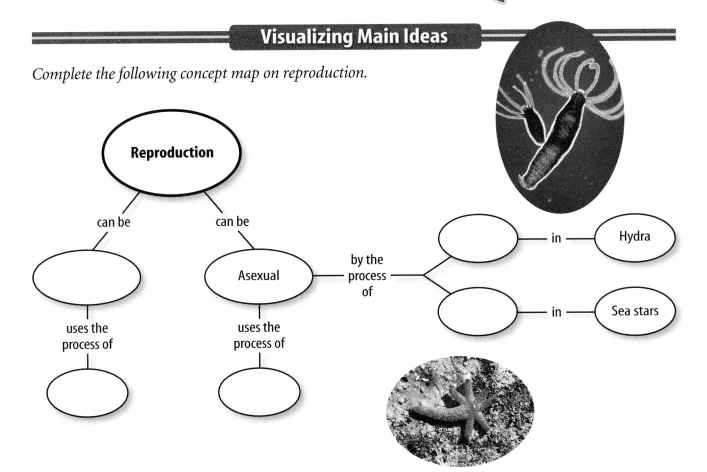

Vocabulary Review

Vocabulary Words

a. asexual reproduction
b. cloning
c. DNA
d. fertilization
e. gene
f. genetics
g. heredity
h. meiosis
i. mitosis
j. mutation
k. sex cell
l. sexual reproduction
m. variation

Using Vocabulary

Explain the differences between the vocabulary words in each of the following sets.

1. mitosis, meiosis
2. asexual reproduction, sexual reproduction
3. cloning, variation
4. fertilization, sexual reproduction
5. mutation, variation
6. gene, DNA
7. asexual reproduction, mitosis
8. sex cells, meiosis
9. genetics, gene
10. DNA, mutation

Study Tip

After each day's lesson, make a practice quiz for yourself. Later, when you're studying for the test, take the practice quizzes that you created.

CHAPTER STUDY GUIDE 437

Chapter 14 Assessment & TEKS Review

Checking Concepts

Choose the word or phrase that best answers the question.

1. Which of these is a process that produces an organism that is genetically identical to another organism?
   A) fertilization
   B) sexual reproduction
   C) budding
   D) mutation

2. What are sperm and eggs?
   A) variations
   B) sex cells
   C) mutations
   D) genes

3. What is the small section of DNA that contains the code for a trait?
   A) a gene
   B) heredity
   C) a variation
   D) a cell

4. What is the study of how traits are passed from parents to offspring?
   A) heredity
   B) variation
   C) genetics
   D) genes

5. Which of these is another name for an observable feature or characteristic of an organism?
   A) sex cell
   B) embryo
   C) trait
   D) gene

6. How are specialized breeds of dogs, cats, horses, and other animals produced?
   A) regeneration
   B) asexual reproduction
   C) selective breeding
   D) budding

7. What is the passing of traits from parent to offspring called?
   A) genetics
   B) variation
   C) heredity
   D) meiosis

8. Which of the following is reproduction that requires male and female sex cells?
   A) asexual reproduction
   B) sexual reproduction
   C) mitosis
   D) heredity

9. What is formed during meiosis?
   A) heredity
   B) sex cells
   C) clones
   D) fertilization

10. What is any change in the DNA of a gene or chromosome called?
    A) an embryo
    B) sex cells
    C) a clone
    D) a mutation

Thinking Critically

11. Explain the relationship among DNA, genes, and chromosomes.

12. Two brown-eyed parents have a baby with blue eyes. Explain how this could happen.

13. This photo shows a picture of a spider plant. Why is this plant an example of asexual reproduction? How could the plant reproduce through sexual reproduction?

14. How is the process of meiosis important in sexual reproduction?

15. Explain how a mutation in a gene could be beneficial to an organism.

Developing Skills

16. **Drawing Conclusions** Some mutations are harmful to organisms. Others are beneficial, and some have no effect at all. Which type of mutation would be least likely to be passed on to future generations?

17. **Recognizing Cause and Effect** What advantage do certain plants have that can reproduce sexually and asexually?

Chapter 14 Assessment

18. **Comparing and Contrasting** Compare and contrast sexual and asexual reproduction.

19. **Recognizing Cause and Effect** What is the role of meiosis and mitosis in the fertilization and development that results in a human baby?

20. **Predicting** A pure, purple-flowered pea plant is crossed with a pure, white-flowered pea plant as illustrated below. What flower color trait would be in the sex cells of each plant? What color of flowers will the resulting pea plant have?

Performance Assessment

21. **Scientific Drawing** Use your imagination and make illustrations for each of the following vocabulary words: *asexual reproduction, genetics,* and *mutation.*

22. **Newspaper Article** Many scientists have reported that it is possible to get DNA from prehistoric creatures. Go to the library and find a newspaper article that describes the discovery of ancient DNA. Write a summary of the article in your Science Journal.

Technology

Go to the Glencoe Science Web site at **tx.science.glencoe.com** or use the **Glencoe Science CD-ROM** for additional chapter assessment.

TAKS Practice

There are a variety of ways that different organisms reproduce. The chart below describes several types of reproduction.
TEKS 6.10 C

Types of Reproduction	
Type	**What is it?**
Cell Division	A cell divides into two identical new cells.
Budding	A new, identical organism grows on the parent and breaks off to live on its own.
Regeneration	A new, identical organism grows from the broken off body part of another organism.
Sexual Reproduction	DNA from two cells is combined to make a new nonidentical organism.

Study the chart and answer the following questions.

1. According to the chart, when a sea star's arm falls off and it grows into a new sea star, it is an example of _____.
   A) cell divison
   B) budding
   C) regeneration
   D) sexual reproduction

2. According to the chart, young birds hatching from eggs is an example of _____.
   F) cell divison
   G) budding
   H) regeneration
   J) sexual reproduction

CHAPTER ASSESSMENT 439

CHAPTER 15

Science TEKS 6.5 A, B; 6.8 B, C; 6.10 B, C; 6.12 A, B, C

Interactions of Living Things

How do Alaskan brown bears and salmon interact? The relationship between these two species is clear to see. However, the Alaskan brown bear also depends on every species of insect and fish that the salmon eats, and many nonliving parts of the environment, too. In this chapter, you will learn how all living things depend on the living and nonliving factors in the environment for survival.

What do you think?

Science Journal Look at the picture below with a classmate. Discuss what you think this might be. Here's a hint: *This species and salmon interact.* Write your answer or best guess in your Science Journal.

Imagine that you are in a crowded elevator. Everyone jostles and bumps each other. The temperature increases and ordinary noises seem louder. What a relief you feel when the doors open and you step out. Like people in an elevator, plants and animals in an area interact. How does the amount of space available to each organism affect its interaction with other organisms?

Measure space

1. Use a meterstick to measure the length and width of the classroom.
2. Multiply the length by the width to find the area of the room in square meters.
3. Count the number of individuals in your class. Divide the area of the classroom by the number of individuals. In your Science Journal, record how much space each person has.

Observe
Write a prediction in your Science Journal about what might happen if the number of students in your classroom doubled.

Before You Read

Making a Cause and Effect Study Fold Make this Foldable to help you understand the cause and effect relationship of biotic and abiotic things.

1. Place a sheet of paper in front of you so the long side is at the top. Fold the paper in half from the left side to the right side. Fold top to bottom and crease. Then unfold.
2. Through the top thickness of paper, cut along the middle fold line to form two tabs as shown. Label the tabs *Biotic*, which means living, and *Abiotic*, which means nonliving, as shown.
3. Before you read the chapter, list examples of biotic and abiotic things around you on the tabs. As you read, write about each under the tabs.

441

SECTION 1

The Environment

As You Read

What You'll Learn
- **Identify** biotic and abiotic factors in an ecosystem.
- **Describe** the different levels of biological organization.
- **Explain** how ecology and the environment are related.

Vocabulary
ecology
abiotic factor
biotic factor
population
community
ecosystem
biosphere

Why It's Important
Abiotic and biotic factors interact to make up your ecosystem. The quality of your ecosystem can affect your health. Your actions can affect the health of the ecosystem.

Ecology

All organisms, from the smallest bacteria to a blue whale, interact with their environment. **Ecology** is the study of the interactions among organisms and their environment. Ecologists, such as the one in **Figure 1,** are scientists who study these relationships. Ecologists divide the environmental factors that influence organisms into two groups. **Abiotic** (ay bi AH tihk) **factors** are the nonliving parts of the environment. Living or once-living organisms in the environment are called **biotic** (bi AH tihk) **factors.**

✓ **Reading Check** *Why is a rotting log considered a biotic factor in the environment?*

Abiotic Factors

In a forest environment, birds, insects, and other living things depend on one another for food and shelter. They also depend on the abiotic factors that surround them, such as water, sunlight, temperature, air, and soil. All of these factors and others are important in determining which organisms are able to live in a particular environment.

Figure 1
Ecologists study biotic and abiotic factors in an environment and the relationships among them. Many times, ecologists must travel to specific environments to examine the organisms that live there.

442 CHAPTER 15 Interactions of Living Things

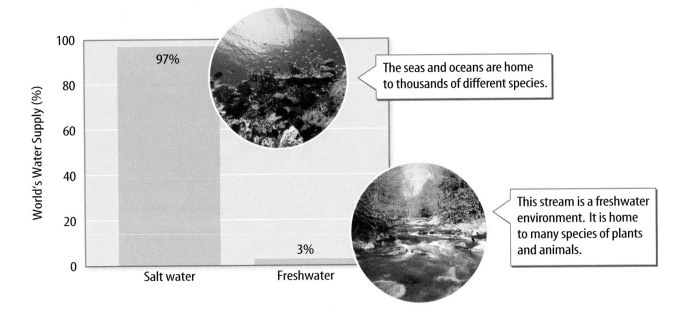

Figure 2
Salt water accounts for 97 percent of the water on Earth. It is found in the seas and oceans. Only three percent of Earth's water is freshwater.

Water All living organisms need water to survive. The bodies of most organisms are 50 percent to 95 percent water. Water is an important part of the cytoplasm in cells and the fluid that surrounds cells. Respiration, photosynthesis, digestion, and other important life processes can only occur in the presence of water.

More than 95 percent of Earth's surface water is found in the oceans. The saltwater environment in the oceans is home to a vast number of species. Freshwater environments, like the one in **Figure 2,** also support thousands of types of organisms.

Light and Temperature The abiotic factors of light and temperature also affect the environment. The availability of sunlight is a major factor in determining where green plants and other photosynthetic organisms live, as shown in **Figure 3.** By the process of photosynthesis, energy from the Sun is changed into chemical energy that is used for life processes. Most green algae live near the water's surface where sunlight can penetrate. On the other hand, little sunlight reaches the forest floor, so very few plants grow close to the forest floor.

Figure 3
Flowers that grow on the forest floor, such as these bluebells, grow during the spring when they receive the most sunlight.

The temperature of a region also determines which plants and animals can live there. Some areas of the world have a fairly consistent temperature year round, but other areas have seasons during which temperatures vary. Water environments throughout the world also have widely varied temperatures. Plant and animal species are found in the freezing cold Arctic, in the extremely hot water near ocean vents, and at almost every temperature in between.

Figure 4
Air pollution can come from many different sources. Air quality in an area affects the health and survival of the species that live there.

Earth Science INTEGRATION

When soil that receives little rain is damaged, a desert can form. This process is called desertification. Use reference materials to find where desertification is occurring in the United States. Record your findings in your Science Journal.

Figure 5
Soil provides a home for many species of animals.

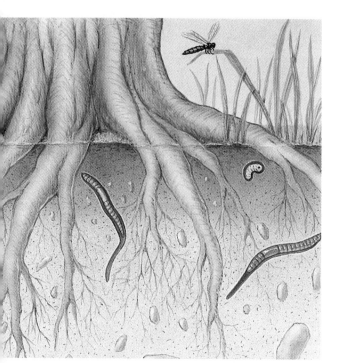

Air Although you can't see the air that surrounds you, it has an impact on the lives of most species. Air is composed of a mixture of gases including nitrogen, oxygen, and carbon dioxide. Most plants and animals depend on the gases in air for respiration. The atmosphere is the layer of gases and airborne particles that surrounds Earth. Polluted air, like the air in **Figure 4,** can cause the species in an area to change, move, or die off.

Clouds and weather occur in the bottom 8 km to 16 km of the atmosphere. All species are affected by the weather in the area where they live. The ozone layer is 20 km to 50 km above Earth's surface and protects organisms from harmful radiation from the Sun. Air pressure, which is the weight of air pressing down on Earth, changes depending on altitude. Higher altitudes have less air pressure. Few organisms live at extreme air pressures.

Reading Check *How does air pollution affect the species in an area?*

Soil From one enviroment to another, soil, as shown in **Figure 5,** can vary greatly. Soil type is determined by the amounts of sand, silt, and clay it contains. Various kinds of soil contain different amounts of nutrients, minerals, and moisture. Different plants need different kinds of soil. Because the types of plants in an area help determine which other organisms can survive in that area, soil affects every organism in an environment.

Biotic Factors

Abiotic factors do not provide everything an organism needs for survival. Organisms depend on other organisms for food, shelter, protection, and reproduction. How organisms interact with one another and with abiotic factors can be described in an organized way.

444 CHAPTER 15 Interactions of Living Things

Levels of Organization The living world is highly organized. Atoms are arranged into molecules, which in turn are organized into cells. Cells form tissues, tissues form organs, and organs form organ systems. Together, organ systems form organisms. Biotic and abiotic factors also can be arranged into levels of biological organization, as shown in **Figure 6**.

Figure 6
The living world is organized in levels.

A Organism An organism is one individual from a population.

B Population All of the individuals of one species that live in the same area at the same time make up a population.

C Community The populations of different species that interact in some way are called a community.

D Ecosystem All of the communities in an area and the abiotic factors that affect them make up an ecosystem.

E Biome A biome is a large region with plants and animals well adapted to the soil and climate of the region.

F Biosphere The level of biological organization that is made up of all the ecosystems on Earth is the biosphere.

Figure 7
Members of a penguin population compete for resources.

Populations All the members of one species that live together make up a **population**. For example, all of the catfish living in a lake at the same time make up a population. Part of a population of penguins is shown in **Figure 7.** Members of a population compete for food, water, mates, and space. The resources of the environment and the ways the organisms use these resources determine how large a population can become.

Communities Most populations of organisms do not live alone. They live and interact with populations of other types of organisms. Groups of populations that interact with each other in a given area form a **community**. For example, a population of penguins and all of the species that they interact with form a community. Populations of organisms in a community depend on each other for food, shelter, and other needs.

Ecosystem In addition to interactions among populations, ecologists also study interactions among populations and their physical surroundings. An **ecosystem** like the one in **Figure 8A** is made up of a biotic community and the abiotic factors that affect it. Examples of ecosystems include coral reefs, forests, and ponds. You will learn more about the interactions that occur in ecosystems later in this chapter.

Biomes Scientists divide Earth into different regions called biomes. A biome (BI ohm) is a large region with plant and animal groups that are well adapted to the soil and climate of the region. Many different ecosystems are found in a biome. Examples of biomes include mountains, as shown in **Figure 8B,** tropical rain forests, and tundra.

Research Visit the Glencoe Science Web site at **tx.science.glencoe.com** for more information about Earth's biomes. Make a poster to communicate to your class what you learn.

Figure 8
Biomes contain many different ecosystems.
A This mountaintop ecosystem is part of the
B mountain biome.

446 CHAPTER 15 Interactions of Living Things

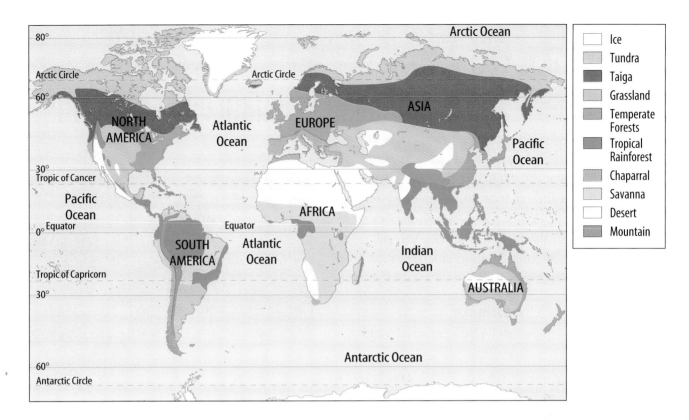

Figure 9
This map shows some of the major biomes of the world. *What biome do you live in?*

The Biosphere Where do all of Earth's organisms live? Living things can be found 11,000 m below the surface of the ocean, 9,000 m high on mountains, and 4.5 km high in Earth's atmosphere. The part of Earth that supports life is the **biosphere** (BI uh sfihr). The biosphere includes the top part of Earth's crust, all the waters that cover Earth's surface, the surrounding atmosphere, and all biomes, including those in **Figure 9.** The biosphere seems huge, but it is only a small part of Earth. If you used an apple as a model of Earth, the thickness of Earth's biosphere could be compared to the thickness of the apple's skin.

Section 1 Assessment

1. What is the difference between an abiotic factor and a biotic factor? Give five examples of each that are in your ecosystem.
2. Contrast a population and a community.
3. What is an ecosystem?
4. How are the terms *ecology* and *environment* related?
5. **Think Critically** Explain how biotic factors change in an ecosystem that has flooded.

Skill Builder Activities

6. **Recording Observations** Each person lives in a population as part of a community. Describe your population and community. **For more help, refer to the** Science Skill Handbook.
7. **Using a Database** Use a database to research biomes. Find the name of the biome that best describes where you live. **For more help, refer to the Technology Skill Handbook.**

SECTION 1 The Environment **447**

Activity

Delicately Balanced Ecosystems

Each year you might visit the same park, but notice little change. However, ecosystems are delicately balanced, and small changes can upset this balance. In this activity, you will observe how small amounts of fertilizer can disrupt an ecosystem.

What You'll Investigate
How do manufactured fertilizers affect pond systems?

Materials
large glass jars of equal size (4)
clear plastic wrap
stalks of *Elodea* (8)
*another aquatic plant
garden fertilzer
*houseplant fertilizer
rubber bands (4)
pond water
triple beam balance
weighing paper
spoon
metric ruler
*Alternate materials

Goals
- **Observe** the effects of manufactured fertilizer on water plants.
- **Predict** the effects of fertilizers on pond and stream ecosystems.

Safety Precautions

Procedure
1. Working in a group, label four jars A, B, C, and D.
2. **Measure** eight *Elodea* stalks to be certain that they are all about equal in length.
3. Fill the jars with pond water and place two stalks of *Elodea* in each jar.
4. Add 5 g of fertilizer to jar B, 10 g to jar C, and 30 g to jar D. Put no fertilizer in jar A.
5. Cover each jar with plastic wrap and secure it with a rubber band. Use your pencil to punch three small holes through the plastic wrap.
6. Place all jars in a well-lit area.
7. Make daily observations of the jars for three weeks. Record your observations in your Science Journal.
8. At the end of the three-week period, remove the *Elodea* stalks. Measure and record the length of each in your Science Journal.

Conclude and Apply
1. **List** the control and variables you used in this experiment.
2. **Compare** the growth of *Elodea* in each jar.
3. **Predict** what might happen to jar A if you added 5 g of fertilizer to it each week.

Communicating Your Data
Compare your results with the results of other students. Research how fertilizer runoff from farms and lawns has affected aquatic ecosystems in your area. **For more help, refer to the** Science Skill Handbook.

SECTION 2
Interactions Among Living Organisms

Characteristics of Populations

You, the person sitting next to you, everyone in your class, and every other organism on Earth is a member of a specific population. Populations can be described by their characteristics such as spacing and density.

Population Size The number of individuals in the population is the population's size, as shown in **Figure 10.** Population size can be difficult to measure. If a population is small and made up of organisms that do not move, the size can be determined by counting the individuals. Usually individuals are too widespread or move around too much to be counted. The population size then is estimated. The number of organisms of one species in a small section is counted and this value is used to estimate the population of the larger area.

Suppose you spent several months observing a population of field mice that live in a pasture. You probably would observe changes in the size of the population. Older mice die. Mice are born. Some are eaten by predators, and some mice wander away to new nests. The size of a population is always changing. The rate of change in population size varies from population to population. In contrast to a mouse population, the number of pine trees in a mature forest changes slowly, but a forest fire could reduce the pine tree population quickly.

As You Read

What You'll Learn
- **Identify** the characteristics that describe populations.
- **Examine** the different types of relationships that occur among populations in a community.
- **Determine** the habitat and niche of a species in a community.

Vocabulary
population density
limiting factor
symbiosis
niche
habitat

Why It's Important
You must interact with other organisms to survive.

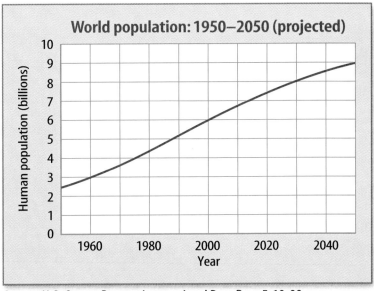

Figure 10
The size of the human population is increasing each year. By the year 2050, the human population is projected to be more than 9 billion.

Source: U.S. Census Bureau, International Data Base 5-10-00.

SECTION 2 Interactions Among Living Organisms **449**

Figure 11
Population density can be shown on a map. This map uses different colors to show varying densities of a population of northern bobwhite birds.

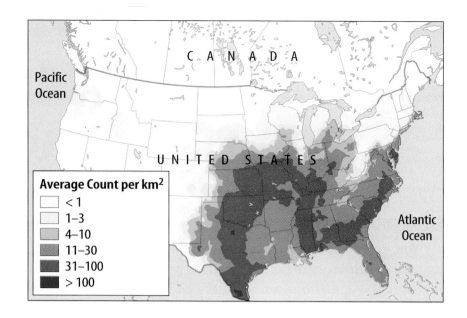

Research Visit the Glencoe Science Web site at **tx.science.glencoe.com** for recent news about the size of the human population. Communicate to your class what you learn.

Figure 12
In some populations, such as creosote bushes in the desert, individuals usually are spaced uniformly throughout the area.

Population Density At the beginning of this chapter, when you figured out how much space is available to each student in your classroom, you were measuring another population characteristic. The number of individuals in a population that occupy a definite area is called **population density**. For example, if 100 mice live in an area of one square kilometer, the population density is 100 mice per square kilometer. When more individuals live in a given amount of space, as seen in **Figure 11,** the population is more dense.

Population Spacing Another characteristic of populations is spacing, or how the organisms are arranged in a given area. They can be evenly spaced, randomly spaced, or clumped together. If organisms have a fairly consistent distance between them, as shown in **Figure 12,** they are evenly spaced. In random spacing, each organism's location is independent of the locations of other organisms in the population. Random spacing of plants usually results when wind or birds disperse seeds. Clumped spacing occurs when resources such as food or living space are clumped. Clumping results when animals gather in herds, flocks, or other groupings.

Limiting Factors Populations cannot continue to grow larger forever. All ecosystems have a limited amount of food, water, living space, mates, nesting sites, and other resources. A **limiting factor,** as shown in **Figure 13,** is any biotic or abiotic factor that limits the number of individuals in a population. A limiting factor also can affect other populations in the community indirectly. For example, a drought might reduce the number of seed-producing plants in a forest clearing. A singular factor means that food can become a limiting factor for deer that eat the plants and for a songbird population that feeds on the seeds of these plants. Food also could become a limiting factor for hawks that feed on the songbirds.

Reading Check *What is an example of a limiting factor?*

Competition is the struggle among organisms to obtain the resources they need to survive and reproduce, as shown in **Figure 14.** As population density increases, so does competition among individuals for the resources in their environment.

Carrying Capacity Suppose a population increases in size year after year. At some point, food, nesting space, or other resources become so scarce that some individuals are not able to survive or reproduce. When this happens, the environment has reached its carrying capacity. Carrying capacity is the largest number of individuals of a species that an environment can support and maintain for a long period of time. If a population gets bigger than the carrying capacity of the environment, some individuals are left without adequate resources. They will die or be forced to move elsewhere.

Figure 13
These antelope and zebra populations live in the grasslands of Africa. *What limiting factors might affect the plant and animal populations shown here?*

What insect populations live in your area? To find out more about insects, see the **Insect Field Guide** at the back of the book.

Figure 14
During dry summers, the populations of animals at existing watering holes increase because some watering holes have dried up. This creates competition for water, a valuable resource.

SECTION 2 Interactions Among Living Organisms

Mini LAB

Observing Symbiosis

Procedure

1. Carefully wash and examine the roots of a **legume plant** and a **nonlegume plant**.
2. Use a **magnifying glass** to examine the roots of the legume plant.

Analysis

1. What differences do you observe in the roots of the two plants?
2. Bacteria and legume plants help one another thrive. What type of symbiotic relationship is this?

Biotic Potential What would happen if a population's environment had no limiting factors? The size of the population would continue to increase. The maximum rate at which a population increases when plenty of food and water are available, the weather is ideal, and no diseases or enemies exist, is its biotic potential. Most populations never reach their biotic potential, or they do so for only a short period of time. Eventually, the carrying capacity of the environment is reached and the population stops increasing.

Symbiosis and Other Interactions

In ecosystems, many species of organisms have close relationships that are needed for their survival. **Symbiosis** (sihm bee OH sus) is any close interaction between two or more different species. Symbiotic relationships can be identified by the type of interaction between organisms. A symbiotic relationship that benefits both species is called mutualism. **Figure 15** shows one example of mutualism.

Commensalism is a form of symbiosis that benefits one organism without affecting the other organism. For example, a species of flatworms benefits by living in the gills of horseshoe crabs, eating scraps of the horseshoe crab's meals. The horseshoe crab is unaffected by the flatworms.

Parasitism is a symbiotic relationship between two species in which one species benefits and the other species is harmed. Some species of mistletoe are parasites because their roots grow into a tree's tissue and take nutrients from the tree.

✓ **Reading Check** *What are some examples of symbiosis?*

Figure 15
The partnership between the desert yucca plant and the yucca moth is an example of mutualism.

The yucca depends on the moth to pollinate its flowers.

The moth depends on the yucca for protected place to lay its eggs and a source of food for its larvae.

Predation One way that population size is regulated is by predation (prih DAY shun). Predation is the act of one organism hunting, killing, and feeding on another organism. Owls are predators of mice, as shown in **Figure 16**. Mice are their prey. Predators are biotic factors that limit the size of the prey population. Availability of prey is a biotic factor that can limit the size of the predator population. Because predators are more likely to capture old, ill, or young prey, the strongest individuals in the prey population are the ones that manage to reproduce. This improves the prey population over several generations.

Habitats and Niches In a community, every species plays a particular role. For example, some are producers and some are consumers. Each also has a particular place to live. The role, or job, of an organism in the ecosystem is called its **niche** (NICH). What a species eats, how it gets its food, and how it interacts with other organisms are all parts of its niche. The place where an organism lives is called its **habitat**. For example, an earthworm's habitat is soil. An earthworm's niche includes loosening, aerating, and enriching the soil.

Figure 16
Owls use their keen senses of sight and hearing to hunt for mice in the dark.

Section 2 Assessment

1. Name three characteristics of populations.
2. Describe how limiting factors can affect the organisms in a population.
3. Explain the difference between a habitat and a niche.
4. Describe and give an example of two symbiotic relationships that occur among populations in a community.
5. **Think Critically** A parasite can obtain food only from its host. Most parasites weaken but do not kill their hosts. Why?

Skill Builder Activities

6. **Drawing Conclusions** Explain how sound could be used to relate the size of the cricket population in one field to the cricket population in another field. **For more help, refer to the Science Skill Handbook.**

7. **Solving One-Step Equations** A 15-m^2 wooded area has the following: 30 ferns, 150 grass plants, and 6 oak trees. What is the population density per m^2 of each species? **For more help, refer to the Math Skill Handbook.**

SECTION 2 Interactions Among Living Organisms

SECTION 3

Matter and Energy

As You Read

What You'll Learn
- **Explain** the difference between a food chain and a food web.
- **Describe** how energy flows through ecosystems.
- **Examine** how materials such as water, carbon, and nitrogen are used repeatedly.

Vocabulary
food chain
food web
water cycle

Why It's Important
You are dependent upon the recycling of matter and the transfer of energy for survival.

Figure 17
These mushrooms are decomposers. Decomposers obtain needed energy for life when they break down or decay other biomass.

Energy Flow Through Ecosystems

Life on Earth is not simply a collection of independent organisms. Even organisms that seem to spend most of their time alone interact with other members of their species. They also interact with members of other species. Most of the interactions among members of different species occur when one organism feeds on another. Food contains nutrients and energy needed for survival. When one organism is food for another organism, some of the energy in the first organism (the food) is transferred to the second organism (the eater).

Producers are organisms that take in and use energy from the Sun or some other source to produce food. Some use the Sun's energy for photosynthesis to produce carbohydrates. For example, plants, algae, and some one-celled, photosynthetic organisms are producers. Consumers are organisms that take in energy when they feed on producers or other consumers. The transfer of energy does not end there. When organisms die, other organisms called decomposers, as shown in **Figure 17**, take in energy as they break down the remains of organisms also known as biomass. This movement of energy through a community can be diagrammed as a food chain or a food web.

Food Chains A **food chain,** as shown in **Figure 18,** is a model, a simple way of showing how energy, in the form of food, passes from one organism to another. When drawing a food chain, arrows between organisms indicate the direction of energy transfer. An example of a pond food chain follows.

small water plants → insects → bluegill → bass

Food chains usually have three or four links. This is because the available energy decreases from one link to the next link. At each transfer of energy, a portion of the energy is lost as heat due to the activities of the organisms. In a food chain, the amount of energy left for the last link is only a small portion of the energy in the first link.

NATIONAL GEOGRAPHIC VISUALIZING A FOOD CHAIN

Figure 18

In nature, energy in food passes from one organism to another in a sequence known as a food chain. All living things are linked in food chains, and there are millions of different chains in the world. Each chain is made up of organisms in a community. The photographs here show a food chain in a North American meadow community.

A The first link in any food chain is a producer—in this case, grass. Grass gets its energy from sunlight.

B The second link of a food chain is usually an herbivore like this grasshopper. Herbivores are animals that feed only on producers.

C The third link of this food chain is a carnivore, an animal that feeds on other animals. This woodhouse toad feeds on grasshoppers.

D The fourth link of this food chain is a garter snake, which feeds on toads.

E The last link in many food chains is a top carnivore, an animal that feeds on other animals, including other carnivores. This great horned owl is a top carnivore.

SECTION 3 Matter and Energy 455

Food Webs Food chains are too simple to describe the many interactions among organisms in an ecosystem. A **food web** is a series of overlapping food chains that exist in an ecosystem. A food web provides a more complete model of the way energy moves through an ecosystem. They also are more accurate models because food webs show how many organisms are part of more than one food chain in an ecosystem.

Humans are a part of many different food webs. Most people eat foods from several different levels of a food chain. Every time you eat a hamburger, an apple, or a tuna fish sandwich, you have become a link in a food web. Can you picture the steps in the food web that led to the food in your lunch?

Problem-Solving Activity

MATH TEKS 6.11 A, 6.12 A

How do changes in Antarctic food webs affect populations?

The food webs in the icy Antarctic Ocean are based on phytoplankton, which are microscopic algae that float near the water's surface. The algae are eaten by tiny, shrimp-like krill, which are consumed by baleen whales, squid, and fish. Toothed whales, seals, and penguins eat the fish and squid. How would changes in any of these populations affect the other populations?

Identifying the Problem

Worldwide, the hunting of baleen whales has been illegal since 1986. It is hoped that the baleen whale population will increase. How will an increase in the whale population affect this food web?

Solving the Problem

1. Populations of seals, penguins, and krill-eating fish increased in size as populations of baleen whales declined. Explain why this occurred.
2. What might happen if the number of baleen whales increases but the amount of krill does not?

Ecological Pyramids Most of the energy in the biosphere comes from the Sun. Producers take in and transform only a small part of the energy that reaches Earth's surface. When an herbivore eats a plant, some of the energy in the plant passes to the herbivore. However, most of it is given off into the atmosphere as heat. The same thing happens when a carnivore eats a herbivore. An ecological pyramid models the number of organisms at each level of a food chain. The bottom of an ecological pyramid represents the producers of an ecosystem. The rest of the levels represent successive consumers.

Chemistry INTEGRATION

Certain bacteria take in energy through a process called chemosynthesis. In chemosynthesis, the bacteria produce food using the energy in chemical compounds. In your Science Journal predict where these bacteria are found.

✔ **Reading Check** *What is an ecological pyramid?*

Energy Pyramid The flow of energy from grass to the hawk in **Figure 19** can be illustrated by an energy pyramid. An energy pyramid compares the energy available at each level of the food chain in an ecosystem. Just as most food chains have three or four links, a pyramid of energy usually has three or four levels. Only about ten percent of the energy at each level of the pyramid is available to the next level. By the time the top level is reached, the amount of energy available is greatly reduced.

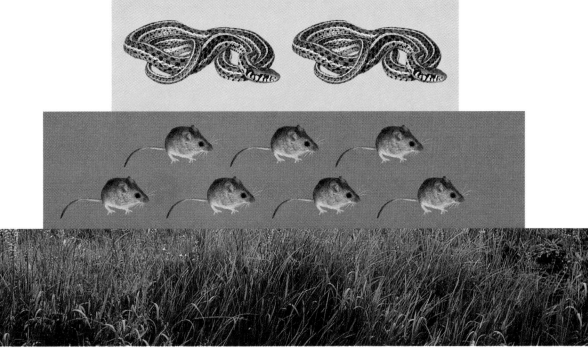

Figure 19
An energy pyramid illustrates that available energy decreases at each successive feeding step. *Why doesn't an energy pyramid have more levels?*

SECTION 3 Matter and Energy **457**

TRY AT HOME Mini LAB

Modeling the Water Cycle

Procedure
1. With a **marker**, make a line halfway up on a **plastic cup**. Fill the cup to the mark with **water**.
2. Cover the top with **plastic wrap** and secure it with a **rubber band or tape**.
3. Put the cup in direct sunlight. Observe the cup for three days. Record your observations.
4. Remove the plastic wrap and observe the cup for seven more days.

Analysis
1. What parts of the water cycle did you observe during this activity?
2. How did the water level in the cup change after the plastic wrap was removed?

The Cycles of Matter

The energy available as food is constantly renewed by plants using sunlight. However, think about the matter that makes up the bodies of living organisms. This is called Earth's biomass. The law of conservation of mass states that matter on Earth is never lost or gained. It is used over and over again. In other words, it is recycled. The carbon atoms in your body might have been on Earth since the planet formed billions of years ago. They have been recycled billions of times. Many important materials that make up your body cycle through the environment. Some of these materials are water, carbon, and nitrogen.

Water Cycle Water molecules on Earth constantly rise into the atmosphere, fall to Earth, and soak into the ground or flow into rivers and oceans. The **water cycle** involves the processes of evaporation, condensation, and precipitation.

Heat from the Sun causes water on Earth's surface to evaporate, or change from a liquid to a gas, and rise into the atmosphere as water vapor. As the water vapor rises, it encounters colder and colder air and the molecules of water vapor slow down. Eventually, the water vapor changes back into tiny droplets of water. It condenses, or changes from a gas to a liquid. These water droplets clump together to form clouds. When the droplets become large and heavy enough, they fall back to Earth as rain or other precipitation. This process is illustrated in **Figure 20**.

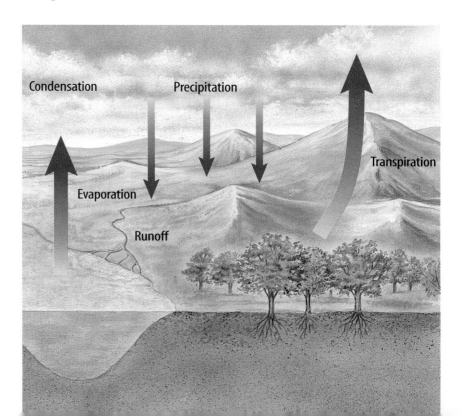

Figure 20
A water molecule that falls as rain can follow several paths through the water cycle.
How many of these paths can you identify in this diagram?

458

Other Cycles in Nature What do you have in common with all organisms? All organisms contain carbon. Earth's atmosphere contains about 0.03 percent carbon in the form of carbon dioxide gas. The movement of carbon through Earth's biosphere is called the carbon cycle, as shown in **Figure 21**.

Nitrogen is an element used by organisms to make proteins and nucleic acids. The nitrogen cycle begins with the transfer of nitrogen from the atmosphere to producers then to consumers. The nitrogen then moves back to the atmosphere or directly into producers again.

Phosphorus, sulfur, and other elements needed by living organisms also are used and returned to the environment. Just as you recycle aluminum, glass, and paper products, the materials that organisms need to live are recycled continuously in the biosphere.

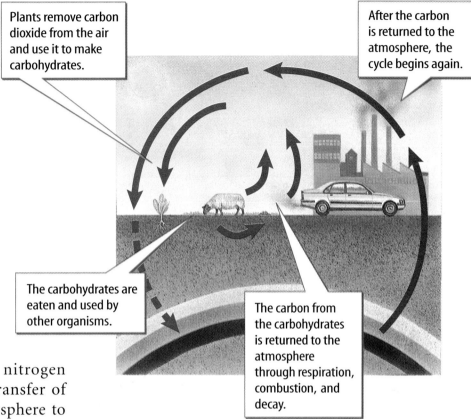

Plants remove carbon dioxide from the air and use it to make carbohydrates.

After the carbon is returned to the atmosphere, the cycle begins again.

The carbohydrates are eaten and used by other organisms.

The carbon from the carbohydrates is returned to the atmosphere through respiration, combustion, and decay.

Figure 21
Carbon can follow several different paths through the carbon cycle. Some carbon is stored in Earth's biomass.

Section Assessment

1. Compare a food chain and a food web.
2. What are the differences among producers, consumers, and decomposers?
3. What is an energy pyramid?
4. How does carbon flow through ecosystems?
5. **Think Critically** Use your knowledge of food chains and the energy pyramid to explain why fewer lions than gazelles live on the African plains.

Skill Builder Activities

6. **Classifying** Look at the food chain in **Figure 18.** Classify each organism as a producer or a consumer. **For more help, refer to the** Science Skill Handbook.
7. **Communicating** In your Science Journal, write a short essay about how the *water cycle, carbon cycle,* and *nitrogen cycle* are important to living organisms. **For more help, refer to the** Science Skill Handbook.

SECTION 3 Matter and Energy

Activity: Design Your Own Experiment

Identifying a Limiting Factor

Organisms depend upon many biotic and abiotic factors in their environment to survive. When these factors are limited or are not available, it can affect an organism's survival. By experimenting with some of these limiting factors, you will see how organisms depend on all parts of their environment.

Recognize the Problem

How do abiotic factors such as light, water, and temperature affect the germination of seeds?

Form a Hypothesis

Based on what you have learned about limiting factors, make a hypothesis about how one specific abiotic factor might affect the germination of a bean seed. Be sure to consider factors that you can change easily.

Safety Precautions

Wash hands after handling soil and seeds.

Goals

- **Observe** the effects of an abiotic factor on the germination and growth of bean seedlings.
- **Design** an experiment that demonstrates whether or not a specific abiotic factor limits the germination of bean seeds.

Possible Materials

bean seeds
small planting containers
soil
water
label
trowel
*spoon
aluminum foil
sunny window
*other light source
refrigerator or oven
*Alternate materials

Using Scientific Methods

Test Your Hypothesis

Plan

1. As a group, agree upon and write out a hypothesis statement.
2. **Decide** on a way to test your group's hypothesis. Keep available materials in mind as you plan your procedure. List your materials.
3. **Design** a data table in your Science Journal for recording data.
4. Remember to test only one variable at a time and use suitable controls.
5. Read over your entire experiment to make sure that all steps are in logical order.
6. **Identify** any constants, variables, and controls in your experiment.
7. Be sure the factor that you will test is measurable.

Do

1. Make sure your teacher approves your plan before you start.
2. Carry out the experiment according to the approved plan.
3. While the experiment is going on, record any observations that you make and complete the data table in your Science Journal.

Analyze Your Data

1. **Compare** the results of this experiment with those of other groups in your class.
2. **Infer** how the abiotic factor you tested affected the germination of bean seeds.
3. **Graph** your results in a bar graph that compares the number of bean seeds that germinated in the experimental container with the number of seeds that germinated in the control container.

Draw Conclusions

1. **Identify** which factor had the greatest effect on the germination of the seeds.
2. **Determine** whether or not you could change more than one factor in this experiment and still have germination of seeds.

Write a set of instructions that could be included on a packet of this type of seeds. Describe the best conditions for seed germination.

ACTIVITY 461

Science and Language Arts

The Solace of Open Spaces
a novel by Gretel Ehrlich

Respond to the Reading

1. From reading this passage, can you guess the occupation of the narrator?
2. Describe the relationship between people and animals in this passage.
3. What words does the author use to indicate that horses are intelligent?

Animals give us their constant, unjaded[1] faces and we burden them with our bodies and civilized ordeals. We're both humbled by and imperious[2] with them. We're comrades who save each other's lives. The horse we pulled from a boghole this morning bucked someone off later in the day; one stock dog refuses to work sheep, while another brings back a calf we had overlooked. . . . What's stubborn, secretive, dumb, and keen[3] in us bumps up against those same qualities in them. . . .

Living with animals makes us redefine our ideas about intelligence. Horses are as mischievous as they are dependable. Stupid enough to let us use them, they are cunning enough to catch us off guard. . . .

We pay for their loyalty; They can be willful, hard to catch, dangerous to shoe and buck on frosty mornings. In turn, they'll work themselves into a lather cutting cows, not for the praise they'll get but for the simple glory of outdodging a calf or catching up with an errant steer. . . .

[1] *Jaded* means "to be weary with fatigue," so *unjaded* means "not to be weary with fatigue."
[2] domineering or overbearing
[3] intellectually smart or sharp

Informative Writing

Understanding Literature

Informative Writing The passage that you have just read is from a work of nonfiction and is based on facts. The passage is informative because it describes the real relationship between people and animals on a ranch in Wyoming. The author speaks from her own point of view, not from the point of view of a disinterested party. She uses her own experience to explain to readers that animals and people depend on each other for survival. For example, she writes, "Living with animals makes us redefine our ideas about intelligence." The language puts her firmly in the story—she is not only telling the story, but living it, too. How might this story have been different if it had been told from the point of view of a visiting journalist?

Science Connection Animals and ranchers are clearly dependent on each other. Ranchers provide nutrition and shelter for animals on the ranch and, in turn, animals provide food and perform work for the ranchers. You might consider the relationship between horses and ranchers to be a symbiotic one. Symbiosis (sihm bee OH sus) is any close interaction among two or more different species.

Linking Science and Writing

Informative Writing Write a short passage about an experience you have had with a pet. In your writing, reflect on how you and the pet are alike and dependent upon each other. Put yourself firmly in the story without overusing the word *I*.

Career Connection

Large-Animal Veterinarian

Dave Garza works to keep horses healthy. Dave spends about 20 percent of his workday in his clinic. He goes there first thing in the morning to perform surgeries and take care of horses that have been brought to him. The rest of the day, he drives to local farms to examine patients. Dave vaccinates horses against rabies, the flu, and the encephalitis virus. He gives them tetanus shots and medication to prevent worms. He also cares for their teeth, replaces their shoes, and helps them deliver their foals in the spring.

SCIENCE *Online* To learn more about careers in veterinary medicine, visit the Glencoe Science Web site at **tx.science.glencoe.com.**

SCIENCE AND LANGUAGE ARTS 463

Chapter 15 Study Guide

Reviewing Main Ideas

Section 1 The Environment

1. Ecology is the study of interactions among organisms and their environment.

2. The nonliving features of the environment are abiotic factors, and the organisms in the environment are biotic factors.

3. Populations and communities make up an ecosystem. *What populations and communities might be present in this ecosystem?*

4. The region of Earth and its atmosphere in which all organisms live is the biosphere.

Section 2 Interactions Among Living Organisms

1. Characteristics that can describe populations include size, spacing, and density.

2. Any biotic or abiotic factor that limits the number of individuals in a population is a limiting factor.

3. A close relationship between two or more species is a symbiotic relationship. A symbiotic relationship that benefits both species is called mutualism. A relationship in which one species benefits and the other is unaffected is called commensalism.

4. The place where an organism lives is its habitat, and its role in the environment is its niche. *How could two similar species of birds live in the same area and nest in the same tree without occupying the same niche?*

Section 3 Matter and Energy

1. Food chains and food webs are models that describe the feeding relationships among organisms in a community.

2. At each level of a food chain, organisms lose energy as heat. Energy on Earth is renewed constantly by sunlight.

3. An ecological pyramid models the number of organisms at each level of a food chain in an ecosystem. *Why is each level of this energy pyramid smaller than the one below it?*

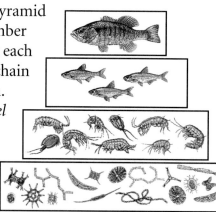

4. Matter on Earth is never lost or gained. It is used over and over again, or recycled.

After You Read

Using your Foldable, explain the cause and effect relationship between specific abiotic and biotic organisms around you.

Chapter 15 Study Guide

Visualizing Main Ideas

Complete the following concept map on the biosphere.

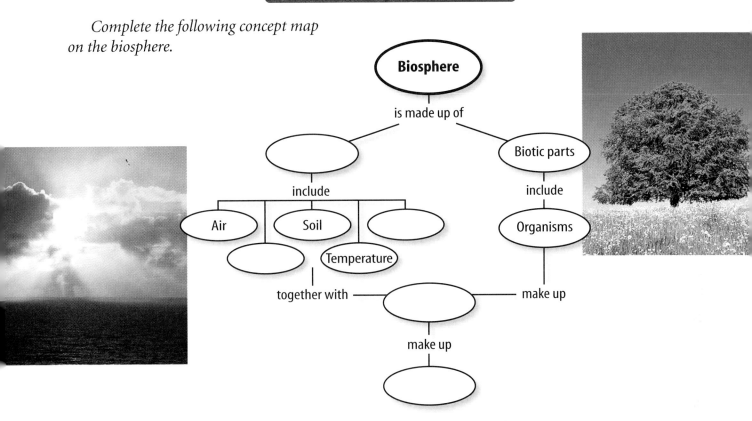

Vocabulary Review

Vocabulary Words

a. abiotic factor
b. biosphere
c. biotic factor
d. community
e. ecology
f. ecosystem
g. food chain
h. food web
i. habitat
j. limiting factor
k. niche
l. population
m. population density
n. symbiosis
o. water cycle

 **Study Tip**

Use tables to organize ideas. For example, put the levels of biological organization in a table. Tables help you review concepts quickly.

Using Vocabulary

Replace the underlined words with the correct vocabulary words.

1. A(n) <u>abiotic factor</u> is any living thing in the environment.

2. A series of overlapping food chains makes up a(n) <u>nitrogen cycle</u>.

3. The size of a population that occupies an area of definite size is its <u>carrying capacity</u>.

4. Where an organism lives in an ecosystem is its <u>niche</u>.

5. The part of Earth that supports life is the <u>limiting factor</u>.

6. Any close relationship between two or more species is <u>habitat</u>.

CHAPTER STUDY GUIDE 465

Chapter 15 Assessment & TEKS Review

Checking Concepts

Choose the word or phrase that best answers the question.

1. Which of the following is NOT cycled in the biosphere?
   A) nitrogen   C) water
   B) soil   D) carbon

2. What are coral reefs, forests, and ponds examples of?
   A) niches   C) populations
   B) habitats   D) ecosystems

3. What is made up of all populations in an area?
   A) niche   C) community
   B) habitat   D) ecosystem

4. What is the term for the total number of individuals in a population occupying a certain area?
   A) clumping   C) spacing
   B) size   D) density

5. Which of the following is an example of a producer?
   A) wolf   C) tree
   B) frog   D) rabbit

6. Which level of the food chain has the most energy?
   A) consumer   C) decomposers
   B) herbivores   D) producers

7. What is a relationship called in which one organism is helped and the other is harmed?
   A) mutualism   C) commensalism
   B) parasitism   D) consumer

8. Which of the following is a model that shows the amount of energy available as it flows through an ecosystem?
   A) niche   C) carrying capacity
   B) energy pyramid D) food chain

9. Which of the following is a biotic factor?
   A) animals   C) sunlight
   B) air   D) soil

10. What are all of the individuals of one species that live in the same area at the same time called?
    A) community   C) biosphere
    B) population   D) organism

Thinking Critically

11. What are two different populations that might be present in a desert biome? Two different ecosystems? Explain.

12. Why are viruses considered parasites?

13. What does carrying capacity have to do with whether or not a population reaches its biotic potential?

14. Why are decomposers vital to the cycling of matter in an ecosystem?

15. Write a paragraph that describes your own habitat and niche.

Developing Skills

16. **Classifying** Classify the following as the result of either evaporation or condensation.
    a. A puddle disappears after a rainstorm.
    b. Rain falls.
    c. A lake becomes shallower.
    d. Clouds form.

17. **Concept Mapping** Use the following information to draw a food web of organisms living in a goldenrod field. *Aphids eat goldenrod sap, bees eat goldenrod nectar, beetles eat goldenrod pollen and goldenrod leaves, stinkbugs eat beetles, spiders eat aphids,* and *assassin bugs eat bees.*

Chapter 15 Assessment

18. **Making and Using Graphs** Use the following data to graph the population density of a deer population over the years. Plot the number of deer on the y-axis and years on the x-axis. Predict what might have happened to cause the changes in the size of the population.

Arizona Deer Population	
Year	Deer Per 400 Hectares
1905	5.7
1915	35.7
1920	142.9
1925	85.7
1935	25.7

19. **Recording Observations** A home aquarium contains water, an air pump, a light, algae, a goldfish, and algae-eating snails. What are the abiotic factors in this environment?

20. **Comparing and Contrasting** Compare and contrast the role of producers, consumers, and decomposers in an ecosystem.

Performance Assessment

21. **Poster** Use your own observations or the results of library research to develop a food web for a nearby park, pond, or other ecosystem. Make a poster display illustrating the food web.

22. **Oral Presentation** Research the steps in the phosphorous cycle. Find out what role phosphorus plays in the growth of algae in ponds and lakes. Present your findings to the class.

Technology

Go to the Glencoe Science Web site at **tx.science.glencoe.com** or use the **Glencoe Science CD-ROM** for additional chapter assessment.

THE PRINCETON REVIEW — TAKS Practice

Biologists want to estimate the total number of fish in a lake. They plan to tow a sampling device from one side of the lake to the other a single time. They discuss the sampling strategies shown below.
TEKS 6.2 B; 6.4 A; 6.12 C

1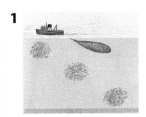

3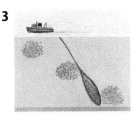

2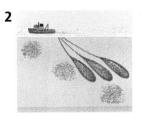

4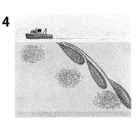

Use the diagrams to answer the following questions.

1. Which strategy is likely to provide the most accurate estimate of the number of fish in the lake?
   A) diagram 1
   B) diagram 2
   C) diagram 3
   D) diagram 4

2. How can the biologists improve their investigation?
   F) tow the sampling device very quickly
   G) tow the sampling device very slowly
   H) tow the sampling device more then once
   J) tow the sampling device around the edge of the lake

CHAPTER 16

Science TEKS 6.11 A; 6.12 A, B, C

Animal Behavior

Eye contact is made, dirt flies, and the silence is shattered. Massive horns clash as two bighorn sheep butt heads. Nearby, a spider spins a web to catch its food. Overhead, the honking of a V-shaped string of geese echoes through the valley. Do organisms learn these actions or do they occur automatically? In this chapter, you will examine the unique behaviors of animals. Also, you'll read about different types of behavior and learn about animal communication.

What do you think?

Science Journal Look at the picture below with a classmate. Discuss what you think this might be or what is happening. Here's a hint: *This instinctive reaction is triggered by their parent's arrival.* Write your answer or best guess in your Science Journal.

One way you communicate is by speaking. Other animals communicate without the use of sound. For example, a gull chick pecks at its parent's beak to get food. Try the activity below to see if you can communicate without speaking.

Observe how humans communicate without using sound

1. Form groups of students. Have one person choose an object and describe that object using gestures.
2. The other students observe and try to identify the object that is being described.
3. Each student in the group should choose an object and describe it without speaking while the others observe and identify the object.

Observe
In your Science Journal, describe how you and the other students were able to communicate without speaking to one another.

Before You Read

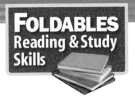

Making a Compare and Contrast Study Fold As you study behaviors, make the following Foldable to help find the similarities and differences between the behaviors of two animals.

1. Place a sheet of paper in front of you so the short side is at the top. Fold the paper in half from the left to the right side. Fold top to bottom but do not crease. Then unfold.
2. Label *Observed Behaviors of Animal 1* and *Observed Behaviors of Animal 2* across the front of the paper, as shown.
3. Through one thickness of paper, cut along the middle fold line to form two tabs, as shown.
4. Before you read the chapter, choose two animals to compare.
5. As you read the chapter, list the behaviors you learn about Animal 1 and Animal 2 under the tabs.

469

SECTION 1

Types of Behavior

As You Read

What You'll Learn
- **Identify** the differences between innate and learned behavior.
- **Explain** how reflexes and instincts help organisms survive.
- **Identify** examples of imprinting and conditioning.

Vocabulary
behavior
innate behavior
reflex
instinct
imprinting
conditioning
insight

Why It's Important
Innate behavior helps you survive on your own.

Behavior

When you come home from school, does your dog run to meet you? Your dog barks and wags its tail as you scratch behind its ears. Sitting at your feet, it watches every move you make. Why do dogs do these things? In nature, dogs are pack animals that generally follow a leader. They have been living with people for about 12,000 years. Domesticated dogs treat people as part of their own pack, as shown in **Figure 1B.**

Animals are different from one another in their behavior. They are born with certain behaviors, and they learn others. **Behavior** is the way an organism interacts with other organisms and its environment. Anything in the environment that causes a reaction is called a stimulus. A stimulus can be external, such as a rival male entering another male's territory, or internal, such as hunger or thirst. You are the stimulus that causes your dog to bark and wag its tail. Your dog's reaction to you is a response.

Figure 1
Dogs are pack animals by nature. **A** This pack of wild dogs must work together to survive. **B** This domesticated dog has accepted a human as its leader.

470 CHAPTER 16 Animal Behavior

Innate Behavior

A behavior that an organism is born with is called an **innate behavior.** These types of behaviors are inherited. They don't have to be learned.

Innate behavior patterns occur the first time an animal responds to a particular internal or external stimulus. For birds like the swallows in **Figure 2A** and the hummingbird in **Figure 2B** building a nest is innate behavior. When it's time for the female weaverbird to lay eggs, the male weaverbird builds an elaborate nest, as shown in **Figure 2C.** Although a young male's first attempt may be messy, the nest is constructed correctly.

The behavior of animals that have short life spans is mostly innate behavior. Most insects do not learn from their parents. In many cases, the parents have died or moved on by the time the young hatch. Yet every insect reacts innately to its environment. A moth will fly toward a light, and a cockroach will run away from it. They don't learn this behavior. Innate behavior allows animals to respond instantly. This quick response often means the difference between life and death.

Reflexes The simplest innate behaviors are reflex actions. A **reflex** is an automatic response that does not involve a message from the brain. Sneezing, shivering, yawning, jerking your hand away from a hot surface, and blinking your eyes when something is thrown toward you are all reflex actions.

In humans a reflex message passes almost instantly from a sense organ along the nerve to the spinal cord and back to the muscles. The message does not go to the brain. You are aware of the reaction only after it has happened. Your body reacts on its own. A reflex is not the result of conscious thinking.

Figure 2
Bird nests come in different sizes and shapes. **A** Cliff swallows build nests out of mud. **B** Hummingbirds build delicate cup-shaped nests on branches of trees. **C** This male weaverbird is knotting the ends of leaves together to secure the nest.

Health
INTEGRATION

A tap on a tendon in your knee causes your leg to stretch. This is known as the knee-jerk reflex. Abnormalities in this reflex tell doctors of a possible problem in the central nervous system. Research other types of reflexes and write a report about them in your Science Journal.

Figure 3
Spiders, like this orb weaver spider, know how to spin webs as soon as they hatch.

Instincts An **instinct** is a complex pattern of innate behavior. Spinning a web like the one in **Figure 3** is complicated, yet spiders spin webs correctly on the first try. Unlike reflexes, instinctive behaviors can take weeks to complete. Instinctive behavior begins when the animal recognizes a stimulus and continues until all parts of the behavior have been performed.

✓ **Reading Check** *What is the difference between a reflex and an instinct?*

Learned Behavior

All animals have innate and learned behaviors. Learned behavior develops during an animal's lifetime. Animals with more complex brains exhibit more behaviors that are the result of learning. However, the behavior of insects, spiders, and other arthropods is mostly instinctive behavior. Fish, reptiles, amphibians, birds, and mammals all learn. Learning is the result of experience or practice.

Learning is important for animals because it allows them to respond to changing situations. In changing environments, animals that have the ability to learn a new behavior are more likely to survive. This is especially important for animals with long life spans. The longer an animal lives, the more likely it is that the environment in which it lives will change.

Learning also can modify instincts. For example, grouse and quail chicks, shown in **Figure 4,** leave their nests the day they hatch. They can run and find food, but they can't fly. When something moves above them, they instantly crouch and keep perfectly still until the danger has passed. They will crouch without moving even if the falling object is only a leaf. Older birds have learned that leaves will not harm them, but they freeze when a hawk moves overhead.

Figure 4
As they grow older, these quail chicks will learn which organisms to avoid. *Why is it important for young quail to react the same toward all organisms?*

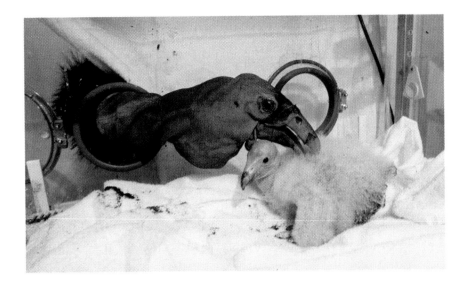

Figure 5
When feeding chicks in captivity, puppets of adult condors are used so the chicks don't associate humans with food.

Research Visit the Glencoe Science Web site at **tx.science.glencoe.com** for the latest information about raising condors to be released into the wild. Communicate to your class what you learn.

Imprinting Learned behavior includes imprinting, trial and error, conditioning, and insight. Have you ever seen young ducks following their mother? This is an important behavior because the adult bird has had more experience in finding food, escaping predators, and getting along in the world. **Imprinting** occurs when an animal forms a social attachment, like the condor in **Figure 5,** to another organism within a specific time period after birth or hatching.

Konrad Lorenz, an Austrian naturalist, developed the concept of imprinting. Working with geese, he discovered that a gosling follows the first moving object it sees after hatching. The moving object, whatever it is, is imprinted as its parent. This behavior works well when the first moving object a gosling sees is an adult female goose. But goslings hatched in an incubator might see a human first and imprint on him or her. Animals that become imprinted toward animals of another species have difficulty recognizing members of their own species.

Figure 6
Were you able to tie your shoes on the first attempt? *What other things do you do every day that required learning?*

Trial and Error Can you remember when you learned to ride a bicycle? You probably fell many times before you learned how to balance on the bicycle. After a while you could ride without having to think about it. You have many skills that you have learned through trial and error such as feeding yourself and tying your shoes, as shown in **Figure 6.**

Behavior that is modified by experience is called trial-and-error learning. Many animals learn by trial and error. When baby chicks first try feeding themselves, they peck at many stones before they get any food. As a result of trial and error, they learn to peck only at food particles.

SECTION 1 Types of Behavior **473**

Mini LAB

Observing Conditioning

Procedure
1. Obtain several **photos of different foods and landscapes** from your teacher.
2. Show each picture to a classmate for 20 s.
3. Record how each photo made your partner feel.

Analysis
1. How did your partner feel after looking at the photos of food?
2. What effect did the landscape pictures have on your partner?
3. Infer how advertising might condition consumers to buy specific food products.

Figure 7
In Pavlov's experiment, a dog was conditioned to salivate when a bell was rung. It associated the bell with food.

Conditioning Do you have an aquarium in your school or home? If you put your hand above the tank, the fish probably will swim to the top of the tank expecting to be fed. They have learned that a hand shape above them means food. What would happen if you tapped on the glass right before you fed them? After a while the fish probably will swim to the top of the tank if you just tap on the glass. Because they are used to being fed after you tap on the glass, they associate the tap with food.

Animals often learn new behaviors by conditioning. In **conditioning,** behavior is modified so that a response to one stimulus becomes associated with a different stimulus. There are two types of conditioning. One type introduces a new stimulus before the usual stimulus. Russian scientist Ivan P. Pavlov performed experiments with this type of conditioning. He knew that the sight and smell of food made hungry dogs secrete saliva. Pavlov added another stimulus. He rang a bell before he fed the dogs. The dogs began to connect the sound of the bell with food. Then Pavlov rang the bell without giving the dogs food. They salivated when the bell was rung even though he did not show them food. The dogs, like the one in **Figure 7,** were conditioned to respond to the bell.

In the second type of conditioning, the new stimulus is given after the affected behavior. Getting an allowance for doing chores is an example of this type of conditioning. You do your chores because you want to receive your allowance. You have been conditioned to perform an activity that you may not have done if you had not been offered a reward.

✔ **Reading Check** *How does conditioning modify behavior?*

Insight How does learned behavior help an animal deal with a new situation? Suppose you have a new math problem to solve. Do you begin by acting as though you've never seen it before, or do you use what you have learned previously in math to solve the problem? If you use what you have learned, then you have used a kind of learned behavior called insight. **Insight** is a form of reasoning that allows animals to use past experiences to solve new problems. In experiments with chimpanzees, as shown in **Figure 8,** bananas were placed out of the chimpanzees' reach. Instead of giving up, they piled up boxes found in the room, climbed them, and reached the bananas. At some time in their lives, the chimpanzees must have solved a similar problem. The chimpanzees demonstrated insight during the experiment. Much of adult human learning is based on insight. When you were a baby, you learned by trial and error. As you grow older, you will rely more on insight.

Figure 8
This illustration shows how chimpanzees may use insight to solve problems.

Section Assessment

1. How is innate behavior different from learned behavior?
2. Compare a reflex with an instinct.
3. What is the difference between an internal and external stimulus?
4. Compare imprinting and conditioning.
5. **Think Critically** Use what you know about conditioning to explain how the term *mouthwatering food* might have come about.

Skill Builder Activities

6. **Researching Information** How are dogs trained to sniff out certain substances? **For more help, refer to the** Science Skill Handbook.
7. **Using an Electronic Spreadsheet** Make a spreadsheet of the behaviors in this section. Sort the behaviors according to whether they are innate or learned behaviors. Then identify the type of innate or learned behavior. **For more help, refer to the** Technology Skill Handbook.

SECTION 1 Types of Behavior

SECTION 2

Behavioral Interactions

As You Read

What You'll Learn
- **Explain** why behavioral adaptations are important.
- **Describe** how courtship behavior increases reproductive success.
- **Explain** the importance of social behavior and cyclic behavior.

Vocabulary
social behavior
society
aggression
courtship behavior
pheromone
cyclic behavior
hibernation
migration

Why It's Important
Organisms must be able to communicate with each other to survive.

Instinctive Behavior Patterns

Complex interactions of innate behaviors between organisms result in many types of animal behavior. For example, courtship and mating within most animal groups are instinctive ritual behaviors that help animals recognize possible mates. Animals also protect themselves and their food sources by defending their territories. Instinctive behavior, just like natural hair color, is inherited.

Social Behavior

Animals often live in groups. One reason, shown in **Figure 9,** is that large numbers provide safety. A lion is less likely to attack a herd of zebras than a lone zebra. Sometimes animals in large groups help keep each other warm. Also, migrating animal groups are less likely to get lost than animals that travel alone.

Interactions among organisms of the same species are examples of **social behavior.** Social behaviors include courtship and mating, caring for the young, claiming territories, protecting each other, and getting food. These inherited behaviors provide advantages that promote survival of the species.

✓ **Reading Check** *Why is social behavior important?*

Figure 9
When several zebras are close together their stripes make it difficult for predators to pick out one individual.

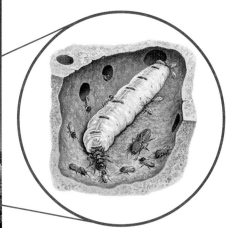

Figure 10
Termites built this large mound in Australia. The mound has a network of tunnels and chambers for the queen to deposit eggs into.

Societies Insects such as ants, bees, and the termites shown in **Figure 10,** live together in societies. A **society** is a group of animals of the same species living and working together in an organized way. Each member has a certain role. Usually a specific female lays eggs, and a male fertilizes them. Workers do all the other jobs in the society.

Some societies are organized by dominance. Wolves usually live together in packs. A wolf pack has a dominant female. The top female controls the mating of the other females. If plenty of food is available, she mates and then allows the others to do so. If food is scarce, she allows less mating. During such times, she is usually the only one to mate.

Territorial Behavior

Many animals set up territories for feeding, mating, and raising young. A territory is an area that an animal defends from other members of the same species. Ownership of a territory occurs in different ways. Songbirds sing, sea lions bellow, and squirrels chatter to claim territories. Other animals leave scent marks. Some animals, like the tiger in **Figure 11,** patrol an area and attack other animals of the same species who enter their territory. Why do animals defend their territories? Territories contain food, shelter, and potential mates. If an animal has a territory, it will be able to mate and produce offspring. Defending territories is an instinctive behavior. It improves the survival rate of an animal's offspring.

Figure 11
A tiger's territory may include several miles. It will confront any other tiger who enters it.

SECTION 2 Behavioral Interactions **477**

Figure 12
Young wolves roll over and make themselves as small as possible to show their submission to adult wolves.

Figure 13
During the waggle dance, if the source is far from the hive, the dance takes the form of a figure eight. The angle of the waggle is equal to the angle from the hive between the Sun and nectar source.

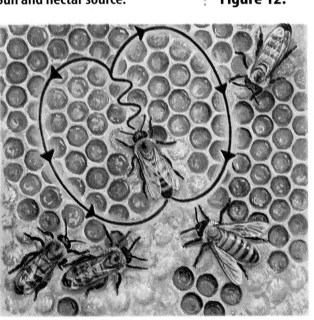

Aggression Have you ever watched as one dog approached another dog that was eating a bone? What happened to the appearance of the dog with the bone? Did its hair on its back stick up? Did it curl its lips and make growling noises? This behavior is aggression. **Aggression** is a forceful behavior used to dominate or control another animal. Fighting and threatening are aggressive behaviors animals use to defend their territories, protect their young, or to get food.

Many animals demonstrate aggression. Some birds let their wings droop below their tail feathers. It may take another bird's perch and thrust its head forward in a pecking motion as a sign of aggression. Cats lay their ears flat, arch their backs, and hiss.

Submission Animals of the same species seldom fight to the death. Teeth, beaks, claws, and horns are used for killing prey or for defending against members of a different species.

To avoid being attacked and injured by an individual of its own species, an animal shows submission. Postures that make an animal appear smaller often are used to communicate surrender. In some animal groups, one individual is usually dominant. Members of the group show submissive behavior toward the dominant individual. This stops further aggressive behavior by the dominant animal. Young animals also display submissive behaviors toward parents or dominant animals, as shown in **Figure 12.**

Communication

In all social behavior, communication is important. Communication is an action by a sender that influences the behavior of a receiver. How do you communicate with the people around you? You may talk, make noises, or gesture like you did in this chapter's Explore Activity. Honeybees perform a dance, as shown in **Figure 13,** to communicate to other bees in the hive where a food source is. Animals in a group communicate with sounds, scents, and actions. Alarm calls, chemicals, speech, courtship behavior, and aggression are forms of communication.

Figure 14
This male Emperor of Germany bird of paradise attracts mates by posturing and fanning its tail.

Courtship Behavior A male bird of paradise, shown in **Figure 14,** spreads its tail feathers and struts. A male sage grouse fans its tail, fluffs its feathers, and blows up its two red air sacs. These are examples of behavior that animals perform before mating. This type of behavior is called **courtship behavior.** Courtship behaviors allow male and female members of a species to recognize each other. These behaviors also stimulate males and females so they are ready to mate at the same time. This helps ensure reproductive success.

In most species the males are more colorful and perform courtship displays to attract a mate. Some courtship behaviors allow males and females to find each other across distances.

Chemical Communication
Ants are sometimes seen moving single file toward a piece of food. Male dogs frequently urinate on objects and plants. Both behaviors are based on chemical communication. The ants have laid down chemical trails that others of their species can follow. The dog is letting other dogs know he has been there. In these behaviors, the animals are using a chemical called a pheromone to communicate. A **pheromone** (FER uh mohn) is a chemical that is produced by one animal to influence the behavior of another animal of the same species. They are powerful chemicals needed only in small amounts. They remain in the environment so that the sender and the receiver can communicate without being in the same place at the same time. They can advertise the presence of an animal to predators, as well as to the intended receiver of the message.

Males and females use pheromones to establish territories, warn of danger, and attract mates. Certain ants, mice, and snails release alarm pheromones when injured or threatened.

Demonstrating Chemical Communication

Procedure
1. Obtain a **sample of perfume or air freshener.**
2. Spray it into the air to leave a scent trail as you move around the house or apartment to a hiding place.
3. Have someone try to discover where you are by following the scent of the substance.

Analysis
1. What was the difference between the first and last room you were in?
2. Would this be an efficient way for humans to communicate? Explain.

SECTION 2 Behavioral Interactions **479**

Figure 15
Many animals use sound to communicate.

A Frogs often croak loud enough to be heard far away.

B Pileated woodpecker calls often can be heard above everything else in the forest.

C Howler monkeys got their name because of the sounds they make.

Sound Communication Male crickets rub one forewing against the other forewing. This produces chirping sounds that attract females. Each cricket species produces several calls that are different from other cricket species. These calls are used by researchers to identify different species. Male mosquitoes have hairs on their antennae that sense buzzing sounds produced by females of their same species. The tiny hairs vibrate only to the frequency emitted by a female of the same species.

Vertebrates use a number of different forms of sound communication. Rabbits thump the ground, gorillas pound their chests, beavers slap the water with their flat tails, and frogs, like the one in **Figure 15,** croak. Do you think that sound communication in noisy environments is useful? Seabirds that live where waves pound the shore rather than in some quieter place must rely on visual signals, not sound, for communication.

Chemistry INTEGRATION

The light produced by fireflies is a particle of visible light that radiates when chemicals produce a high-energy state and then return to their normal state. Hypothesize how this helps fireflies survive. Write your hypothesis in your Science Journal.

Light Communication Certain kinds of flies, marine organisms, and beetles have a special form of communication called bioluminescence. Bioluminescence, shown in **Figure 16,** is the ability of certain living things to give off light. This light is produced through a series of chemical reactions in the organism's body. Probably the most familiar bioluminescent organisms in North America are fireflies. They are not flies, but beetles. The flash of light is produced on the underside of the last abdominal segments and is used to locate a prospective mate. Each species has its own characteristic flashing. Males fly close to the ground and emit flashes of light. Females must flash an answer at exactly the correct time to attract males.

480 CHAPTER 16 Animal Behavior

NATIONAL GEOGRAPHIC VISUALIZING BIOLUMINESCENCE

Figure 16

Many marine organisms use bioluminescence as a form of communication. This visible light is produced by a chemical reaction and often confuses predators or attracts mates. Each organism on this page is shown in its normal and bioluminescent state.

▼ **KRILL** The blue dots shown below this krill are all that are visible when krill bioluminesce. The krill may use bioluminescence to confuse predators.

▲ **JELLYFISH** This jellyfish lights up like a neon sign when it is threatened.

◀ **BLACK DRAGONFISH** The black dragonfish lives in the deep ocean where light doesn't penetrate. It has light organs under its eyes that it uses like a flashlight to search for prey.

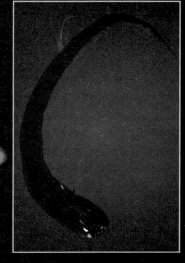

▲ **DEEP-SEA SEA STAR** The sea star uses light to warn predators of its unpleasant taste.

Uses of Bioluminescence Many bioluminescent animals are found deep in oceans where sunlight does not reach. The ability to produce light may serve several functions. One species of fish dangles a special luminescent organ in front of its mouth. This lures prey close enough to be caught and eaten. Deep-sea shrimp secrete clouds of a luminescent substance when disturbed. This helps them escape their predators. Patterns of luminescence on an animal's body may serve as marks of recognition similar to the color patterns of animals that live in sunlit areas.

Cyclic Behavior

Why do most songbirds rest at night while some species of owls rest during the day? Some animals like the owl in **Figure 17** show regularly repeated behaviors such as sleeping in the day and feeding at night.

A **cyclic behavior** is innate behavior that occurs in a repeating pattern. It often is repeated in response to changes in the environment. Behavior that is based on a 24-hour cycle is called a circadian rhythm. Most animals come close to this 24-hour cycle of sleeping and wakefulness. Experiments show that even if animals can't tell whether it is night or day, they continue to behave in a 24-hour cycle.

Animals that are active during the day are diurnal (dy UR nul). Animals that are active at night are nocturnal. Owls are nocturnal. They have round heads, big eyes, and flat faces. Their flat faces reflect sound and help them navigate at night. Owls also have soft feathers that make them almost silent while flying.

✓ **Reading Check** *What is a diurnal behavior?*

Science Online

Research Visit the Glencoe Science Web site at **tx.science.glencoe.com** for more information about owl behavior. Communicate to your class what you learn.

Figure 17
Barn owls usually sleep during the day and hunt at night.
What type of behavior does the owl exhibit?

Hibernation Some cyclic behaviors also occur over long periods of time. **Hibernation** is a cyclic response to cold temperatures and limited food supplies. During hibernation, an animal's body temperature drops to near that of its surroundings, and its breathing rate is greatly reduced. Animals in hibernation, such as the bats in **Figure 18,** survive on stored body fat. The animal remains inactive until the weather becomes warm in the spring. Some mammals and many amphibians and reptiles hibernate.

Animals that live in desert like environments also go into a state of reduced activity. This period of inactivity is called estivation. Desert animals sometimes estivate due to extreme heat, lack of food, or periods of drought.

Figure 18
Many bats find a frost-free place like this abandoned coal mine to hibernate for the winter when food supplies are low.

Problem-Solving Activity

How can you determine which animals hibernate?

Many animals hibernate in the winter. During this period of inactivity, they survive on stored body fat. While they are hibernating, they undergo several physical changes. Heart rate slows down and body temperature decreases. The degree to which the body temperature decreases varies among animals. Scientists have disagreed about whether some animals truly hibernate or if they just reduce their activity and go into a light sleep. Usually, a true hibernator's body temperature will decrease significantly while it is hibernating.

Identifying the Problem
The table on the right shows the difference between the normal body temperature and the hibernating body temperature of several animals. What similarities do you notice?

Average Body Temperatures of Hibernating Animals		
Animal	Normal Body Temperature (°C)	Hibernating Body Temperature (°C)
Woodchuck	37	3
Squirrel	32	4
Grizzly Bear	32–37	27–32
Whippoorwill	40	18
Hoary Marmot	37	10

Solving the Problem
1. Which animals would you classify as true hibernators and which would you classify as light sleepers? Explain.
2. Some animals such as snakes and frogs also hibernate. Why would it be difficult to record their normal body temperature on this table?
3. Which animal has the least amount of change in body temperature?

SECTION 2 Behavioral Interactions

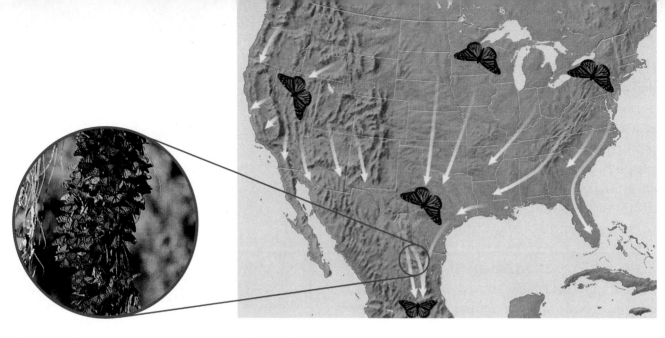

Figure 19
Many monarch butterflies travel from the United States to Mexico for the winter.

Migration Instead of hibernating, many birds and mammals move to new locations when the seasons change. This instinctive seasonal movement of animals is called **migration.** Most animals migrate to find food or reproduce in an environment that is more favorable for the survival of its offspring. Many species of birds fly for hours or days without stopping. The blackpoll warbler flies more than 4,000 km nonstop from North America to its winter home in South America. The trip takes nearly 90 hours. Monarch butterflies, shown in **Figure 19,** can migrate as much as 2,900 km. Gray whales swim from cold arctic waters to the waters off the coast of northern Mexico. After the young are born, they make the return trip.

Section Assessment

1. What are some examples of courtship behavior? How does this behavior help organisms survive?
2. How are cyclic behaviors, such as hibernation, a response to stimuli in the environment?
3. Give two reasons why animals migrate.
4. What is the difference between hibernation and migration?
5. **Think Critically** Suppose a species of frog lives close to a loud waterfall. It often waves a bright blue foot in the air. What might the frog be doing?

Skill Builder Activities

6. **Testing a Hypothesis** Design an experiment that tests the hypothesis that ants leave chemical trails to show other ants where food can be found. **For more help, refer to the** Science Skill Handbook.
7. **Solving One-Step Equations** Some cicadas emerge from the ground every 17 years. The population of one type of caterpillar peaks every five years. If the peak cycle of the caterpillars and the emergence of cicadas coincided in 1990, in what year will they coincide again? **For more help, refer to the** Math Skill Handbook.

Activity

Observing Earthworm Behavior

Earthworms often can be seen wriggling across sidewalks, driveways, and yards on moist nights. Why don't you see many earthworms during the day?

What You'll Investigate
How do earthworms respond to light?

Materials
scissors
shoe box with lid
flashlight
tape
paper
moist paper towels
earthworms
timer

Goals
- **Predict** how earthworms will behave in the presence of light.

Safety Precautions

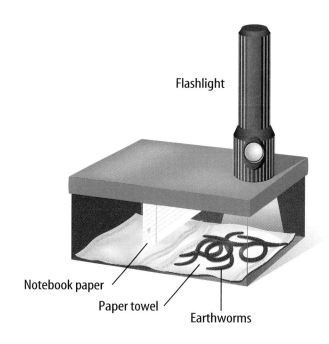

Flashlight

Notebook paper
Paper towel
Earthworms

Procedure
1. Cut a round hole, smaller than the end of the flashlight, near one end of the lid.
2. Tape a sheet of paper to the lid so it hangs just above the bottom of the box and about 10 cm away from the end with the hole in it.
3. Place the moist paper towels in the bottom of the box.
4. Place the earthworms in the end of the box that has the hole in it.
5. Hold the flashlight over the hole and turn it on.
6. Leave the box undisturbed for 30 minutes, then open the lid and observe the worms.
7. **Record** the results of your experiment in your Science Journal.

Conclude and Apply
1. Which direction did the earthworms move when the light was turned on?
2. Based on your observations, what can you infer about earthworms?
3. What type of behavior did the earthworms exhibit? Explain.
4. **Predict** where you would need to go to find earthworms during the day.

Communicating Your Data

Write a story that describes a day in the life of an earthworm. List activities, dangers, and problems an earthworm can face. Include a description of its habitat. **For more help, refer to the** Science Skill Handbook.

ACTIVITY **485**

Activity: Model and Invent

Animal Habitats

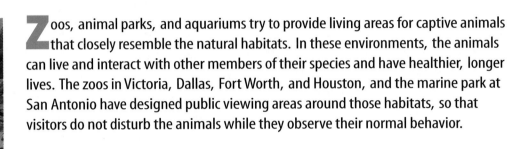

Zoos, animal parks, and aquariums try to provide living areas for captive animals that closely resemble the natural habitats. In these environments, the animals can live and interact with other members of their species and have healthier, longer lives. The zoos in Victoria, Dallas, Fort Worth, and Houston, and the marine park at San Antonio have designed public viewing areas around those habitats, so that visitors do not disturb the animals while they observe their normal behavior.

Recognize the Problem

What types of environments are best suited for raising animals in captivity?

Thinking Critically

How can the habitats provided at an animal park affect the behavior of animals?

Goals

- **Research** the natural habitat and basic needs of one animal.
- **Design** and model an appropriate zoo, animal park, or aquarium environment for this animal. Working cooperatively with your classmates, design an entire zoo or animal park.

Possible Materials

poster board
markers or colored pencils
materials that can be used to make a scale model

Data Source

SCIENCE *Online* Go to the Glencoe Science Web site at **tx.science.glencoe.com** for more information about existing zoos, animal parks, and aquariums.

486 CHAPTER 16 Animal Behavior

Using Scientific Methods

Planning the Model

1. Choose an animal to research. Find out where this animal is found in nature. What does it eat? What are its natural predators? Does it exhibit unique territorial, courtship, or other types of behavior? How is this animal adapted to its natural environment?
2. **Design** a model of a proposed habitat in which this animal can live successfully. Don't forget to include all of the things, such as shelter, food, and water, that your animal will need to survive. Will there be any other organisms in the habitat?

Check the Model Plans

1. **Research** how zoos, animal parks, or aquariums provide habitats for animals. Information may be obtained by viewing the Glencoe Science Web site and contacting scientists who work at zoos, animal parks, and aquariums.
2. **Present** your design to your class in the form of a poster, slide show, or video. Compare your proposed habitat with that of the animal's natural environment. Make sure you include a picture of your animal in its natural environment.

Making the Model

1. Using all of the information you have gathered, create a model exhibit area for your animal.
2. Indicate what other plants and animals may be present in the exhibit area.

Analyzing and Applying Results

1. **Decide** whether all of the animals studied in this activity can coexist in the same zoo or wildlife preserve.
2. **Predict** which animals could be grouped together in exhibit areas.
3. **Determine** how large your zoo or wildlife preserve needs to be. Which animals require a large habitat?
4. Using the information provided by the rest of your classmates, design an entire zoo or aquarium that could include the majority of animals studied.
5. **Analyze** problems that might exist in your design. Suggest some ways you might want to improve your design.

Give an oral presentation to another class on the importance of providing natural habitats for captive animals. **For more help, refer to the** Science Skill Handbook.

Oops! Accidents in SCIENCE
SOMETIMES GREAT DISCOVERIES HAPPEN BY ACCIDENT!

Going to the Dogs

A simple and surprising stroll showed that dogs really are humans' best friends

You've probably seen visually impaired people walking with their trusted and gentle four-legged guides—or "seeing-eye" dogs. The specially trained dogs serve as eyes for people who can't see, making it possible for them to lead independent lives. But what you probably didn't know is that about 80 years ago, a doctor and his patient discovered this canine ability entirely by accident!

Many people were killed or injured during World War I. Near the end of that war, Dr. Gerhard Stalling and his dog strolled with a patient—a German soldier who had been blinded—around hospital grounds in Germany.

German shepherds make excellent guide dogs.

A dog safely guides its owner across a street.

While they were walking, the doctor was briefly called away. The dog and the soldier stayed outside. A few moments later, when the doctor returned, the dog and the soldier were gone! Searching the paths frantically, Dr. Stalling made an astonishing discovery. His pet had led the soldier safely around the hospital grounds. And together the two strolled peacefully back toward the doctor.

School for Dogs

Inspired by what his dog could do, Dr. Stalling set up the first school in the world dedicated to training dogs as guides. Dorothy Eustis, an American woman working as a dog trainer for the International Red Cross in Switzerland, traveled to Stalling's school about ten years later. A report of her visit and study of the way Stalling trained dogs appeared in a New York City newspaper in 1927.

Hearing the story, Morris Frank, a visually impaired American, became determined to get himself a guide dog. He wrote to Dorothy Eustis and asked that she train a dog for him. She accepted his request on one condition.

She wanted Frank to join her in Switzerland for the training process. Frank and his guide dog Buddy returned to New Jersey in 1928. Within a year, Frank set up a training facility in New Jersey, "The Seeing Eye, Inc."

German shepherds, golden retrievers, and Labrador retrievers seem to make the best guide dogs. They learn hand gestures and simple commands to lead visually impaired people across streets and safely around obstacles. This is what scientists call "learned behavior." Animals gain learned behavior through experience. Learning happens gradually and in steps. In fact, scientists say that learning is a somewhat permanent change in behavior due to experience. But, a guide dog not only learns to respond to special commands, it must also know when *not* to obey. If its human owner urges the dog to cross the street and the dog sees that a car is approaching and refuses, the dog has learned to disobey the command. This trait, called "intelligent disobedience," ensures the safety of the owner and the dog—a sure sign that dogs are still humans' best friends.

This girl gets to help train a future guide dog for The Seeing Eye, Inc.

CONNECTIONS Write Lead a blindfolded partner around the classroom. Help your partner avoid obstacles. Then trade places. Write in your Science Journal about your experience leading and being led.

SCIENCE *Online*
For more information, visit
tx.science.glencoe.com

Chapter 16 Study Guide

Reviewing Main Ideas

Section 1 Types of Behavior

1. Behavior that an animal has when it's born is innate behavior. Other animal behaviors are learned through experience. *In the figure below, what type of behavior is the dog exhibiting?*

2. Reflexes are simple innate behaviors. An instinct is a complex pattern of innate behavior.

3. Learned behavior includes imprinting, in which an animal forms a social attachment immediately after birth.

4. Behavior modified by experience is learning by trial and error.

5. Conditioning occurs when the response to one stimulus becomes associated with another. Insight uses past experiences to solve new problems.

Section 2 Behavioral Interactions

1. Behavioral adaptations such as defense of territory, courtship behavior, and social behavior help species of animals survive and reproduce.

2. Courtship behaviors allow males and females to recognize each other and prepare to mate.

3. Interactions among members of the same species are social behaviors. *What type of social behavior is this male peacock displaying?*

4. Communication among organisms occurs in several ways including chemical, sound, and light. *How will other ants, like the one shown, be able to locate food that is far from their nest?*

5. Cyclic behaviors are behaviors that occur in repeating patterns. Animals that are active during the day are diurnal. Animals that are active at night are nocturnal.

After You Read

FOLDABLES Reading & Study Skills Compare and contrast the behaviors of Animal 1 and Animal 2 listed in your foldable. How many of the behaviors you listed were innate? Learned?

Chapter 16 Study Guide

Visualizing Main Ideas

Complete the following concept map on types of behavior.

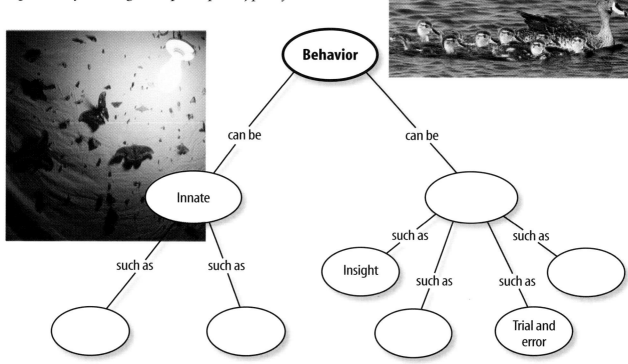

Vocabulary Review

Vocabulary Words

a. aggression
b. behavior
c. conditioning
d. courtship behavior
e. cyclic behavior
f. hibernation
g. imprinting
h. innate behavior
i. insight
j. instinct
k. migration
l. pheromone
m. reflex
n. social behavior
o. society

Study Tip
THE PRINCETON REVIEW

Take good notes, even during lab. Lab experiments reinforce key concepts, and looking back on these notes can help you better understand what happened and why.

Using Vocabulary

Explain the differences between the vocabulary words given below. Then explain how the words are related.

1. conditioning, imprinting
2. innate behavior, social behavior
3. insight, instinct
4. social behavior, society
5. instinct, reflex
6. hibernation, migration
7. courtship behavior, pheromone
8. cyclic behavior, migration
9. aggression, social behavior
10. behavior, reflex

CHAPTER STUDY GUIDE 491

Chapter 16 Assessment & TEKS Review

Checking Concepts

Choose the word or phrase that best answers the question.

1. What is an instinct an example of?
   A) innate behavior   C) imprinting
   B) learned behavior D) conditioning

2. What is a spider spinning a web an example of?
   A) conditioning   C) learned behavior
   B) imprinting   D) an instinct

3. Which animals depend least on instinct and most on learning?
   A) birds   C) mammals
   B) fish   D) amphibians

4. What is an area that an animal defends from other members of the same species called?
   A) society   C) migration
   B) territory D) aggression

5. What is a forceful act used to dominate or control?
   A) courtship C) aggression
   B) reflex   D) hibernation

6. Which of the following is NOT an example of courtship behavior?
   A) fluffing feathers
   B) taking over a perch
   C) singing songs
   D) releasing pheromones

7. What is an organized group of animals doing specific jobs called?
   A) community C) society
   B) territory D) circadian rhythm

8. What is the response of inactivity and slowed metabolism that occurs during cold conditions?
   A) hibernation C) migration
   B) imprinting   D) circadian rhythm

9. Which of the following is a reflex?
   A) writing   C) sneezing
   B) talking   D) riding a bicycle

10. What are behaviors that occur in repeated patterns called?
    A) cyclic   C) reflex
    B) imprinting D) society

Thinking Critically

11. Explain the type of behavior involved when the bell rings at the end of class.

12. Discuss the advantages and disadvantages of migration as a means of survival.

13. Explain how a habit such as tying your shoes, is different from a reflex.

14. Use one example to explain how behavior increases an animal's chance for survival.

15. Hens lay more eggs in the spring when the number of daylight hours increases. How can farmers use this knowledge of behavior to their advantage?

Developing Skills

16. **Testing a Hypothesis** Design an experiment to test a hypothesis about a specific response to a stimulus from an animal.

17. **Recording Observations** Make observations of a dog, cat, or bird for a week. Record what you see. How did the animal communicate with other animals and with you?

Chapter 16 Assessment

18. Forming a Hypothesis Make a hypothesis about how frogs communicate with each other. How could you test your hypothesis?

19. Classifying Make a list of 25 things that you do regularly. Classify each as an innate or learned behavior. Which behaviors do you have more of?

20. Concept Mapping Complete the following concept map about communication. Use these words: *sound, chirping, bioluminescence,* and *buzzing.*

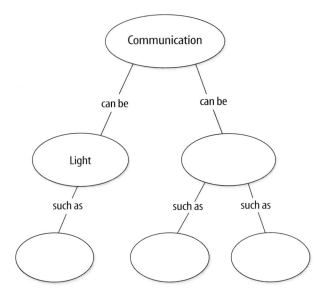

Performance Assessment

21. Poster Draw a map showing the migration route of monarch butterflies, gray whales, or blackpoll warblers.

Technology

Go to the Glencoe Science Web site at **tx.science.glencoe.com** or use the **Glencoe Science CD-ROM** for additional chapter assessment.

THE PRINCETON REVIEW TAKS Practice

A biologist is given illustrations of different behaviors. The different types of behaviors are listed below. **TEKS 8.2 B, C**

Study the table and answer the following questions.

1. A reflex is an automatic response to a stimulus. Which one of the behaviors in the table is an example of a reflex?

   A) one   **C)** three
   B) two   **D)** four

2. Trial and error is a type of learned behavior that is modified by experience. Which of the behaviors in the table is an example of a trial-and-error behavior?

   F) one   **H)** three
   G) two   **J)** four

Reading Comprehension

Read the passage. Then read each question that follows the passage. Decide which is the best answer to each question.

Antoine Laurent Lavoisier

Many scientists assisted in the development of modern chemistry. Among them was a French scientist named Antoine Laurent Lavoisier (1743–1794). He is known as the father of modern chemistry because of his theories of combustion, his development of a new system of chemical nomenclature, and his writing of the first modern textbook of chemistry. He served his country in many ways, including as a tax collector, an economist, and the director of France's gunpowder-making facility. Many considered him to be a great patriot to his nation.

Lavoisier's experiments were some of the first chemistry experiments that used measurements to support a hypothesis. He performed chemical reactions inside sealed flasks. In his most famous experiment, a sealed flask was filled with air and mercury. Lavoisier weighed the flask and then heated it for several days until the mercury turned red. The altered appearance of the mercury indicated that the air and mercury had reacted chemically to produce a different substance. Once he observed this color change, Lavoisier weighed the flask again and found that it weighed the same as it did before.

Lavoisier's experiment showed that the matter that exists at the start of a chemical reaction exists at the end of the chemical reaction, even if the matter's properties have changed. Because the flask weighed the same before and after the chemical reaction, Lavoisier knew that nothing had left or entered the flask. The heat caused the atoms trapped inside to rearrange and form a new substance. Although it took many years for Lavoisier's explanation of what he observed to be accepted by everyone, it was eventually recognized as correct. His work helped scientists everywhere to understand chemical reactions better.

Test-Taking Tip As you read the passage, underline or circle key words. Refer back to these key words as you answer the questions.

The two flasks were weighed before and after the chemical reaction occurred.

1. The passage best supports which of the following conclusions? *Reading TEKS 7.10 H*
    A) Energy is neither created nor destroyed.
    B) Matter is neither created nor destroyed.
    C) Only heat and time are needed to create any kind of matter.
    D) Glass and heat always create new substances.

2. You can tell from the passage that a patriot is someone who _____. *Reading TEKS 7.9 B*
    F) makes gunpowder
    G) reacts chemically
    H) convinces other people
    J) serves a country

TAKS Practice

Reasoning and Skills

Read each question and choose the best answer.

1. Water exists in different states. Cells contain mostly water. Which of the following best represents the way that water is found in cells? *Science TEKS 7.3 C*

   A)   C)

   B)    D)

   Test-Taking Tip   Think about what you would find in your own cells.

Some of the Elements that Make up Organisms			
Element	Bacteria	Alfalfa	Human
Carbon	12%	11%	19%
Nitrogen	3%	1%	3%
Oxygen	74%	78%	65%
Phosphorus	1%	1%	1%
Hydrogen	10%	9%	10%

2. Elements come together in different combinations to make the many substances in your body. According to the information above, which element is most abundant in humans? *Science TEKS 7.2 C*

   F) carbon
   G) nitrogen
   H) oxygen
   J) phosphorus

Test-Taking Tip   Use key words from the question, such as *in humans*, to direct your attention to the correct column in the table.

Experiments in Plant Growth			
Plant	Days Grown	Type of Light Bulb (watts)	Height (cm)
1	68	100	35
2	68	100	27.5

3. Above is a picture showing how an experiment on plant growth was performed. Data from the experiment also could have included the _____. *Science TEKS 7.2 C*

   A) type of flower pot that was used
   B) size of the laboratory
   C) day of week that the data were recorded
   D) amount of water given to plants

   Test-Taking Tip   Think about which of the answer choices can affect a plant's growth.

Consider this question carefully before writing your answer on a separate sheet of paper.

4. Governments have started to talk about reducing the pollution their countries release into oceans and air. Why would pollution from one country bother another, faraway country? *Science TEKS 7.14 C*

   Test-Taking Tip   Think about how pollutants might move through water or air. Then, carefully write your answer.

Grade 5 TEKS Review

On the pages that follow, you will find a review of the major science concepts that you learned in Grade 5. These materials on these pages reviews all the science concepts covered in the Grade 5 TEKS. Use this review as you study the science concepts required by the Grade 6 TEKS.

CONTENTS

TEKS 5.5	Systems	497
TEKS 5.6	Change Occurs in Cycles	499
TEKS 5.7	Physical Properties of Matter	502
TEKS 5.8	Forms of Energy	505
TEKS 5.9	Species Adaptations	508
TEKS 5.10	Characteristics of Offspring	511
TEKS 5.11	Impact of Past Events	513
TEKS 5.12	The Natural World of Earth and Sky	517

TEKS 5.5 Review

Systems

What You'll Review

TEKS 5.5
A system is a collection of cycles, structures, and processes that interact.

5.5 A describe some cycles, structures, and processes that are found in a simple system;

5.5 B describe some interactions that occur in a simple system.

Directions: *Review these concepts from Grade 5. They will help you to study Chapters 1, 7, 8, and 15 about different types of systems, a topic required by the Grade 6 TEKS.*

For 5.5 A, recall that:

Systems are composed of different structures, processes, and cycles. For example, an ecosystem includes the living community and the nonliving parts of the environment. One part of the nonliving environment is water. Water can take many forms—lakes, rivers, oceans, and springs are a few examples. These forms are all structures. They are the way water is organized in the environment.

These structures, however, don't stay the same. They change through different processes. Streams change as erosion and deposition occur. As a stream flows along, it picks up and moves Earth materials from one place to another. In some places, water flows swiftly, and rock can be eroded to form canyons and waterfalls. In other places, water drops the sediment that it has been carrying. When sediment is deposited, sand bars and deltas might form. Streams, like many other systems, are constantly changing.

Many of the processes that affect water are all part of a large cycle called the water cycle. In the water cycle, water moves between Earth and the atmosphere. You can trace the water cycle in **Figure 1.**

The example of water shows that a system is built of structures, processes, and cycles. Each system has many such examples. Structures in an ecosystem range from microscopic cells to giant mountains. Processes like erosion and evolution shape Earth and living things, and elements like carbon and nitrogen cycle through the ecosystem in a way similar to water.

Figure 1
Start anywhere on the water cycle and trace your way through the cycle back to a starting point.

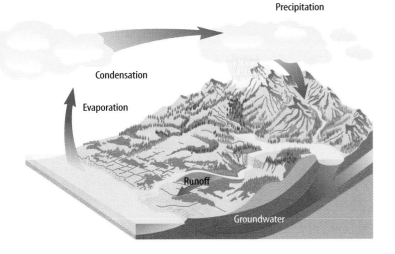

TEKS 5.5 Review

TEKS 5.5 A Assessment

1. What are the names of two processes by which water can change a stream?
2. Name three cycles in an ecosystem.
3. **Thinking Critically** Do cycles have a beginning and an end?

For 5.5 B, recall that:

A system is made up of structures, processes, and cycles. But there is another aspect to a system—the interactions among its various parts. Important interactions exist among living things and between living things and the nonliving parts of the environment in which they exist.

Living things within an ecosystem interact in different ways. One way is the predator and prey relationship. Owls prey on mice. If the mice disappear, the owls won't be able to get the food they need to live. If the owls disappear, the population of mice may then grow too large for the environment to sustain. Therefore, the health of owls and mice depend on one another.

Living things also interact with the nonliving parts of the environment. Different species of fish will live in whitewater rapids rather than in a calm, quiet pool in a stream. Some animals, like the polar bear in **Figure 2,** are adapted to survive the arctic cold. Black bears, however, prefer warmer climates. Any changes in climate will affect the species that are able to thrive in an area.

The relationship between predator and prey and the relationship between species and climate show the interdependence of the parts of a system. Changes in one part affect the other parts. If you understand the system, you can predict how one change will affect other parts of a system.

Figure 2
Polar bears thrive in the icy cold.

TEKS 5.5 B Assessment

1. What do you expect would happen if the mice in an area suddenly disappeared due to disease?
2. What bear would you expect to find in Greenland, where the weather is cold and icy?
3. **Thinking Critically** Almost all bears depend to some extent on plants in their diets. What species of bear would you expect to depend the least on plants? Why?

TEKS 5.6 Review

Change Occurs in Cycles: Past Events Affect Present and Future Events

What You'll Review

TEKS 5.6
Change Occurs in Cycles

5.6 A identify events and describe changes that occur on a regular basis such as in daily, weekly, lunar, and seasonal cycles;

5.6 B identify the significance of the water, carbon, and nitrogen cycles;

5.6 C describe and compare life cycles of plants and animals.

Directions: *Review these concepts from Grade 5. They will help you to study Chapters 6, 10, 12, 13, and 15, which cover the Grade 6 TEKS about how change occurs in cycles and how past events affect present and future events.*

For 5.6 A, recall that:

Many of Earth's changes occur in cycles, or a set of events that take place regularly. One change is that the Sun rises every morning and sets every night. This cycle takes one day, the time it takes Earth to spin once on its axis.

Other changes are due to the pull of gravity. Earth's orbit around the Sun takes one year. The tilt of Earth's axis as it orbits causes the cycle of seasons.

Some places have a cycle of four seasons. Winter days are short and cold. Spring days are cool to warm with much rain, allowing plants, such as the tulips in **Figure 1,** to grow. Summer days are long and hot. Autumn days are warm to cool with some rain. During this time, plants die off in many parts of the country.

The lunar cycle follows the Moon's orbit around Earth. It, in part, causes the ocean tides to rise and fall.

Figure 1
Tulip bulbs are planted in autumn, but they develop and bloom only after they have been exposed to cold temperatures.

TEKS 5.6 A Assessment

1. What is a cycle?
2. Which of Earth's cycles takes one day to complete? One year?
3. **Thinking Critically** Why do you think some places on Earth do not experience four distinct seasons?

TEKS REVIEW 499

TEKS 5.6 Review

For 5.6 B, recall that:

Earth's land, water, and air also follow cycles. They can combine, change form, break apart, and start the cycle all over again.

The Sun is involved in the water cycle, as shown in **Figure 2,** when its heat beats down on ocean water and other wet areas. The water evaporates, or changes to gas, and enters the atmosphere. There, water condenses, or changes back to liquid. When it returns to land as precipitation—rain or snow—the cycle continues.

The carbon cycle moves carbon dioxide through Earth's atmosphere and all living things. Animals and humans breathe out carbon dioxide. Plants use this gas to produce food that is a source of energy. Then, some animals eat plants and continue the cycle.

Nitrogen is an element that makes up 78 percent of the gases in the atmosphere. It is important because it is found in proteins and both plants and animals depend on protein. However, while nitrogen is plentiful in the air you breathe, it is only useful when you eat foods that contain it. Nitrogen compounds pass from the air into the soil or may enter a plant in the form of fertilizer.

There it helps plants to grow—as when they are fertilized. Nitrogen may also enter small growths—or nodules—on the roots of certain plants, such as peas. Bacteria in the nodules change the nitrogen into a form that plant cells can use.

Bacteria in the soil help decay dead plants and animals. This decaying matter adds more nitrogen to the soil and returns some to the air.

Figure 2
Water moves from Earth's surface to the atmosphere and back again in the water cycle.

TEKS 5.6 B Assessment

1. What forms does water take in its cycle?
2. Why is the carbon cycle so important to living things?
3. **Thinking Critically** Plants and animals that live in rivers, seas, and oceans also need nitrogen. How might they get their supply in the nitrogen cycle?

For 5.6 C, recall that:

The life cycles of plants and animals range from short to long. Adult male mosquitoes, for example, live only seven to ten days. Some redwood trees live for more than a thousand years.

The life cycles of plants have two basic stages. One stage produces spores. The other stage produces male and female sex cells. In some plants, fruits containing seeds are produced during the plant's life cycle. In other types of plants, such as pine trees, cones are produced.

Animals reproduce in one of two ways, similar to the plant stages. In the life cycle of simple animals, such as the sea anemone in **Figure 3,** a young bud forms on the parent's side. After the bud grows, it breaks off and grows on its own. Most animals, however, reproduce when a male cell fertilizes a female egg cell.

Some animals change forms at each stage as they develop into adults. A butterfly's egg changes into a caterpillar, which then forms a hard-shelled pupa. After days or months, depending on the kind, an adult butterfly emerges.

TEKS 5.6 C Assessment

1. What are the two stages of a plant's life cycle?
2. How are animal and plant reproduction alike?
3. **Thinking Critically** Write a statement about the variety of ways plants and animals reproduce.

Figure 3
A sea anemone can reproduce in three different ways—by forming buds, releasing eggs, or dividing itself into two parts.

TEKS 5.7 Review

Physical Properties of Matter

 You'll Review

TEKS 5.7
Matter has physical properties.

5.7 A classify matter based on its physical properties including magnetism, physical state, and the ability to conduct or insulate heat, electricity, and sound;

5.7 B demonstrate that some mixtures maintain the physical properties of their ingredients;

5.7 C identify changes that can occur in the physical properties of the ingredients of solutions such as dissolving sugar in water;

5.7 D observe and measure characteristic properties of substances that remain constant, such as boiling points and melting points.

Figure 1
Salt is a solid. Its particles vibrate but don't move out of their positions completely.

Directions: *Review these concepts from Grade 5. They will help you to study Chapter 3 on the physical properties of matter, a topic required by the Grade 6 TEKS.*

For 5.7 A, recall that:

Matter is anything that has mass and takes up space. Scientists classify matter by its physical properties. One obvious property of matter is its physical state. Matter can be in the form of a solid, liquid, gas or plasma. The attraction between the particles of solids is strong and locks them tightly in place. In solids, the particles can vibrate, but they can't leave their positions. **Figure 1** shows a common solid. In liquids, the particles have more energy of motion and are able to flow. In gases, they can move freely and far apart from each other.

Another property, magnetism, is a force that certain objects can exert even when they are not in direct contact with one another. Some magnetic materials occur in nature. Others can be magnetized.

Yet another property of matter is the ability to conduct or insulate. Materials that offer little resistance to the flow of energy are called conductors, and materials that are poor conductors are called insulators. For heat, insulators include vegetable fibers, cork, and foamed plastics. By contrast, metals including aluminum, copper, and steel are excellent conductors. Electric current is composed of flowing electrons, so conductors offer little resistance to moving electrons, while insulators do not allow electrons to flow easily. Copper is an excellent conductor of electricity, and plastic, rubber, and glass are insulators.

Matter transmits sound. Sound needs an elastic medium through which to travel; it cannot exist in a vacuum. Porous materials, like cork and some foams, tend to absorb sound.

TEKS 5.7 A Assessment

1. Name three physical properties of matter.
2. Give an example of matter that does not conduct electricity but that does transmit sound.
3. **Thinking Critically** What might you use to construct a hot pad on which to set a casserole dish straight from the oven?

For 5.7 B, recall that:

Mixtures are combinations of two or more substances that can be separated by physical means. In mixtures, the components maintain their physical properties and do not change. In other words, they do not combine chemically.

Mixtures can be composed of solids, liquids, gases, or any combination of the three. Examples include mixing sugar and salt, iron filings and sand, and water and oil. In heterogeneous mixtures like the one in **Figure 2,** it is easy to tell the ingredients apart. These mixtures can be mixed evenly and still have parts that are distinguishable.

Figure 2
A mixture contains at least two different substances that keep their physical characteristics.

TEKS 5.7 B Assessment

1. What are mixtures?
2. Give an example of a heterogeneous mixture.
3. **Thinking Critically** How might you separate a mixture of iron filings and sand?

For 5.7 C, recall that:

The composition of a solution, also called a homogeneous mixture, is the same throughout. The substances that make it up cannot be distinguished, but they still can be separated by physical means. Examples of homogeneous mixtures include carbonated beverages, vinegar, salt water, and sugar water, or the solution shown in **Figure 3.**

Figure 3
Once mixed, the compounds in homogeneous mixtures cannot be distinguished from each other.

TEKS REVIEW 503

TEKS 5.7 Review

Examples In salt water and sugar water, a solid is mixed with a liquid. The solid dissolves in the water. You can taste the solid, but you can't see it—not even with a microscope.

Carbonated beverages are a mixture of a gas and a liquid. The gas is carbon dioxide. It forms the bubbles that you see in your drink.

In vinegar, the mixture is a liquid and a liquid. Vinegar is approximately 95 percent water, 4 percent acetic acid, and 1 percent other.

TEKS 5.7 C Assessment

1. What is a solution?
2. Give an example of a homogeneous mixture.
3. **Thinking Critically** How might you separate a solution of salt water?

For 5.7 D, recall that:

Adding heat to a solid causes its particles to move faster. As particles move faster and faster, the temperature of the substance increases until a point where the substance changes from a solid to a liquid. The temperature at which a substance changes from a solid to a liquid is its melting point. **Figure 4** shows the difference between particles in a solid and particles in a liquid. Once all the particles have turned to liquid, the temperature of the substance increases again until it reaches a point where the substance changes from a liquid to a gas. The temperature at which liquid changes to gas is the boiling point.

Boiling points and melting points are characteristics of matter that do not change under constant conditions. For example, the boiling point of water is 100°C and the melting point of water is 0°C. Other substances have different and distinct boiling and melting points.

Figure 4
Liquid particles can move around each other but can't break away completely. Solid particles have even less energy and merely vibrate in place.

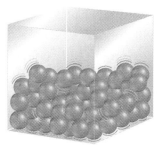

Liquid

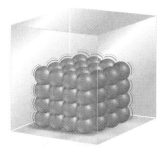

Solid

TEKS 5.7 D Assessment

1. What is the boiling point of water?
2. What is the difference between a boiling point and a melting point?
3. **Thinking Critically** Which will be higher: boiling points or melting points? Why?

TEKS 5.8 Review

Forms of Energy

 You'll Review

TEKS 5.8
Energy occurs in many forms.

5.8 A differentiate among forms of energy including light, heat, electrical, and solar energy;

5.8 B identify and demonstrate everyday examples of how light is reflected, such as from tinted windows, and refracted, such as in cameras, telescopes, and eyeglasses;

5.8 C demonstrate that electricity can flow in a circuit and can produce heat, light, sound, and magnetic effects;

5.8 D verify that vibrating an object can produce sound.

Directions: *Review these concepts from Grade 5. They will help you to study Chapter 5 on forms of energy, a topic required by the Grade 6 TEKS.*

For 5.8 A, recall that:

Energy is the ability to cause change. It cannot be created or destroyed, but can only be transformed. There are many different forms of energy, including light, heat, electrical, and solar.

Light energy is a special type of wave called an electromagnetic wave. It does not need a medium in which to travel. It even can travel in a vacuum. When light waves strike an object, some of that light is reflected and some is absorbed. The colors you see depend upon the light waves the object reflects.

Heat energy is generated by the continuous motion of molecules in matter. Heat is transferred from warmer objects to colder ones. One source of heat is the geothermal energy produced by Earth's molten interior. Heat is different from temperature; heat is the flow of energy from an object with higher temperature to an object with lower temperature.

Electrical charge can flow as a current. As a current, electrical charge usually flows through a closed path called a circuit. A simple circuit, shown in **Figure 1**, includes a source of electrical energy like a battery and a conductor that provides the path in which the current flows. A switch stops and starts the current. Electrical energy can be changed into other forms of energy such as heat and light.

Solar energy comes from the Sun. It is nonpolluting, renewable, and plentiful, but it is available only when the Sun is shining. Potentially, the Sun could provide enough energy to supply all of Earth's needs. However, humans do not know how to harness all of it.

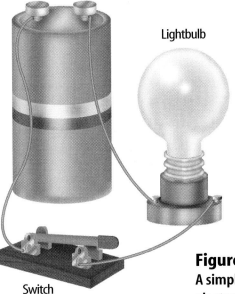

Figure 1
A simple circuit has a closed path for electrons to follow. The switch allows a person to open or close the circuit.

TEKS REVIEW 505

TEKS 5.8 Review

TEKS 5.8 A Assessment

1. What are three different forms of energy? How do they differ from one another?
2. Choose one form of energy and describe it.
3. **Thinking Critically** Which form of energy might be most effective for powering a car?

For 5.8 B, recall that:

When light energy strikes matter, some of it is absorbed and some of it is reflected. Some of it also could be refracted, or bent, as it travels through the new substance. What happens depends on the makeup of the matter.

Reflection is the process of light striking an object and bouncing off of it. When you look in a mirror, you are seeing the reflected light. If the light were not reflected, you would be unable to see your image. As shown in **Figure 2**, the smooth surface of a window can have a similar effect. It reflects light, so you can see a reflected image.

Refraction is the bending of a light wave when it changes speed as it moves from one material to another. A magnifying glass refracts light. It produces a larger image because it bends the light that is traveling through it. Cameras, telescopes, and eyeglasses also bend light.

Figure 2
The building at the left has a rough surface that scatters light, and the building at the right has a smooth surface that reflects light like a mirror.

TEKS 5.8 B Assessment

1. What is the difference between refraction and reflection?
2. Give an example of both refraction and reflection.
3. **Thinking Critically** Is a smooth surface necessary for a reflection to occur?

For 5.8 C, recall that:

Electricity can flow in circuits, closed loops through which electrons flow, like the ones shown in **Figure 3.** This is the way electricity is used most often. For example, a lightning bolt can deliver a huge amount of energy all at once, but appliances, like refrigerators and televisions, need a smaller, steady flow of electricity. Just as a continuous flow of water can do work, so can a continuous flow of electricity.

Figure 3
A series circuit has only one path for the electric current to follow. A parallel circuit has more than one.

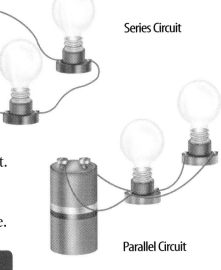

Series Circuit

Parallel Circuit

Toasters and electric ovens convert electricity to heat and light. Stereos use electricity to produce sound. An electric fan converts electricity to movement or mechanical energy. A lamp converts electricity to light and heat. Electricity and magnetism are related. Devices called electromagnets create strong magnetic fields when an electric current is passed through a conducting wire coiled around an iron core.

TEKS 5.8 C Assessment

1. What is a circuit?
2. Electricity can be transformed into what forms of energy?
3. **Thinking Critically** What happens to a lightbulb if its circuit is left open?

For 5.8 D, recall that:

When an object vibrates, it causes the air around it to vibrate. This produces energy in the form of sound waves. When those waves reach your ears, they stimulate nerves that send impulses to your brain. As a result, you hear sound. **Figure 4** illustrates this process. You can hear many sounds, but some waves are beyond the range of human hearing.

In addition to air, sound can travel through solids, liquids, or other gases. However, it cannot exist in a vacuum. Sound waves need a medium to travel through.

Not all sounds are the same in loudness and pitch. A sound's loudness is determined by the amount of energy the wave carries. A sound's pitch is determined by the frequency of the vibration. Frequency measures how many wavelengths pass a point in 1 s.

TEKS 5.8 D Assessment

1. How is sound created?
2. Explain how sound travels from a vibrating object to your ear.
3. **Thinking Critically** Two sound waves have the same frequency, but carry different amounts of energy. What is the same about the two sounds and what is different?

Figure 4
A vibrating object produces sound waves that travel to your ears.

TEKS 5.9 Review

Species Adaptations

What You'll Review

TEKS 5.9
Adaptations may increase the survival of the members of a species.

5.9 A compare the adaptive characteristics of species that improve their ability to survive and reproduce in an ecosystem;

5.9 B analyze and describe adaptive characteristics that result in an organism's unique niche in an ecosystem;

5.9 C predict some adaptive characteristics required for survival and reproduction by an organism in an ecosystem.

Directions: *Review these concepts from Grade 5. They will help you to study Chapters 14, 15, and 16 on adaptations, a topic required by the Grade 6 TEKS.*

For 5.9 A, recall that:

An adaptation is any structure or behavior that helps an organism survive in its environment. Adaptations include how a species gets its food, how it protects itself, or even how it looks. Sometimes these traits change over a long period of time to increase the species' chance of survival. Changes usually occur as the environment in which the species lives is altered.

Look closely at an animal or plant population and you will notice the adaptations. A wolf in a cold climate will have fur which insulates it from cold temperatures. The cactus in **Figure 1** has a thick, fleshy stem which stores water in the dry desert where it lives. Other plants, in temperate climates where water is more plentiful, do not have water-storage stems.

Figure 1
The cactus survives in the desert in part because of its waxy cuticle, which prevents moisture loss, and its thick stem, which stores water.

Some plants now found on land evolved from aquatic green algae ancestors. Over time, a lack of water caused some of the algae to dry up and die. Other algae adapted to less water and formed cellulose, an organic compound made up of long chains of glucose. The cellulose gave the plant structure and support. Cell walls helped reduce water loss. This is an example of an adaptation that occurred over time.

TEKS 5.9 A Assessment

1. What is an adaptation?
2. Name an adaptation of a species that helps it survive in its environment.
3. **Thinking Critically** Which would thrive in the Sahara Desert: a Siberian wolf or a gecko lizard? Why?

For 5.9 B, recall that:

Each organism has a unique niche, or role that it plays, in the ecosystem. Scientists often classify animals and plants by this role.

One part of an organism's niche is the way that it gets its food. Some animals eat plants. Scientists called them herbivores. The animals that eat other animals are called carnivores, or predators. The animals that eat both plants and animals are known as omnivores. Bacteria and fungi that feed on dead organisms are known as decomposers. Parasites, like ticks and fleas, are organisms that feed off of living organisms, harming the host in the process. Plants that make their own food are known as autotrophs, or producers.

Animals and plants have specific characteristics that allow them to play the role they do. These are adaptations that have evolved over time. For example, a hairy woodpecker has a long, thin beak that enables it to dig insects out of wood and a foot structure that allows it to climb trees. These adaptations allow the woodpecker to get food. Roadrunners have a foot structure which allows them to run and escape predators. **Figure 2** shows a variation that could be a disadvantage. These sorts of variations tend to decrease in an animal population over time.

Figure 2
Some variations tend to decrease in the population over time. What might happen to an albino squirrel?

TEKS 5.9 B Assessment

1. Name an adaptive characteristic of an animal and how it benefits that animal.
2. What happens to a characteristic that is a disadvantage in an animal population?
3. **Thinking Critically** Why would a duck have webbed feet, waterproof plumage, and toothlike notches on the inner edge of its jaw?

TEKS 5.9 Review

For 5.9 C, recall that:

Knowing the environment, you can predict adaptive characteristics that might be necessary for survival in an ecosystem. An effective way to do so is by looking at the climate. You can find different species in each of the six climate classifications—tropical, mild, dry, continental, polar, and high elevation. Species in each of these climates adapted to the temperature and environmental conditions. For example, freezing temperatures in the arctic require adaptations such as warm fur for insulation. The extremely dry climate of a desert requires adaptations for water storage, such as a cactus with a thick, fleshy stem.

Sometimes a species displays a behavioral adaptation that helps it survive. For example, chipmunks hibernate in winter when the weather is cold and food is scarce.

Reproduction is another behavior that varies depending on the species and the environment in which it lives. Female frogs lay hundreds of jellylike eggs in the water, which male frogs fertilize. They hatch into larvae, which eventually change into tadpoles and, finally, adult frogs. By contrast, most mammals give birth to live young after a period of development inside the mother. Some mammals are nearly helpless when born, but others—like deer—are able to stand and move shortly after birth. A deer native to Texas is shown in **Figure 3.**

TEKS 5.9 C Assessment

1. What characteristics might a plant living in a desert need?
2. Give an example of a behavioral adaptation.
3. **Thinking Critically** Why might a young deer need to be able to run soon after it is born?

Figure 3
The white-tailed deer is one of the animals that has adapted to Texas's climate and is found throughout the state.

TEKS 5.10 Review

Characteristics of Offspring

What You'll Review

TEKS 5.10
Likenesses between offspring and parents can be inherited or learned.

5.10 A identify traits that are inherited from parent to offspring in plants and animals;

5.10 B give examples of learned characteristics that result from the influence of the environment.

Directions: *Review these concepts from Grade 5. They will help you study Chapter 14 on characteristics of offspring, a topic required by the Grade 6 TEKS.*

For 5.10 A, recall that:

Parents and offspring often resemble each other. That is because the parent passes certain traits, or distinguishing characteristics, to the offspring. For example, a baby duck and its adult mother probably will have the same beak, webbed feet, and eye color. Human babies often look like one of their parents. As they grow, more and more of the resemblance can be seen.

Like animals, plants pass characteristics to their offspring. For plants, one good example is flower color. Pea plants studied by Gregor Mendel in the 1800s had either purple or white flowers, as in **Figure 1,** depending on the parents' traits. These traits are controlled in large part by genes.

The genes that control a trait are either dominant or recessive in form. A person inherits two copies of every gene. You may inherit two copies of a dominant form of a gene, two copies of the recessive form, or one of each form, depending on what forms your parents have. Dimples are a trait controlled by a dominant gene. A child born with dimples has at least one parent with the dominant gene. That offspring, when grown, can pass a gene for dimples to his or her child. In this way, the traits are inherited from generation to generation—in plants and in animals.

Figure 1
Gregor Mendel, considered the father of genetics, learned about passing down traits by studying garden peas such as these.

TEKS 5.10 A Assessment

1. What is an inherited trait?
2. Give examples of traits passed from animal parent to offspring, and from plant parent to offspring.
3. **Think Critically** How might inheritance of traits play a role in passing along disease?

TEKS 5.10 Review

For 5.10 B, recall that:

Some characteristics are passed down genetically, and others are passed from parent to offspring in a different way. They are not part of the organism's genetic makeup. They are behaviors learned from the parent and suited to the needs of the environment.

Setting up a territory for feeding, mating, and raising young is one such behavior. Songbirds sing to mark their territories, whereas squirrels chatter and sea lions bellow. Other animals leave scent marks or patrol their areas to keep intruders away.

Aggression, a forceful act of dominating another animal, is another form of behavior. Fighting and growling are aggressive behaviors. Animals use aggression to defend themselves, their territory, and their food supply.

If you've ever seen a male sage grouse fan his tail, fluff his feathers, and blow up his two air sacs, you've witnessed courtship behavior. Courtship behaviors allow males and females of the same species to recognize each other. They are done prior to mating, which helps the species reproduce, and they are modeled by adults to offspring.

Social behavior and communication are also examples of behaviors adults model for offspring. Social behavior includes caring for the young, protecting each other, and living and working in an organized way. The penguins in **Figure 2** are working together to stay warm. Communication includes sounds and actions. For example, rabbits also thump the ground, gorillas pound their chests, and prairie dogs bark.

Figure 2
Emperor penguins learn to huddle together for warmth and protection.

TEKS 5.10 B Assessment

1. What is a learned trait?
2. Give examples of traits passed down from parent to offspring this way.
3. **Thinking Critically** Male fireflies produce patterns in their flashes when they want to mate. How might this trait improve chances for survival of the species?

TEKS 5.11 Review

Impact of Past Events

 You'll Review

TEKS 5.11
Certain past events affect present and future events.

5.11 A identify and observe actions that require time for changes to be measurable, including growth, erosion, dissolving, weathering, and flow;

5.11 B draw conclusions about "what happened before" using data such as from tree-growth rings and sedimentary rock sequences;

5.11 C identify past events that led to the formation of the Earth's renewable, non-renewable, and inexhaustible resources.

Directions: *Review these concepts from Grade 5. They will help you to study Chapters 6, 7, 8, and 9 on the impact of past events, a topic required by the Grade 6 TEKS.*

For 5.11 A, recall that:

Some physical changes happen quickly, like the melting of ice. Other physical changes take place much more slowly. In fact, you might not even notice these changes until a lot of time has passed—maybe even months or years.

Growth is one such change. The human body takes years to grow from a child into an adult. Trees also take years to grow from saplings into their full size. Flowers grow more quickly from seed to plant, but even then growth happens too slowly to sit and observe it.

Erosion and weathering happen when wind and water interact with the earth. Over time, wind, water, and ice break rocks into smaller pieces. This is weathering. In erosion, particles are picked up and carried to new locations. Gravity, glaciers, wind, and water are all agents of erosion. After many years, the original rock formation looks markedly changed. **Figure 1** shows a canyon that has experienced weathering and erosion.

Figure 1
Water eroded this area in Yellowstone National Park's Upper Grand Canyon.

TEKS 5.11 Review

Sometimes minerals dissolve in water; dissolved minerals can then form new compounds and new minerals. The new minerals might be totally different from the minerals that were dissolved originally.

TEKS 5.11 A Assessment

1. Give an example of a physical change that happens over time.
2. What is the difference between erosion and weathering?
3. **Think Critically** Which would take less time: erosion of rock or soil? Why?

For 5.11 B, recall that:

Scientists use clues to draw conclusions about physical changes that occurred in the past. They use what they know about growth, weathering, erosion, dissolving, and flow.

Scientists know that 75 percent of the rocks at Earth's surface are sedimentary. These are rocks that form when sediments become cemented and pressed together. Sediments are loose materials—minerals, bits of shell, and pieces of rock—that have been moved by erosion.

Figure 2
Layers of sedimentary rock form a canyon wall in Arizona.

Sedimentary rock forms as layers like those shown in **Figure 2.** They are often different colors, with the oldest rocks on the bottom and the most recently formed rocks at the top. This is true unless Earth's forces disturb the order, such as moving rocks along faults.

Because scientists know the order in which the layers were deposited, they can determine the relative ages of fossils within the rocks and learn about the evolution of life on Earth.

In a similar way, scientists can determine the age of a tree. If the inside of the trunk is exposed by a horizontal cut, you can see the tree rings. A tree grows one ring of new wood each year just under the bark. That means the older wood is close to the center and the newer wood is near the outside. In wet years, the growth ring is wider. Scientists can study these rings to determine the tree's age and past environmental conditions.

TEKS 5.11 B Assessment

1. Where is the oldest rock in a sedimentary rock sequence? Where is the oldest wood in a tree?
2. Explain how a scientist might use a clue in the environment to learn what has happened in the past.
3. **Thinking Critically** How can sedimentary rock layers be used to learn about past life?

For 5.11 C, recall that:

Earth has many natural resources, which are materials in the environment that living things use to meet their needs. They fall into two basic categories—renewable and nonrenewable. A renewable resource is one like water, which is recycled or replaced by ongoing processes. A nonrenewable resource is one that is available in limited amounts, like petroleum or other fossil fuels—it either cannot be replaced, or it can be replaced only slowly. Sometimes a resource can be described as inexhaustible. An inexhaustible resource is one that is vast and unlimited, such as energy from the Sun.

Past events formed Earth's nonrenewable resources. Oil and gas take millions of years to form. Algae and tiny ocean animals called plankton must die, fall to the floor of the sea, and accumulate in sediment. Over time, thick layers of sand and mud bury this sediment. The combination of pressure and heat causes the organic matter to form oil and natural gas. **Figure 3** shows layers of rock covering oil and natural gas deposits.

Figure 3
Engineers drill deep into rock to reach deposits of oil and natural gas.

TEKS 5.11 Review

Coal Formation Coal also takes millions of years to form. Huge, fernlike plants died and fell into swamps where they were covered by mud, sand, and other dead plants. After layers of sediment piled up, pressure was greater. The combination of heat and pressure changed the materials into coal.

Compared to the processes that develop nonrenewable resources, the processes that result in renewable resources take place quickly. One example is hydroelectric power. Hydroelectric power is produced by transforming the energy of flowing water into electricity. This water is on the surface of Earth, and it is part of the water cycle. Water that evaporates from Earth's surface falls back again as rain or snow. This means the flowing waters that hydroelectric plants use are replenished continually.

TEKS 5.11 C Assessment

1. What is the difference between a renewable resource and a nonrenewable one?
2. Give an example of a renewable, nonrenewable, and inexhaustible resource.
3. **Thinking Critically** What happens if humans use up a nonrenewable resource, such as oil?

TEKS 5.12 Review

The Natural World of Earth and Sky

What You'll Review

TEKS 5.12
The natural world includes earth materials and objects in the sky.

5.12 A interpret how land forms are the result of a combination of constructive and destructive forces such as deposition of sediment and weathering;

5.12 B describe processes responsible for the formation of coal, oil, gas, and minerals;

5.12 C identify the physical characteristics of the Earth and compare them to the physical characteristics of the moon;

5.12 D identify gravity as the force that keeps planets in orbit around the Sun and the moon in orbit around the Earth.

Directions: *Review these concepts from Grade 5. They will help you to study Chapters 4, 7, 8, and 12 on the natural world of Earth and sky, a topic required by the Grade 6 TEKS.*

For 5.12 A, recall that:

According to many scientists, the land that we see today looks different from the landscape as it was millions of years ago. Over time, different forces have acted upon the land to create different landforms.

Our land is made up of many different materials. These materials include rocks, soil, and sediment. The forces that act upon these materials help create and destroy landforms, such as plains, mountains, and plateaus. **Figure 1** shows the three basic landforms.

Figure 1
Three basic types of landforms are plains, plateaus, and mountains.

TEKS 5.12 Review

Erosion Weathering is an example of a destructive force. In weathering, wind, rain, and ice wear down the rock in landforms and break it apart.

Have you ever seen a pothole in a road at the end of winter? A pothole is a place where water has frozen, expanded and broken some of the road apart. This is a type of weathering. The four main agents of erosion are wind, water, gravity, and glaciers. Erosion changes the face of Earth, and it can make a very big difference in the way things look. You may think that the Rocky Mountains in Colorado are very high, but at one time, the Appalachian mountains in the eastern United States were even higher. Over millions of years, the Appalachians have been worn down by erosion.

But, constructive forces exist in nature, too. Pieces of rock and sediment are moved to new places, where they can accumulate to build new landforms, such as deltas and fans. The dropping of sediments in new places is called deposition.

Sediment moved by erosion piles up and becomes compacted under pressure. If the sediment is small, it can form solid rock. If it is bigger, then it must be cemented together by a solution of water and minerals.

TEKS 5.12 A Assessment

1. Explain the difference between a constructive force and a destructive force.
2. How does sedimentary rock form?
3. Name four agents of erosion.
4. **Thinking Critically** What causes the uneven surfaces of mountaintops?

For 5.12 B, recall that:

Oil, natural gas, and coal are referred to as fossil fuels. That is because they began as organisms (both plants and animals) that died and accumulated. It takes millions of years for oil and natural gas to form. Billions of algae and other microscopic ocean organisms called plankton die and fall to the floor of the sea and accumulate in deep layers. Sand and mud bury the layers. Over millions of years, the pressure and heat from these layers cause oil and natural gas to form.

Figure 2
This area in Healy, Alaska, shows a deposit of coal.

Coal forms in a similar manner. Mud and sand covered huge fernlike plants that died and fell into swamps. A combination of heat and pressure over millions of years changed the material into coal. **Figure 2** shows a coal deposit. Coal is made up almost entirely of the element carbon. For years coal has been mined and used as fuel. It is probably the most commonly used fossil fuel across the world.

Minerals occur throughout Earth. Even Earth's mantle consists of minerals. Minerals occur naturally, but they never were alive, and they are not by-products of life processes. Instead, minerals form from processes like the cooling of molten rock or the reactions of substances dissolved in water. Almost all nonliving substances used by humans are made from minerals—including the lead in the pencil you may be holding. Gems are minerals, too. Gems are valued because they are rare and because of their beauty. Examples of gems are rubies, sapphires, emeralds, and diamonds.

TEKS 5.12 B Assessment

1. Describe how coal is formed.
2. How is the process similar to the formation of oil and natural gas?
3. **Thinking Critically** What can you tell coal miners about how their workplace may have appeared millions of years ago?

TEKS 5.12 Review

For 5.12 C, recall that:

Even a quick glance at **Figure 3** reveals Earth is a sphere. Earth is sphere shaped, but it is not a perfect sphere. It bulges around the equator and is flatter around the poles. The poles are the areas at the north and south ends of Earth's axis, the imaginary line around which Earth spins. Spinning on an axis is referred to as rotation. It takes about one day, or 24 h, for Earth to complete one rotation. Earth's rotation causes day and night.

Earth is made up of different layers of materials. The very center, or inner core, of Earth is solid, and it is made mostly of iron with small amounts of nickel. The outer core also is composed of iron and nickel, but it is in the liquid phase. The thickest layer, called the mantle, comes next. It is more than 2,800 km thick. The outer layer is called the crust. Silicon and oxygen are the most common elements in Earth's crust.

The Moon's exterior looks nothing like Earth. It is barren with craters and mountains. Unlike Earth, the Moon has no atmosphere to help protect it from extreme temperature changes. But like Earth, the Moon is spherical. It also is made up of layers of rock and minerals. Under its crust, the Moon has a solid mantle that could be as deep as 1,000 km. Below that might be a partly molten mantle, and then a solid, iron-rich core.

As in Earth's crust, the elements oxygen, silicon, and aluminum are the most abundant elements in the Moon's crust.

Figure 3
Earth's nickname is the blue planet. *What feature of Earth gives it this nickname?*

TEKS 5.12 C Assessment

1. What physical similarities do Earth and the Moon share?
2. What physical differences are there?
3. **Thinking Critically** Why might the lack of an atmosphere contribute to the craters on the Moon?

For 5.12 D, recall that:

Gravity is the attraction between all matter. It causes things to fall to the ground on Earth. Its pull accelerates things down hills and gives you weight. Gravity is a force that all objects in the universe exert on all the other objects. The size of this force depends on the masses of the objects and the distance between them.

Some scientists think that gravity helped form Earth's solar system about 4.6 billion years ago. The solar system started with a cloud of gas, ice, and dust that was rotating slowly in space. A nearby star might have exploded, causing shock waves that compressed the cloud. As it contracted or pulled in toward its center, the cloud's density became greater. The resulting increase in gravitational pull drew more gas and dust toward the center. This caused the cloud to spin even faster and the temperature to increase.

Eventually, nuclear fusion occurred to form the beginning of the Sun. The remaining gas, ice, and dust particles collided and stuck together to form the planets and other objects.

Gravity is the force that keeps the planets in Earth's solar system in orbit around the Sun. It also keeps the Moon in orbit around Earth and the astronaut in **Figure 4** on the surface of the Moon.

Figure 4
Astronauts had to learn how to move in the weaker gravity of the Moon.

TEKS 5.12 D Assessment

1. What is gravity?
2. What role does gravity play in the planets' orbits?
3. **Thinking Critically** What might be the effects if gravity on Earth disappeared?

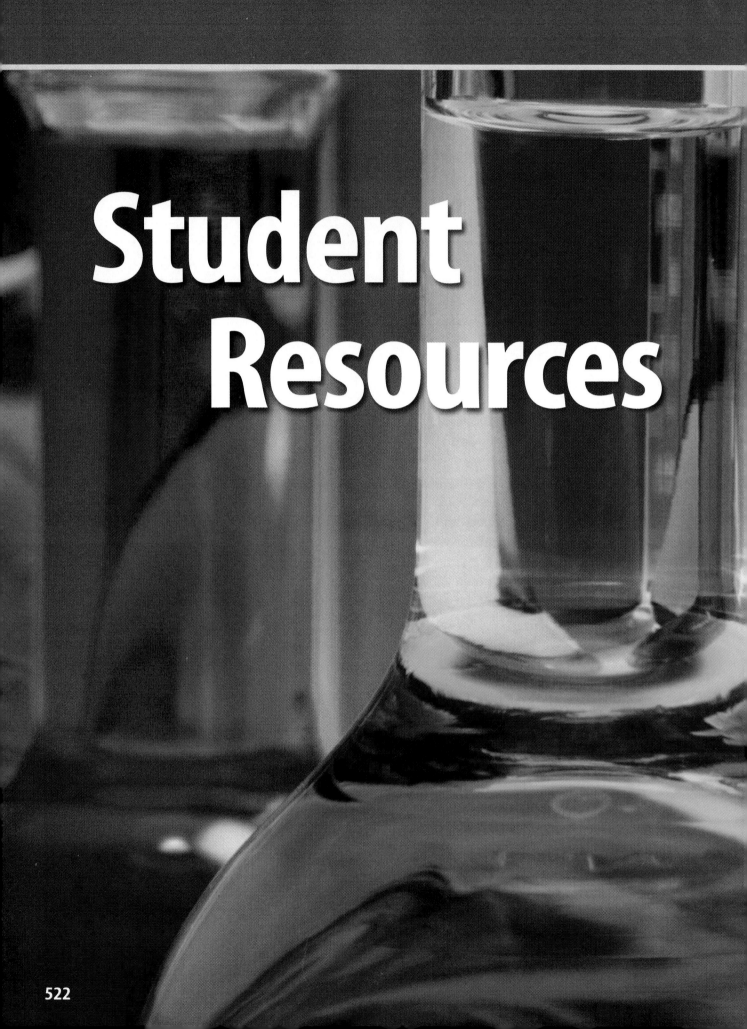

Student Resources

Student Resources

CONTENTS

Field Guides — 524

Amusement Park Rides Field Guide 524
Building Stones Field Guide 528
Living in Space Field Guide 532
Insects Field Guide 536

Skill Handbooks — 540

Science Skill Handbook 540
- **Organizing Information** 540
  - Researching Information 540
  - Evaluating Print and Nonprint Sources 540
  - Interpreting Scientific Illustrations .. 541
  - Venn Diagram 541
  - Concept Mapping 541
  - Writing a Paper 543
- **Investigating and Experimenting** 544
  - Identifying a Question 544
  - Forming Hypotheses 544
  - Predicting 544
  - Testing a Hypothesis 544
  - Identifying and Manipulating Variables and Controls 545
  - Collecting Data 546
  - Measuring in SI 547
  - Making and Using Tables 548
  - Recording Data 549
  - Recording Observations 549
  - Making Models 549
  - Making and Using Graphs 550
- **Analyzing and Applying Results** 551
  - Analyzing Results 551
  - Forming Operational Definitions ... 551
  - Classifying 551
  - Comparing and Contrasting 552
  - Recognizing Cause and Effect 552
  - Interpreting Data 552
  - Drawing Conclusions 552
  - Evaluating Others' Data and Conclusions 553
  - Communicating 553

Technology Skill Handbook 554
- Using a Word Processor 554
- Using a Database 554
- Using an Electronic Spreadsheet 555
- Using a Computerized Card Catalog .. 556
- Using Graphics Software 556
- Developing Multimedia Presentations 557

Math Skill Handbook 558
- Converting Units 558
- Using Fractions 559
- Calculating Ratios 560
- Using Decimals 560
- Using Percentages 561
- Using Precision and Significant Digits .. 561
- Solving One-Step Equations 562
- Using Proportions 563
- Using Statistics 564

Reference Handbook — 565

A. Safety in the Science Classroom 565
B. Periodic Table 566
C. SI/Metric to English, English to Metric Conversions 568
D. Care and Use of a Microscope 569
E. Diversity of Life 570
F. Weather Map Symbols 574
G. Topographic Map Symbols 575
H. Minerals 576
I. Rocks 578

English Glossary 579
Spanish Glossary 589
Index 601

Field GUIDE

Amusement Park Rides

If you like smooth, gentle rides, don't expect to get one at an amusement park. Amusement park rides are designed to provide thrills—plummeting down hills at 160 km/h, whizzing around curves so fast you think you'll fall out of your seat, zooming upside down, plunging over waterfalls, dropping so fast and far that you feel weightless. It's all part of the fun.

May the Force Be with You

What you might not realize as you're screaming with delight is that amusement park rides are lessons in physics. You can apply Newton's laws of motion to everything from the water slide and the bumper cars to the roller coasters. Amusement park ride designers know how to use the laws of motion to jolt, bump, and jostle you enough to make you scream, while still keeping you safe from harm. They don't just plan how the laws of motion cause these rides to move, they also plan how you will move when you are on the rides. These designers also use Newton's law of motion when they design and build the rides to make the structures safe and lasting. Look at the forces at work on some popular amusement park rides.

Free-fall ride

Free Fall

Slowly you rise up, up, up. Gravity is pulling you downward, but your seat exerts an upward force on you. Then, in an instant, you're plummeting toward the ground at speeds of more than 100 km/h. When you fall, your seat falls at the same rate and no longer exerts a force on you. Because you don't feel your seat pushing upward, you have the feeling of being weightless—at least for a few seconds.

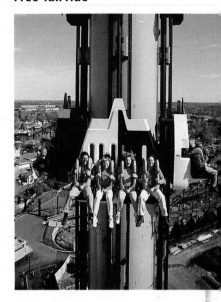

Field Activity

The next time you're at an amusement park, watch the rides. When you return home, make drawings of the rides using arrows to show how they move. Group the rides according to their movements. Compare your drawings and observations to the information in this field guide.

524 STUDENT RESOURCES

Field Guide

Roller Coaster: Design

The biggest coasters—some as tall as a 40-story building—are made of steel. Steel roller coasters are stronger and sway less than wooden roller coasters. This allows for more looping, more hills, and faster speeds.

Roller Coaster: The Coaster's Motion

Roller coasters are gravity-powered trains. Some coasters have motor-driven chains that move the cars to the top of the first hill. Then, gravity keeps it going.

The first hill is the highest point on the track. As the coaster rolls down the first hill, it converts potential energy to kinetic energy that sends it up the next hill. With each hill it climbs, it loses a little energy due to friction. That is why each hill is generally lower than the one before it.

Roller Coaster: Your Ride

Inertia is at work when you sweep around curves on a roller coaster. Inertia is the tendency for a body that's moving in a certain direction to keep moving in the same direction. For example, when the coaster swings right, inertia tries to keep you going in a straight line at a constant speed. As a result, you are pushed to the left side of your car.

Inertia tends to keep bodies moving in a straight line.

Bumper Cars: The Car's Motion

You control your bumper car's acceleration with the accelerator pedal. When the car you're in bumps head-on into another car, your car comes to an abrupt stop. The big rubber bumper around the bottom of the car diffuses the force of the collision by prolonging the impact.

Bumper Cars: Your Ride

When you first accelerate in a bumper car, you feel as though you are being pushed back in your seat. This sensation and the jolt you feel when you hit another car are due to inertia. On impact, your car stops, but your inertia makes you continue to move forward. It's the same jolt you feel in a car when someone slams on the brakes.

In a bumper-car collision, inertia keeps each rider moving forward.

Swing Ride: Design

Some of the more powerful swing rides make about eight revolutions around the central pole each minute. These swing rides are capable of moving their riders at speeds of close to 50 km/h.

Swing Ride: Forces

As the swings rotate, your inertia wants to fling you outward, but the chain that connects your seat to the ride's central pole prevents you from being flung into the air. You can see the changes in force as the swing ride changes speeds. As the ride speeds up and the forces exerted on the chain increase, your swing rises, moves outward, and travels almost parallel to the ground. As the ride slows, these forces on the chains decrease, returning the swings slowly to their original position.

The arrows show the forces at work in a swing ride.

Field Guide

Building Stones

Since ancient times, people have used naturally available materials such as stone and wood to construct their homes, places of worship, palaces, and other buildings and monuments. As early as 2500 B.C., the Egyptians had learned how to cut and transport large blocks of limestone from mountainsides for use in the construction of the pyramids. The Maya also used stone to construct their magnificent cities. Since those times, stone has remained a popular building material not only for its beauty but also for its durability and strength.

Types of Building Stone and Some Famous Stone Monuments

The choice of stone used in construction depends on the purpose, availability, cost, and properties of the particular stone. On the following pages are some of the most common building stones, one or more of which was used in the construction of many of the famous buildings in the world. By using this Field Guide, you can learn which stones were used to build these structures. You also can learn to identify the stones used to construct other monuments, buildings, and structures in your neighborhood or town.

Granite

Granite's mineral composition determines its color. For example, some granites are pink because they contain the mineral potassium feldspar. The minerals in granite resist wear and tear caused by wind and precipitation.

The pink color of granite in Enchanted Rock is caused by the presence of microcline, a variety of potassium feldspar.

Marischal College

Built in the late 1890s in Aberdeen, Scotland, Marischal College is one of the largest granite buildings in the world. Constructed in a gothic style, the building has an ornate façade and represents unusually intricate sculpture work for granite.

Field Activity

Take a walk around your town and find a public building or structure made of stone. Then use this field guide to identify the stones used to construct that building. Record your findings in your Science Journal. Visit the Glencoe Science Web site at **tx.science.glencoe.com** to find out which stones were used to build other famous buildings.

528 STUDENT RESOURCES

Limestone

Limestone commonly ranges in color from white to gray, but other colors are possible. Compared to other stones, limestone is not resistant when exposed to air and water in humid climates. However, it still is used as a building stone, especially where it is available locally.

Indiana limestone

Marble

In its purest form marble is white, but impurities produce other colors. The impurities often weave throughout marble, producing a beautiful mosaic appearance. Many marbles are composed mainly of calcite or dolomite. The softness of these minerals allows marble to be carved easily.

Italian marble

Lincoln Memorial

The Lincoln Memorial in Washington, D.C., was built to honor Abraham Lincoln. The exterior of the building, which is constructed in the Greek style, is made of marble quarried from Colorado and Tennessee. The walls and columns inside are made of Indiana limestone, and the ceiling, floor, and platform are made of marble from Tennessee and Alabama. Lincoln's statue was carved using white marble from Georgia.

Field Guide

Sandstone

The sand grains in sandstone give it a coarse, rustic appearance. Although sandstone often is white or tan, mineral cements can color the rock vivid shades of orange or red. When sand grains are well-cemented and composed mainly of resistant minerals such as quartz, sandstones can be excellent building materials.

Uluru (OO lew rew), the world's largest outcropping of sandstone, was previously called Ayers Rock.

Piece of sandstone

The Red Fort

Built by the Shah Jahan in the seventeenth century, the Red Fort in Delhi, India, is made of red sandstone. The octagonal, or eight-sided, fort contains palaces, gardens, army barracks, and other buildings, including a stable for the king's horses and elephants. The entire fort is surrounded by a large wall.

Field Guide

Slate

The color of slate varies, but it is usually gray, black, green, or red. The color is determined by the stones' organic and mineral content. Slate contains fine-grained, well-compacted clay minerals and mica which make it water-tight and easy to separate into layers. These properties make slate ideal to use for roofing and paving.

The Gloddfa Ganol Slate Mine in Wales is the world's largest slate mine.

Mud Island Park

The flat and compact nature of slate makes it a natural choice for this River Walk in Mud Island Park, Memphis, Tennessee. A street map of Memphis was etched into the surface of the slate.

Mud Island Park River Walk

FIELD GUIDE 531

Field Guide

Living in Space

Early astronauts were crammed into tiny space capsules where they could barely move in their seats. Food was a tasteless paste squeezed from a tube or a hard, bite-sized cube. Today, space shuttle astronauts have a two-level cabin with sleeping bunks, a galley for preparing food, and exercise equipment. Living in space isn't what it used to be.

Living in Orbit

Although conditions on a spacecraft are better now than in the past, the problems astronauts face are the same. They still go about their daily life, but space has no air, food, or water. This makes it hard to prepare meals and wash dishes afterward. It complicates how you drink beverages out of a glass. Due to these challenges, space shuttle crews must carry everything they need with them to survive in space.

By far the biggest challenge for astronauts is still the lack of gravity. Imagine eating a meal as part of it floats away, or sleeping in a bed that drifts into walls. NASA scientists have found ways to overcome such problems. This field guide offers a look at some of them.

Life-Support System

People need oxygen to breathe. The shuttle carries canisters of super-cold liquid oxygen and pressurized nitrogen to create an atmosphere in the crew compartment that is similar to Earth's— 79 percent nitrogen and 21 percent oxygen. The shuttle also circulates air through canisters of lithium hydroxide and activated charcoal, removing carbon dioxide and odors from it. Crew members must change one of the two canisters every 12 h.

Field Activity

Read a science-fiction description of people living and working in space. In your Science Journal, describe how people performed daily tasks such as eating, sleeping, and getting around. Go to the Glencoe Science Web site at **tx.science.glencoe.com** and click on the NASA link to find out more about living and working in space. Compare what you wrote with what you learn in this field guide.

Electricity

Fuel cells generate electricity by chemically combining hydrogen and oxygen. As a by-product, they produce 3 kg of water each hour—some of which is used to prepare food.

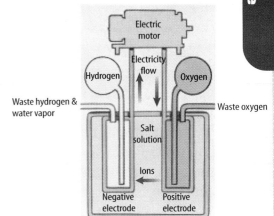

A Typical Menu

Astronauts eat three meals per day, chosen for them from a list of 70 foods and 20 beverages. They eat foods such as sausage, eggs, bread, fruits, vegetables, rice, and even turkey with gravy.

Food Preservation

Foods are not refrigerated. Some foods are freeze-dried, so water is added before they are eaten. Some foods are heated to kill bacteria and sealed in airtight foil packets. Irradiated food, such as bread and some meat, has been exposed to radiation to kill bacteria.

Food Preparation

Astronauts prepare and eat their food in the galley. A different person serves each meal, which takes about 20 min to prepare. The astronaut injects water into dried or powdered foods that need it, and puts hot dishes into the oven to warm them. Some foods can be eaten right out of the pouches.

These astronauts are enjoying a meal together.

Field Guide

Working Out

To help prevent bone and muscle deterioration due to space's weightless environment, astronauts exercise for 15 min each day on 7-day to 14-day missions. They work out for 30 min daily on 30-day missions. They can use a treadmill, a rowing machine, or an exercise bike. Even with this exercise, astronauts can lose more than one percent of their bone density for each month they are in space.

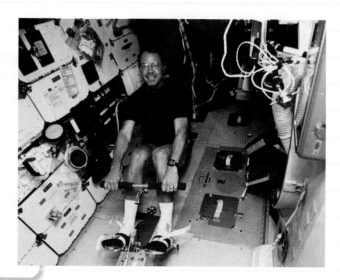

Using a rowing machine

Exercise Equipment

The base of the treadmill hooks into the floor or walls. An astronaut can stand on the treadmill with rubber bungee cords attached to a belt and shoulder harness. The cord is tightened to increase resistance.

Using a treadmill

Getting Some Sleep

Weightless astronauts can sleep in unusual places. Each astronaut's sleep station contains a bed made up of a padded board with a fireproof sleeping bag attached. Two astronauts sleep on bunks facing up. One sleeps on the underside of the lower bunk, facing the floor. The fourth sleeps vertically against the wall.

Sleeping compartments

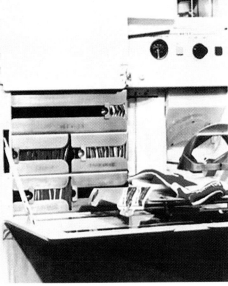

This is their hand-washing station.

Cleaning Up

After 8 h of sleep, astronauts have 45 min for morning hygiene. There aren't any showers or baths in space. To keep clean, astronauts just wipe themselves (and their hair) off with a wet cloth. They also can wash their hands at the hand-washing station. Water is air-blasted at their hands and then immediately sucked up.

This astronaut uses a wet cloth to keep clean.

Waste Management

Astronauts have a special toilet they use in space. It utilizes air instead of water to remove bodily wastes. The waste is then held in a tank until the spacecraft returns to Earth.

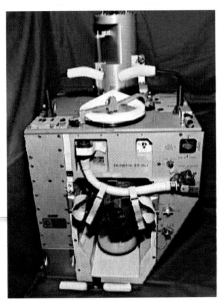

Here is a space shuttle toilet.

FIELD GUIDE 535

Field Guide

It's brown and creepy, and it has wings and six legs. If you call it a bug, you might be correct, but if you call it an insect, you are definitely correct. Insects belong to a large group of animals called arthropods. They are related to shrimp, spiders, lobsters, and centipedes. More insect species exist than all other animal species on Earth. Insects are found from the tropics to the tundra. Some live in water all or part of their lives, and some insects even live inside other animals. Insects play important roles in the environment. Many are helpful, but others are destructive.

How Insects Are Classified

An insect's body is divided into three parts—head, thorax, and abdomen. The head has a pair of antennae and eyes and paired mouthparts. Three pairs of jointed legs and, sometimes, wings are attached to the thorax. The abdomen has neither wings nor legs. Insects have a hard covering over their entire body. They shed this covering, then replace it as they grow. Insects are classified into smaller groups called orders. By observing an insect and recognizing certain features, you can identify the order it belongs to. This field guide presents ten insect orders.

Insects

Insect Orders

Convergent ladybug beetle

Coleoptera

Beetles

This is the largest order of insects. Many sizes, shapes, and colors of beetles can be found. All beetles have a pair of thick, leathery wings that meet in a straight line and cover another pair of wings, the thorax, and all or most of the abdomen. Some beetles are considered to be serious pests. Other beetles feed on insects or eat dead and decaying organisms. Not all beetles are called beetles. For example, fireflies, June bugs, and weevils are types of beetles.

Male stag beetle

Field Activity

For a week, use this field guide to help identify insect orders. Look for insects in different places and at different times. Visit the Glencoe Science Web site at **tx.science.glencoe.com** to view other insects that might not be found in your city. In your Science Journal, record the order of insect found, along with the date, time, and place.

Field Guide

Dermaptera

Earwigs

The feature that quickly identifies this brown, beetlelike insect is the pair of pincerlike structures that extend from the end of the abdomen. Earwigs usually are active at night and hide under litter or in any dark, protected place during the day. They can damage plants.

Earwig

Diptera

Flies and Mosquitoes

These are small insects with large eyes. They have two pair of wings but only one pair can be seen when the insect is at rest and the wings are folded. Their mouths are adapted for piercing and sucking, or scraping and lapping. Many of these insects are food for larger animals. Some spread diseases, others are pests, and some eat dead and decaying organisms. They are found in many different environments.

Common housefly

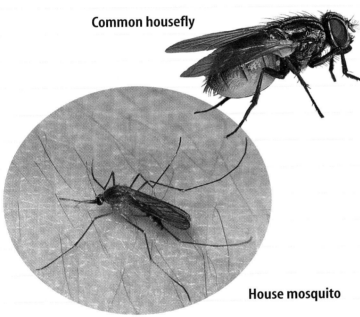

House mosquito

Odonata

Dragonflies and Damselflies

These insects have two pairs of transparent, multi-veined wings that are nearly equal in size and cannot be folded against the insect's body. The head has a pair of large eyes on it, and the abdomen is long and thin. These insects are usually seen near bodies of water. Most members of this group hunt during flight and catch small insects, such as mosquitoes.

Dragonfly

FIELD GUIDE 537

Field Guide

Isoptera

Termites

Adult termites are small, dark brown or black, and can have wings. Immature forms of this insect are small, soft bodied, pale yellow or white, and wingless. The adults are sometimes confused with ants. The thorax and abdomen of a termite look like one body part, but a thin waist separates the thorax and abdomen of an ant. Termites live in colonies in the ground or in wood.

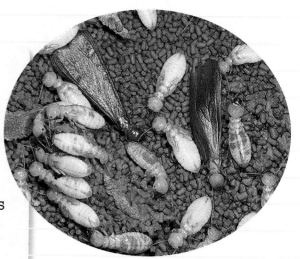
Pacific coast termites

Dictyoptera

Cockroaches and Mantises

These insects have long, thin antennae on the head. In species with wings, the back wings are thin and fan-like when they are opened, and are larger than the front wings. In the mantis, the front legs are adapted for grasping. The other two pairs of legs are similar to those of a cockroach. Praying mantises are beneficial because they eat other, often harmful, insects. Most cockroaches are pests.

American cockroach

Carolina praying mantis

Hymenoptera

Ants, Bees, and Wasps

Members of this order can be so small that they're visible only with a magnifier. Others may be nearly 35 mm long. These insects have two pairs of transparent wings, if present. They are found in many different environments, in colonies or alone. They are important because they pollinate flowers, and some prey on harmful insects. Honeybees make honey and wax.

Paper wasp

American bumblebee

Black carpenter ant

Field Guide

Lepidoptera

Butterflies and Moths

Butterflies and moths have two pairs of wings with colorful patterns created by thousands of tiny scales. The antennae of most moths are feathery. A butterfly's antennae are thin, and each has a small knob on the tip. Adult's mouthparts are adapted as a long, coiled tube for drinking nectar. Most moths are active at night, but some tropical moths are active during the day.

Yellow woolly bear moth

Buckeye butterfly

Periodic cicada

Water boatman

Hemiptera

Bugs

The prefix of this order, "*Hemi-*", means "half" and describes the front pair of wings. Near the insect's head, the front wings are thick and leathery, and are thin at the tip. Wing tips usually overlap when they are folded over the insect's back and cover a smaller pair of thin wings. Some bugs live on land and others are aquatic.

Orthoptera

Grasshoppers, Crickets, and Katydids

These insects have large hind legs adapted for leaping. They usually have two pairs of wings. The outer pair is hard and covers a transparent pair. Many of these insects make singing noises by rubbing one body part against another. Males generally make these sounds. Many of these insects are considered pests because swarms of them can destroy a farmer's crops in a few days.

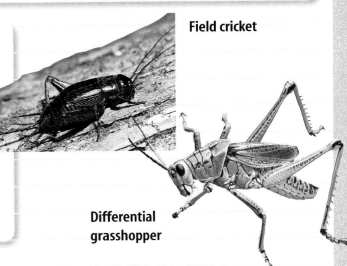

Field cricket

Differential grasshopper

FIELD GUIDE 539

Science Skill Handbook

Organizing Information

As you study science, you will make many observations and conduct investigations and experiments. You will also research information that is available from many sources. These activities will involve organizing and recording data. The quality of the data you collect and the way you organize it will determine how well others can understand and use it. In **Figure 1,** the student is obtaining and recording information using a thermometer.

Putting your observations in writing is an important way of communicating to others the information you have found and the results of your investigations and experiments.

Researching Information

Scientists work to build on and add to human knowledge of the world. Before moving in a new direction, it is important to gather the information that already is known about a subject. You will look for such information in various reference sources. Follow these steps to research information on a scientific subject:

Step 1 Determine exactly what you need to know about the subject. For instance, you might want to find out about one of the elements in the periodic table.

Step 2 Make a list of questions, such as: Who discovered the element? When was it discovered? What makes the element useful or interesting?

Step 3 Use multiple sources such as textbooks, encyclopedias, government documents, professional journals, science magazines, and the Internet.

Step 4 List where you found the sources. Make sure the sources you use are reliable and the most current available.

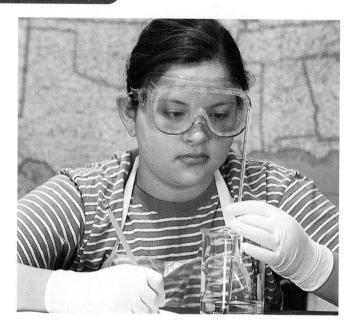

Figure 1
Making an observation is one way to gather information directly.

Evaluating Print and Nonprint Sources

Not all sources of information are reliable. Evaluate the sources you use for information, and use only those you know to be dependable. For example, suppose you want to find ways to make your home more energy efficient. You might find two Web sites on how to save energy in your home. One Web site contains "Energy-Saving Tips" written by a company that sells a new type of weatherproofing material you put around your door frames. The other is a Web page on "Conserving Energy in Your Home" written by the U.S. Department of Energy. You would choose the second Web site as the more reliable source of information.

In science, information can change rapidly. Always consult the most current sources. A 1985 source about saving energy would not reflect the most recent research and findings.

Science Skill Handbook

Interpreting Scientific Illustrations

As you research a science topic, you will see drawings, diagrams, and photographs. Illustrations help you understand what you read. Some illustrations are included to help you understand an idea that you can't see easily by yourself. For instance, you can't see the tiny particles in an atom, but you can look at a diagram of an atom as labeled in **Figure 2** that helps you understand something about it. Visualizing a drawing helps many people remember details more easily. Illustrations also provide examples that clarify difficult concepts or give additional information about the topic you are studying.

Most illustrations have a label or caption. A label or caption identifies the illustration or provides additional information to better explain it. Can you find the caption or labels in **Figure 2**?

Figure 2
This drawing shows an atom of carbon with its six protons, six neutrons, and six electrons.

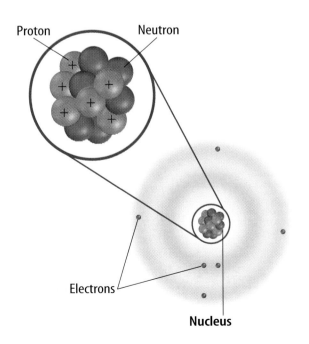

Venn Diagram

Although it is not a concept map, a Venn diagram illustrates how two subjects compare and contrast. In other words, you can see the characteristics that the subjects have in common and those that they do not.

The Venn diagram in **Figure 3** shows the relationship between two different substances made from the element carbon. However, due to the way their atoms are arranged, one substance is the gemstone diamond, and the other is the graphite found in pencils.

Concept Mapping

If you were taking a car trip, you might take some sort of road map. By using a map, you begin to learn where you are in relation to other places on the map.

A concept map is similar to a road map, but a concept map shows relationships among ideas (or concepts) rather than places. It is a diagram that visually shows how concepts are related. Because a concept map shows relationships among ideas, it can make the meanings of ideas and terms clear and help you understand what you are studying.

Overall, concept maps are useful for breaking large concepts down into smaller parts, making learning easier.

Figure 3
A Venn diagram shows how objects or concepts are alike and how they are different.

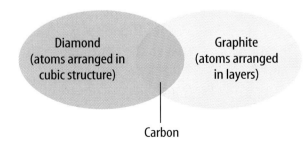

SCIENCE SKILL HANDBOOK **541**

Science Skill Handbook

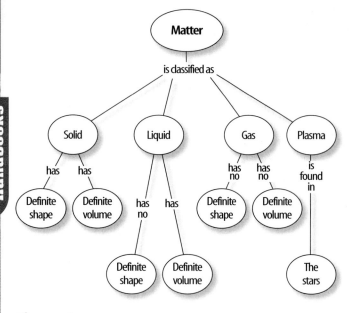

Figure 4
A network tree shows how concepts or objects are related.

Network Tree Look at the network tree in **Figure 4,** that describes the different types of matter. A network tree is a type of concept map. Notice how some words are in ovals while others are written across connecting lines. The words inside the ovals are science terms or concepts. The words written on the connecting lines describe the relationships between the concepts.

When constructing a network tree, write the topic on a note card or piece of paper. Write the major concepts related to that topic on separate note cards or pieces of paper. Then arrange them in order from general to specific. Branch the related concepts from the major concept and describe the relationships on the connecting lines. Continue branching to more specific concepts. If necessary, write the relationships between the concepts on the connecting lines until all concepts are mapped. Then examine the network tree for relationships that cross branches and add them to the network tree.

Events Chain An events chain is another type of concept map. It models the order, or sequence, of items. In science, an events chain can be used to describe a sequence of events, the steps in a procedure, or the stages of a process.

When making an events chain, first find the one event that starts the chain. This event is called the initiating event. Then, find the next event in the chain and continue until you reach an outcome. Suppose you are asked to describe why and how a sound might make an echo. You might draw an events chain such as the one in **Figure 5.** Notice that connecting words are not necessary in an events chain.

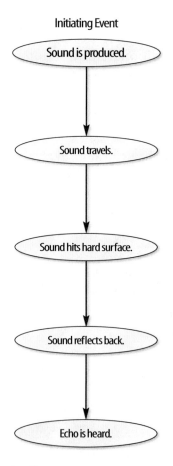

Figure 5
Events chains show the order of steps in a process or event.

542 STUDENT RESOURCES

Science Skill Handbook

Cycle Map A cycle concept map is a specific type of events chain map. In a cycle concept map, the series of events does not produce a final outcome. Instead, the last event in the chain relates back to the beginning event.

You first decide what event will be used as the beginning event. Once that is decided, you list events in order that occur after it. Words are written between events that describe what happens from one event to the next. The last event in a cycle concept map relates back to the beginning event. The number of events in a cycle concept varies, but is usually three or more. Look at the cycle map, as shown in **Figure 6.**

Figure 7
A spider map allows you to list ideas that relate to a central topic but not necessarily to one another.

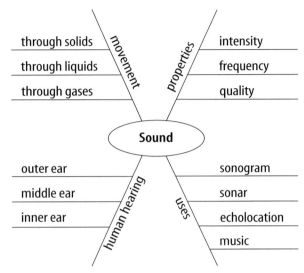

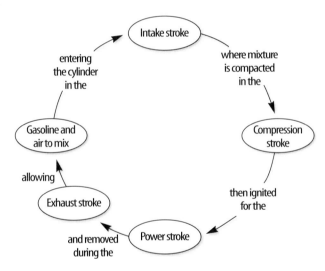

Figure 6
A cycle map shows events that occur in a cycle.

Spider Map A type of concept map that you can use for brainstorming is the spider map. When you have a central idea, you might find you have a jumble of ideas that relate to it but are not necessarily clearly related to each other. The spider map on sound in **Figure 7** shows that if you write these ideas outside the main concept, then you can begin to separate and group unrelated terms so they become more useful.

Writing a Paper

You will write papers often when researching science topics or reporting the results of investigations or experiments. Scientists frequently write papers to share their data and conclusions with other scientists and the public. When writing a paper, use these steps.

Step 1 Assemble your data by using graphs, tables, or a concept map. Create an outline.

Step 2 Start with an introduction that contains a clear statement of purpose and what you intend to discuss or prove.

Step 3 Organize the body into paragraphs. Each paragraph should start with a topic sentence, and the remaining sentences in that paragraph should support your point.

Step 4 Position data to help support your points.

Step 5 Summarize the main points and finish with a conclusion statement.

Step 6 Use tables, graphs, charts, and illustrations whenever possible.

SCIENCE SKILL HANDBOOK 543

Science Skill Handbook

Investigating and Experimenting

You might say the work of a scientist is to solve problems. When you decide to find out why your neighbor's hydrangeas produce blue flowers while yours are pink, you are problem solving, too. You might also observe that your neighbor's azaleas are healthier than yours are and decide to see whether differences in the soil explain the differences in these plants.

Scientists use orderly approaches to solve problems. The methods scientists use include identifying a question, making observations, forming a hypothesis, testing a hypothesis, analyzing results, and drawing conclusions.

Scientific investigations involve careful observation under controlled conditions. Such observation of an object or a process can suggest new and interesting questions about it. These questions sometimes lead to the formation of a hypothesis. Scientific investigations are designed to test a hypothesis.

Identifying a Question

The first step in a scientific investigation or experiment is to identify a question to be answered or a problem to be solved. You might be interested in knowing how beams of laser light like the ones in **Figure 8** look the way they do.

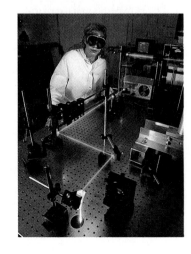

Figure 8
When you see lasers being used for scientific research, you might ask yourself, "Are these lasers different from those that are used for surgery?"

Forming Hypotheses

Hypotheses are based on observations that have been made. A hypothesis is a possible explanation based on previous knowledge and observations.

Perhaps a scientist has observed that certain substances dissolve faster in warm water than in cold. Based on these observations, the scientist can make a statement that he or she can test. The statement is a hypothesis. The hypothesis could be: *A substance dissolves in warm water faster.* A hypothesis has to be something you can test by using an investigation. A testable hypothesis is a valid hypothesis.

Predicting

When you apply a hypothesis to a specific situation, you predict something about that situation. First, you must identify which hypothesis fits the situation you are considering. People use predictions to make everyday decisions. Based on previous observations and experiences, you might form a prediction that if substances dissolve in warm water faster, then heating the water will shorten mixing time for powdered fruit drinks. Someone could use this prediction to save time in preparing a fruit punch for a party.

Testing a Hypothesis

To test a hypothesis, you need a procedure. A procedure is the plan you follow in your experiment. A procedure tells you what materials to use, as well as how and in what order to use them. When you follow a procedure, data are generated that support or do not support the original hypothesis statement.

544 STUDENT RESOURCES

Science Skill Handbook

For example, premium gasoline costs more than regular gasoline. Does premium gasoline increase the efficiency or fuel mileage of your family car? You decide to test the hypothesis: "If premium gasoline is more efficient, then it should increase the fuel mileage of my family's car." Then you write the procedure shown in **Figure 9** for your experiment and generate the data presented in the table below.

Figure 9
A procedure tells you what to do step by step.

> **Procedure**
> 1. Use regular gasoline for two weeks.
> 2. Record the number of kilometers between fill-ups and the amount of gasoline used.
> 3. Switch to premium gasoline for two weeks.
> 4. Record the number of kilometers between fill-ups and the amount of gasoline used.

Gasoline Data			
Type of Gasoline	Kilometers Traveled	Liters Used	Liters per Kilometer
Regular	762	45.34	0.059
Premium	661	42.30	0.064

These data show that premium gasoline is less efficient than regular gasoline in one particular car. It took more gasoline to travel 1 km (0.064) using premium gasoline than it did to travel 1 km using regular gasoline (0.059). This conclusion does not support the hypothesis.

Are all investigations alike? Keep in mind as you perform investigations in science that a hypothesis can be tested in many ways. Not every investigation makes use of all the ways that are described on these pages, and not all hypotheses are tested by investigations. Scientists encounter many variations in the methods that are used when they perform experiments. The skills in this handbook are here for you to use and practice.

Identifying and Manipulating Variables and Controls

In any experiment, it is important to keep everything the same except for the item you are testing. The one factor you change is called the independent variable. The factor that changes as a result of the independent variable is called the dependent variable. Always make sure you have only one independent variable. If you allow more than one, you will not know what causes the changes you observe in the dependent variable. Many experiments also have controls—individual instances or experimental subjects for which the independent variable is not changed. You can then compare the test results to the control results.

For example, in the fuel-mileage experiment, you made everything the same except the type of gasoline that was used. The driver, the type of automobile, and the type of driving were the same throughout. In this way, you could be sure that any mileage differences were caused by the type of fuel—the independent variable. The fuel mileage was the dependent variable.

If you could repeat the experiment using several automobiles of the same type on a standard driving track with the same driver, you could make one automobile a control by using regular gasoline over the four-week period.

SCIENCE SKILL HANDBOOK 545

Science Skill Handbook

Collecting Data

Whether you are carrying out an investigation or a short observational experiment, you will collect data, or information. Scientists collect data accurately as numbers and descriptions and organize it in specific ways.

Observing Scientists observe items and events, then record what they see. When they use only words to describe an observation, it is called qualitative data. For example, a scientist might describe the color, texture, or odor of a substance produced in a chemical reaction. Scientists' observations also can describe how much there is of something. These observations use numbers, as well as words, in the description and are called quantitative data. For example, if a sample of the element gold is described as being "shiny and very dense," the data are clearly qualitative. Quantitative data on this sample of gold might include "a mass of 30 g and a density of 19.3 g/cm^3." Quantitative data often are organized into tables. Then, from information in the table, a graph can be drawn. Graphs can reveal relationships that exist in experimental data.

When you make observations in science, you should examine the entire object or situation first, then look carefully for details. If you're looking at an element sample, for instance, check the general color and pattern of the sample before using a hand lens to examine its surface for any smaller details or characteristics. Remember to record accurately everything you see.

Scientists try to make careful and accurate observations. When possible, they use instruments such as microscopes, metric rulers, graduated cylinders, thermometers, and balances. Measurements provide numerical data that can be repeated and checked.

Sampling When working with large numbers of objects or a large population, scientists usually cannot observe or study every one of them. Instead, they use a sample or a portion of the total number. To *sample* is to take a small, representative portion of the objects or organisms of a population for research. By making careful observations or manipulating variables within a portion of a group, information is discovered and conclusions are drawn that might apply to the whole population.

Estimating Scientific work also involves estimating. To estimate is to make a judgment about the size or the number of something without measuring or counting every object or member of a population. Scientists first measure or count the amount or number in a small sample. A geologist, for example, might remove a 10-g sample from a large rock that is rich in copper ore, as in **Figure 10**. Then a chemist would determine the percentage of copper by mass and multiply that percentage by the total mass of the rock to estimate the total mass of copper in the large rock.

Figure 10
Determining the percentage of copper by mass that is present in a small piece of a large rock, which is rich in copper ore, can help estimate the total mass of copper ore that is present in the rock.

Science Skill Handbook

Measuring in SI

The metric system of measurement was developed in 1795. A modern form of the metric system, called the International System, or SI, was adopted in 1960. SI provides standard measurements that all scientists around the world can understand.

The metric system is convenient because unit sizes vary by multiples of 10. When changing from smaller units to larger units, divide by a multiple of 10. When changing from larger units to smaller, multiply by a multiple of 10. To convert millimeters to centimeters, divide the millimeters by 10. To convert 30 mm to centimeters, divide 30 by 10 (30 mm equal 3 cm).

Prefixes are used to name units. Look at the table below for some common metric prefixes and their meanings. Do you see how the prefix *kilo-* attached to the unit *gram* is *kilogram*, or 1,000 g?

Metric Prefixes			
Prefix	Symbol	Meaning	
kilo-	k	1,000	thousand
hecto-	h	100	hundred
deka-	da	10	ten
deci-	d	0.1	tenth
centi-	c	0.01	hundredth
milli-	m	0.001	thousandth

Now look at the metric ruler shown in **Figure 11.** The centimeter lines are the long, numbered lines, and the shorter lines are millimeter lines.

When using a metric ruler, line up the 0-cm mark with the end of the object being measured, and read the number of the unit where the object ends, in this instance it would be 4.5 cm.

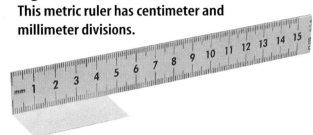

Figure 11
This metric ruler has centimeter and millimeter divisions.

Liquid Volume The unit that is used to measure liquids is the liter. A liter has the volume of 1,000 cm^3. The prefix *milli-* means "thousandth (0.001)." A milliliter is one thousandth of 1 L, and 1 L has the volume of 1,000 mL. One milliliter of liquid completely fills a cube measuring 1 cm on each side. Therefore, 1 mL equals 1 cm^3.

Beakers and graduated cylinders are used to measure liquid volume. The surface of liquids is always curved when viewed in a glass cylinder. This curved surface is the *meniscus*. A meniscus must be looked at along a horizontal line of sight as in Figure 12. A graduated cylinder is marked from bottom to top in milliliters. This one contains 79 mL of a liquid.

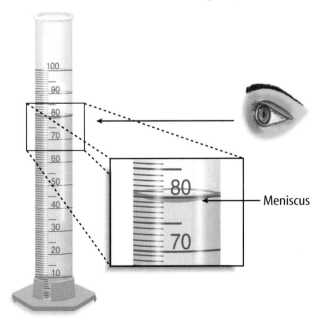

Figure 12
Graduated cylinders measure liquid volume.

SCIENCE SKILL HANDBOOK 547

Science Skill Handbook

Mass Scientists measure mass in grams. You might use a beam balance similar to the one shown in **Figure 13.** The balance has a pan on one side and a set of beams on the other side. Each beam has a rider that slides on the beam.

Before you find the mass of an object, slide all the riders back to the zero point. Check the pointer on the right to make sure it swings an equal distance above and below the zero point. If the swing is unequal, find and turn the adjusting screw until you have an equal swing.

Place an object on the pan. Slide the largest rider along its beam until the pointer drops below zero. Then move it back one notch. Repeat the process on each beam until the pointer swings an equal distance above and below the zero point. Sum the masses on each beam to find the mass of the object. Move all riders back to zero when finished.

Figure 13
A triple beam balance is used to determine the mass of an object.

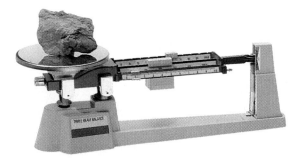

You should never place a hot object on the pan or pour chemicals directly onto the pan. Instead, find the mass of a clean container. Remove the container from the pan, then place the chemicals in the container. Find the mass of the container with the chemicals in it. To find the mass of the chemicals, subtract the mass of the empty container from the mass of the filled container.

Making and Using Tables

Browse through your textbook and you will see tables in the text and in the activities. In a table, data, or information, are arranged so that they are easier to understand. Activity tables help organize the data you collect during an activity so results can be interpreted.

Making Tables To make a table, list the items to be compared in the first column and the characteristics to be compared in the first row. The title should clearly indicate the content of the table, and the column or row heads should tell the reader what information is found in there. The table below lists materials collected for recycling on three weekly pick-up days. The inclusion of kilograms in parentheses also identifies for the reader that the figures are mass units.

Recyclable Materials Collected During Week			
Day of Week	Paper (kg)	Aluminum (kg)	Glass (kg)
Monday	5.0	4.0	12.0
Wednesday	4.0	1.0	10.0
Friday	2.5	2.0	10.0

Using Tables How much paper, in kilograms, is being recycled on Wednesday? Locate the column labeled "Paper (kg)" and the row "Wednesday." The information in the box where the column and row intersect is the answer. Did you answer "4.0"? How much aluminum, in kilograms, is being recycled on Friday? If you answered "2.0," you understand how to read the table. How much glass is collected for recycling each week? Locate the column labeled "Glass (kg)" and add the figures for all three rows. If you answered "32.0," then you know how to locate and use the data provided in the table.

Science Skill Handbook

Recording Data

To be useful, the data you collect must be recorded carefully. Accuracy is key. A well-thought-out experiment includes a way to record procedures, observations, and results accurately. Data tables are one way to organize and record results. Set up the tables you will need ahead of time so you can record the data right away.

Record information properly and neatly. Never put unidentified data on scraps of paper. Instead, data should be written in a notebook like the one in **Figure 14.** Write in pencil so information isn't lost if your data get wet. At each point in the experiment, record your information and label it. That way, your data will be accurate and you will not have to determine what the figures mean when you look at your notes later.

Figure 14
Record data neatly and clearly so they are easy to understand.

Recording Observations

It is important to record observations accurately and completely. That is why you always should record observations in your notes immediately as you make them. It is easy to miss details or make mistakes when recording results from memory. Do not include your personal thoughts when you record your data. Record only what you observe to eliminate bias. For example, when you record the time required for five students to climb the same set of stairs, you would note which student took the longest time. However, you would not refer to that student's time as "the worst time of all the students in the group."

Making Models

You can organize the observations and other data you collect and record in many ways. Making models is one way to help you better understand the parts of a structure you have been observing or the way a process for which you have been taking various measurements works.

Models often show things that are too large or too small for normal viewing. For example, you normally won't see the inside of an atom. However, you can understand the structure of the atom better by making a three-dimensional model of an atom. The relative sizes, the positions, and the movements of protons, neutrons, and electrons can be explained in words. An atomic model made of a plastic-ball nucleus and pipe-cleaner electron shells can help you visualize how the parts of the atom relate to each other.

Other models can be devised on a computer. Some models, such as those that illustrate the chemical combinations of different elements, are mathematical and are represented by equations.

Science Skill Handbook

Making and Using Graphs

After scientists organize data in tables, they might display the data in a graph that shows the relationship of one variable to another. A graph makes interpretation and analysis of data easier. Three types of graphs are the line graph, the bar graph, and the circle graph.

Line Graphs A line graph like in **Figure 15** is used to show the relationship between two variables. The variables being compared go on two axes of the graph. For data from an experiment, the independent variable always goes on the horizontal axis, called the *x*-axis. The dependent variable always goes on the vertical axis, called the *y*-axis. After drawing your axes, label each with a scale. Next, plot the data points.

A data point is the intersection of the recorded value of the dependent variable for each tested value of the independent variable. After all the points are plotted, connect them.

Figure 15
This line graph shows the relationship between distance and time during a bicycle ride lasting several hours.

Bar Graphs Bar graphs compare data that do not change continuously. Vertical bars show the relationships among data.

To make a bar graph, set up the *y*-axis as you did for the line graph. Draw vertical bars of equal size from the *x*-axis up to the point on the *y*-axis that represents the value of *x*.

Figure 16
The amount of aluminum collected for recycling during one week can be shown as a bar graph or circle graph.

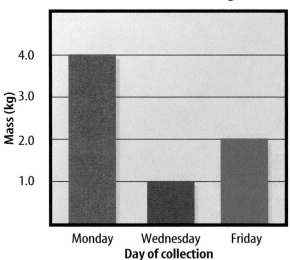

Circle Graphs A circle graph uses a circle divided into sections to display data as parts (fractions or percentages) of a whole. The size of each section corresponds to the fraction or percentage of the data that the section represents. So, the entire circle represents 100 percent, one-half represents 50 percent, one-fifth represents 20 percent, and so on.

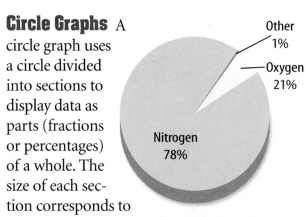

550 STUDENT RESOURCES

Science Skill Handbook

Analyzing and Applying Results

Analyzing Results

To determine the meaning of your observations and investigation results, you will need to look for patterns in the data. You can organize your information in several of the ways that are discussed in this handbook. Then you must think critically to determine what the data mean. Scientists use several approaches when they analyze the data they have collected and recorded. Each approach is useful for identifying specific patterns in the data.

Forming Operational Definitions

An operational definition defines an object by showing how it functions, works, or behaves. Such definitions are written in terms of how an object works or how it can be used; that is, they describe its job or purpose.

For example, a ruler can be defined as a tool that measures the length of an object (how it can be used). A ruler also can be defined as something that contains a series of marks that can be used as a standard when measuring (how it works).

Classifying

Classifying is the process of sorting objects or events into groups based on common features. When classifying, first observe the objects or events to be classified. Then select one feature that is shared by some members in the group but not by all. Place those members that share that feature into a subgroup. You can classify members into smaller and smaller subgroups based on characteristics.

How might you classify a group of chemicals? You might first classify them by state of matter, putting solids, liquids, and gases into separate groups. Within each group, you could then look for another common feature by which to further classify members of the group, such as color or how reactive they are.

Remember that when you classify, you are grouping objects or events for a purpose. For example, classifying chemicals can be the first step in organizing them for storage. Both at home and at school, poisonous or highly reactive chemicals should all be stored in a safe location where they are not easily accessible to small children or animals. Solids, liquids, and gases each have specific storage requirements that may include waterproof, airtight, or pressurized containers. Are the dangerous chemicals in your home stored in the right place? Keep your purpose in mind as you select the features to form groups and subgroups.

Figure 17
Color is one of many characteristics that are used to classify chemicals.

SCIENCE SKILL HANDBOOK **551**

Science Skill Handbook

Comparing and Contrasting

Observations can be analyzed by noting the similarities and differences between two or more objects or events that you observe. When you look at objects or events to see how they are similar, you are comparing them. Contrasting is looking for differences in objects or events. The table below compares and contrasts the characteristics of two elements.

Elemental Characteristics		
Element	Aluminum	Gold
Color	silver	gold
Classification	metal	metal
Density (g/cm^3)	2.7	19.3
Melting Point (°C)	660	1064

Recognizing Cause and Effect

Have you ever heard a loud pop right before the power went out and then suggested that an electric transformer probably blew out? If so, you have observed an effect and inferred a cause. The event is the effect, and the reason for the event is the cause.

When scientists are unsure of the cause of a certain event, they design controlled experiments to determine what caused it.

Interpreting Data

The word *interpret* means "to explain the meaning of something." Look at the problem originally being explored in an experiment and figure out what the data show. Identify the control group and the test group so you can see whether or not changes in the independent variable have had an effect. Look for differences in the dependent variable between the control and test groups.

These differences you observe can be qualitative or quantitative. You would be able to describe a qualitative difference using only words, whereas you would measure a quantitative difference and describe it using numbers. If there are differences, the independent variable that is being tested could have had an effect. If no differences are found between the control and test groups, the variable that is being tested apparently had no effect.

For example, suppose that three beakers each contain 100 mL of water. The beakers are placed on hot plates, and two of the hot plates are turned on, but the third is left off for a period of 5 min. Suppose you are then asked to describe any differences in the water in the three beakers. A qualitative difference might be the appearance of bubbles rising to the top in the water that is being heated but no rising bubbles in the unheated water. A quantitative difference might be a difference in the amount of water that is present in the beakers.

Inferring Scientists often make inferences based on their observations. An inference is an attempt to explain, or interpret, observations or to indicate what caused what you observed. An inference is a type of conclusion.

When making an inference, be certain to use accurate data and accurately described observations. Analyze all of the data that you've collected. Then, based on everything you know, explain or interpret what you've observed.

Drawing Conclusions

When scientists have analyzed the data they collected, they proceed to draw conclusions about what the data mean. These conclusions are sometimes stated using words similar to those found in the hypothesis formed earlier in the process.

Science Skill Handbook

Conclusions To analyze your data, you must review all of the observations and measurements that you made and recorded. Recheck all data for accuracy. After your data are rechecked and organized, you are almost ready to draw a conclusion such as "salt water boils at a higher temperature than freshwater."

Before you can draw a conclusion, however, you must determine whether the data allow you to come to a conclusion that supports a hypothesis. Sometimes that will be the case, other times it will not.

If your data do not support a hypothesis, it does not mean that the hypothesis is wrong. It means only that the results of the investigation did not support the hypothesis. Maybe the experiment needs to be redesigned, but very likely, some of the initial observations on which the hypothesis was based were incomplete or biased. Perhaps more observation or research is needed to refine the hypothesis.

Avoiding Bias Sometimes drawing a conclusion involves making judgments. When you make a judgment, you form an opinion about what your data mean. It is important to be honest and to avoid reaching a conclusion if no supporting evidence for it exists or if it was based on a small sample. It also is important not to allow any expectations of results to bias your judgments. If possible, it is a good idea to collect additional data. Scientists do this all the time.

For example, the *Hubble Space Telescope* was sent into space in April, 1990, to provide scientists with clearer views of the universe. *Hubble* is the size of a school bus and has a 2.4-m-diameter mirror. *Hubble* helped scientists answer questions about the planet Pluto.

For many years, scientists had only been able to hypothesize about the surface of the planet Pluto. *Hubble* has now provided pictures of Pluto's surface that show a rough texture with light and dark regions on it. This might be the best information about Pluto scientists will have until they are able to send a space probe to it.

Evaluating Others' Data and Conclusions

Sometimes scientists have to use data that they did not collect themselves, or they have to rely on observations and conclusions drawn by other researchers. In cases such as these, the data must be evaluated carefully.

How were the data obtained? How was the investigation done? Was it carried out properly? Has it been duplicated by other researchers? Were they able to follow the exact procedure? Did they come up with the same results? Look at the conclusion, as well. Would you reach the same conclusion from these results? Only when you have confidence in the data of others can you believe it is true and feel comfortable using it.

Communicating

The communication of ideas is an important part of the work of scientists. A discovery that is not reported will not advance the scientific community's understanding or knowledge. Communication among scientists also is important as a way of improving their investigations.

Scientists communicate in many ways, from writing articles in journals and magazines that explain their investigations and experiments, to announcing important discoveries on television and radio, to sharing ideas with colleagues on the Internet or presenting them as lectures.

Technology Skill Handbook

People who study science rely on computers to record and store data and to analyze results from investigations. Whether you work in a laboratory or just need to write a lab report with tables, good computer skills are a necessity.

Using a Word Processor

Suppose your teacher has assigned a written report. After you've completed your research and decided how you want to write the information, you need to put all that information on paper. The easiest way to do this is with a word processing application on a computer.

A computer application that allows you to type your information, change it as many times as you need to, and then print it out so that it looks neat and clean is called a word processing application. You also can use this type of application to create tables and columns, add bullets or cartoon art to your page, include page numbers, and check your spelling.

Helpful Hints

- If you aren't sure how to do something using your word processing program, look in the help menu. You will find a list of topics there to click on for help. After you locate the help topic you need, just follow the step-by-step instructions you see on your screen.
- Just because you've spellchecked your report doesn't mean that the spelling is perfect. The spell-check feature can't catch misspelled words that look like other words. If you've accidentally typed *cold* instead of *gold*, the spell checker won't know the difference. Always reread your report to make sure you didn't miss any mistakes.

Figure 18
You can use computer programs to make graphs and tables.

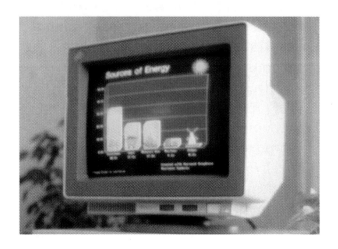

Using a Database

Imagine you're in the middle of a research project busily gathering facts and information. You soon realize that it's becoming more difficult to organize and keep track of all the information. The tool to use to solve information overload is a database. Just as a file cabinet organizes paper records, a database organizes computer records. However, a database is more powerful than a simple file cabinet because at the click of a mouse, the contents can be reshuffled and reorganized. At computer-quick speeds, databases can sort information by any characteristics and filter data into multiple categories.

Helpful Hints

- Before setting up a database, take some time to learn the features of your database software by practicing with established database software.
- Periodically save your database as you enter data. That way, if something happens such as your computer malfunctions or the power goes off, you won't lose all of your work.

Technology Skill Handbook

Doing a Database Search

When searching for information in a database, use the following search strategies to get the best results. These are the same search methods used for searching internet databases.

- Place the word *and* between two words in your search if you want the database to look for any entries that have both the words. For example, "gold *and* silver" would give you information that mentions both gold and silver.
- Place the word *or* between two words if you want the database to show entries that have at least one of the words. For example "gold *or* silver" would show you information that mentions either gold or silver.
- Place the word *not* between two words if you want the database to look for entries that have the first word but do not have the second word. For example, "gold *not* jewelry" would show you information that mentions gold but does not mention jewelry.

In summary, databases can be used to store large amounts of information about a particular subject. Databases allow biologists, Earth scientists, and physical scientists to search for information quickly and accurately.

Using an Electronic Spreadsheet

Your science fair experiment has produced lots of numbers. How do you keep track of all the data, and how can you easily work out all the calculations needed? You can use a computer program called a spreadsheet to record data that involve numbers. A spreadsheet is an electronic mathematical worksheet.

Type your data in rows and columns, just as they would look in a data table on a sheet of paper. A spreadsheet uses simple math to do data calculations. For example, you could add, subtract, divide, or multiply any of the values in the spreadsheet by another number. You also could set up a series of math steps you want to apply to the data. If you want to add 12 to all the numbers and then multiply all the numbers by 10, the computer does all the calculations for you in the spreadsheet. Below is an example of a spreadsheet that records test-car data.

Helpful Hints

- Before you set up the spreadsheet, identify how you want to organize the data. Include any formulas you will need to use.
- Make sure you have entered the correct data into the correct rows and columns.
- You also can display your results in a graph. Pick the style of graph that best represents the data with which you are working.

Figure 19
A spreadsheet allows you to display large amounts of data and do calculations automatically.

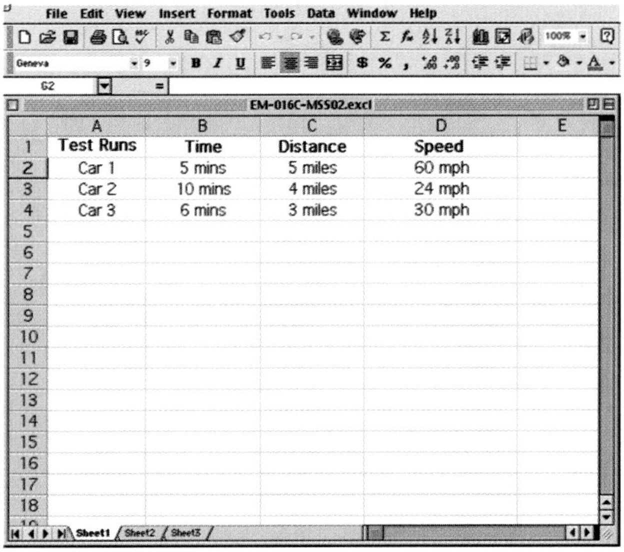

	A	B	C	D
1	Test Runs	Time	Distance	Speed
2	Car 1	5 mins	5 miles	60 mph
3	Car 2	10 mins	4 miles	24 mph
4	Car 3	6 mins	3 miles	30 mph

Technology Skill Handbook

Using a Computerized Card Catalog

When you have a report or paper to research, you probably go to the library. To find the information you need in the library, you might have to use a computerized card catalog. This type of card catalog allows you to search for information by subject, by title, or by author. The computer then will display all the holdings the library has on the subject, title, or author requested.

A library's holdings can include books, magazines, databases, videos, and audio materials. When you have chosen something from this list, the computer will show whether an item is available and where in the library to find it.

Helpful Hints

- Remember that you can use the computer to search by subject, author, or title. If you know a book's author but not the title, you can search for all the books the library has by that author.
- When searching by subject, it's often most helpful to narrow your search by using specific search terms, such as *and*, *or*, and *not*. If you don't find enough sources, you can broaden your search.
- Pay attention to the type of materials found in your search. If you need a book, you can eliminate any videos or other resources that come up in your search.
- Knowing how your library is arranged can save you a lot of time. The librarian will show you where certain types of materials are kept and how to find specific holdings.

Using Graphics Software

Are you having trouble finding that exact piece of art you're looking for? Do you have a picture in your mind of what you want but can't seem to find the right graphic to represent your ideas? To solve these problems, you can use graphics software. Graphics software allows you to create and change images and diagrams in almost unlimited ways. Typical uses for graphics software include arranging clipart, changing scanned images, and constructing pictures from scratch. Most graphics-software applications work in similar ways. They use the same basic tools and functions. Once you master one graphics application, you can use any other graphics application relatively easily.

Figure 20
Graphics software can use your data to draw bar graphs.

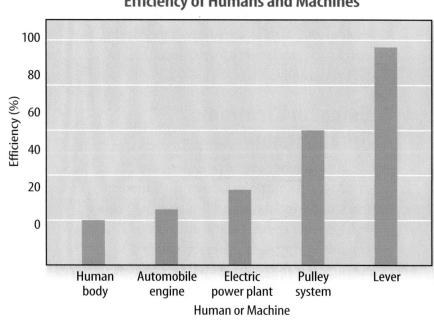

Technology Skill Handbook

Figure 21
Graphics software can use your data to draw circle graphs.

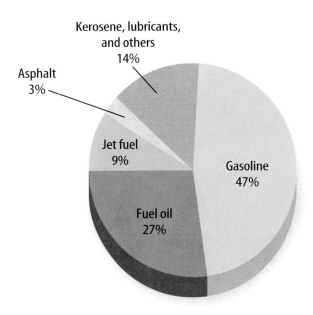

Helpful Hints

- As with any method of drawing, the more you practice using the graphics software, the better your results will be.
- Start by using the software to manipulate existing drawings. Once you master this, making your own illustrations will be easier.
- Clip art is available on CD-ROMs and the Internet. With these resources, finding a piece of clip art to suit your purposes is simple.
- As you work on a drawing, save it often.

Developing Multimedia Presentations

It's your turn—you have to present your science report to the entire class. How do you do it? You can use many different sources of information to get the class excited about your presentation. Posters, videos, photographs, sound, computers, and the Internet can help show your ideas.

First, determine what important points you want to make in your presentation. Then, write an outline of what materials and types of media would best illustrate those points. Maybe you could start with an outline on an overhead projector, then show a video, followed by something from the Internet or a slide show accompanied by music or recorded voices. You might choose to use a presentation builder computer application that can combine all these elements into one presentation. Make sure the presentation is well constructed to make the most impact on the audience.

Figure 22
Multimedia presentations use many types of print and electronic materials.

Helpful Hints

- Carefully consider what media will best communicate the point you are trying to make.
- Make sure you know how to use any equipment you will be using in your presentation.
- Practice the presentation several times.
- If possible, set up all of the equipment ahead of time. Make sure everything is working correctly.

Math Skill Handbook

Use this Math Skill Handbook to help solve problems you are given in this text. You might find it useful to review topics in this Math Skill Handbook first.

Converting Units

In science, quantities such as length, mass, and time sometimes are measured using different units. Suppose you want to know how many miles are in 12.7 km.

Conversion factors are used to change from one unit of measure to another. A conversion factor is a ratio that is equal to one. For example, there are 1,000 mL in 1 L, so 1,000 mL equals 1 L, or:

$$1{,}000 \text{ mL} = 1 \text{ L}$$

If both sides are divided by 1 L, this equation becomes:

$$\frac{1{,}000 \text{ mL}}{1 \text{ L}} = 1$$

The **ratio** on the left side of this equation is equal to one and is a conversion factor. You can make another conversion factor by dividing both sides of the top equation by 1,000 mL:

$$1 = \frac{1 \text{ L}}{1{,}000 \text{ mL}}$$

To **convert units**, you multiply by the appropriate conversion factor. For example, how many milliliters are in 1.255 L? To convert 1.255 L to milliliters, multiply 1.255 L by a conversion factor.

Use the **conversion factor** with new units (mL) in the numerator and the old units (L) in the denominator.

$$1.255 \text{ L} \times \frac{1{,}000 \text{ mL}}{1 \text{ L}} = 1{,}255 \text{ mL}$$

The unit L divides in this equation, just as if it were a number.

Example 1 There are 2.54 cm in 1 inch. If a meterstick has a length of 100 cm, how long is the meterstick in inches?

Step 1 Decide which conversion factor to use. You know the length of the meterstick in centimeters, so centimeters are the old units. You want to find the length in inches, so inch is the new unit.

Step 2 Form the conversion factor. Start with the relationship between the old and new units.

$$2.54 \text{ cm} = 1 \text{ inch}$$

Step 3 Form the conversion factor with the old unit (centimeter) on the bottom by dividing both sides by 2.54 cm.

$$1 = \frac{2.54 \text{ cm}}{2.54 \text{ cm}} = \frac{1 \text{ inch}}{2.54 \text{ cm}}$$

Step 4 Multiply the old measurement by the conversion factor.

$$100 \text{ cm} \times \frac{1 \text{ inch}}{2.54 \text{ cm}} = 39.37 \text{ inches}$$

The meterstick is 39.37 inches long.

Example 2 There are 365 days in one year. If a person is 14 years old, what is his or her age in days? (Ignore leap years)

Step 1 Decide which conversion factor to use. You want to convert years to days.

Step 2 Form the conversion factor. Start with the relation between the old and new units.

$$1 \text{ year} = 365 \text{ days}$$

Step 3 Form the conversion factor with the old unit (year) on the bottom by dividing both sides by 1 year.

$$1 = \frac{1 \text{ year}}{1 \text{ year}} = \frac{365 \text{ days}}{1 \text{ year}}$$

Step 4 Multiply the old measurement by the conversion factor:

$$14 \text{ years} \times \frac{365 \text{ days}}{1 \text{ year}} = 5{,}110 \text{ days}$$

The person's age is 5,110 days.

Practice Problem A book has a mass of 2.31 kg. If there are 1,000 g in 1 kg, what is the mass of the book in grams?

Math Skill Handbook

Using Fractions

A **fraction** is a number that compares a part to the whole. For example, in the fraction $\frac{2}{3}$, the 2 represents the part and the 3 represents the whole. In the fraction $\frac{2}{3}$, the top number, 2, is called the numerator. The bottom number, 3, is called the denominator.

Sometimes fractions are not written in their simplest form. To determine a fraction's **simplest form,** you must find the greatest common factor (GCF) of the numerator and denominator. The greatest common factor is the largest factor that is common to the numerator and denominator.

For example, because the number 3 divides into 12 and 30 evenly, it is a common factor of 12 and 30. However, because the number 6 is the largest number that evenly divides into 12 and 30, it is the **greatest common factor.**

After you find the greatest common factor, you can write a fraction in its simplest form. Divide both the numerator and the denominator by the greatest common factor. The number that results is the fraction in its **simplest form.**

Example Twelve of the 20 chemicals used in the science lab are in powder form. What fraction of the chemicals used in the lab are in powder form?

Step 1 Write the fraction.

$$\frac{part}{whole} = \frac{12}{20}$$

Step 2 To find the GCF of the numerator and denominator, list all of the factors of each number.

Factors of 12: 1, 2, 3, 4, 6, 12 (the numbers that divide evenly into 12)

Factors of 20: 1, 2, 4, 5, 10, 20 (the numbers that divide evenly into 20)

Step 3 List the common factors.

1, 2, 4.

Step 4 Choose the greatest factor in the list of common factors.

The GCF of 12 and 20 is 4.

Step 5 Divide the numerator and denominator by the GCF.

$$\frac{12 \div 4}{20 \div 4} = \frac{3}{5}$$

In the lab, $\frac{3}{5}$ of the chemicals are in powder form.

Practice Problem There are 90 rides at an amusement park. Of those rides, 66 have a height restriction. What fraction of the rides has a height restriction? Write the fraction in simplest form.

Math Skill Handbook

Calculating Ratios

A **ratio** is a comparison of two numbers by division.

Ratios can be written 3 to 5 or 3:5. Ratios also can be written as fractions, such as $\frac{3}{5}$. Ratios, like fractions, can be written in simplest form. Recall that a fraction is in **simplest form** when the greatest common factor (GCF) of the numerator and denominator is 1.

Example A chemical solution contains 40 g of salt and 64 g of baking soda. What is the ratio of salt to baking soda as a fraction in simplest form?

Step 1 Write the ratio as a fraction. $\frac{\text{salt}}{\text{baking soda}} = \frac{40}{64}$

Step 2 Express the fraction in simplest form.
The GCF of 40 and 64 is 8.
$$\frac{40}{64} = \frac{40 \div 8}{64 \div 8} = \frac{5}{8}$$

The ratio of salt to baking soda in the solution is $\frac{5}{8}$.

Practice Problem Two metal rods measure 100 cm and 144 cm in length. What is the ratio of their lengths in simplest fraction form?

Using Decimals

A **decimal** is a fraction with a denominator of 10, 100, 1,000, or another power of 10. For example, 0.854 is the same as the fraction $\frac{854}{1,000}$.

In a decimal, the decimal point separates the ones place and the tenths place. For example, 0.27 means twenty-seven hundredths, or $\frac{27}{100}$, where 27 is the **number of units** out of 100 units. Any fraction can be written as a decimal using division.

Example Write $\frac{5}{8}$ as a decimal.

Step 1 Write a division problem with the numerator, 5, as the dividend and the denominator, 8, as the divisor. Write 5 as 5.000.

Step 2 Solve the problem.

$$\begin{array}{r} 0.625 \\ 8\overline{)5.000} \\ \underline{48} \\ 20 \\ \underline{16} \\ 40 \\ \underline{40} \\ 0 \end{array}$$

Therefore, $\frac{5}{8} = 0.625$.

Practice Problem Write $\frac{19}{25}$ as a decimal.

Math Skill Handbook

Using Percentages

The word *percent* means "out of one hundred." A **percent** is a ratio that compares a number to 100. Suppose you read that 77 percent of Earth's surface is covered by water. That is the same as reading that the fraction of Earth's surface covered by water is $\frac{77}{100}$. To express a fraction as a percent, first find an equivalent decimal for the fraction. Then, multiply the decimal by 100 and add the percent symbol. For example, $\frac{1}{2} = 1 \div 2 = 0.5$. Then $0.5 \times 100 = 50 = 50\%$.

Example Express $\frac{13}{20}$ as a percent.

Step 1 Find the equivalent decimal for the fraction.

$$\begin{array}{r} 0.65 \\ 20\overline{)13.00} \\ \underline{120} \\ 100 \\ \underline{100} \\ 0 \end{array}$$

Step 2 Rewrite the fraction $\frac{13}{20}$ as 0.65.

Step 3 Multiply 0.65 by 100 and add the % sign.

$0.65 \cdot 100 = 65 = 65\%$

So, $\frac{13}{20} = 65\%$.

Practice Problem In one year, 73 of 365 days were rainy in one city. What percent of the days in that city were rainy?

Using Precision and Significant Digits

When you make a **measurement**, the value you record depends on the precision of the measuring instrument. When adding or subtracting numbers with different precision, the answer is rounded to the smallest number of decimal places of any number in the sum or difference. When multiplying or dividing, the answer is rounded to the smallest number of significant figures of any number being multiplied or divided. When counting the number of **significant figures**, all digits are counted except zeros at the end of a number with no decimal such as 2,500, and zeros at the beginning of a decimal such as 0.03020.

Example The lengths 5.28 and 5.2 are measured in meters. Find the sum of these lengths and report the sum using the least precise measurement.

Step 1 Find the sum.

$$\begin{array}{rl} 5.28 \text{ m} & \text{2 digits after the decimal} \\ + \; 5.2 \text{ m} & \text{1 digit after the decimal} \\ \hline 10.48 \text{ m} & \end{array}$$

Step 2 Round to one digit after the decimal because the least number of digits after the decimal of the numbers being added is 1.

The sum is 10.5 m.

Practice Problem Multiply the numbers in the example using the rule for multiplying and dividing. Report the answer with the correct number of significant figures.

MATH SKILL HANDBOOK

Math Skill Handbook

Solving One-Step Equations

An **equation** is a statement that two things are equal. For example, $A = B$ is an equation that states that A is equal to B.

Sometimes one side of the equation will contain a **variable** whose value is not known. In the equation $3x = 12$, the variable is x.

The equation is solved when the variable is replaced with a value that makes both sides of the equation equal to each other. For example, the solution of the equation $3x = 12$ is $x = 4$. If the x is replaced with 4, then the equation becomes $3 \cdot 4 = 12$, or $12 = 12$.

To solve an equation such as $8x = 40$, divide both sides of the equation by the number that multiplies the variable.

$$8x = 40$$
$$\frac{8x}{8} = \frac{40}{8}$$
$$x = 5$$

You can check your answer by replacing the variable with your solution and seeing if both sides of the equation are the same.

$$8x = 8 \cdot 5 = 40$$

The left and right sides of the equation are the same, so $x = 5$ is the solution.

Sometimes an equation is written in this way: $a = bc$. This also is called a **formula**. The letters can be replaced by numbers, but the numbers must still make both sides of the equation the same.

Example 1 Solve the equation $10x = 35$.

Step 1 Find the solution by dividing each side of the equation by 10.

$$10x = 35 \quad \frac{10x}{10} = \frac{35}{10} \quad x = 3.5$$

Step 2 Check the solution.

$$10x = 35 \quad 10 \times 3.5 = 35 \quad 35 = 35$$

Both sides of the equation are equal, so $x = 3.5$ is the solution to the equation.

Example 2 In the formula $a = bc$, find the value of c if $a = 20$ and $b = 2$.

Step 1 Rearrange the formula so the unknown value is by itself on one side of the equation by dividing both sides by b.

$$a = bc$$
$$\frac{a}{b} = \frac{bc}{b}$$
$$\frac{a}{b} = c$$

Step 2 Replace the variables a and b with the values that are given.

$$\frac{a}{b} = c$$
$$\frac{20}{2} = c$$
$$10 = c$$

Step 3 Check the solution.

$$a = bc$$
$$20 = 2 \times 10$$
$$20 = 20$$

Both sides of the equation are equal, so $c = 10$ is the solution when $a = 20$ and $b = 2$.

Practice Problem In the formula $h = gd$, find the value of d if $g = 12.3$ and $h = 17.4$.

Math Skill Handbook

Using Proportions

A **proportion** is an equation that shows that two ratios are equivalent. The ratios $\frac{2}{4}$ and $\frac{5}{10}$ are equivalent, so they can be written as $\frac{2}{4} = \frac{5}{10}$. This equation is an example of a proportion.

When two ratios form a proportion, the **cross products** are equal. To find the cross products in the proportion $\frac{2}{4} = \frac{5}{10}$, multiply the 2 and the 10, and the 4 and the 5. Therefore $2 \cdot 10 = 4 \cdot 5$, or $20 = 20$.

Because you know that both proportions are equal, you can use cross products to find a missing term in a proportion. This is known as **solving the proportion.** Solving a proportion is similar to solving an equation.

Example The heights of a tree and a pole are proportional to the lengths of their shadows. The tree casts a shadow of 24 m at the same time that a 6-m pole casts a shadow of 4 m. What is the height of the tree?

Step 1 Write a proportion.

$$\frac{\text{height of tree}}{\text{height of pole}} = \frac{\text{length of tree's shadow}}{\text{length of pole's shadow}}$$

Step 2 Substitute the known values into the proportion. Let h represent the unknown value, the height of the tree.

$$\frac{h}{6} = \frac{24}{4}$$

Step 3 Find the cross products.

$$h \cdot 4 = 6 \cdot 24$$

Step 4 Simplify the equation.

$$4h = 144$$

Step 5 Divide each side by 4.

$$\frac{4h}{4} = \frac{144}{4}$$
$$h = 36$$

The height of the tree is 36 m.

Practice Problem The ratios of the weights of two objects on the Moon and on Earth are in proportion. A rock weighing 3 N on the Moon weighs 18 N on Earth. How much would a rock that weighs 5 N on the Moon weigh on Earth?

Math Skill Handbook

Using Statistics

Statistics is the branch of mathematics that deals with collecting, analyzing, and presenting data. In statistics, there are three common ways to summarize the data with a single number—the mean, the median, and the mode.

The **mean** of a set of data is the arithmetic average. It is found by adding the numbers in the data set and dividing by the number of items in the set.

The **median** is the middle number in a set of data when the data are arranged in numerical order. If there were an even number of data points, the median would be the mean of the two middle numbers.

The **mode** of a set of data is the number or item that appears most often.

Another number that often is used to describe a set of data is the range. The **range** is the difference between the largest number and the smallest number in a set of data.

A **frequency table** shows how many times each piece of data occurs, usually in a survey. The frequency table below shows the results of a student survey on favorite color.

Color	Tally	Frequency
red	IIII	4
blue	HHI	5
black	II	2
green	III	3
purple	HHI II	7
yellow	HHI I	6

Based on the frequency table data, which color is the favorite?

Example The speeds (in m/s) for a race car during five different time trials are 39, 37, 44, 36, and 44.

To find the mean:
Step 1 Find the sum of the numbers.

$$39 + 37 + 44 + 36 + 44 = 200$$

Step 2 Divide the sum by the number of items, which is 5.

$$200 \div 5 = 40$$

The mean measure is 40 m/s.

To find the median:
Step 1 Arrange the measures from least to greatest.

36, 37, <u>39</u>, 44, 44

Step 2 Determine the middle measure.

The median measure is 39 m/s.

To find the mode:
Step 1 Group the numbers that are the same together.

44, 44, 36, 37, 39

Step 2 Determine the number that occurs most in the set.

<u>44, 44</u>, 36, 37, 39

The mode measure is 44 m/s.

To find the range:
Step 1 Arrange the measures from largest to smallest.

44, 44, 39, 37, 36

Step 2 Determine the largest and smallest measures in the set.

<u>44</u>, 44, 39, 37, <u>36</u>

Step 3 Find the difference between the largest and smallest measures.

$$44 - 36 = 8$$

The range is 8 m/s.

Practice Problem Find the mean, median, mode, and range for the data set 8, 4, 12, 8, 11, 14, 16.

Reference Handbook A

Safety in the Science Classroom

1. Always obtain your teacher's permission to begin an investigation.

2. Study the procedure. If you have questions, ask your teacher. Be sure you understand any safety symbols shown on the page.

3. Use the safety equipment provided for you. Goggles and a safety apron should be worn during most investigations.

4. Always slant test tubes away from yourself and others when heating them or adding substances to them.

5. Never eat or drink in the lab, and never use lab glassware as food or drink containers. Never inhale chemicals. Do not taste any substances or draw any material into a tube with your mouth.

6. Report any spill, accident, or injury, no matter how small, immediately to your teacher, then follow his or her instructions.

7. Know the location and proper use of the fire extinguisher, safety shower, fire blanket, first aid kit, and fire alarm.

8. Keep all materials away from open flames. Tie back long hair and tie down loose clothing.

9. If your clothing should catch fire, smother it with the fire blanket, or get under a safety shower. NEVER RUN.

10. If a fire should occur, turn off the gas then leave the room according to established procedures.

Follow these procedures as you clean up your work area

1. Turn off the water and gas. Disconnect electrical devices.

2. Clean all pieces of equipment and return all materials to their proper places.

3. Dispose of chemicals and other materials as directed by your teacher. Place broken glass and solid substances in the proper containers. Make sure never to discard materials in the sink.

4. Clean your work area. Wash your hands thoroughly after working in the laboratory.

First Aid	
Injury	Safe Response ALWAYS NOTIFY YOUR TEACHER IMMEDIATELY
Burns	Apply cold water.
Cuts and Bruises	Stop any bleeding by applying direct pressure. Cover cuts with a clean dressing. Apply ice packs or cold compresses to bruises.
Fainting	Leave the person lying down. Loosen any tight clothing and keep crowds away.
Foreign Matter in Eye	Flush with plenty of water. Use eyewash bottle or fountain.
Poisoning	Note the suspected poisoning agent.
Any Spills on Skin	Flush with large amounts of water or use safety shower.

REFERENCE HANDBOOK B

PERIODIC TABLE OF THE ELEMENTS

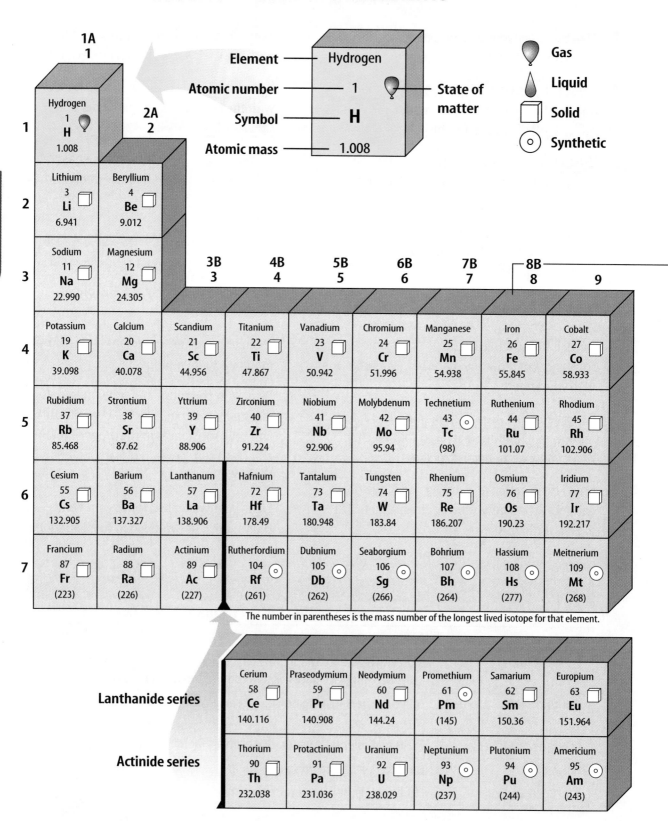

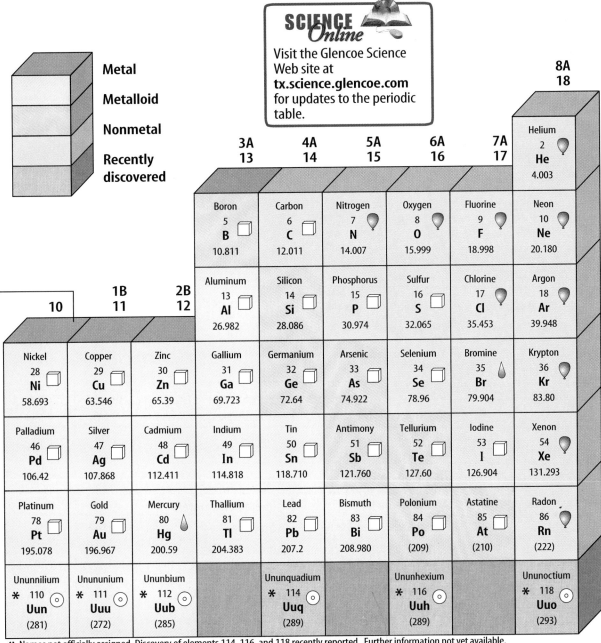

Reference Handbook C

SI—Metric/English, English/Metric Conversions

	When you want to convert:	To:	Multiply by:
Length	inches	centimeters	2.54
	centimeters	inches	0.39
	yards	meters	0.91
	meters	yards	1.09
	miles	kilometers	1.61
	kilometers	miles	0.62
Mass and Weight*	ounces	grams	28.35
	grams	ounces	0.04
	pounds	kilograms	0.45
	kilograms	pounds	2.2
	tons (short)	tonnes (metric tons)	0.91
	tonnes (metric tons)	tons (short)	1.10
	pounds	newtons	4.45
	newtons	pounds	0.22
Volume	cubic inches	cubic centimeters	16.39
	cubic centimeters	cubic inches	0.06
	liters	quarts	1.06
	quarts	liters	0.95
	gallons	liters	3.78
Area	square inches	square centimeters	6.45
	square centimeters	square inches	0.16
	square yards	square meters	0.83
	square meters	square yards	1.19
	square miles	square kilometers	2.59
	square kilometers	square miles	0.39
	hectares	acres	2.47
	acres	hectares	0.40
Temperature	To convert °Celsius to °Fahrenheit		°C × 9/5 + 32
	To convert °Fahrenheit to °Celsius		5/9 (°F − 32)

*Weight is measured in standard Earth gravity.

Reference Handbook D

Care and Use of a Microscope

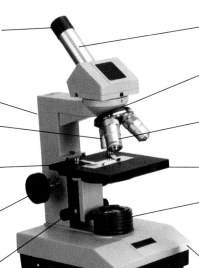

Eyepiece Contains magnifying lenses you look through.

Arm Supports the body tube.

Low-power objective Contains the lens with the lowest power magnification.

Stage clips Hold the microscope slide in place.

Fine adjustment Sharpens the image under high magnification.

Coarse adjustment Focuses the image under low power.

Body tube Connects the eyepiece to the revolving nosepiece.

Revolving nosepiece Holds and turns the objectives into viewing position.

High-power objective Contains the lens with the highest magnification.

Stage Supports the microscope slide.

Light source Provides light that passes upward through the diaphragm, the specimen, and the lenses.

Base Provides support for the microscope.

Caring for a Microscope

1. Always carry the microscope holding the arm with one hand and supporting the base with the other hand.

2. Don't touch the lenses with your fingers.

3. The coarse adjustment knob is used only when looking through the lowest-power objective lens. The fine adjustment knob is used when the high-power objective is in place.

4. Cover the microscope when you store it.

Using a Microscope

1. Place the microscope on a flat surface that is clear of objects. The arm should be toward you.

2. Look through the eyepiece. Adjust the diaphragm so light comes through the opening in the stage.

3. Place a slide on the stage so the specimen is in the field of view. Hold it firmly in place by using the stage clips.

4. Always focus with the coarse adjustment and the low-power objective lens first. After the object is in focus on low power, turn the nosepiece until the high-power objective is in place. Use ONLY the fine adjustment to focus with the high-power objective lens.

Making a Wet-Mount Slide

1. Carefully place the item you want to look at in the center of a clean, glass slide. Make sure the sample is thin enough for light to pass through.

2. Use a dropper to place one or two drops of water on the sample.

3. Hold a clean coverslip by the edges and place it at one edge of the water. Slowly lower the coverslip onto the water until it lies flat.

4. If you have too much water or a lot of air bubbles, touch the edge of a paper towel to the edge of the coverslip to draw off extra water and draw out unwanted air.

STUDENT RESOURCES 569

Reference Handbook E

Diversity of Life: Classification of Living Organisms

A six-kingdom system of classification of organisms is used today. Two kingdoms—Kingdom Archaebacteria and Kingdom Eubacteria—contain organisms that do not have a nucleus and that lack membrane-bound structures in the cytoplasm of their cells. The members of the other four kingdoms have a cell or cells that contain a nucleus and structures in the cytoplasm, some of which are surrounded by membranes. These kingdoms are Kingdom Protista, Kingdom Fungi, Kingdom Plantae, and Kingdom Animalia.

Kingdom Archaebacteria
one-celled; some absorb food from their surroundings; some are photosynthetic; some are chemosynthetic; many found in extremely harsh environments including salt ponds, hot springs, swamps, and deep-sea hydrothermal vents

Kingdom Eubacteria
one-celled; most absorb food from their surroundings; some are photosynthetic; some are chemosynthetic; many are parasites; many are round, spiral, or rod-shaped; some form colonies

Kingdom Protista
Phylum Euglenophyta one-celled; photosynthetic or take in food; most have one flagellum; euglenoids

Phylum Bacillariophyta one-celled; photosynthetic; have unique double shells made of silica; diatoms

Phylum Dinoflagellata one-celled; photosynthetic; contain red pigments; have two flagella; dinoflagellates

Phylum Chlorophyta one-celled, many-celled, or colonies; photosynthetic; contain chlorophyll; live on land, in freshwater, or salt water; green algae

Phylum Rhodophyta most are many-celled; photosynthetic; contain red pigments; most live in deep, saltwater environments; red algae

Phylum Phaeophyta most are many-celled; photosynthetic; contain brown pigments; most live in saltwater environments; brown algae

Phylum Rhizopoda one-celled; take in food; are free-living or parasitic; move by means of pseudopods; amoebas

Kingdom Eubacteria
Bacillus anthracis

Phylum Chlorophyta
Desmids

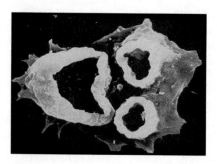

Amoeba

Phylum Zoomastigina one-celled; take in food; free-living or parasitic; have one or more flagella; zoomastigotes

Phylum Ciliophora one-celled; take in food; have large numbers of cilia; ciliates

Phylum Sporozoa one-celled; take in food; have no means of movement; are parasites in animals; sporozoans

Phyla Myxomycota and Acrasiomycota one- or many-celled; absorb food; change form during life cycle; cellular and plasmodial slime molds

Phylum Oomycota many-celled; are either parasites or decomposers; live in freshwater or salt water; water molds, rusts and downy mildews

Kingdom Fungi

Phylum Zygomycota many-celled; absorb food; spores are produced in sporangia; zygote fungi; bread mold

Phylum Ascomycota one- and many-celled; absorb food; spores produced in asci; sac fungi; yeast

Phylum Basidiomycota many-celled; absorb food; spores produced in basidia; club fungi; mushrooms

Phylum Deuteromycota members with unknown reproductive structures; imperfect fungi; *Penicillium*

Mycophycota organisms formed by symbiotic relationship between an ascomycote or a basidiomycote and green alga or cyanobacterium; lichens

Phylum Myxomycota
Slime mold

Phylum Oomycota
Phytophthora infestans

Lichens

Kingdom Plantae

Divisions Bryophyta (mosses), **Anthocerophyta** (hornworts), **Hepatophyta** (liverworts), **Psilophyta** (whisk ferns) many-celled nonvascular plants; reproduce by spores produced in capsules; green; grow in moist, land environments

Division Lycophyta many-celled vascular plants; spores are produced in conelike structures; live on land; are photosynthetic; club mosses

Division Sphenophyta vascular plants; ribbed and jointed stems; scalelike leaves; spores produced in conelike structures; horsetails

Division Pterophyta vascular plants; leaves called fronds; spores produced in clusters of sporangia called sori; live on land or in water; ferns

Division Ginkgophyta deciduous trees; only one living species; have fan-shaped leaves with branching veins and fleshy cones with seeds; ginkgoes

Division Cycadophyta palmlike plants; have large, featherlike leaves; produces seeds in cones; cycads

Division Coniferophyta deciduous or evergreen; trees or shrubs; have needlelike or scalelike leaves; seeds produced in cones; conifers

Division Gnetophyta shrubs or woody vines; seeds are produced in cones; division contains only three genera; gnetum

Division Anthophyta dominant group of plants; flowering plants; have fruits with seeds

Kingdom Animalia

Phylum Porifera aquatic organisms that lack true tissues and organs; are asymmetrical and sessile; sponges

Phylum Cnidaria radially symmetrical organisms; have a digestive cavity with one opening; most have tentacles armed with stinging cells; live in aquatic environments singly or in colonies; includes jellyfish, corals, hydra, and sea anemones

Phylum Platyhelminthes bilaterally symmetrical worms; have flattened bodies; digestive system has one opening; parasitic and free-living species; flatworms

Division Bryophyta
Liverwort

Division Anthophyta
Tomato plant

Phylum Platyhelminthes
Flatworm

Reference Handbook E

Phylum Chordata

Phylum Nematoda round, bilaterally symmetrical body; have digestive system with two openings; free-living forms and parasitic forms; roundworms

Phylum Mollusca soft-bodied animals, many with a hard shell and soft foot or footlike appendage; a mantle covers the soft body; aquatic and terrestrial species; includes clams, snails, squid, and octopuses

Phylum Annelida bilaterally symmetrical worms; have round, segmented bodies; terrestrial and aquatic species; includes earthworms, leeches, and marine polychaetes

Phylum Arthropoda largest animal group; have hard exoskeletons, segmented bodies, and pairs of jointed appendages; land and aquatic species; includes insects, crustaceans, and spiders

Phylum Echinodermata marine organisms; have spiny or leathery skin and a water-vascular system with tube feet; are radially symmetrical; includes sea stars, sand dollars, and sea urchins

Phylum Chordata organisms with internal skeletons and specialized body systems; most have paired appendages; all at some time have a notochord, nerve cord, gill slits, and a postanal tail; include fish, amphibians, reptiles, birds, and mammals

Reference Handbook F

Weather Map Symbols

Sample Station Model

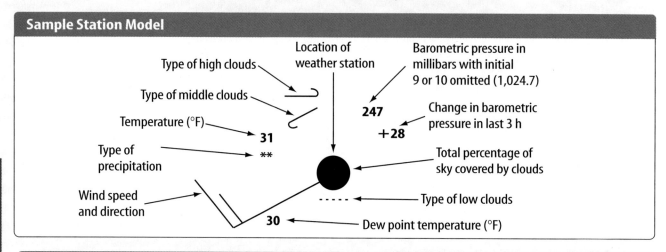

Sample Plotted Report at Each Station

Precipitation		Wind Speed and Direction		Sky Coverage		Some Types of High Clouds	
≡	Fog	○	0 calm	○	No cover		Scattered cirrus
★	Snow	/	1-2 knots	⊙	1/10 or less		Dense cirrus in patches
●	Rain	⌄	3-7 knots	◔	2/10 to 3/10		Veil of cirrus covering entire sky
⚡	Thunderstorm	⌄	8-12 knots	◔	4/10		Cirrus not covering entire sky
,	Drizzle	⌄	13-17 knots	◐	—		
▽	Showers	⌄	18-22 knots	◕	6/10		
		⌄	23-27 knots	◕	7/10		
		⌄	48-52 knots	◉	Overcast with openings		
		1 knot = 1.852 km/h		●	Completely overcast		

Some Types of Middle Clouds		Some Types of Low Clouds		Fronts and Pressure Systems	
∠	Thin altostratus layer	⌒	Cumulus of fair weather	(H) or High (L) or Low	Center of high- or low-pressure system
⫽	Thick altostratus layer	⌣	Stratocumulus	▲▲▲▲	Cold front
⌒	Thin altostratus in patches	-----	Fractocumulus of bad weather	⌒⌒⌒⌒	Warm front
⌒	Thin altostratus in bands	—	Stratus of fair weather	▲⌒▲⌒	Occluded front
				⌒▲⌒▲	Stationary front

574 STUDENT RESOURCES

Reference Handbook G

Topographic Map Symbols

━━━━━	Primary highway, hard surface	～～	Index contour
━▬━▬━	Secondary highway, hard surface	⋯⋯⋯	Supplementary contour
═══════	Light-duty road, hard or improved surface	～	Intermediate contour
=========	Unimproved road	⌒	Depression contours
┼┼┼┼┼	Railroad: single track		
╫╫╫╫╫	Railroad: multiple track	━ ━ ━	Boundaries: national
┿┿┿┿┿	Railroads in juxtaposition	━ ━ ━	State
		━ ━ ━	County, parish, municipal
▪▟▓	Buildings	━ ━ ━	Civil township, precinct, town, barrio
♁ ✝ cem	Schools, church, and cemetery	━ ━ ━	Incorporated city, village, town, hamlet
▫ ▨	Buildings (barn, warehouse, etc)	━ ･ ━	Reservation, national or state
∘ ∘	Wells other than water (labeled as to type)	━ ━ ━	Small park, cemetery, airport, etc.
•••⦰	Tanks: oil, water, etc. (labeled only if water)	━ ･･ ━	Land grant
∘ ⚑	Located or landmark object; windmill	━━━━━	Township or range line, U.S. land survey
✕ ✕	Open pit, mine, or quarry; prospect	━ ━ ━	Township or range line, approximate location
⫲	Marsh (swamp)	～～	Perennial streams
⫲	Wooded marsh	→━━←	Elevated aqueduct
▢	Woods or brushwood	∘ ⌒	Water well and spring
⋮⋮	Vineyard	～	Small rapids
⋮⋮	Land subject to controlled inundation	≈≈≈	Large rapids
⋮⋮	Submerged marsh	▨	Intermittent lake
⋮⋮	Mangrove	～	Intermittent stream
⋮⋮	Orchard	→====←	Aqueduct tunnel
⋮⋮	Scrub	▨	Glacier
▨	Urban area	～	Small falls
x7369	Spot elevation	▨	Large falls
670	Water elevation	▨	Dry lake bed

Reference Handbook H

Minerals

Mineral (formula)	Color	Streak	Hardness	Breakage Pattern	Uses and Other Properties
Graphite (C)	black to gray	black to gray	1–1.5	basal cleavage (scales)	pencil lead, lubricants for locks, rods to control some small nuclear reactions, battery poles
Galena (PbS)	gray	gray to black	2.5	cubic cleavage perfect	source of lead, used for pipes, shields for X rays, fishing equipment sinkers
Hematite (Fe_2O_3)	black or reddish-brown	reddish-brown	5.5–6.5	irregular fracture	source of iron; converted to pig iron, made into steel
Magnetite (Fe_3O_4)	black	black	6	conchoidal fracture	source of iron, attracts a magnet
Pyrite (FeS_2)	light, brassy, yellow	greenish-black	6–6.5	uneven fracture	fool's gold
Talc ($Mg_3Si_4O_{10}(OH)_2$)	white, greenish	white	1	cleavage in one direction	used for talcum powder, sculptures, paper, and tabletops
Gypsum ($CaSO_4 \cdot 2H_2O$)	colorless, gray, white, brown	white	2	basal cleavage	used in plaster of paris and dry wall for building construction
Sphalerite (ZnS)	brown, reddish-brown, greenish	light to dark brown	3.5–4	cleavage in six directions	main ore of zinc; used in paints, dyes, and medicine
Muscovite ($KAl_3Si_3O_{10}(OH)_2$)	white, light gray, yellow, rose, green	colorless	2–2.5	basal cleavage	occurs in large, flexible plates; used as an insulator in electrical equipment, lubricant
Biotite ($K(Mg,Fe)_3(AlSi_3O_{10})(OH)_2$)	black to dark brown	colorless	2.5–3	basal cleavage	occurs in large, flexible plates
Halite ($NaCl$)	colorless, red, white, blue	colorless	2.5	cubic cleavage	salt; soluble in water; a preservative

Reference Handbook H

Minerals					
Mineral (formula)	**Color**	**Streak**	**Hardness**	**Breakage Pattern**	**Uses and Other Properties**
Calcite ($CaCO_3$)	colorless, white, pale blue	colorless, white	3	cleavage in three directions	fizzes when HCl is added; used in cements and other building materials
Dolomite ($(CaMg(CO_3)_2)$)	colorless, white, pink, green, gray, black	white	3.5–4	cleavage in three directions	concrete and cement; used as an ornamental building stone
Fluorite (CaF_2)	colorless, white, blue, green, red, yellow, purple	colorless	4	cleavage in four directions	used in the manufacture of optical equipment; glows under ultraviolet light
Hornblende $((CaNa)_{2-3}(Mg,Al,Fe)_5\text{-}(Al,Si)_2 Si_6O_{22}(OH)_2)$	green to black	gray to white	5–6	cleavage in two directions	will transmit light on thin edges; 6-sided cross section
Feldspar $(KAlSi_3O_8)$ $(NaAlSi_3O_8),$ $(CaAl_2Si_2O_8)$	colorless, white to gray, green	colorless	6	two cleavage planes meet at 90° angle	used in the manufacture of ceramics
Augite $((Ca,Na)(Mg,Fe,Al)(Al,Si)_2 O_6)$	black	colorless	6	cleavage in two directions	square or 8-sided cross section
Olivine $((Mg,Fe)_2 SiO_4)$	olive, green	none	6.5–7	conchoidal fracture	gemstones, refractory sand
Quartz (SiO_2)	colorless, various colors	none	7	conchoidal fracture	used in glass manufacture, electronic equipment, radios, computers, watches, gemstones

Reference Handbook I

Rocks		
Rock Type	**Rock Name**	**Characteristics**
Igneous (intrusive)	Granite	Large mineral grains of quartz, feldspar, hornblende, and mica. Usually light in color.
	Diorite	Large mineral grains of feldspar, hornblende, and mica. Less quartz than granite. Intermediate in color.
	Gabbro	Large mineral grains of feldspar, augite, and olivine. No quartz. Dark in color.
Igneous (extrusive)	Rhyolite	Small mineral grains of quartz, feldspar, hornblende, and mica, or no visible grains. Light in color.
	Andesite	Small mineral grains of feldspar, hornblende, and mica or no visible grains. Intermediate in color.
	Basalt	Small mineral grains of feldspar, augite, and olivine or no visible grains. No quartz. Dark in color.
	Obsidian	Glassy texture. No visible grains. Volcanic glass. Fracture looks like broken glass.
	Pumice	Frothy texture. Floats in water. Usually light in color.
Sedimentary (detrital)	Conglomerate	Coarse grained. Gravel or pebble size grains.
	Sandstone	Sand-sized grains 1/16 to 2 mm.
	Siltstone	Grains are smaller than sand but larger than clay.
	Shale	Smallest grains. Often dark in color. Usually platy.
Sedimentary (chemical or organic)	Limestone	Major mineral is calcite. Usually forms in oceans, lakes, and caves. Often contains fossils.
	Coal	Occurs in swampy areas. Compacted layers of organic material, mainly plant remains.
Sedimentary (chemical)	Rock Salt	Commonly forms by the evaporation of seawater.
Metamorphic (foliated)	Gneiss	Banding due to alternate layers of different minerals, of different colors. Parent rock often is granite.
	Schist	Parallel arrangement of sheetlike minerals, mainly micas. Forms from different parent rocks.
	Phyllite	Shiny or silky appearance. May look wrinkled. Common parent rocks are shale and slate.
	Slate	Harder, denser, and shinier than shale. Common parent rock is shale.
Metamorphic (non-foliated)	Marble	Calcite or dolomite. Common parent rock is limestone.
	Soapstone	Mainly of talc. Soft with greasy feel.
	Quartzite	Hard with interlocking quartz crystals. Common parent rock is sandstone.

English Glossary

This glossary defines each key term that appears in bold type in the text. It also shows the chapter, section, and page number where you can find the word used.

A

abiotic (ay bi AHT ihk) **factor:** any nonliving part of the environment, such as water, sunlight, temperature, and air. (Chap. 15, Sec. 1, p. 442)

abrasion: erosion that occurs when wind blows sediments into rocks, makes pits in the rocks, and produces smooth, polished surfaces. (Chap. 8, Sec. 2, p. 238)

acceleration: change in velocity divided by the amount of time over which the change occurs; can take place when an object changes speed, direction, or both. (Chap. 4, Sec. 1, p. 109)

accuracy: compares a measurement to the true value. (Chap. 2, Sec. 1, p. 45)

aerosols (ER uh sahls): in the atmosphere, solids such as dust, salt, and pollen, and liquid droplets such as acids. (Chap. 10, Sec. 1, p. 285)

aggression: forceful behavior, such as fighting, used by an animal to control or dominate another animal to protect their young, defend territory, or get food. (Chap. 16, Sec. 2, p. 478)

air mass: large body of air that develops over a particular region of Earth's surface. (Chap. 10, Sec. 3, p. 298)

alternative resources: any renewable and inexhaustible sources of energy to generate electricity, including solar energy, wind, and geothermal energy. (Chap. 5, Sec. 3, p. 149)

aquifer: permeable rock layer or sediment through which water flows freely; acts as a groundwater reservoir. (Chap. 9, Sec. 1, p. 254)

artesian well: well drilled into a pressurized aquifer that supplies freshwater, usually without pumping. (Chap. 9, Sec. 1, p. 257)

asexual (ay SEK shul) **reproduction:** a type of reproduction, such as budding or regeneration, in which a new organism is produced from a part of another organism by the process of mitosis. (Chap. 14, Sec. 1, p. 419)

astronomical (as truh NAHM ih kul) **unit:** unit of measure that equals 150 million km, which is the mean distance from Earth to the Sun. (Chap. 12, Sec. 2, p. 359)

atmosphere: layer of gases surrounding Earth that protects living things against harmful doses of ultraviolet radiation and X-ray radiation and absorbs and distributes warmth. (Chap. 10, Sec. 1, p. 284)

B

balanced forces: forces that cancel each other out because they are equal and acting in opposite directions. (Chap. 4, Sec. 2, p. 111)

bar graph: a type of graph that uses bars of varying sizes to show relationships among variables. (Chap. 2, Sec. 3, p. 58)

behavior: the way in which an organism interacts with other organisms and its environment; can be innate or learned. (Chap. 16, Sec. 1, p. 470)

bioremediation: process that uses living organisms to remove pollutants. (Chap. 9, Sec. 2, p. 267)

biosphere (BI uh sfihr): part of Earth that supports life—the top part of Earth's crust, all the waters covering Earth's surface, and the surrounding atmosphere; includes all biomes, ecosystems, communities, and populations. (Chap. 15, Sec. 1, p. 447)

biotic (bi AHT ihk) **factor:** any living or once-living organism in the environment. (Chap. 15, Sec. 1, p. 442)

English Glossary

boiling point: temperature at which a substance in a liquid state becomes a gas. (Chap. 3, Sec. 1, p. 79)

C

cave: underground chamber that opens to the surface and is formed when slightly acidic groundwater dissolves compounds in rock. (Chap. 9, Sec. 3, p. 269)

cell membrane: protective outer covering of all cells that is made up of a double layer of fatlike molecules and regulates the interaction between the cell and the environment. (Chap. 13, Sec. 1, p. 388)

cell theory: states that all organisms are made up of one or more cells, the cell is the basic unit of life, and all cells come from other cells. (Chap. 13, Sec. 2, p. 401)

cell wall: rigid structure that encloses, supports, and protects the cells of plants, algae, fungi, and most bacteria. (Chap. 13, Sec. 1, p. 389)

chemical change: change in which the identity of a substance changes due to its chemical properties and forms a new substance or substances. (Chap. 3, Sec. 2, p. 85)

chemical energy: energy that is stored in chemicals. (Chap. 5, Sec. 1, p. 135)

chemical property: any characteristic, such as the ability to burn, that allows a substance to undergo a change that results in a new substance. (Chap. 3, Sec. 2, p. 84)

chemical weathering: process in which the chemical composition of rocks is changed by agents such as natural acids and oxygen. (Chap. 8, Sec. 1, p. 228)

chloroplast: green, chlorophyll-containing, plant-cell organelle that converts sunlight, carbon dioxide, and water into sugar. (Chap. 13, Sec. 1, p. 392)

circle graph: a type of graph that shows the parts of a whole; sometimes called a pie graph, each piece of which represents a percentage of the total. (Chap. 2, Sec. 3, p. 58)

cloning: making copies of organisms, each of which is a clone that receives DNA from only one parent cell. (Chap. 14, Sec. 1, p. 420)

comet: large body of frozen ice and rock that travels toward the center of the solar system, might originate in the Oort Cloud, and develops a bright, glowing tail as it approaches the Sun. (Chap. 12, Sec. 2, p. 364)

community: all of the populations of different species in a given area that interact in some way and depend on one another for food, shelter, and other needs. (Chap. 15, Sec. 1, p. 446)

conditioning: occurs when the response to a stimulus becomes associated with another stimulus. (Chap. 16, Sec. 1, p. 474)

constant: variable that is not changed in an experiment. (Chap. 1, Sec. 2, p. 18)

constellation (kan stuh LAY shun): group of stars that forms a pattern in the sky and can be named after a real or imaginary animal, object, or person. (Chap. 12, Sec. 3, p. 366)

controlled experiment: involves observing the effect of one thing while keeping all other things constant. (Chap. 1, Sec. 2, p. 18)

courtship behavior: behavior that allows males and females of the same species to recognize each other and get ready to mate. (Chap. 16, Sec. 2, p. 479)

critical thinking: involves using knowledge and thinking skills to evaluate evidence and explanations. (Chap. 1, Sec. 4, p. 27)

crust: Earth's outermost layer, which is thinnest under the oceans and thickest through the mountains and contains all features of Earth's surface. (Chap. 7, Sec. 1, p. 200)

crystal: solid material with atoms arranged in a repeating pattern. (Chap. 6, Sec. 1, p. 168)

cyclic behavior: behavior that occurs in repeated patterns. (Chap. 16, Sec. 2, p. 482)

English Glossary

cytoplasm: constantly moving gel-like mixture inside the cell membrane that contains heredity material and is the location of most of a cell's life processes. (Chap. 13, Sec. 1, p. 388)

D

deflation: erosion of land that occurs when wind blows across loose sediments and carries them away, often leaving behind particles that are too heavy to move. (Chap. 8, Sec. 2, p. 238)

density: measurable physical property that can be found by dividing the mass of an object by its volume. (Chap. 3, Sec. 1, p. 76)

dew point: temperature at which air is saturated and condensation can occur. (Chap. 10, Sec. 2, p. 292)

displacement: measures the direction and distance between the end position and starting position. (Chap. 4, Sec. 1, p. 106)

DNA: deoxyribonucleic acid—a chemical inside cells that contains hereditary information and controls how an organism will look and function by controlling which proteins a cell produces. (Chap. 14, Sec. 1, p. 416)

dripstone: deposits of calcium carbonate, such as stalagmites and stalactites, that are left behind when groundwater drips and evaporates inside caves. (Chap. 9, Sec. 3, p. 271)

E

Earth science: study of Earth systems and systems in space, including weather and climate systems and the study of nonliving things such as rocks, oceans, and planets. (Chap. 1, Sec. 1, p. 10)

eclipse (ih KLIHPS): event that occurs when the Moon moves between the Sun and Earth (solar eclipse), or when Earth moves between the Sun and the Moon (lunar eclipse) and casts a shadow. (Chap. 12, Sec. 1, p. 355)

ecology: study of all the interactions among organisms and their environment. (Chap. 15, Sec. 1, p. 442)

ecosystem: all of the communities in a given area and the abiotic factors that affect them. (Chap. 15, Sec. 1, p. 446)

electrical energy: energy carried by electric current that comes out of batteries and wall sockets, is generated at large power plants, and is readily transformed into other types of energy. (Chap. 5, Sec. 1, p. 136)

electromagnetic spectrum: arrangement of electromagnetic waves according to their wavelengths. (Chap. 11, Sec. 1, p. 321)

endoplasmic reticulum (ER): cytoplasmic organelle that moves materials around in a cell and is made up of a complex series of folded membranes; can be rough (with attached ribosomes) or smooth (without attached ribosomes). (Chap. 13, Sec. 1, p. 393)

energy: the ability to cause change. (Chap. 5, Sec. 1, p. 132)

erosion: wearing away and removal of rock material that occurs by agents such as gravity, ice, wind, and water. (Chap. 8, Sec. 2, p. 233)

estimation: method of making an educated guess at a measurement. (Chap. 2, Sec. 1, p. 43)

extrusive (ehk STREW sihv): igneous rocks that have small or no crystals and form when melted rock cools quickly on Earth's surface. (Chap. 6, Sec. 2, p. 175)

F

fault: large fracture in rock along which movement occurs. (Chap. 7, Sec. 1, p. 203)

fault-block mountains: sharp, jagged mountains made of huge, tilted blocks of rock that are separated from surrounding rock by faults and form because of pulling forces. (Chap. 7, Sec. 2, p. 210)

fertilization: process in which sperm and egg join, resulting in a new organism. (Chap. 14, Sec. 1, p. 421)

English Glossary

first law of motion: states that an object at rest will remain at rest or moving in a straight line at a constant speed unless a force acts on it. (Chap. 4, Sec. 3, p. 115)

folded mountain: mountain that forms by the folding of rock layers caused by compressive forces. (Chap. 7, Sec. 2, p. 211)

foliated: metamorphic rocks with visible layers of minerals. (Chap. 6, Sec. 3, p. 184)

food chain: model that describes how energy in the form of food passes from one organism to another. (Chap. 15, Sec. 3, p. 454)

food web: model that describes how energy in the form of food moves through a community; a series of overlapping food chains. (Chap. 15, Sec. 3, p. 456)

force: a push or a pull that has a size and a direction. (Chap. 4, Sec. 2, p. 110)

friction: force that resists motion between two touching surfaces. (Chap. 4, Sec. 2, p. 112)

front: boundary that develops where air masses of different temperatures collide; can be cold, warm, stationary, or occluded. (Chap. 10, Sec. 3, p. 299)

G

galaxy (GAL uk see): group of stars, gas, and dust held together by gravity. (Chap. 12, Sec. 3, p. 369)

gem: rare, valuable mineral that can be cut and polished. (Chap. 6, Sec. 1, p. 172)

gene: small section of DNA on a chromosome that carries information about a trait. (Chap. 14, Sec. 2, p. 426)

generator: device that transforms kinetic energy into electrical energy. (Chap. 5, Sec. 2, p. 142)

genetics (juh NET ihks): study of how traits are passed from parent to offspring. (Chap. 14, Sec. 2, p. 425)

geyser: hot spring that erupts periodically and shoots water and steam into the air. (Chap. 9, Sec. 1, p. 258)

Golgi bodies: organelles that package cellular materials and transport them within the cell or out of the cell. (Chap. 13, Sec. 1, p. 393)

graph: used to collect, organize, and summarize data in a visual way, making it easy to use and understand. (Chap. 2, Sec. 3, p. 57)

gravity: pull that all objects with mass exert on other matter. (Chap. 4, Sec. 2, p. 113)

groundwater: freshwater contained in the open spaces of soil and rock; one of Earth's most important resources for drinking, irrigation, and washing. (Chap. 9, Sec. 1, p. 252)

H

habitat: place where an organism lives. (Chap. 15, Sec. 2, p. 453)

heredity (huh RED uh tee): passing on of traits from parents to offspring. (Chap. 14, Sec. 2, p. 425)

hibernation: cyclic response of inactivity and slowed metabolism that occurs during periods of cold temperatures and limited food supplies. (Chap. 16, Sec. 2, p. 483)

host cell: living cell in which a virus can actively reproduce or in which a virus can hide until activated by environmental stimuli. (Chap. 13, Sec. 3, p. 402)

humidity: amount of water vapor in the atmosphere. (Chap. 10, Sec. 2, p. 292)

hurricane: large storm, up to 970 km in diameter, that begins as a low-pressure area over tropical oceans, has sustained winds that can reach 250 km/h and gusts up to 300 km/h. (Chap. 10, Sec. 3, p. 303)

hypothesis: reasonable guess that can be tested and is based on what is known and observed. (Chap. 1, Sec. 2, p. 14)

English Glossary

I

igneous (IHG nee us) **rock:** intrusive or extrusive rock that is produced when melted rock from inside Earth cools and hardens. (Chap. 6, Sec. 2, p. 175)

imprinting: occurs when an animal forms a social attachment to another organism during a specific period following birth or hatching. (Chap. 16, Sec. 1, p. 473)

inertia: measures an object's tendency to resist changing its motion and depends on the amount of mass an object has. (Chap. 4, Sec. 2, p. 111)

inexhaustible resources: a source of energy which cannot be used up by humans. (Chap. 5, Sec. 3, p. 149)

infer: to draw a conclusion based on observation. (Chap. 1, Sec. 2, p. 16)

innate behavior: behavior that an organism is born with and that does not have to be learned, such as a reflex or instinct. (Chap. 16, Sec. 1, p. 471)

inner core: solid, innermost layer of Earth's interior that is the hottest part of Earth and experiences the greatest amount of pressure. (Chap. 7, Sec. 1, p. 199)

insight: form of reasoning that allows animals to use past experiences to solve new problems. (Chap. 16, Sec. 1, p. 475)

instinct: complex pattern of innate behavior, such as spinning a web, that can take weeks to complete. (Chap. 16, Sec. 1, p. 472)

intrusive (ihn trew sihv): igneous rocks that have large crystals and form when melted rock cools slowly and hardens underneath Earth's surface. (Chap. 6, Sec. 2, p. 175)

isostasy: principle stating that Earth's crust and lithosphere float on the upper part of the mantle. (Chap. 7, Sec. 2, p. 214)

K

kelvin (K): SI unit for temperature. (Chap. 2, Sec. 2, p. 54)

kilogram (kg): SI unit for mass. (Chap. 2, Sec. 2, p. 53)

kinetic energy: energy an object has due to its motion. (Chap. 5, Sec. 1, p. 133)

L

law of conservation of energy: states that energy can change its form but it is never created or destroyed. (Chap. 5, Sec. 2, p. 138)

law of conservation of mass: states that the mass of the products of a chemical change is always the same as the mass of what you started with. (Chap. 3, Sec. 2, p. 89)

life science: study of living systems and how they interact. (Chap. 1, Sec. 1, p. 9)

light-year: about 9.5 million km—the distance that light travels in one year; used to measure large distances between stars or galaxies. (Chap. 12, Sec. 3, p. 372)

limiting factor: any biotic or abiotic factor that limits the number of individuals in a population. (Chap. 15, Sec. 2, p. 451)

line graph: a type of graph used to show the relationship between two variables that are numbers on an x-axis and a y-axis. (Chap. 2, Sec. 3, p. 57)

lithosphere: Earth's mantle and crust. (Chap. 7, Sec. 1, p. 202)

M

mantle: largest layer of Earth's interior that lies above the outer core and is solid yet flows slowly. (Chap. 7, Sec. 1, p. 200)

mass: amount of matter in an object, which is measured in kilograms. (Chap. 2, Sec. 2, p. 53)

mass movement: occurs when gravity alone causes rock or sediment to move down a slope. (Chap. 8, Sec. 2, p. 233)

matter: anything that has mass and takes up space. (Chap. 3, Sec. 1, p. 75)

English Glossary

measurement: way to describe objects and events with numbers; for example, length, volume, mass, weight, and temperature. (Chap. 2, Sec. 1, p. 42)

mechanical weathering: process that breaks rocks down into smaller pieces without changing them chemically. (Chap. 8, Sec. 1, p. 226)

meiosis (mi OH sus): process in which sex cells are formed in the reproductive organs; involves two divisions of the nucleus, producing four sex cells, each having half the number of chromosomes as the original cell. (Chap. 14, Sec. 1, p. 421)

melting point: temperature at which a solid becomes a liquid. (Chap. 3, Sec. 1, p. 79)

metamorphic (me tuh MOR fihk) **rock:** new rock that forms when existing rock is heated or squeezed. (Chap. 6, Sec. 3, p. 183)

meteorite: any space fragment that survives its plunge through the atmosphere and lands on Earth's surface. (Chap. 12, Sec. 2, p. 365)

meter (m): SI unit for length. (Chap. 2, Sec. 2, p. 51)

migration: instinctive seasonal movement of animals to find food or to reproduce in better conditions. (Chap. 16, Sec. 2, p. 484)

mineral: inorganic solid material found in nature that always has the same chemical makeup, atoms arranged in an orderly pattern, and properties such as cleavage and fracture, color, hardness, and streak and luster. (Chap. 6, Sec. 1, p. 166)

mitochondrion: cell organelle that breaks down lipids and carbohydrates and releases energy. (Chap. 13, Sec. 1, p. 392)

mitosis (mi TOH sus): cell division process in which DNA in the nucleus is duplicated and the nucleus divides into two nuclei that contain the same genetic information. (Chap. 14, Sec. 1, p. 418)

model: any representation of an object or an event that is used as a tool for understanding the natural world; can communicate observations and ideas, test predictions, and save time, money, and lives. (Chap. 1, Sec. 3, p. 21)

mutation: change in a gene or chromosome that can result from something in the environment or an error in mitosis or meiosis, can be harmful, neutral, or beneficial, and adds variation to the genes of a species. (Chap. 14, Sec. 2, p. 430)

N

niche (NICH): role of an organism in the ecosystem, including what it eats, how it interacts with other organisms, and how it gets its food. (Chap. 15, Sec. 2, p. 453)

nonfoliated: metamorphic rocks that lack distinct layers or bands. (Chap. 6, Sec. 3, p. 184)

nonrenewable resources: any energy sources that eventually will run out, such as coal and oil. (Chap. 5, Sec. 3, p. 146)

nuclear energy: energy stored in atomic nuclei that can be transformed into other forms of energy by complex power plants. (Chap. 5, Sec. 1, p. 136)

nucleus: organelle that controls all the activities of a cell and contains hereditary material made of proteins and DNA. (Chap. 13, Sec. 1, p. 390)

O

observatory: building that can house an optical telescope; often has a dome-shaped roof that can be opened for viewing. (Chap. 11, Sec. 1, p. 322)

orbit: curved path followed by a satellite as it revolves around an object. (Chap. 11, Sec. 2, p. 329; Chap. 12, Sec. 1, p. 353)

ore: mineral that contains enough of a useful metal that it can be mined and sold at a profit. (Chap. 6, Sec. 1, p. 173)

organ: structure, such as the heart, made up of different types of tissues that all work together. (Chap. 13, Sec. 1, p. 395)

English Glossary

organelle: structure in the cytoplasm of a eukaryotic cell that can act as a storage site, process energy, move materials, or manufacture substances. (Chap. 13, Sec. 1, p. 390)

outer core: layer of Earth that lies above the inner core and is thought to be composed mostly of molten metal. (Chap. 7, Sec. 1, p. 200)

P

permeable: describes rock that allows groundwater to flow through it because it contains many well-connected pores or cracks. (Chap. 9, Sec. 1, p. 253)

pheromone (FER uh mohn): powerful chemical produced by an animal to influence the behavior of another animal of the same species. (Chap. 16, Sec. 2, p. 479)

photovoltaic: device that transforms radiant energy directly into electrical energy. (Chap. 5, Sec. 3, p. 150)

physical change: change in which the properties of a substance change but the identity of the substance always remains the same. (Chap. 3, Sec. 1, p. 75)

physical property: any characteristic of a material, such as state, color, and volume, that can be observed or measured without changing or attempting to change the material. (Chap. 3, Sec. 1, p. 74)

physical science: study of matter, which is anything that takes up space and has mass, and the study of energy, which is the ability to cause change. (Chap. 1, Sec. 1, p. 10)

plate: section of Earth's crust and rigid, upper mantle that moves slowly around on the asthenosphere. (Chap. 7, Sec. 1, p. 202)

pollution: contamination of the environment by introducing something harmful. (Chap. 9, Sec. 2, p. 260)

population: all of the individuals of one species that live in the same space at the same time. (Chap. 15, Sec. 1, p. 446)

population density: number of individuals in a population that occupies an area of limited size. (Chap. 15, Sec. 2, p. 450)

porosity: volume of pore space divided by the volume of a rock or soil sample. (Chap. 9, Sec. 1, p. 253)

potential energy: energy stored in an object due to its position. (Chap. 5, Sec. 1, p. 134)

precipitation: occurs when drops of water or crystals of ice become too large to be suspended in a cloud and fall in the form of rain, freezing rain, sleet, snow, or hail. (Chap. 10, Sec. 2, p. 294)

precision: describes how closely measurements agree with each other and how carefully measurements were made. (Chap. 2, Sec. 1, p. 44)

Project Apollo: final stage in the U.S. program to reach the Moon in which Neil Armstrong was the first human to step onto the Moon's surface. (Chap. 11, Sec. 2, p. 334)

Project Gemini: second stage in the U.S. program to reach the Moon in which an astronaut team connected with another spacecraft in orbit. (Chap. 11, Sec. 2, p. 333)

Project Mercury: first step in the U.S. program to reach the Moon that orbited a piloted spacecraft around Earth and brought it back safely. (Chap. 11, Sec. 2, p. 333)

R

radiant energy: energy of light. (Chap. 5, Sec. 1, p. 135)

radio telescope: collects and records radio waves traveling through space; can be used day or night under most weather conditions. (Chap. 11, Sec. 1, p. 325)

rate: the amount of change in a given amount of time. (Chap. 2, Sec. 2, p. 54)

reflecting telescope: optical telescope that uses a concave mirror to focus light and form an image at the focal point. (Chap. 11, Sec. 1, p. 322)

English Glossary

reflex: simple innate behavior, such as yawning or blinking, that is an automatic response and does not involve a message to the brain. (Chap. 16, Sec. 1, p. 471)

refracting telescope: optical telescope that uses a double convex lens to bend light and form an image at the focal point. (Chap. 11, Sec. 1, p. 322)

relative humidity: measure of the amount of water vapor in the air compared with the amount that could be held at a specific temperature. (Chap. 10, Sec. 2, p. 292)

renewable resources: any energy sources that are replenished continually. (Chap. 5, Sec. 3, p. 148)

revolution (rev uh LEW shun): movement of Earth around the Sun, which takes a year to complete. (Chap. 12, Sec. 1, p. 353)

ribosome: small structure on which cells make their own proteins. (Chap. 13, Sec. 1, p. 392)

rock: solid inorganic material that usually is made of two or more minerals and can be metamorphic, sedimentary, or igneous. (Chap. 6, Sec. 1, p. 166)

rock cycle: diagram that shows the slow, continuous process of rocks changing from one type to another. (Chap. 6, Sec. 3, p. 185)

rocket: special motor that can work in space and burns liquid or solid fuel. (Chap. 11, Sec. 2, p. 327)

rotation (roh TAY shun): spinning of Earth on its axis, which occurs once every 24 h; produces day and night and causes the planets and stars to appear to rise and set. (Chap. 12, Sec. 1, p. 352)

runoff: water that flows over Earth's surface. (Chap. 8, Sec. 2, p. 239)

S

sanitary landfill: landfill lined with plastic or concrete, or located in clay-rich soil; reduces the chance of hazardous wastes leaking into the surrounding soil and groundwater. (Chap. 9, Sec. 2, p. 262)

satellite: any natural or artificial object that revolves around another object. (Chap. 11, Sec. 2, p. 329)

science: way of learning more about the natural world that provides possible explanations to questions and involves using a collection of skills. (Chap. 1, Sec. 1, p. 6)

scientific law: a rule that describes a pattern in nature but does not try to explain why something happens. (Chap. 1, Sec. 1, p. 7)

scientific theory: a possible explanation for repeatedly observed patterns in nature. (Chap. 1, Sec. 1, p. 7)

second law of motion: states that net force acting on an object causes the object to accelerate in the direction of the force. (Chap. 4, Sec. 3, p. 116)

sedimentary rock: a type of rock made from pieces of other rocks, dissolved minerals, or plant and animal matter that collects to form rock layers. (Chap. 6, Sec. 2, p. 179)

sex cells: specialized cells—female eggs and male sperm—that are produced by the process of meiosis, carry DNA, and join in sexual reproduction. (Chap. 14, Sec. 1, p. 420)

sexual reproduction: a type of reproduction in which a new organism is produced from the DNA of two sex cells (egg and sperm). (Chap. 14, Sec. 1, p. 420)

SI: International System of Units, related by multiples of ten, that allows quantities to be measured in the exact same way throughout the world. (Chap. 2, Sec. 2, p. 50)

sinkhole: depression formed when a cave's roof is no longer able to support the land above it and the land collapses into the cave. (Chap. 9, Sec. 3, p. 272)

social behavior: interactions among members of the same species, including courtship and mating, getting food, caring for young, and protecting each other. (Chap. 16, Sec. 2, p. 476)

English Glossary

society: a group of animals of the same species that live and work together in an organized way, with each member doing a specific job. (Chap. 16, Sec. 2, p. 477)

soil: mixture of weathered rock, organic matter, water, and air that evolves over time and supports the growth of plant life. (Chap. 8, Sec. 1, p. 230)

solar system: system of nine planets and numerous other objects that orbit the Sun, all held in place by the Sun's gravity. (Chap. 12, Sec. 2, p. 358)

space probe: instrument that travels far into the solar system, gathers data, and sends them back to Earth. (Chap. 11, Sec. 2, p. 330)

space shuttle: reusable spacecraft that can carry cargo, astronauts, and satellites to and from space. (Chap. 11, Sec. 3, p. 335)

space station: large facility with living quarters, work and exercise areas, and equipment and support systems for humans to live and work in space and conduct research not possible on Earth. (Chap. 11, Sec. 3, p. 336)

speed: distance traveled divided by the amount of time that is needed to travel that distance. (Chap. 4, Sec. 1, p. 106)

state of matter: physical property that is dependent on temperature and pressure and occurs in four forms—solid, liquid, gas, or plasma. (Chap. 3, Sec. 1, p. 77)

subduction: a type of plate movement that occurs when one plate sinks beneath another plate. (Chap. 7, Sec. 1, p. 205)

subsidence: occurs when water no longer fills the pores in an aquifer and the land above the aquifer sinks. (Chap. 9, Sec. 2, p. 268)

supernova: bright explosion of the outer part of a supergiant that takes place after its core collapses. (Chap. 12, Sec. 3, p. 369)

symbiosis (sihm bee OH sus): any close interaction among two or more different species, including mutualism, commensalism, and parasitism. (Chap. 15, Sec. 2, p. 452)

system: collection of structures, cycles, and processes that relate to and interact with each other. (Chap. 1, Sec. 1, p. 8)

T

table: presents information in rows and columns, making it easier to read and understand. (Chap. 2, Sec. 3, p. 57)

technology: use of science to help people in some way. (Chap. 1, Sec. 1, p. 11)

thermal energy: energy that all objects have; increases as the object's temperature increases. (Chap. 5, Sec. 1, p. 134)

third law of motion: states that when a force is applied on an object, an equal force is applied by the object in the opposite direction. (Chap. 4, Sec. 3, p. 119)

tissue: group of similar cells that work together to do one job. (Chap. 13, Sec. 1, p. 395)

topography: the surface features of an area. (Chap. 8, Sec. 1, p. 230)

tornado: violent, whirling wind, usually less than 200 m in diameter, that travels in a narrow path over land and can be highly destructive. (Chap. 10, Sec. 3, p. 302)

troposphere (TROH puh sfihr): layer of the atmosphere that is closest to Earth's surface and contains nearly all of its clouds and weather. (Chap. 10, Sec. 1, p. 286)

turbine: set of steam-powered fan blades that spins a generator at a power plant. (Chap. 5, Sec. 2, p. 142)

U

upwarped mountain: mountain that forms when forces inside Earth push up the crust. (Chap. 7, Sec. 2, p. 211)

English Glossary

V

variable: factor that can be changed in an experiment. (Chap. 1, Sec. 2, p. 18)

variations: different ways that a trait can appear—for example, differences in height, hair color, or weight. (Chap. 14, Sec. 2, p. 429)

velocity: measure of speed in a particular direction. (Chap. 4, Sec. 1, p. 108)

virus: structure that contains heredity material surrounded by a protein coat and makes copies of itself only inside living organisms. (Chap. 13, Sec. 3, p. 402)

volcanic mountain: mountain that forms when magma is forced upward and flows onto Earth's surface. (Chap. 7, Sec. 2, p. 212)

W

water cycle: continuous cycle of water molecules on Earth as they rise into the atmosphere, fall back to Earth as rain or other precipitation, and flow into rivers and oceans through the processes of evaporation, condensation, and precipitation. (Chap. 10, Sec. 1, p. 289; Chap. 15, Sec. 3, p. 458)

water table: top surface of the zone of saturation; also the surface of lakes and rivers. (Chap. 9, Sec. 1, p. 254)

weather: current condition of the atmosphere including cloud cover, temperature, wind speed and direction, humidity, and air pressure. (Chap. 10, Sec. 2, p. 290)

weathering: natural mechanical or chemical process that causes rocks to change by breaking them down and causing them to crumble. (Chap. 8, Sec. 1, p. 226)

Z

zone of saturation: in an aquifer, the zone where the pores are full of water. (Chap. 9, Sec. 1, p. 254)

Spanish Glossary

Este glossario define cada término clave que aparece en negrillas en el texto. También muestra el capítulo y el número de página en donde se usa dicho término.

A

abiotic factor / factor abiótico: cualquier parte inanimada o sin vida en un medio ambiente, como por ejemplo, el agua, la luz solar, la temperatura y el aire. (Cap. 15, Sec. 1, pág. 442)

abrasion / abrasión: erosión que ocurre cuando el viento sopla sedimentos en las rocas, forma hoyos en las rocas y produce superficies lisas y pulidas. (Cap. 8, Sec. 2, pág. 238)

acceleration / aceleración: cambio en la velocidad dividido entre la cantidad de tiempo en que ocurre tal cambio; puede llevarse a cabo cuando un objeto cambia de rapidez o dirección o las dos. (Cap. 4, Sec. 1, pág. 109)

accuracy / exactitud: compara una medida con el verdadero valor. (Cap. 2, Sec. 1, pág. 45)

aerosols / aerosoles: sólidos presentes en la atmósfera, como el polvo, la sal y el polen; también pueden ser gotitas líquidas como los ácidos. (Cap. 10, Sec. 1, pág. 285)

aggression / agresión: comportamiento enérgico, como las peleas, que usa un animal para controlar o dominar a otro animal con el propósito de proteger sus crías, defender su territorio u obtener alimento. (Cap. 16, Sec. 2, pág. 478)

air mass / masa de aire: extenso flujo de aire que se desarrolla sobre una región en particular de la superficie terrestre. (Cap. 10, Sec. 3, pág. 298)

alternative resources / recursos alternos: toda fuente de energía, tanto renovable como inagotable, que se utiliza para generar electricidad; incluye la energía solar, la energía eólica y la energía geotérmica. (Cap. 5, Sec. 3, pág. 149)

aquifer / acuífero: capa rocosa o sedimento permeable por el cual el agua corre libremente; actúa como un reservorio de aguas subterráneas. (Cap. 9, Sec. 1, pág. 254)

artesian well / pozo artesiano: pozo perforado en un acuífero presurizado que provee agua dulce, generalmente, sin ser bombeada. (Cap. 9, Sec. 1, pág. 257)

asexual reproduction / reproducción asexual: un tipo de reproducción, como la gemación o la regeneración, en la cual un nuevo organismo se produce a partir de una parte de otro organismo mediante el proceso de mitosis. (Cap. 14, Sec. 1, pág. 419)

astronomical unit / unidad astronómica: unidad de medida equivalente a 150 millones de kilómetros, lo cual es la distancia promedio de la Tierra al Sol. (Cap. 12, Sec. 2, pág. 359)

atmosphere / atmósfera: capa de gases que rodea la Tierra y protege a los seres vivos contra dosis dañinas de radiación ultravioleta y radiación de rayos X; absorbe y distribuye el calor. (Cap. 10, Sec. 1, pág. 284)

B

balanced forces / fuerzas equilibradas: fuerzas que se anulan entre sí porque las dos son iguales y porque actúan en direcciones opuestas. (Cap. 4, Sec. 2, pág. 111)

bar graph / gráfica de barras: tipo de gráfica que usa barras de distintos tamaños para mostrar relaciones entre variables. (Cap. 2, Sec. 3, pág. 58)

behavior / comportamiento: la interacción de un organismo con otro organismo y su ambiente; puede ser innato o adquirido. (Cap. 16, Sec. 1, pág. 470)

SPANISH GLOSSARY

Spanish Glossary

bioremediation / biorremediación: proceso que usa organismos vivos para eliminar contaminantes. (Cap. 9, Sec. 2, pág. 267)

biosphere / biosfera: parte de la Tierra que sustenta la vida: la parte superior de la corteza terrestre, toda el agua que cubre la superficie de la Tierra y la atmósfera circundante; incluye todos los biomas, ecosistemas, comunidades y poblaciones. (Cap. 15, Sec. 1, pág. 447)

biotic factor / factor biótico: cualquier organismo vivo o que alguna vez vivió en el medio ambiente. (Cap. 15, Sec. 1, pág. 442)

boiling point / punto de ebullición: temperatura a la cual una sustancia en estado líquido se transforma en un gas. (Cap. 3, Sec. 1, pág. 79)

C

cave / caverna: cámara subterránea que se abre a la superficie y que se forma cuando el agua subterránea, ligeramente ácida, disuelve compuestos en las rocas. (Cap. 9, Sec. 3, pág. 269)

cell membrane / membrana celular: cubierta externa protectora de todas las células; formada por una capa doble de moléculas adiposas y controla la interacción entre la célula y el ambiente. (Cap. 13, Sec. 1, pág. 388)

cell theory / teoría celular: establece que todos los organismos están formados por una o más células, la célula es la unidad básica de la vida y todas las células provienen de otras células. (Cap. 13, Sec. 2, pág. 401)

cell wall / pared celular: estructura rígida que encierra, sostiene y protege las células vegetales, las células de las algas, de los hongos y de la mayoría de las bacterias. (Cap. 13, Sec. 1, pág. 389)

chemical change / cambio químico: cambio en el cual la identidad de una sustancia cambia debido a sus propiedades químicas y forma una nueva sustancia o sustancias. (Cap. 3, Sec. 2, pág. 85)

chemical energy / energía química: energía almacenada en sustancias químicas. (Cap. 5, Sec. 1, pág. 135)

chemical property / propiedad química: cualquier característica, como la capacidad de quemarse, que permite que una sustancia sufra un cambio, el cual da como resultado una nueva sustancia. (Cap. 3, Sec. 2, pág. 84)

chemical weathering / meteorización química: proceso en el cual los agentes como los ácidos naturales y el oxígeno transforman la composición química de las rocas. (Cap. 8, Sec. 1, pág. 228)

chloroplast / cloroplasto: organelo de las células vegetales, de color verde y que contiene clorofila, que convierte la luz solar, el dióxido de carbono y el agua en azúcar. (Cap. 13, Sec. 1, pág. 392)

circle graph / gráfica circular: tipo de gráfica que muestra partes de un todo; cada parte es un sector que representa un porcentaje del total. (Cap. 2, Sec. 3, pág. 58)

cloning / clonación: hacer copias de un organismo, cada una de la cuales es un clon que recibe DNA de solo una célula progenitora. (Cap. 14, Sec. 1, pág. 420)

comet / cometa: astro extenso formado por hielo congelado y roca que viaja hacia el centro del sistema solar, es posible que provenga de la nube de Oort; desarrolla una cola brillante e incandescente a medida que se acerca al Sol. (Cap. 12, Sec. 2, pág. 364)

community / comunidad: todas las poblaciones de diferentes especies en un área dada que interactúan de alguna manera y que dependen entre sí para obtener alimento, refugio y otras necesidades. (Cap. 15, Sec. 1, pág. 446)

conditioning / condicionamiento: ocurre cuando la respuesta a un estímulo se asocia con otro estímulo. (Cap. 16, Sec. 1, pág. 474)

Spanish Glossary

constant / constante: variable que no se cambia en un experimento. (Cap. 1, Sec. 2, pág. 18)

constellation / constelación: grupo de estrellas que forma un patrón en el firmamento y puede recibir su nombre de un animal, una persona o un objeto real o imaginario. (Cap. 12, Sec. 3, pág. 366)

controlled experiment / experimento controlado: implica la observación del efecto que produce una cosa mientras se mantienen constantes las demás cosas. (Cap. 1, Sec. 2, pág. 18)

courtship behavior / comportamiento de cortejo: tipo de comportamiento que permite que machos y hembras de una especie se reconozcan mutuamente y se preparen para el apareo. (Cap. 16, Sec. 2, pág. 479)

critical thinking / pensamiento crítico: implica el uso del conocimiento y las destrezas del pensamiento para evaluar pruebas y explicaciones. (Cap. 1, Sec. 4, pág. 27)

crust / corteza: capa más externa de la Tierra que es más delgada en los océanos y más gruesa a través de montañas y contiene todos los relieves de la superficie terrestre. (Cap. 7, Sec. 1, pág. 200)

crystal / cristal: material sólido cuyos átomos están ordenados en un patrón repetitivo. (Cap. 6, Sec. 1, pág. 168)

cyclic behavior / comportamiento cíclico: comportamiento que ocurre en forma de patrones repetidos. (Cap. 16, Sec. 2, pág. 482)

cytoplasm / citoplasma: mezcla gelatinosa en continuo movimiento dentro de la membrana celular que contiene material hereditario y en la cual se lleva a cabo la mayoría de los procesos de una célula. (Cap. 13, Sec. 1, pág. 388)

D

deflation / deflación: erosión del terreno que ocurre cuando el viento sopla y transporta sedimentos sueltos, a menudo dejando atrás partículas demasiado pesadas que no puede mover. (Cap. 8, Sec. 2, pág. 238)

density / densidad: propiedad física que se puede medir; se puede calcular dividiendo la masa de un cuerpo entre su volumen. (Cap. 3, Sec. 1, pág. 76)

dew point / punto de condensación o de rocío: temperatura a la cual el aire se satura y puede ocurrir la condensación. (Cap. 10, Sec. 2, pág. 292)

displacement / desplazamiento: mide la dirección y distancia entre la posición final y la posición inicial. (Cap. 4, Sec. 1, pág. 106)

DNA / DNA: ácido desoxirribonucleico: una sustancia química presente en las células que contiene información hereditaria y controla la apariencia y el funcionamiento de un organismo al controlar las proteínas que produce una célula. (Cap. 14, Sec. 1, pág. 416)

dripstone / carbonato cálcico: puede ser una estalactita o una estalagmita; estos depósitos quedan atrás cuando el agua subterránea gotea y se evapora dentro de cavernas o grutas. (Cap. 9, Sec. 3, pág. 271)

E

Earth science / ciencias terrestres: estudio de los sistemas terrestres y los espaciales, entre ellos, los sistemas del tiempo y del clima, y el estudio de las cosas sin vida como las rocas, los océanos y los planetas. (Cap. 1, Sec. 1, pág. 10)

eclipse / eclipse: fenómeno que ocurre cuando la Luna se mueve entre el Sol y la Tierra (eclipse solar) o cuando la Tierra se mueve entre el Sol y la Luna (eclipse lunar) y el astro proyecta una sombra. (Cap. 12, Sec. 1, pág. 355)

ecology / ecología: estudio de todas las interacciones entre los organismos y su ambiente. (Cap. 15, Sec. 1, pág. 442)

ecosystem / ecosistema: todas las comunidades en un área dada y los factores abióticos que las afectan. (Cap. 15, Sec. 1, pág. 446)

Spanish Glossary

electrical energy / energía eléctrica: energía transportada por la corriente eléctrica que sale de las pilas y de los enchufes de pared, se genera en centrales eléctricas grandes y se transforma fácilmente en otros tipos de energía. (Cap. 5, Sec. 1, pág. 136)

electromagnetic spectrum / espectro electromagnético: arreglo de ondas electromagnéticas según sus longitudes de onda. (Cap. 11, Sec. 1, pág. 321)

endoplasmic reticulum (ER) / retículo endoplásmico (RE): organelo citoplásmico que mueve materiales dentro de una célula y que está formado por una serie compleja de membranas plegadas; puede ser áspero (con ribosomas adheridos) o liso (sin ribosomas adheridos). (Cap. 13, Sec. 1, pág. 393)

energy / energía: la capacidad de causar cambios. (Cap. 5, Sec. 1, pág. 132)

erosion / erosión: desgaste y transporte de material rocoso causado por agentes como la gravedad, el hielo, el viento y el agua. (Cap. 8, Sec. 2, pág. 233)

estimation / estimación: método de hacer una conjetura razonada de una medida. (Cap. 2, Sec. 1, pág. 43)

extrusive / extrusivas: rocas ígneas con o sin cristales que se forman cuando la roca fundida se enfría rápidamente en la superficie terrestre. (Cap. 6, Sec. 2, pág. 175)

F

fault / falla: fractura grande a lo largo de la cual ocurre movimiento(Cap. 7, Sec. 1, pág. 203)

fault-block mountains / montañas de bloques de falla: montañas abruptas y rugosas formadas por enormes bloques inclinados de rocas que están separados de las rocas circundantes por fallas y que se forman debido a las fuerzas de tracción. (Cap. 7, Sec. 2, pág. 210)

fertilization / fecundación: unión de un espermatozoide y un óvulo, de la cual se origina un nuevo organismo. (Cap. 14, Sec. 1, pág. 421)

first law of motion / primera ley del movimiento: establece que un cuerpo en reposo permanece en reposo o se mueve en línea recta a una velocidad constante a menos que una fuerza actúe sobre él. (Cap. 4, Sec. 3, pág. 115)

folded mountain / montaña plegada: montaña que se forma debido al plegamiento de capas rocosas provocado por fuerzas de compresión. (Cap. 7, Sec. 2, pág. 211)

foliated / foliadas: rocas metamórficas que poseen capas visibles de minerales. (Cap. 6, Sec. 3, pág. 184)

food chain / cadena alimenticia: modelo que describe la manera en que la energía pasa de un organismo a otro en forma de alimento. (Cap. 15, Sec. 3, pág. 454)

food web / red alimenticia: modelo que describe cómo la energía (en forma de alimento) se mueve por una comunidad; una serie de cadenas alimenticias superpuestas. (Cap. 15, Sec. 3, pág. 456)

force / fuerza: un empuje o un halón que posee tanto un tamaño como una dirección. (Cap. 4, Sec. 2, pág. 110)

friction / fricción: fuerza que resiste el movimiento entre dos superficies en contacto. (Cap. 4, Sec. 2, pág. 112)

front / frente: límite que se desarrolla en el punto donde chocan dos masas de aire con temperaturas diferentes; puede ser un frente frío, cálido, estacionario u ocluido. (Cap. 10, Sec. 3, pág. 299)

G

galaxy / galaxia: grupo de estrellas, gases y polvo que se mantienen unidos gracias a la gravedad. (Cap. 12, Sec. 3, pág. 369)

gem / gema: mineral precioso y valioso que se puede cortar y pulir. (Cap. 6, Sec. 1, pág. 172)

Spanish Glossary

gene / gene: pequeña sección de DNA en un cromosoma que transporta información sobre un rasgo. (Cap. 14, Sec. 2, pág. 426)

generator / generador: dispositivo que transforma la energía cinética en energía eléctrica. (Cap. 5, Sec. 2, pág. 142)

genetics / genética: estudio de la transmisión de los rasgos de los progenitores a la progenie. (Cap. 14, Sec. 2, pág. 425)

geyser / géiser: agua termal que hace erupción periódicamente y que arroja agua y vapor al aire. (Cap. 9, Sec. 1, pág. 258)

Golgi bodies / cuerpos de Golgi: organelos que almacenan materiales celulares y los transportan dentro o fuera de la célula. (Cap. 13, Sec. 1, pág. 393)

graph / gráfica: instrumento que se utiliza para recopilar, organizar y resumir datos de una manera visual, facilitando de esta manera su uso y comprensión. (Cap. 2, Sec. 3, pág. 57)

gravity / gravedad: halón que todo cuerpo con masa ejerce sobre otro cuerpo. (Cap. 4, Sec. 2, pág. 113)

groundwater / agua subterránea: agua dulce en los espacios abiertos de suelo y roca; uno de los recursos terrestres más importantes para obtener agua potable, irrigar y lavar. (Cap. 9, Sec. 1, pág. 252)

H

habitat / hábitat: morada de un organismo. (Cap. 15, Sec. 2, pág. 453)

heredity / herencia: transmisión de rasgos genéticos de una generación a la siguiente. (Cap. 14, Sec. 2, pág. 425)

hibernation / hibernación: respuesta cíclica de inactividad y disminución del metabolismo que ocurre durante períodos de temperaturas frías y abastecimientos limitados de alimentos. (Cap. 16, Sec. 2, pág. 483)

host cell / célula huésped: célula viva en la cual un virus se puede reproducir activamente o en la cual un virus puede ocultarse hasta que los estímulos ambientales lo activen. (Cap. 13, Sec. 3, pág. 402)

humidity / humedad: cantidad de vapor de agua presente en la atmósfera. (Cap. 10, Sec. 2, pág. 292)

hurricane / huracán: tormenta de gran alcance, hasta de 970 km de diámetro, que comienza como un área de baja presión sobre los océanos tropicales, tiene vientos sostenidos que pueden alcanzar 250 km/h y ráfagas de hasta 300 km/h. (Cap. 10, Sec. 3, pág. 303)

hypothesis / hipótesis: conjetura razonable que se puede poner a prueba y que se basa en lo que se sabe y lo observable. (Cap. 1, Sec. 2, pág. 14)

I

igneous rock / roca ígnea: roca intrusiva o extrusiva que se produce cuando la roca fundida del interior de la Tierra se enfría y se endurece. (Cap. 6, Sec. 2, pág. 175)

imprinting / impronta: ocurre cuando un animal forma un vínculo social con otro organismo durante un período específico después del nacimiento o de salir del cascarón. (Cap. 16, Sec. 1, pág. 473)

inertia / inercia: mide la tendencia de un cuerpo a resistir un cambio en su movimiento y depende de la cantidad de masa que tiene tal cuerpo. (Cap. 4, Sec. 2, pág. 111)

inexhaustible resources / recursos inagotables: fuente energética que no podemos agotar los seres humanos. (Cap. 5, Sec. 3, pág. 149)

infer / inferir: sacar una conclusión basándose en una observación. (Cap. 1, Sec. 2, pág. 16)

innate behavior / comportamiento innato: comportamiento con que nace un organismo y el cual no tiene que ser aprendido como un reflejo o un instinto. (Cap. 16, Sec. 1, pág. 471)

inner core / núcleo interno: capa sólida en lo más profundo del interior de la Tierra; es la parte más caliente del planeta y soporta la mayor cantidad de presión. (Cap. 7, Sec. 1, pág. 199)

Spanish Glossary

insight / discernimiento: forma de razonamiento que permite a los animales usar las experiencias previas para resolver nuevos problemas. (Cap. 16, Sec. 1, pág. 475)

instinct / instinto: patrón complejo de comportamiento innato, como por ejemplo, tejer una telaraña y el que puede demorar semanas en completarse. (Cap. 16, Sec. 1, pág. 472)

intrusive / intrusivas: rocas ígneas que poseen cristales grandes y que se forman cuando la roca fundida se enfría lentamente y se endurece debajo de la superficie terrestre. (Cap. 6, Sec. 2, pág. 175)

isostasy / isostasia: principio según el cual la corteza y la litosfera terrestres flotan en la parte superior del manto terrestre. (Cap. 7, Sec. 2, pág. 214)

K

kelvin (K) / kelvin (K): unidad de temperatura del SI. (Cap. 2, Sec. 2, pág. 54)

kilogram (kg) / kilogramo (kg): unidad de masa del SI. (Cap. 2, Sec. 2, pág. 53)

kinetic energy / energía cinética: energía que tiene un cuerpo debido a su movimiento. (Cap. 5, Sec. 1, pág. 133)

L

law of conservation of energy / ley de conservación de la energía: establece que la energía puede transformarse pero nunca se crea ni se destruye. (Cap. 5, Sec. 2, pág. 138)

law of conservation of mass / ley de conservación de la masa: establece que la masa de los productos de un cambio químico es siempre la misma que la masa con que se empezó. (Cap. 3, Sec. 2, pág. 89)

life science / ciencias biológicas: estudio de los sistemas vivos y sus interacciones. (Cap. 1, Sec. 1, pág. 9)

light-year / año luz: equivale a aproximadamente 9.5 millones de km, o sea, la distancia que la luz viaja en un año. El año luz se utiliza para medir grandes distancias entre estrellas o galaxias. (Cap. 12, Sec. 3, pág. 372)

limiting factor / factor limitativo: cualquier factor biótico o abiótico que limita el número de individuos en una población. (Cap. 15, Sec. 2, pág. 451)

line graph / gráfica lineal: tipo de gráfica que se utiliza para mostrar la relación entre dos variables, en forma de números, en un eje x y un eje y. (Cap. 2, Sec. 3, pág. 57)

lithosphere / litosfera: el manto y la corteza de la Tierra (Cap. 7, Sec. 1, pág. 202)

M

mantle / manto: capa más extensa del interior de la Tierra que se halla encima del núcleo externo; el manto es sólido pero flota lentamente. (Cap. 7, Sec. 1, pág. 200)

mass / masa: cantidad de materia que posee un cuerpo, la cual se mide en kilogramos. (Cap. 2, Sec. 2, pág. 53)

mass movement / movimiento de masas: se presenta cuando la gravedad por sí sola hace que las rocas o sedimentos desciendan por una pendiente. (Cap. 8, Sec. 2, pág. 233)

matter / materia: todo lo que posee masa y ocupa espacio. (Cap. 3, Sec. 1, pág. 75)

measurement / medida: manera de describir objetos y eventos con números; por ejemplo: longitud, volumen, masa, peso y temperatura. (Cap. 2, Sec. 1, pág. 42)

mechanical weathering / meteorización mecánica: proceso que rompe las rocas en fragmentos más pequeños sin alterarlas químicamente. (Cap. 8, Sec. 1, pág. 226)

meiosis / meiosis: proceso en el cual las células sexuales se forman en los órganos reproductores; implica dos divisiones del núcleo que producen cuatro células sexuales, cada una con la mitad del número de cromosomas que la célula original. (Cap. 14, Sec. 1, pág. 421)

Spanish Glossary

melting point / punto de fusión: temperatura a la cual un sólido se convierte en un líquido. (Cap. 3, Sec. 1, pág. 79)

metamorphic rock / roca metamórfica: roca nueva que se forma cuando la roca existente se calienta o se comprime. (Cap. 6, Sec. 3, pág. 183)

meteorite / meteorito: cualquier fragmento espacial que sobrevive su caída a través de la atmósfera y que llega a la superficie terrestre. (Cap. 12, Sec. 2, pág. 365)

meter (m) / metro (m): unidad de longitud del SI. (Cap. 2, Sec. 2, pág. 51)

migration / migración: movimiento instintivo de ciertos animales de mudarse a lugares nuevos cuando cambian las estaciones, en busca de alimentos o para encontrar condiciones más propicias para el apareo. (Cap. 16, Sec. 2, pág. 484)

mineral / mineral: material sólido inorgánico que se halla en la naturaleza y que siempre posee la misma composición química: átomos arreglados en un patrón ordenado y propiedades como crucero, fractura, color, dureza y veta y brillo. (Cap. 6, Sec. 1, pág. 166)

mitochondrion / mitocondria: organelo celular que descompone lípidos y carbohidratos y libera energía. (Cap. 13, Sec. 1, pág. 392)

mitosis / mitosis: proceso de división celular en el cual se duplica el DNA presente en el núcleo y el núcleo se divide en dos núcleos que contienen la misma información genética. (Cap. 14, Sec. 1, pág. 418)

model / modelo: cualquier representación de un objeto o un fenómeno que se utiliza como instrumento para comprender el mundo natural; puede comunicar observaciones e ideas, probar predicciones y ahorrar tiempo, dinero y salvar vidas. (Cap. 1, Sec. 3, pág. 21)

mutation / mutación: cambio en un gene o cromosoma que puede resultar de algo en el medio ambiente o de un error en la mitosis o en la meiosis; puede ser dañino, neutro o beneficioso y además puede añadir variación a los genes de una especie. (Cap. 14, Sec. 2, pág. 430)

N

niche / nicho: papel que tiene un organismo en el ecosistema en el cual se incluye lo que come, su manera de interactuar con otros organismos y de conseguir alimento. (Cap. 15, Sec. 2, pág. 453)

nonfoliated / no foliadas: rocas metamórficas que carecen de capas o bandas distintivas. (Cap. 6, Sec. 3, pág. 184)

nonrenewable resources / recursos no renovables: toda fuente de energía que se agota a la larga, como el carbón y el petróleo. (Cap. 5, Sec. 3, pág. 146)

nuclear energy / energía nuclear: energía almacenada en los núcleos atómicos que se puede transformar en otras formas de energía en centrales eléctricas complejas. (Cap. 5, Sec. 1, pág. 136)

nucleus / núcleo: organelo que controla todas las actividades de una célula y contiene material hereditario compuesto por proteínas y DNA. (Cap. 13, Sec. 1, pág. 390)

O

observatory / observatorio: centro que puede albergar un telescopio óptico; tiene a menudo un techo en forma de domo que se puede abrir para observar el espacio. (Cap. 11, Sec. 1, pág. 322)

orbit / órbita: trayectoria curva que sigue un satélite a medida que gira alrededor de un cuerpo. (Cap. 11, Sec. 2, pág. 329; Cap. 12, Sec. 1, pág. 353)

ore / mena: mineral que contiene suficiente cantidad de un metal útil como para que se pueda minar y vender con fines de lucro. (Cap. 6, Sec. 1, pág. 173)

Spanish Glossary

organ / órgano: estructura, como el corazón, compuesta por tipos diferentes de tejidos que funcionan en conjunto. (Cap. 13, Sec. 1, pág. 395)

organelle / organelo: estructura en el citoplasma de una célula eucariota que puede actuar como lugar de almacenamiento, puede procesar energía, mover materiales o elaborar sustancias. (Cap. 13, Sec. 1, pág. 390)

outer core / núcleo externo: capa de la Tierra ubicada sobre el núcleo interno; se piensa que está compuesto en su mayor parte de metal fundido. (Cap. 7, Sec. 1, pág. 200)

P

permeable / permeable: describe la roca que permite la filtración del agua subterránea debido a que contiene muchos poros o grietas bien conectados. (Cap. 9, Sec. 1, pág. 253)

pheromone / feromona: poderosa sustancia química producida por un animal para influir sobre el comportamiento de otro animal de la misma especie. (Cap. 16, Sec. 2, pág. 479)

photovoltaic / célula fotovoltaica: dispositivo que transforma la energía radiante directamente en energía eléctrica. (Cap. 5, Sec. 3, pág. 150)

physical change / cambio físico: cambio en el cual las propiedades de un sustancia cambian pero la identidad de la sustancia permanece siempre igual. (Cap. 3, Sec. 1, pág. 75)

physical property / propiedad física: cualquier característica de un material, como el estado, el color y el volumen, que se puede observar o medir sin alterar o intentar alterar el material. (Cap. 3, Sec. 1, pág. 74)

physical science / ciencias físicas: estudio de la materia, que es todo lo que ocupa espacio y posee masa, y el estudio de la energía, la cual es la capacidad de producir cambios. (Cap. 1, Sec. 1, pág. 10)

plate / placa: sección de la corteza terrestre y del manto superior rígido que se mueve lentamente en la astenosfera. (Cap. 7, Sec. 1, pág. 202)

pollution / contaminación: alteración nociva del medio ambiente debido a la introducción de elementos dañinos. (Cap. 9, Sec. 2, pág. 260)

population / población: todos los individuos de una especie que viven en el mismo espacio al mismo tiempo. (Cap. 15, Sec. 1, pág. 446)

population density / densidad demográfica: número de individuos en una población que ocupan un área de tamaño limitado. (Cap. 15, Sec. 2, pág. 450)

porosity / porosidad: volumen del espacio poroso dividido entre el volumen de una muestra de roca o de suelo. (Cap. 9, Sec. 1, pág. 253)

potential energy / energía potencial: energía almacenada en un cuerpo debido a su posición. (Cap. 5, Sec. 1, pág. 134)

precipitation / precipitación: se presenta cuando gotas de agua o cristales de hielo alcanzan un tamaño demasiado grande como para estar suspendidos en una nube y caen en forma de lluvia, lluvia congelada, cellisca, nieve o granizo. (Cap. 10, Sec. 2, pág. 294)

precision / precisión: describe el grado de aproximación de las medidas entre sí y el grado de exactitud con que se tomaron tales medidas. (Cap. 2, Sec. 1, pág. 44)

Project Apollo / Proyecto Apolo: etapa final del programa espacial de EE.UU. para llegar a la Luna, en la cual el astronauta Neil Armstrong fue el primer ser humano en poner pie sobre la superficie lunar. (Cap. 11, Sec. 2, pág. 334)

Project Gemini / Proyecto Géminis: segunda etapa del programa espacial de EE.UU. para llegar a la Luna, en la cual un equipo de astronautas se conectó con otra astronave en órbita. (Cap. 11, Sec. 2, pág. 333)

Spanish Glossary

Project Mercury / Proyecto Mercurio: primer paso en el programa espacial de EE.UU. para llegar a la Luna que orbitó una astronave piloteada alrededor de la Tierra, la cual regresó a salvo. (Cap. 11, Sec. 2, pág. 333)

R

radiant energy / energía radiante: energía de la luz. (Cap. 5, Sec. 1, pág. 135)

radio telescope / radiotelescopio: recopila y registra ondas radiales que viajan por el espacio; se puede usar de día o de noche bajo casi cualquier condición meteorológica. (Cap. 11, Sec. 1, pág. 325)

rate / tasa: razón de dos clases distintas de medidas. (Cap. 2, Sec. 2, pág. 54)

reflecting telescope / telescopio reflector: telescopio óptico que usa un espejo cóncavo para enfocar la luz y formar una imagen en el punto focal. (Cap. 11, Sec. 1, pág. 322)

reflex / reflejo: comportamiento innato simple, como bostezar o parpadear, que es una respuesta automática y que no involucra el envío de un mensaje al encéfalo. (Cap. 16, Sec. 1, pág. 471)

refracting telescope / telescopio refractor: telescopio óptico que usa una lente convexa doble para doblar la luz y formar una imagen en el punto focal. (Cap. 11, Sec. 1, pág. 322)

relative humidity / humedad relativa: medida de la cantidad de vapor de agua presente en el aire comparada con la cantidad que éste puede sostener a una temperatura dada. (Cap. 10, Sec. 2, pág. 292)

renewable resources / recursos renovables: toda fuente de energía que se regenera continuamente. (Cap. 5, Sec. 3, pág. 148)

revolution / revolución: movimiento de la Tierra alrededor del Sol, el cual se demora un año en completarse. (Cap. 12, Sec. 1, pág. 353)

ribosome / ribosoma: estructura pequeña en la cual las células producen sus propias proteínas. (Cap. 13, Sec. 1, pág. 392)

rock / roca: material sólido inorgánico que por lo general está compuesto de dos o más minerales; puede ser metamórfica, sedimentaria o ígnea. (Cap. 6, Sec. 1, pág. 166)

rock cycle / ciclo de las rocas: diagrama que muestra el proceso lento y continuo de las rocas en el cual éstas cambian de un tipo a otro. (Cap. 6, Sec. 3, pág. 185)

rocket / cohete: motor especial que funciona en el espacio y que quema combustible líquido o sólido. (Cap. 11, Sec. 2, pág. 327)

rotation / rotación: movimiento giratorio de la Tierra alrededor de su eje, el cual ocurre una vez cada 24 horas, produciendo el día y la noche y hace aparecer los planetas y estrellas como si saliesen y se pusiesen. (Cap. 12, Sec. 1, pág. 352)

runoff / escorrentía: agua que corre sobre la superficie terrestre. (Cap. 8, Sec. 2, pág. 239)

S

sanitary landfill / vertedero controlado: vertedero forrado con plástico o concreto, o ubicado en suelo arcilloso; reduce la posibilidad de que los desechos peligrosos se filtren por el suelo y lleguen a las aguas subterráneas circundantes. (Cap. 9, Sec. 2, pág. 262)

satellite / satélite: cualquier astro natural o artificial que gira alrededor de otro astro. (Cap. 11, Sec. 2, pág. 329)

science / ciencia: manera de aprender más acerca de la naturaleza que ofrece posibles explicaciones a preguntas e implica el uso de un número de destrezas. (Cap. 1, Sec. 1, pág. 6)

scientific law / ley científica: una regla que describe un patrón en la naturaleza pero que no intenta explicar por qué suceden las cosas. (Cap. 1, Sec. 1, pág. 7)

Spanish Glossary

scientific theory / teoría científica: una posible explicación para patrones que se observan repetidamente en la naturaleza. (Cap. 1, Sec. 1, pág. 7)

second law of motion / segunda ley del movimiento: establece que la fuerza neta que actúa sobre un cuerpo es la causa de que tal cuerpo acelere en la dirección de la fuerza. (Cap. 4, Sec. 3, pág. 116)

sedimentary rock / roca sedimentaria: un tipo de roca formada por fragmentos de otras rocas, minerales disueltos o materia vegetal y animal que se congregan para formar capas rocosas. (Cap. 6, Sec. 2, pág. 179)

sex cells / células sexuales: células especializadas (óvulos femeninos y espermatozoides masculinos) que se producen mediante meiosis, son portadoras de DNA y se unen en la reproducción sexual. (Cap. 14, Sec. 1, pág. 420)

sexual reproduction / reproducción sexual: tipo de reproducción en la cual un nuevo organismo se produce a partir del DNA de dos células sexuales (óvulo y espermatozoide). (Cap. 14, Sec. 1, pág. 420)

SI / SI: Sistema internacional de unidades, relacionado por múltiplos de diez, que permite que las cantidades se midan de la misma manera exacta en todo el mundo. (Cap. 2, Sec. 2, pág. 50)

sinkhole / dolina: depresión que se forma cuando el techo de una caverna no puede seguir soportando la corteza terrestre encima de él provocando su colapso hacia el interior de la caverna. (Cap. 9, Sec. 3, pág. 272)

social behavior / comportamiento social: interacciones entre los miembros de la misma especie; incluye el comportamiento de cortejo, el apareo, la obtención de alimentos, el cuidado de las crías y la protección mutua. (Cap. 16, Sec. 2, pág. 476)

society / sociedad: grupo de animales de la misma especie que viven y trabajan juntos de manera organizada en la que cada cual realiza una tarea específica. (Cap. 16, Sec. 2, pág. 477)

soil / suelo: mezcla de roca meteorizada, materia orgánica, agua y aire que se forma con el tiempo y sustenta el crecimiento de la vida vegetal. (Cap. 8, Sec. 1, pág. 230)

solar system / sistema solar: sistema compuesto por nueve planetas y numerosos otros cuerpos celestes que giran alrededor de nuestro Sol y que se mantienen unidos gracias a la gravedad solar. (Cap. 12, Sec. 2, pág. 358)

space probe / sonda espacial: instrumento que viaja a gran distancia en el sistema solar, recopila datos y los envía a la Tierra. (Cap. 11, Sec. 2, pág. 330)

space shuttle / transbordador espacial: astronave reutilizable que puede transportar cargamento, astronautas y satélites hacia y desde el espacio. (Cap. 11, Sec. 3, pág. 335)

space station / estación espacial: instalaciones con zonas de habitación, de trabajo y de ejercicio y equipo y sistemas de apoyo para que los seres humanos vivan y trabajen en el espacio y efectúen investigación que no es posible llevar a cabo en la Tierra. (Cap. 11, Sec. 3, pág. 336)

speed / rapidez: distancia recorrida dividida entre la cantidad de tiempo que se necesita para recorrer esa distancia. (Cap. 4, Sec. 1, pág. 106)

state of matter / estado de la materia: propiedad física que depende de la temperatura y la presión y que se presenta en cuatro formas: sólido, líquido, gas o plasma. (Cap. 3, Sec. 1, pág. 77)

subduction / subducción: un tipo de movimiento de placas que ocurre cuando una placa se hunde debajo de otra. (Cap. 7, Sec. 1, pág. 205)

subsidence / subsidencia: se presenta cuando el agua deja de llenar los poros de un acuífero provocando el hundimiento de la corteza terrestre que se halla encima del acuífero. (Cap. 9, Sec. 2, pág. 268)

supernova / supernova: explosión brillante de la parte externa de una supergigante que se lleva a cabo después del colapso de su núcleo. (Cap. 12, Sec. 3, pág. 369)

symbiosis / simbiosis: cualquier interacción estrecha entre dos o más especies diferentes; incluye el mutualismo, el comensalismo y el parasitismo. (Cap. 15, Sec. 2, pág. 452)

system / sistema: conjunto de estructuras, ciclos y procesos que se relacionan e interactúan entre sí. (Cap. 1, Sec. 1, pág. 8)

T

table / tabla: despliega información en hileras y columnas facilitando la lectura y comprensión de los datos. (Cap. 2, Sec. 3, pág. 57)

technology / tecnología: uso de la ciencia para ayudar a las personas de alguna manera. (Cap. 1, Sec. 1, pág. 11)

thermal energy / energía térmica: energía que tienen todos los cuerpos; aumenta conforme aumenta la temperatura del cuerpo. (Cap. 5, Sec. 1, pág. 134)

third law of motion / tercera ley del movimiento: establece que cuando se le aplica una fuerza a un cuerpo, éste aplica una fuerza semejante en dirección contraria. (Cap. 4, Sec. 3, pág. 119)

tissue / tejido: grupo de células semejantes que funcionan juntas para efectuar una tarea. (Cap. 13, Sec. 1, pág. 395)

topography / topografiá: el relieve superficial de un area. (Cap. 8, Sec. 1, pág. 230)

tornado / tornado: viento violento y arremolinado, por lo general con un diámetro menor de 200 m, que viaja en una trayectoria estrecha sobre tierra y puede ser sumamente destructivo. (Cap. 10, Sec. 3, pág. 302)

troposphere / troposfera: capa de la atmósfera que se halla más próxima a la superficie terrestre y que contiene casi todas sus nubes y condiciones meteorológicas. (Cap. 10, Sec. 1, pág. 286)

turbine / turbina: conjunto de álabes accionados a vapor que hace girar un generador en una central eléctrica. (Cap. 5, Sec. 2, pág. 142)

U

upwarped mountain / montaña plegada anticlinal: montaña que se forma cuando las fuerzas internas de la Tierra empujan la corteza hacia arriba. (Cap. 7, Sec. 2, pág. 211)

V

variable / variable: factor que se puede cambiar en un experimento. (Cap. 1, Sec. 2, pág. 18)

variations / variaciones: diferentes maneras en que puede darse un rasgo; por ejemplo, las diferencias en estatura, color del cabello o peso. (Cap. 14, Sec. 2, pág. 429)

velocity / velocidad: medida de la rapidez en una dirección en particular. (Cap. 4, Sec. 1, pág. 108)

virus / virus: estructura que contiene material hereditario rodeado por un revestimiento proteico y que sólo se desarrolla en el interior de organismos vivos. (Cap. 13, Sec. 3, pág. 402)

volcanic mountain / montaña volcánica: montaña que se forma cuando el magma es forzado a ascender y fluye sobre la superficie terrestre. (Cap. 7, Sec. 2, pág. 212)

W

water cycle / ciclo del agua: ciclo interminable en el cual el agua circula entre la superficie terrestre y la atmósfera, a través de los procesos de evaporación, transpiración, precipitación y condensación. (Cap. 10, Sec. 1, pág. 289; Cap. 15, Sec. 3, pág. 458)

Spanish Glossary

water table / capa freática: nivel superior de la zona de saturación: también es la superficie de lagos y ríos. (Cap. 9, Sec. 1, pág. 254)

weather / tiempo (atmosférico): estado de la atmósfera; incluye las nubes, la temperatura, la velocidad y dirección del viento, la humedad y la presión atmosférica. (Cap. 10, Sec. 2, pág. 290)

weathering / meteorización: proceso mecánico o químico natural que rompe y desintegra las rocas. (Cap. 8, Sec. 1, pág. 226)

Z

zone of saturation / zona de saturación: en un acuífero, es la zona donde los poros están llenos de agua. (Cap. 9, Sec. 1, pág. 254)

Index

The index for *Texas Science Grade 6* will help you locate major topics in the book quickly and easily. Each entry in the index is followed by the number of the pages on which the entry is discussed. A page number given in boldfaced type indicates the page on which that entry is defined. A page number given in italic type indicates a page on which the entry is used in an illustration or photograph. The abbreviation *act.* indicates a page on which the entry is used in an activity.

A

Abiotic factors, 442–444, *443, 444;* air, 444, *444;* light, 443, *443;* soil, 444, *444;* temperature, 443; water, 443, *443*
Abrasion, 238
Acceleration, 109; calculating, 117; force and, 116–119, *118;* and gravity, *act.* 5; mass and, 116–119, *118*
Accuracy, 43, *43,* 44, **45**–47, *45, 46*
Acid(s): and cave formation, 270, *270;* natural, 228–229, *229;* weathering and, 228–229, *229*
Action: and reaction, 119–120, *119, 120*
Active viruses, 403, *403*
Activities, 31, 32–33, 55, 60–61, 91, 92–93, 114, 122–123, 144, 152–153, 187, 188–189, 208, 216–217, 232, 242–243, 259, 274–275, 305, 306–307, 326, 342–343, 357, 374–375, 396, 406–407, 424, 432–433, 448, 460–461, 485, 486–487
Aerosols, 285, *285*
Age-progression models, 23, *24*
Aggression, 478
Agriculture: and groundwater pollution, 265
Air: as abiotic factor, 444; effects of temperature on, *act.* 283; temperature of, 290–291, *290, 291;* weight of, 284, *284*
Air mass, 298, *298, act.* 305
Air pollution: and environment, 444, *444*

Air resistance, 112
Aldrin, Edwin, 334
Algae: green, 443
Allele, 426–427
Alternative resources, 149–151, *150, 151*
Altitude: and atmospheric pressure, 444
Altocumulus clouds, 293
Altostratus clouds, 293
Animal(s): aggression in, 478; communication of, *act.* 469, 478–482, *478, 480, 481;* competition among, 451, *451;* conditioning of, 474, *474;* courtship behavior of, 479, *479;* cyclic behavior of, 482–485, *482, 483, 484, act.* 485; feedlots for, 265; habitats of, 453, *act.* 486–487, *486, 487;* hibernation of, 483, *483;* identifying, 81; imprinting of, 473, *473;* innate behavior of, 470–471, *470, 471;* instincts of, 472, *472,* 476; learned behavior of, 472–475, *472, 473, 474, 475;* migration of, 484, *484;* reflexes of, 471; social behavior of, 476–477, *476, 477;* submission in, 478, *478;* territorial behavior of, 477, *477;* weathering, 227
Animal cell, *391*
Antarctica: food webs in, 456, *456*
Apatite, 167, 170
Appalachian Mountains, 211, *211*
Apparent magnitude, 367
Aquatic ecosystems: freshwater, 443
Aquifer, 254, *254,* 267, *267,* 268, *268, act.* 274–275
Armstrong, Neil, 334
Artesian springs, 257
Artesian wells, 257, *257, act.* 259
Ash, 85, 88–89
Asking questions. *see* all activities, 13
Asteroid, 362
Asteroid belt, 362
Astrolabe, 343
Astronauts, 318, 333, *333,* 334, *334,* 336
Astronomical unit (AU), 359
Astronomy Integrations, 51
Atmosphere, 282–307, **284;** carbon dioxide in, 285; gases in, 285, *285;* layers of, 286–287, *286, 287;* nitrogen in, 285; oxygen in, 285; ozone layer in, 286; temperature in, *act.* 283, 286, *286, 287;* weather change, 290–297
Atmospheric pressure, 284, 291, 295, 300, 301, *act.* 306
Average speed, 107–108, *108*

B

Bacteria: chemosynthesis and, 457; nitrogen and, 459; shapes of, *388*
Bacteriophage, *386,* 404
Balanced forces, 111, *111*
Bar graph, 58, *58, 59*
Basalt, *176*
Basaltic rock, 175

Index

Bats, hibernation of, *483*
Bear, *440*
Behavior, 470, *470;* conditioned, 474, *474;* courtship **479,** *479;* cyclic, 482–485, *482, 483, 484,* act. 485; innate, 470–**471,** *470, 471;* learned, 472–475, *472, 473, 474, 475;* of packs, 470, *470,* 477; social, **476**–**477,** *476, 477;* territorial, 477, *477*
Betelgeuse, 367
Big Dipper, 366
Bioluminescence, 480–482, *481*
Biomass: transforming chemical energy, 139, 454; cycles of matter, 458
Biomes, *445,* **446,** *446, 447*
Bioremediation, 267, *267*
Biosphere, 447, *447*
Biotic factors, 442, 444–447, *445, 446*
Biotic potential, 452
Birds: courtship behavior of, 479, *479;* cyclic behavior of, 482, *482;* innate behavior of, 471, *471;* learned behavior of, 472, *472,* 473, *473;* sound communication of, 480, *480*
Black hole, 369, *369*
Boiling points, 79, *79*
Burning, 85, 88–90
Butterflies, 484, *484*

C

Carbon cycle, 459, *459*
Carbon dioxide: in atmosphere, 285
Carbonic acid, 228, 270
Carrying capacity, 451, *451*
Cave(s), 269; formation of, 269–271, *269, 270, 271;* karst areas, 273, *273*
Cave bacon, *271*
Cave pearls, *271*
Cell(s), 386–405; animal, *391;* comparing, 388, *388,* act. 396; eukaryotic, 389, *389;* host, **402,** *403,* 404, *404;* magnifying, 397–400, *397, 398*–*399;* nucleus of, 390, *391;* organization of, 389–390, *390, 391;* plant, *391;* prokaryotic, 389, *389;* structure of, 388–395, *388*
Cell membrane, 388, *390, 390*
Cell theory, 401
Cell wall, 389, *389*
Celsius, 54
Chalk, 180, 181, *181,* 228
Changes. *see* Chemical changes; Physical changes
Charon (moon of Pluto), 364
Charts, construct, 289, act. 433; analyze, 60, 61, 114; examine, 60, 61, 114, 143, act. 432–433; evaluate, 143, 207
Chemical changes, 85–86, *85, 86, 87,* act. 92–93
Chemical communication, 479
Chemical energy, 135, *135,* 139, *139, 140*
Chemical properties, 84–90; of building materials, 88; classifying according to, 88; common, 85–86, *85;* examples of, *84*
Chemical rocks, 180, *180*
Chemical weathering, 228–229, *228, 229*
Chemist, *10,* 11
Chemistry Integrations, 43, 200, 255, 330, 457–458, 480
Chemosynthesis, 457
Chimpanzees, 475, *475*
Chlorophyll, 392, *392*
Chloroplasts, 392, *392*
Chromosome(s), 390
Circle graph, 58, *58*
Circumference, of Earth, act. 342–343
Cirques, 237, *237*
Cirrus clouds, *293,* 294
Classification: according to chemical properties, 88; of coins, act. 73; of igneous rock, 175; of minerals, 171, act. 188–189; of parts of a system, 8; of soils, act. 232
Cleavage, 168, *168*
Climate, and soil formation, 230–231
Clouds, *292,* 293–294, *293, 294*
Coal, 146, 180
Coins, classifying, act. 73
Cold front, 299, *299*
Collins, Michael, 334
Colony: space, act. 374–375
Color: of minerals, 169, *169;* as physical property, 75, *75*
Combustion, 85, 88–90
Comets, 364, *364*
Communication: of animals, act. 469, 478–482, *478, 480, 481;* chemical, 479; of data, 56–59; light, 480–482, *481;* in science, 17, *17;* sound, 480, *480;* through models, 25
Communities, *445,* **446;** symbiosis in, 452, *452*
Compression, 205
Computer models, 22, *22*
Concave mirror, 322, *322*
Conclusion, 16
Condensation, 288, 289
Conditioning, 474, *474*
Conduction, 291, *291*
Conglomerate, *179,* 180, *185*
Conservation of energy, 138, 145; of mass, 88–90, **89,** *90*
Constant, 18
Constellation, 366, *366,* 367
Construction materials, 88
Consumers, 454
Continent(s), 205
Continental glaciers, 236, *236*
Continental plates, 205, *205*
Convection, 206, *206,* 291, *291*
Convex lens, 322, *322*
Copper, 173, *173, 174;* ductility of, 80; reaction with oxygen, 85
Copper oxide, 85
Core(s): inner, **199,** *200–201;* outer, **200,** *200–201*
Coriolis effect, 296, 297, 300
Corrosion, 85
Corundum, 170
Courtship behavior, 479, *479*
Crash-test dummies, 25, *25*
Crater(s), *212*

Index

Creep, *234,* 235, *235*
Critical thinking, 27
Crust: of Earth, 168, *168,* **200,** *200–201,* 209–217, *act.* 216–217; oceanic, 205, *205;* uplift of, 209–217, *act.* 216–217
Crystal, 167, *167,* **168,** *168,* 178, *178*
Cubic meter, 52, *52*
Cumulonimbus clouds, 293, *294*
Cumulus clouds, 293, *293*
Cycles: carbon, 459, *459;* life cycle of stars, 368–369, *368;* lunar, 354, *354;* nitrogen, 459; rock, **185**–186, *185;* water, 288, 289, **458,** *458*
Cyclic behavior, 482–485, *482, 483, 484, act.* 485
Cytoplasm, 388, 390
Cytoskeleton, 390, *390*

D

Data, collect by observing, *act.* 387, 395, *act.* 396, 398, *act.* 441, *act.* 448, 448, 452, 458, *act.* 460–461; communicating, 17, *17,* 56–59; organizing, 15
Data tables, 57
Decomposers, 454, *454*
Deflation, 238, *238*
Deimos (moon of Mars), 361
Delta, 241, *241*
Density, 76, 77
Deoxyribonucleic acid (DNA), 390
Deposition: by glaciers, 237, *237;* of sediment, 237, *237,* 241, *241*
Desert(s): wind erosion in, 238
Detrital rocks, 179–180
Dew point, 292
Diamond, 170, 172, *172*
Dichotomous key, 81
Digits: number of, 48; significant, 49, *49*
Diseases: and contaminated water, 263

Displacement, 106, *106*
Distance: and motion, 105, *105;* in space, 358–359, 363, 372
DNA (deoxyribonucleic acid), 390
Dogs: behavior of, 470, *470,* 474, *474*
Dolomite, 173, 228
Drawings: scale, *act.* 55; as scientific illustrations, 56, *56*
Dripstone, 271, *271*
Ductility, 80, *80*
Dune(s), 238, *238*

E

Earth, 361, *361;* circumference of, *act.* 342–343; crust of, **200,** *200–201,* 209–217, *act.* 216–217; inner core of, 199, *200–201;* interior of, 198–200, *199, 200–201;* layers of, 199–200, *199, 200–201;* mantle of, 200, *200–201;* minerals in crust, 173, *173;* motion through space, 103, *104;* moving plates of, 202–208, *203, 205, 206, 207, act.* 208; orbit of, 353, *353;* outer core of, 200, *200–201;* revolution of, 353, *353;* rotation of, **352,** *352;* structure of, 199–201, *199, 200–201;* water on, 287–289, *288*
Earthquakes: Earth's plates and, 203, *203,* 206, *206;* faults and, 203, 206, *206;* seismic waves and 198, *198,* 199
Earth science, 10, *10*
Earth Science Integrations, 22, 79, 116, 119, 146, 430, 444
Earth systems: rock cycle, 185; groundwater, 254, 255; surface water, 254, 255
Earthworms: behavior of, *act.* 485
Easterlies, 296, 297
Eclipses, 355; lunar, 356, *356;* solar, 355, *355*
Ecological pyramids, 457

Ecologist, 442, *442*
Ecology, 442
Ecosystems, 445, **446,** *446;* balance of, *act.* 448; energy flow through, 454–457, *454, 455, 456, 457;* habitats in, 453; limiting factors in, 451, *451, act.* 460–461; populations in, *445,* 446, *446*
Electrical energy, 136, 141, *141,* 142–143, *142, 143*
Electricity: consumption of, 148, *act.* 152–153; generating, 142–143, *142, 143,* 148, *149*
Electrolysis, 86
Electromagnetic spectrum, 320–321, **321**
Electron microscopes, 399–400
Elliptical galaxy, 369, *370*
Endangered species, 57–59, *57, 58, 59*
Endoplasmic reticulum (ER), 392, **393,** *393*
Energy, 130–153, *act.* 131, **132,** *132;* alternative sources of, 149–151, *150, 151;* chemical, **135,** *135,* 139, *139, 140;* conservation of, **138,** 145; consumption of, 148, *act.* 152–153; electrical, **136,** 141, *141,* 142–143, *142, 143;* flow-through ecosystems, 454–457, *454, 455, 456, 457;* in food chain, 454, *455;* forms of, 134–136, *134, 135, 136;* from fossil fuels, 146, *146;* geothermal, 150; kinetic, **133,** *133,* 138, *138,* 139; matter and, 454–459; of motion, 133, *133;* nuclear, **136,** *136,* 137, *137;* photosynthesis and, 454; potential, **134,** *134,* 138, *138,* 139; radiant, **135,** *135;* solar, 130, 149–150, *150;* sources of, 145–153, *144, act.* 152–153; thermal, **134,** *134,* 141–142, *142;* transfer of, 132; using, 145; wind, 151, *151*
Energy-producing organelles, 392, *392*

INDEX 603

Index

Energy pyramids, 457, *457*
Energy transformations, 137–143, *137, act.* 144; chemical energy, 139, *139, 140;* electrical energy, 141, *141;* between kinetic and potential energy, 138, *138,* 139; thermal energy, 141–142, *142;* tracking, 137
Energy types, research, *act.* 152–153; describe, 146, 147, 151
Environment, 266–267, 442–448; abiotic factors in, 442–444, *443, 444;* biotic factors in, 442, 444–447, *445, 446;* biotic potential of, 452; carrying capacity of, 451, *451;* population and, *445, 446, 446*
Environmental Protection Agency, 266, *281*
Environmental Science Integrations, 394
Erosion, 224, *act.* 225, **233**–243; agents of, 233; by glaciers, 236–237, *236, 237;* gravity and, 233–235, *233, 234, 235;* gully, 240, *240;* mass movement and, **233**–235, *233, 234, 235;* rill, 240; of soil, *act.* 242–243; stream, 240–241, *240;* by water, 239–241, *239, 240;* by wind, 238, *238*
Estimation, 43–44, *44*
Estivation, 483
Eukaryotic cell, 389, *389*
Euphausiids, *440*
Europa (moon of Jupiter), 332, *332,* 362
Evaluation, 27–30; of evidence, 27–28, *27;* of promotional materials, 30; of scientific explanation, 28–29, *29*
Evaporation, 288, 289
Evidence, evaluating, 27–28, *27;* direct evidence, 88, *act.* 91–92, 108, 114, 129, 202, *act.* 396, *act.* 448, *act.* 460–461, *act.* 424; indirect evidence, 52, 198, 199–201, 214, 216–217, 357
Exosphere, *286,* 287
Experiments: controlled, 18, *18*
Explanations: scientific, 28–29, *29*
Explore Activities, 5, 41, 73, 101, 131, 165, 197, 225, 251, 283, 319, 351, 387, 415, 441, 469
External stimuli, 470–475, 476–484, *act.* 485
Extrusive igneous rock, 175, 177, *177, 178*

F

Fahrenheit, 54
Farming. *see* Agriculture
Fault(s), **203,** 206, *206*
Fault-block mountains, 210, *210*
Feedlots, 265
Feldspar, 170, 171, *228*
Field Guides: Amusement Park Rides Field Guide 524–527, Building Stones Field Guide 528–531, Living in Space Field Guide 532–535, Insects Field Guide 536–539
Field investigations, 19, 20, *act.* 60, *act.* 342
Fissure, 176
Flammability, 85, 88–90
Flint, 168, *168*
Fluorite, *167,* 170
Foldables, 5, 41, 73, 101, 131, 165, 197, 225, 251, 283, 319, 351, 387, 415, 441, 469
Folded mountains, 211, *211*
Foliated rocks, 184, *184, act.* 187
Food chains, 454, *455*
Food web, 456, *456*
Fool's gold (pyrite), 168, 169, *169,* 170
Force(s), **110**–113; acceleration and, 116–119, *118;* balanced, **111,** *111;* changing motion with, 110–112, *110, 111,* 116, *116;* compression, 205; of gravity, 113; shaping Earth, 209–211, 212, 233, 270; tension, 203, *203,* 205, *210;* unbalanced, 112
Force pairs, 120
Fossey, Dian, 9, *9*
Fossil(s), in rocks, 180, 181, *181*
Fossil fuels, as source of energy, 146, *146*
Fracture, 168, *168*
Frequency, 428, *act.* 433
Freshwater ecosystems, 443
Friction, **112,** *112*
Fronts, 299–300, *299, 300, act.* 305
Full Moon, 354, *354*
Funnel cloud, 302, *302*
Fusion, 368

G

Gabbro, *177*
Gagarin, Yuri A., 333
Galaxies, 323, **369**–373, *373;* elliptical, 369, *370;* irregular, 369, *370;* spiral, 369, *370*
Galena, 173
Galilei, Galileo, 284, 324, 371
Ganymede (moon of Jupiter), 362
Garnet, *172*
Gas(es), 78, *78;* in atmosphere, 285, *285*
Gasoline, 145, 146, 263
Gasoline engines, 139, *139*
Gaspra (asteroid), 362
Gems, 172, *172*
Gene, 426–427
Gene therapy, 405
Generator, 142, *142*
Genetics, **425**–431; *act.* 432–433
Geothermal energy, 150
Geyser, 180, *180,* **258,** *258*
Glaciers: continental, 236, *236;* deposition by, 237, *237;* erosion by, 236–237, *236, 237;* valley, 236, *236*
Glass, *166*
Glenn, John, 333, *333*
Gneiss, 184, *185, act.* 187
Gold: identifying, 81; 169, 170
Golgi bodies, 393, *393*

Index

Gram, 50
Grand Canyon, 233
Granite, *177, 183,* 185
Granitic rock, 175
Graph(s), 57–59, *57, 58, 59;* of motion, 108, *108;* construct, *act.* 61, 289, 412, *act.* 461, *act.* 531; analyze, *act.* 61, 289, 412, *act.* 461, *act.* 531; examine, 412, *act.* 433, *act.* 461; evaluate, 108, 292, *act.* 433
Graphite, *166*
Gravity, 113; and acceleration of objects, *act.* 5; adjusting to, 215, *215;* erosion by, 233–235, *233, 234, 235;* mass movement and, 233–235, *233, 234, 235;* specific, 170
Great Red Spot (Jupiter), 362
Green algae, 443
Groundwater, 250–275, **252;** aquifers, 254, *254,* 267, *267,* 268, *268, act.* 574–575; cave formation and, 269–271, *269, 270, 271;* cleanup of, 266–267, *266, 267;* disappearing streams and, 272, *272;* flow of, 255; geysers, 258, *258;* importance of, 252, *252;* karst areas and, 272–273, *273;* pollution of, 260–265, *260, 261, 262, 263, 264,* 268, *268, act.* 274–275; shortages of, 267, *267;* sinkholes and, 272, *272;* springs, 256, *256,* 257; subsidence and, 268; water table and, 254, *255;* wells, 256, *256,* 257, *act.* 259
Guessing, 13
Gully erosion, 240, *240*
Gypsum, 170

H

Habitats, 453, *act.* 486–487, *486, 487*
Hail, 294, *294*
Halite, *168,* 170, 171
Hardness, 170
Hawaiian Islands: volcanoes in, 213, *213*

Health Integrations, 47, 85, 263, 286, 321, 421, 471
Hematite, 173
Heredity, 425
Hibernation, 483, *483*
High-pressure center, 300
Himalaya, *196,* 209, *209*
Hooke, Robert, 401
Host cell, 402, *403,* 404, *404*
Hot springs, 256
Hubble Space Telescope, 323, *323,* 340, 373, *373*
Humidity, 292, *292;* relative, 292
Humus, 231, *231*
Hurricane, 303, *303*
Hydroelectric power, 148, *149*
Hypothesis, 14; stating, 14; testing, 15; formulating. *see all* Design Your Own Activities, *act.* 60–61, 306–307, *act.* 406–407, *act.* 432–433, *act.* 460–461; analyze, 14, 15, *act.* 31, 216, 217; review, 15, 32; critique, 16, 32, 206, 207

I

Ice. *see* Glaciers; melting point of, 79
Ice wedging, 227, *227*
Idea models, 23
Igneous rock, 175–178, *175, 178;* classifying, 175; extrusive, **171,** 172, *172,* 173; intrusive, **175,** *176,* 177, *177*
Illustrations: scientific, 56–57, *56*
Impact of research on: history of science, 26, 337
Impact of: scientists, 34, 35, *act.* 326; contributions of, 34, 35, *act.* 326
Imprinting, 473, *473*
Industrial wastes, 264, *264*
Inertia, 111, *111*
Inexhaustible resources, 149, 151, *act.* 152–153, 154
Inference: from promotional materials, 30; products, 30, *act.* 152–153; services, 30; **16,** 29
Inflammability, 85
Innate behavior, 470–471, *470, 471*
Inner core, 199, *200–201*
Inner planets, 360–361; Earth, 361, *361;* Mars, 361, *361;* Mercury, 360, *360;* Venus, 360, *360*
Insects: communication among, 478, *478;* migration of, 484, *484*
Insight, 475, *475*
Instinct, 472, *472,* 476
Interactions: among living things, 449–453; space and, *act.* 441, 450, *450*
Interior: of Earth, 198–200, *199, 200–201*
Internal stimuli, 446, 451, 453, 474
International System of Units (SI), 50–54
Intrusive igneous rock, 175, *176,* 177, *177*
Investigations. *see all* Activities
Investigate procedures: plan, *see all* Activities, 6, 7, 11, 13, 15, 16, *act.* 60–61, *act.* 242–243, *act.* 406–407, *act.* 460–461, *act.* 432–433; implement, 60, 61, 244, *act.* 406–407, *act.* 460–461, *act.* 432–433
Io (moon of Jupiter), 362
Ionosphere, *286,* 287, *287*
Iron: in ore, 173; rusting of, 85, *85*
Iron oxide, 85
Irregular galaxy, 369, *370*
Isostasy, 214, *215, act.* 216–217

J

Jenner, Edward, 404
Jet streams, 297, *297*
Jupiter, 362, *362;* exploration of, 330, 332; moons of, 332, *332*

Index

K

Kaolinite, *228*
Karst areas, 272–273, *273*
Keck telescopes, 324, *324*, 340
Kelvin, **54**, *54*
Kilogram, 50, **53**
Kilometer, 51, *51*
Kimberlite, 172
Kinetic energy, **133**; mass and, 133, *133*; speed and, 133, *133*; transforming chemical energy to, 139, *139*, *140*; transforming to and from potential energy, 138, *138*, 139
Kuiper Belt, 364

L

Laboratory investigations, 20, 305, 326, 333, *act.* 415, *act.* 424, *act.* 448, *act.* 460–461
Laboratory safety, 19–20, *19*, *74*, *74*
Landfills: sanitary, **262**, *262*
Landslides, *233*
Latent viruses, 403
Lava, 176, 186, *186*, 212, *212*, 213
Law(s): of conservation of energy, **138**, 145; of conservation of mass, 88–90, **89**, *90*; Newton's first law of motion, **115**–116, *115*, *116*; Newton's second law of motion, **116**–119, *118*; Newton's third law of motion, **119**–120, *119*, *120*; scientific, **7**
Lead, 173
Learned behavior, 472–475, *472*, *473*, *474*, *475*
Leeuwenhoek, Antonie van, 397
Length: measuring, 76, *76*, *act.* 41, 47, 50, 51, *51*
Lens, 322, *322*
Life science, **9**, *9*
Life Science Integrations, 13, 141, 167, 291, 360
Light: energy of, 135, *135*; and environment, 443, *443*; speed of, 321, 372, *372*; visible, *act.* 319
Light communication, 480–482, *481*
Lightning, 301, *301*
Light pollution, 324
Light-year, 372
Limestone, 86, 171, 180, *183*, 228
Limiting factors, **451**, *451*, *act.* 460–461
Line graph, 57, *57*
Liquid(s), 78, *78*; boiling point of, 79, *79*; layers of, *act.* 91
Lithosphere, **200**, 203
Living things: interactions among, 449–453; levels of organization of, 445–447, *445*
Local Group, 372
Loess, 238
Louse, *386*
Low-pressure center, 300, 301
Lunar cycle, 354, *354*
Lunar eclipse, 356, *356*
Luster, 79, 169

M

Magellan mission, 330
Magma, 172, *172*, 177, 212, *212*, 213
Magma chamber, 212
Magnesium, 173
Magnetic properties, 80, *80*, 170
Magnetite, 170
Magnifying glass, *act.* 387, 400
Magnitude: apparent, 367
Malleability, 80
Mantle: of Earth, **200**, *200–201*
Manure, 265
Maps: construct, 181, 207, 223; analyze, 181, 207, 223; examine, 55, 223; evaluate, 223, 275
Marble, *183*, 184, 228
Marble launch: analyzing, *act.* 131
Mariner 2 mission, 330
Mars, 361, *361*; exploration of, 330, 338, *338*

Mass, **53**, *53*; acceleration and, 116–119, *118*; conservation of, 88–90, **89**, *90*; inertia and, 111, *111*; and kinetic energy, 133, *133*; as physical property, 76, 77; weight and, 113
Mass movement, **233**–235, *233*, *234*, *235*
Materials: disposal of, 262
Math Skill Handbook, 558–564
Math Skills Activities, 48, 107, 117, 295, 328, 394, 429
Matter, **75**; cycles of, 458–459, *458*, *459*; define matter, 10; define energy, 132; energy and, 454–459; physical changes in, 75, 87, *87*, *act.* 92–93; interactions between water cycle, 288, 289, 458; biomass, 454, 458; states of. *see* States of matter
Measuring, 40–61, **42**; accuracy of, 40, 43, 44, 45–47, *46*, 51; estimation vs., 43–44, *44*; of length, 76, *76*, *act.* 41, 47, 50, 51, *51*; of mass, **53**, *53*; of motion, *act.* 114; of pore space, *act.* 251, 253; precision of, 44–48, *45*, *46*; rounding, 47, 48; in SI, 50–54, *act.* 55; of small object, *act.* 387; of soil erosion, *act.* 242–243; in space, 359, 363, 372; of space, *act.* 441; of speed, 54, *act.* 60–61; of temperature, 54, *54*; of time, 40, *40*, *45*, 54; units of, 50–54; of volume, 52, *52*; of weight, 53, *53*
Measuring porosity, 253
Mechanical weathering, **226**–227, *226*, *227*
Melting points, 79
Mercury, 360, *360*
Mesosphere, **286**, *286–287*
Metal(s): properties of, 79–80, *80*; uses of, 80, *80*
Metamorphic rocks, *183*, **183**–184, *184*, *act.* 187

Index

Meteorite, 360, **365,** *365*
Meter, 51
Mica, 168, *168*
Mice, 453, *453*
Microscopes, 397–400, *397, 398–399, act.* 406–407, 569
Mid-ocean ridges, 207, *207*
Migration, 484, *484*
Milky Way, 371, *371,* 372
Mineral(s), 166–174, *166;* classifying, 171, *act.* 188–189; cleavage of, 168, *168;* color of, 169, *169;* common, 171–174, *172, 173, 174;* in Earth's crust, 171; formation of, 167, *167;* fracture of, 168, *168;* gems, **172,** *172;* hardness of, 170; identifying, 167–170, *168, 169, 170;* luster of, 169; magnetic properties of, 170; physical properties of, 167–170, *168, 169, 170;* rock-forming, 171; specific gravity of, 170; streak test of, 169, *169;* unique properties of, 170, *170*
MiniLABS, Thinking Like A Scientist 23, Measuring Accurately 44, Observing Yeast 88, Demonstrating the Third Law of Motion 120, Building a Solar Collector 149, Classifying Minerals 171, Modeling Mountains 211, Rock Dissolving Acids 229, Measuring Porosity 253, Creating a Low-Pressure Center 301, Modeling a Satellite 333, Observing Distance and Size 355, Modeling Cytoplasm 390, Observing Yeast Budding 419, Observing Symbiosis 452, Observing Conditioning 474
Mining, 173, *173*
Mining wastes, 264
Mitochondria, 392, *392*
Model(s), 21–26; limitations of, 26, *26;* making, 23, *23,* 24; need for, 21, *21;* types of, 22–23, *22;* using, 25, *25*
Mohs, Friedrich, 170
Mohs scale of hardness, 170
Moon(s): eclipse of, 356, *356;* exploration of, 333–334, *334, 339,* 339; of Jupiter, 332, *332;* 362; of Mars, 361; movement of, 354–356, *355, 356;* of Neptune, 364; of Pluto, 364; of Saturn, 340, 363; of Uranus, 363
Moon phases, 354, *354, act.* 357
Motion, 102–123; acceleration and, 109, 116–119, *118;* air resistance and, 112; balanced forces and, 111, *111;* changing, 110–112, *110, 111,* 116, *116;* changing position and, 105–106, *105, 106;* displacement and, 106, *106;* distance and, 105, *105;* of Earth through space, 103, *104;* energy of, 133, *133;* forces and, 110–113, *110, 111;* friction and, 112, *112;* graphing, 108, *108;* gravity and, 113; laws of, 115–121, *121;* measuring, *act.* 114; Newton's first law of, **115**–116, *115, 116;* Newton's second law of, **116**–119, *118;* Newton's third law of, **119**–120, *119, 120;* reference points and, 103, *103, 104;* as relative, 102–103, *102, 103;* speed and, 106–108, *108, act.* 114, *act.* 122–123; unbalanced forces and, 112; velocity and, 108–109, *108*
Mount St. Helens (Washington State), 212
Mountains, *196,* 209; fault-block, **210,** *210;* folded, **211,** *211;* formation of, 203, 209–213, *209, 210, 211, 212, 213;* modeling, 211; upwarped, **211;** volcanic, **212**–213, *212, 213*
Mudflow, 224, 234, 235
Muscle cell, 388
Muscles: transforming chemical energy to kinetic energy in, 139, *140*
Mutualism, 452, *452*

N

National Aeronautics and Space Administration (NASA), 330, 335, 337, 338, 340, 341
National Geographic Unit Openers: How are arms and centimeters connected? 2–3, How are charcoal and celebrations connected? 70–71, How are rocks and fluorescent lights connected? 162–163, How are Inuit and astronauts connected? 316–317, How are seaweed and cell cultures connected? 384–385
National Geographic Visualizing: The Modeling of King Tut 24, Precision and Accuracy 46, Dichotomous Keys 82, Earth's Motion 104, Energy Transformations 140, Igneous Rock Features 178, Rift Valleys 204, Mass Movements 234, Sources of Groundwater Pollution 261, The Water Cycle 288, Space Probes 331, Galaxies 370, Microscopes 398, Human Reproduction 422, A Food Chain 455, Bioluminescence 481
National Weather Service, 304
Neptune, 364, *364*
Nerve cells, 388, *388*
Nest building, 471, *471*
Neutron star, 369
New Millennium Program (NMP), 339
New moon, 354, *354*
Newton, 53

Index

Newton, Isaac, 115
Newton's first law of motion, 115–116, *115, 116*
Newton's second law of motion, 116–119, *118*
Newton's third law of motion, 119–120, *119, 120*
Next Generation Space Telescope, 340, *340*
Niche, 453
Nimbostratus clouds, 293, *294*
Nitrogen: in atmosphere, 285; boiling point of, 79, *79*
Nitrogen cycle, 459
Nonfoliated rocks, **184**, *184*
Nonrenewable resources, **146**, *146*
North Star, 342, *act.* 342
Nuclear energy, **136**, *136*
Nuclear fusion, 368
Nucleus, 390, *391*
Nutrients: vitamins, 86

O

Observation, 13, *13*, 27–28, *27*
Observatories, 322
Occluded front, 300, *300*
Ocean(s): and environment, 443, *443*
Oceanic plates, 205, *205*
Ocean water: groundwater pollution from, 268, *268*
Oil, 146
Old Faithful, 258, *258*
Oops! Accidents in Science, 94–95, 190–191, 488–489
Oort Cloud, 364
Optical telescopes, 322–324, *323, 324, act.* 326
Orbit, 329, *329*, 353, *353*
Ore, 173, *173*
Organ(s), **395**, *395*; systems, 395
Organelle(s), 390–394; energy-producing, 392, *392*; manufacturing, 392–393; recycling, 394; storage, 393; transport, 393, *393*
Organic rocks, 180
Organisms: responses from internal stimuli, 471; responses from external stimuli, *act.* 485; responses from components of ecosystem, 442, 451
Ouachita Mountains, 209, *209*
Outer core, **200**, *200–201*
Outer planets, 362–365; Jupiter, 362, *362*; Neptune, 364, *364*; Pluto, 364, *364*; Saturn, 363, *363*; Uranus, 363, *363*
Outwash, 237
Owl, 453, *453*, 482, *482*
Oxidation, 229, *229*
Oxygen: in atmosphere, 285; and chemical weathering, 229, *229*
Ozone, 287
Ozone layer, 286

P

Pack behavior, 470, *470*, 477
Pan balance, 53, *53*
Percent, *act.* 242, 253, 258, 281, 429
Permeable rock, 253
Phases of the Moon, 354, *354, act.* 357
Pheromone, 479
Phobos (moon of Mars), 361
Phosphorus, 459
Photographs: scientific, 57
Photosynthesis, 454
Photovoltaic collector, 150, *150*
Phyllite, 184
Physical changes, **75**, 87, *87, act.* 92–93
Physical models, 22
Physical properties, 74–83, 88; boiling points, 79, *79*; cleavage, 168, *168*; color, 75, *75*, 169, *169*; density, 76, 77, *77*; fracture, 168, *168*; hardness, 170; identification by, 81–83, *81, 82, 83*; length, 76, *76*; luster, 79, 169; magnetic, 80, *80*; mass, 76, 77; melting points, 79; metallic, 79–80, *80*; of minerals, 167, 168, 169, 170; senses and, 74, *74*; shape, 75, *75*; size, 77, *77*; streak, 169, *169*; volume, 76, 77
Physical science, 10–11, *10*
Physicist, *10*, 11
Physics Integrations, 177, 214, 235, 372, 400
Pipe, *212*
Planets: colonizing, *act.* 374–375; inner, 360–361, *360, 361*; outer, 362–364, *362, 363, 364*
Plant(s): cell walls in, 389, *389*; photosynthesis in, 454; roots of, 229; soil formation and, 231; weathering and, 227, *227*, 229, *229*
Plant cell, *391*
Plate(s), **202**, *202*; collisions of, 205, *205*; earthquakes and, 203, *203*, 206, *206*; movement of, 202–208, *203, 205, 206*; ridge-push, 207; slab-pull, 207, *207, act.* 208
Plate boundaries, 203; transform, 206, *206*
Pluto, 364, *364*
Plutonium, 147
Polar easterlies, 297
Polaris, 342, *act.* 342
Pollution, 260; of air, 444, *444*; cleanup of, 266–267, *266, 267*; and fossil fuels, 146; of groundwater, 260–265, *260, 261, 262, 263, 264*, 268, *268, act.* 274–275; light, 324
Ponds: polluted, 264, *264*
Population(s), **446**; characteristics of, 449–452; competition and, 451, *451*; environment and, 445, 446, *446*; food webs and, 456, *456*; size of, 449, *449*; spacing of, *act.* 441, 450, *450*
Population density, 450, *450*
Pore space: measuring, *act.* 251, 253; variations in, 253, *253*
Porosity, 253
Position: changing, 105–106, *105, 106*
Potential energy, **134**, *134*, 138, *138*, 139
Power: hydroelectric, 148, *149*
Power plants, 142–143, *142*

Index

Precipitation, 289, **294;** fronts and, 299, *299, 300;* temperature and, 294, *294*
Precision, 44–47, *45, 46,* 48
Predators, 453, *453*
Prediction, 14, 25
Prefixes: in SI, 50
Pressure: atmospheric, 284, 291, 295, 300, 301, *act.* 306, 444
Prevailing westerlies, 297
Prey, 453, *453*
Problem-Solving Activities, 17, 89, 148, 171, 214, 239, 265, 363, 456, 483
Producers, 454
Project Apollo, 334, *334*
Project Gemini, 333
Project Mercury, 333, *333*
Prokaryotic cell, 389, *389*
Promotional materials: evaluating, 30
Properties: chemical. *see* Chemical properties; of gems, **170,** *172;* magnetic, 80, *80,* 170; of metals, 79–80, *80;* physical. *see* Physical properties
Pyrite (fool's gold), 168, 169, 170

Q

Quartz, *166,* 168, 170, 171
Quartzite, 183, 184
Questioning, 13, *13*

R

Rabies vaccinations, 405
Radiant energy, 135, *135*
Radiation: from space, 320–325; from the Sun, 286, 287; ultraviolet, 286
Radio, 141, *141, act.* 144
Radioactive wastes, 147
Radio telescopes, 325, *325*
Radio waves, 287, *287,* 321
Rain, 294, *294. see also* Precipitation
Range: 123
Rate, 54
Reaction: and action, 119–120, *119, 120*
Recycling organelle, 394
Red blood cell, 388, *388*
Red shift, 372
Reference Handbook: Safety in the Science Classroom 565, Periodic Table 566–567, SI/Metric to English, English to Metric Conversions 568, Care and Use of a Microscope 569, Diversity of Life 570–573, Weather Map Symbols 574, Topographic Map Symbols 575, Minerals 576–577, Rocks 578
Reference points, 103, *103,* 104
Reflecting telescopes, 322, *322*
Reflex, 471
Refracting telescopes, 322, *322*
Refraction: double, 170, *170*
Relative humidity, 292
Renewable resources, 148, *149*
Reproduction: of viruses, 402, *403*
Research, 9, 18, 47, 58, 75, 113, 138, 172, 212, 236, 240, 254, 257, 265, 271, 297, 299, 303, 337, 359, 403, 417, 431, 446, 450, 473, 482
Research, impact of: on scientific thought, 116, 284, 373; on society, 35, 94, *act.* 152, 191, 265, *act.* 275, 304, 309, 341, 345, 435, 489; on environment, 233–235, 266–268
Resistance, 112
Resources: alternative, 149–151, *150, 151;* of, 151, 268; nonrenewable, **146,** *146;* renewable, **148,** *149*
Revolution, 353, *353*
Rhyolite, *177*
Ribosome, 392, 393
Rift valleys, 203, *204*
Rill erosion, 240
Rock(s), 166, 175–187; chemical, 180, *180;* chemical weathering of, 227–229, *228, 229;* detrital, 179–180; foliated, **184,** *184, act.* 187; formation of, 175, *175,* 176, 177, *176, 177,* 178, *178,* 179, 182, *182;* fossils in, 180, 181, *181;* ice wedging and, 227, *227;* igneous, **175**–179, *175, 176,* 177, *178;* from lava, 176; from magma, *176–177,* 177; mechanical weathering of, 226–227, *226, 227;* metamorphic, **183**–184; non-foliated, **184,** *184;* observing, *act.* 165; organic, 180; permeable, **253;** pore space in. *see* Pore space; sedimentary, **179**–181, *179, 180, 181;* soil formation and, 230–231
Rock cycle, 185–186, *185*
Rocket car, 120
Rockets, 327, *327,* 335
Rock-forming minerals, 171
Rock slides, *234,* 235
Root(s): as agent of weathering, 229
Rotation: of Earth, **352,** *352*
Rounding, 47, 48
Runoff, 239, *239*
Rust, 85, *85,* 229, *229*

S

Safe Practices: field investigation, 19, 20, *act.* 60, *act.* 342; laboratory investigations, 20, 305, 326, 333, *act.* 415, *act.* 424, *act.* 448, *act.* 460–461
Safety: in laboratory, 19–20, *19, 20,* 74, *74;* and severe weather, 304
Salt. *see* Halite; identifying, 81; as source of groundwater pollution, *260,* 263
San Andreas Fault, 206, *206*
Sand dunes, 238, *238*
Sandstone, *179,* 180, *183*
Sanitary landfills, 262, *262*
Satellites, 329, *329,* 335
Saturation: zone of, **254**
Saturn, 340, 363, *363*
Scale, 47, 53, *53*
Scale drawing, *act.* 55
Schist, 184
Schleiden, Matthias, 401

Index

Science, 4, **6**–33; history of, 34–35, 244–245, 408–409; branches of, 9–11; communication in, 17, *17*; evaluation in, 27–30, *27*, *29*; experiments in, 18, *18*; inference in, 16; models in, 21–26, *21*, *22*, *23*, *24*, *25*, *26*; observation in, 13, *13*, 27–28, *27*; safety in, 19–20, *19*, *20*; skills in, 12–17, *12*; systems in, 8–9, *8*; technology and, 11, *11*
Science and Language Arts, 124–125, 376–377, 462–463
Science Online: Collect Data, 76, 118, 184, 332, 353, 368, 404; Data Update, 148, 203, 339; Research. *see* Research
Science Skill Handbook, 540–553
Scientific explanations: evaluating, 28–29, *29*
Scientific illustrations, 56–57, *56*
Scientific law, 7
Scientific methods, 12
Scientific theory, 7
Scientific thought, 24, 401, 402
Scientists: contributions of 9–11, 24, 34–35, 115–116, 333–334, 397–401, 474, 488
Seasons, 353, *353*
Sediment(s): deposition of, 237, *237*, 241, *241*
Sedimentary rocks, 179–181, *179*, *180*, *181*
Seismic waves, 198, *198*, *199*
Selecting equipment, 44, *act.* 406–407, 460–461
Selective breeding, 431
Septic systems, 263, *263*
Shale, 180
Shape: as physical property, 75, *75*
Shearing, 206
Sheep, *468*
Sheet flow, 239, *239*
Shepard, Alan B., 333
SI (International System of Units), 50–54, *act.* 55

Significant digits, 49, *49*
Silica, 175
Silicates, 171
Silt, 238
Siltstone, *179*, 180
Silver tarnish, 85, *85*
Sinkholes, 272, *272*
Sirius, 367
Size: as physical property, 77, *77*
Slump, *234*, **235**
Smallpox, 404
Smelting, 173
Snow, 294. *see also* Precipitation
Soapstone, 184
Social behavior, 476–477, *476*, *477*
Society, 405, **477**
Soil, 230; as abiotic factor in environment, 444, *444*; classifying, *act.* 232; formation of, 230–231, *230*, *231*; in pore space. *see* Pore space
Soil erosion: measuring, *act.* 242–243
Soil profile, 230, *231*
Solar collector, 149–150, *150*
Solar eclipse, 355, *355*
Solar energy, *130*, 149–150, *150*
Solar system, 350, *350*, **358**–365, *358*–*359*; asteroids in, 362, *362*; comets in, 364, *364*; distances in, 358–359, *363*; inner planets of, 360–361, *360*, *361*; models of, 21, 26, *26*; outer planets of, 362–364, *362*, *363*, *364*
Solar wind, 364, *364*
Solid(s), 78, *78*; melting point of, 79
Sound communication, 480, *480*
Space: distance in, 358–359, 363, 372; measuring, *act.* 441
Space colony, *act.* 374–375
Space exploration, 318, *act.* 319, 327–341; early missions, 327–334; international cooperation in, 336–337, *336*, *337*; of Jupiter, 330, 332; of Mars, 330, 338, *338*, *361*; of Moon, 333–334, *334*, 339, *339*; of

Saturn, 340; of Venus, 330
Space probes, 330–331, *330*, *331*; *Cassini*, 340, *340*; *Galileo*, 332; *Global Surveyor*, 338; *Pioneer 10*, 330, 332; *Voyager*, 332
Space shuttle, 327, *335*, **335**, 336, *act.* 122–123
Space stations, 318, **336**; *International*, 336, 337, *337*; *Mir*, 336, *336*; *Skylab*, 336, *336*
Space suit, 341, *341*
Space travel: equipment, 336; types of transportation, 333–334, 335–337
Spacing: and populations, *act.* 441, 450, *450*
Species: changes in traits, 426–431; endangered, 57–59, *57*, *58*, *59*; natural occurrence, 427–428; selective breeding, 431, 436; genetic materials, 388, 390; role of genes, 426
Specific gravity, 170
Spectrum: electromagnetic, *320*–*321*, **321**
Speed, 106–108; average, 107–108, *108*; calculating, 295; and kinetic energy, 133, *133*; measuring, 54, *act.* 60–61, 114; of space shuttle, *act.* 122–123; velocity and, 108–109, *108*
Speed of light, 372, *372*
Spiders, 472, *472*
Spiral galaxy, 369, *370*
Springs, 256, *256*, 257
Sputnik I, 329, 333
Stalactites, 271, *271*
Stalagmites, 271, *271*
Standardized Test Practices, 39, 67, 99, 129, 159, 195, 223, 249, 281, 313, 349, 381, 413, 439, 493, 674
Star(s), 366–369; apparent magnitude of, 367; colors of, 367; constellations of, 366,

Index

366; estimating number of, *act.* 351; fusion reaction in, 368; life cycle of, 368–369, *368*; navigation by, *act.* 342; neutron, 369
States of matter, 77–79, *78, 79;* changes of, 78, *79*
Stationary front, 300, *300*
Stereotactic Radiotherapy (SRT), 47
Stimuli: external, 470–475, 476–484, *act.* 485; internal, 446, 451, 453, 474
Storage tanks: as source of ground-water pollution, 263
Stratosphere, 286, *286*, 287
Stratus clouds, 293, *293*
Streak, 169, *169*
Stream(s): disappearing, 272, *272*
Stream erosion, 240–241, *240*
Structure and function, 388–395; cells, *388, act.* 396, 401; organs, 395; organ systems, 395; organisms, 395; populations, 446, 449–453
Subduction, 205
Submission, 478, *478*
Subsidence, 268
Substances: creating, 90; comparing properties of new and original, 86; physical properties, 75–77, 167–170; chemical properties, 88
Sugar: reaction with sulfuric acid, *86*
Sulfur, 459
Sulfuric acid, *86*
Sun: apparent magnitude of, 367; color of, 367; eclipse of, 355, *355;* life cycle of, 368; location of, 371, *371;* radiation from, 286, 287
Sundial, *45*
Supergiants, 369
Supernova, 369, *369*
Surface water, 239–241
Symbiosis, 452, *452*
System(s), 8–9, *8;* formed as a result of two or more systems, 8; identify, 8, 297, 298–300, 369, 395; describe, *act.* 8, 8–9

T

Table, 57; construct, 17, 114, *act.* 433; analyze 114
TAKS Practice, 68-69, 160-161, 314-315, 382-383, 494-495
Talc, 170
Tarnish, 85, *85*
Technology, 11. *see* Telescopes. *see also* Space probes; astrolabe, *343;* gasoline engines, 139, *139;* generator, 142, *142;* pan balance, 53, *53;* photovoltaic collector, 150, *150;* radio, 141, *141, act.* 144; rockets, 327, *327,* 335; satellites, 329, *329,* 335; scale, 47, 53, *53;* science and, 11, *11;* solar collector, 149–150, *150;* space shuttle, 327, 335, *335,* 336, *act.* 122–123; space suit, 341, *341;* Stereotactic Radiotherapy (SRT), 47; turbine, 142, *142;* windmill, 151, *151*
Technology Skill Handbook, 554–557
Telescopes, 322–326; *Hubble,* 322, 323, *323,* 340, 373, *373;* Keck, 324, *324,* 340; Next Generation, 340, *340;* optical, 322–324, *322, 323, 324, act.* 326; radio, **325,** *325;* reflecting, **322,** *322;* refracting, **322,** *322*
Temperature: of air, *act.* 283, 290–291, *290, 291;* in atmosphere, *act.* 283, 286, *286,* 287; and environment, 443; humidity and, 292, *292;* measuring, 54, *54;* precipitation and, 294, *294*
Temperature scales, 54, *54*
Tension, 203, *203,* 205, 210
Termites, 477, *477*
Territory, **477,** *477*
Teton Range (Wyoming), 210, *210*
Theories: analyze, 7; review, 7, 206; critique, 7; strengths and weaknesses, 30, 136–137, 202–207, 368–369
Theory: scientific, **7**

The Princeton Review, *see* TAKS practice; Standardized Test Practice
Thermal energy, 134, *134,* 141–142, *142*
Thermometer, 290
Thermosphere, *286,* 287
Thunderstorms, 301, *301*
Till, 237, *237*
Time: measuring, 40, *40,* 45, 54
TIME Science and History, 34–35, 244–245, 408–409
TIME Science and Society, 308–309, 344–345, 434–435
Tissue, **395,** *395*
Titan (moon of Saturn), 340, 363
Titanium, 88
Tools: beakers, *act.* 171, *act.* 251, *act.* 274, *act.* 424; petri dishes, 406; metersticks, *act.* 44, 54, 55, *act.* 114, *act.* 306; graduated cylinders, *act.* 44, *act.* 251, *act.* 274; weather instruments, 305–306; hot plates, 88, 289, 301; test tubes, 23, 208; safety goggles, 19, 305, *act.* 424, *act.* 448; spring scales, 53; balances, 53, *act.* 242–243, *act.* 448; microscopes, 8, 232, *act.* 396, 402, *act.* 406, *act.* 424; telescopes, 326; thermometers, *act.* 305–306, 307; calculators, 17, 18, 19, 49, 109, 114, 117, *act.* 242; field equipment, 324, *act.* 343; computers, all internet activities, 59, 121, 402, *act.* 433, 446, 447, 450; computer probes, *act.* 44, *act.* 149, *act.* 306, 307; timing devices, 114, 149, *act.* 242–243; magnets, 171; compasses, *act.* 343
Topography: and soil formation, **230**
Tornado, 302, *302*
Trade winds, 296, *296*
Traits, 425–431; *act.* 432–433
Transform boundary, 206, *206*
Transpiration, 289

INDEX 611

Index

Trial and error, 473, *473*
Triton (moon of Neptune), 364
Troposphere, 286, *286*
Try at Home MiniLabs: Classifying Parts of a System 8, Forming a Hypothesis 14, Measuring Volume 52, Changing Density 77, Modeling Acceleration 109, Analyzing Energy Transformations 139, Modeling How Fossils Form Rocks 180, Modeling Tension and Compression 205, Analyzing Soils 230, Modeling Groundwater Pollution 264, Observing Condensation and Evaporation 289, Observing Effects of Light Pollution 324, Modeling Constellations 367, Observing Magnified Objects 400, Modeling Probability 427, Modeling the Water Cycle 458, Demonstrating Chemical Communication 479
Turbine, 142, *142*

Ultraviolet radiation, 286
Unbalanced forces, 112
Universe, 373, *373*
Uplifting, 209–215
Upwarped mountains, 211
Uranus, 363, *363*
Ursa Major, 366, *366*
Using equipment, *act.* 242–243, *act.* 274–275, *act.* 306–307, *act.* 406–407, *act.* 432–433, *act.* 460–461

Vaccines, 404, 405
Vacuoles, 393
Valley glaciers, 236, *236*
Variables, 18
Velocity, 108–109, *108*
Vent, 212

Venus, 360, *360;* exploration of, 330
Viking I **mission,** 330
Virchow, Rudolf, 401
Virus(es), 402–405; active, 403, *403;* effects of, 404; fighting, 404–405; latent, 403; reproduction of, 402, *403;* shapes of, 402, *402*
Vitamin(s), 86
Volcanic mountains, 212–213, *212, 213*
Volcanoes: and Earth's plates, 203, 213; effect on atmosphere, 285, *285;* and formation of igneous rock, 171, *171*
Volume: measuring, 52, *52, act.* 52; as physical property, 76, 77
Volcanologist, 10, *10*

Waning, 354, *354*
Warm front, 299, *299*
Waste(s): industrial, 264, *264;* mining, 264; radioactive, 147; spills of, 262
Waste disposal. *see* Sanitary landfills; Septic systems
Water: boiling point of, 79; and environment, 443, *443;* erosion by, 239–241, *239, 240;* hard vs. soft, 255; movement of, 239–241; surface water, 239–241. *see also* Groundwater
Water cycle, 56, *56,* 288, 289, **458,** *458*
Water pollution: cleanup of, 266–267, *266, 267;* of groundwater, 260–265, *260, 261, 262, 263, 264,* 268, *268, act.* 274–275; sources of, 260–265, *260, 261, 262, 263, 264,* 268, *268*
Watershed: *act.* 254
Water table, 254, *255*
Water vapor: in atmosphere, 285, 292, *292*

Wave(s): electromagnetic, 320–321; radio, 287, *287,* 321; seismic, 198, *198,* 199
Wavelength, *320–321,* 321
Waxing, 354, *354*
Weather, 290–297; air masses and, 298, *298, act.* 305; atmospheric pressure and, 284, 291, 295, 300, 301, *act.* 306; clouds and, 293–294, *293, 294;* forecasting, 297; fronts and, 299–300, *299, 300, act.* 305; humidity and, 292, *292;* precipitation and, 289, 294, *294,* 299, *299, 300;* safety and, 304; severe, 301–304, *301, 302, 303, 304;* temperature and, 290–291, *290, 291,* 292, *292,* 294, *294;* wind and, 295–297, *296, 297,* 302, *302,* 303, *303*
Weathering, 224, **226**–231; chemical, **228**–229, *228, 229;* mechanical, **226**–227, *226, 227*
Weight: and mass, 113; measuring, 53, *53*
Well(s): artesian, **257,** *257, act.* 259; drilling, 256, *256*
Westerlies, 297
White dwarf, 368
Wind, 295–297, *296, 297;* hurricanes and, 303, *303;* solar, 364, *364;* tornadoes and, **302,** *302*
Wind energy, 151, *151*
Wind erosion, 238, *238*
Windmill, 151, *151*
Wind tunnel, 25

Yeast, *act.* 88, *act.* 419

Zone of saturation, 254

Art Credits

Glencoe would like to acknowledge the artists and agencies who participated in illustrating this program: Absolute Science Illustration; Andrew Evansen; Argosy; Articulate Graphics; Craig Attebery represented by Frank & Jeff Lavaty; CHK America; Gagliano Graphics; Pedro Julio Gonzalez represented by Melissa Turk & The Artist Network; Robert Hynes represented by Mendola Ltd.; Morgan Cain & Associates; JTH Illustration; Laurie O'Keefe; Matthew Pippin represented by Beranbaum Artist's Representative; Precision Graphics; Publisher's Art; Rolin Graphics, Inc.; Wendy Smith represented by Melissa Turk & The Artist Network; Kevin Torline represented by Berendsen and Associates, Inc.; WILDlife ART; Phil Wilson represented by Cliff Knecht Artist Representative; Zoo Botanica.

Photo Credits

Abbreviation Key: AA=Animals Animals; AH=Aaron Haupt; AP=Amanita Pictures; CB=CORBIS; BD=Bob Daemmrich; DM=Doug Martin; DRK=DRK Photo; ES=Earth Scenes; FI=First Image; FP=Fundamental Photographs; GB=Geoff Butler; IS=Index Stock; KS=KS Studios; LA=Liaison Agency; MM=Matt Meadows; PA=Peter Arnold, Inc.; PD=PhotoDisc; PE=PhotoEdit; PQ=PictureQuest; PR=Photo Researchers; PT=PhotoTake NYC; RM=Richard Megna; SB=Stock Boston; SPL=Science Photo Library; SS=SuperStock; TIW=The Image Works; TSA=Tom Stack & Associates; TSM=The Stock Market; VU=Visuals Unlimited.

Cover (background)Paul Ruben Archives, (l)PD, (r)Laurence Parent; **ix** David Ducros/SPL/PR; **xiii** AFP/CB; **xv** NASA/Science Source/PR; **xvi** Gil Lopez-Espina/VU; **xvii** Joe McDonald/VU; **xviii** Stephen J. Krasemann/PA; **xix** NASA; **xx** Leonard Lee Rue/PR; **1** NASA/SPL/PR; **2** PD; **2-3** Wolfgang Kaehler; **3** PD; **4** National Marine Mammal Lab; **4-5** Kennan Ward/CB; **5** Richard Hutchings; **6** (l)Jack Star/Photo-link/PD, (c)Rudi VonBriel, (r)Richard T. Nowitz/CB; **8** Mary Kate Denny/PE; **9** National Geographic; **10** (l)G. Brad Lewis/Stone, (c)Roger Ball/TSM, (r)Will & Deni McIntyre/PR; **11** (t)AFP/CB, (b)Reuters NewMedia, Inc./CB; **13 14** Richard Hutchings; **15** MM; **16** Icon Images; **17** Richard Hutchings/PE/PQ; **18** Rudi VonBriel; **19** BD; **20** Glasheen Graphics/IS; **21** (t)David Young-Wolff/PE, (c)John Bavosi/SPL/PR, (bl)A. Ramey/PE, (br)Donald C. Johnson/TSM; **22** CB Images/PQ; **23** Todd Gipstein/CB; **24** (tl c br)Betty Pat Gatliff, (tr)Richard Nowitz/Words & Pictures/PQ, (bl)Michael O'Brian/Mud Island, Inc.; **25** (tl)CB, (tr)Tom Wurl/SB/PQ, (b)Jim Sugar Photography/CB; **26** (l) Stock Montage, (r) North Wind Picture Archives; **27** Digital Art/CB; **28** SS; **29** (t)Lester V. Bergman/CB, (b)Bob Handelman/Stone; **31** John Evans; **32** (t)AH, (b)MM; **34** Benainous-Deville/LA; **35** (t)UPI/Bettmann/CB, (b)Reuters/CB; **36** Robert Glusic/PD; **38** Tim Courlas; **39** Charles D. Winters/PR; **40** FI; **40-41** Brent Jones/SB; **41** MM; **42** Paul Almasy/CB; **43** AFP/CB; **44** David Young-Wolff/PE; **45** (tl)Lowell D. Franga, (tr)The Purcell Team/CB, (b)Len Delessio/IS; **46** photo by Richard T. Nowitz, imaging by Janet Dell Russell Johnson; **47** MM; **49** FI; **51** Tom Prettyman/PE; **53** (t)Michael Dalton/FP, (cl)David Young-Wolff/PE, (cr)Dennis Potokar/PR, (b)MM; **55** Michael Newman/PE; **57** John Cancalosi/SB; **60 61** Richard Hutchings; **62** (t)Fletcher & Baylis/PR, (b)Owen Franken/CB; **63** CMCD/PD; **64** Fred Bavendam/Stone; **65** FI; **69** MM; **70-71** PD; **71** Stephen Frisch/SB/PQ; **72** Michael Newman; **72-73** AP Photo/Steve Nutt; **73** Mark Burnett; **75 76** MM; **77** AH; **79** David Taylor/SPL/PR; **80** (t)SS, (b)Ray Pfortner/PA; **81** AP; **82** (tl)Steve Kaufman/CB, (tr)Tom McHugh/PR, (b)Fred Bavendam/Minden Pictures; **83** Don Tremain/PD; **84** (l)RM/FP, (cl)John Lund/Stone, (cr)Richard Pasley/SB, (r)T.J. Florian/Rainbow/PQ; **85** (l)Philippe Colombi/PD, (c)Michael Newman/PE, (r)Roger K. Burnard; **86** MM; **87** (t)Ralph Cowan/FPG, (bl br)AH; **89** Jeff J. Daly/VU; **90** Timothy Fuller; **91** John Evans; **92** (t)AH, (b)MM; **93** (l)John A. Rizzo/PD, (c)AH, (r)John A. Rizzo/PD; **94** Adam Woolfitt/CB; **95** (t)Ted Thai/Timepix, (b)Dave G. Houser/CB; **96** (tl)SS, (tr)S. Solum/Photolinks/PD, (bl)AP, (br)Pat Lacroix/TIB; **97** (tl)file photo, (tr c)Siede Preis/PD, (bl)John Evans, (br)AH; **98** Elaine Shay; **100** Tyler Stableford/TIB/PQ; **101** David Young-Wolff/PE; **101-102** SS; **102** Michael Newman/PE; **103** BD; **104** Stephen R. Wagner; **110** (l)Tom Pantages, (r)Jodi Jacobson/PA; **111** BD; **112** (t)Kenneth Jarecke/Contact Press Images/PQ, (c)Stock South/PQ, (b)DM; **114** file photo; **115** Doug Menuez/PD; **116** Globus Bros./TSM; **117** Ed Bock/TSM; **118** (l)Duomo, (r)Amwell/Stone; **119** Richard Hutchings; **120** DiMaggio/Kalish/TSM; **121** file photo; **122 123** NASA; **124** Tony Arruza/CB; **125** Courtesy of Nadia Roberts; **126** (t)S.R. Maglione/PR, (c)David Madison, (b)Davies & Starr/Stone; **127** (l)BD, (r)Stock South/PQ; **130** Charles Krebs/Stone; **130-131** Roger Ressmeyer/CB; **131** MM; **132** (lc)file photo, (r)Mark Burnett; **133** (t)BD, (c)Al Tielemans/Duomo, (b)BD; **134** KS; **135** (tl tr)BD, (b)Andrew McClenaghan/SPL/PR;

Credits

136 Mark Burnett/PR; **137** Lori Adamski Peek/Stone; **138** Richard Hutchings; **139** Ron Kimball Photography; **140** (tl)Judy Lutz, (tc tr bl)Stephen R. Wagner, (br)Lennart Nilsson; **142 144** KS; **150** (t)Dr. Jeremy Burgess/SPL/PR, (b)John Keating/PR; **151** Billy Hustace/Stone; **152** SS; **153** Roger Ressmeyer/CB; **154** (l)Reuters NewMedia Inc./CB, (r)PD; **155** (l)KS, (r)Dominic Oldershaw; **156** (t)James Blank/FPG, (c)Robert Torres/Stone, (bl br)SS; **157** (l)Lowell Georgia/CB, (r)Mark Richards/PE; **158** Reuters NewMedia Inc./CB; **162-163** Mike Zens/CB; **163** Mark A. Schneider/VU; **164** Carr Clifton/Minden Pictures; **164-165** David Muench/CB; **165** Mark Burnett; **166** (l)DM, (r)Mark Burnett; **167** Mark A. Schneider/VU; **168** (t)Manuel Sanchis Calvete/CB, (c)Jose Manuel Sanchis Calvete/CB, (bl)DM, (br)Mark A. Schneider/VU; **169** (t)Albert J. Copley/VU, (b)FP; **170** Tim Courlas; **172** (tl)Ryan McVay/PD, (tr)Lester V. Bergman/CB, (b)Margaret Courtney-Clarke/PR; **173** (t)Walter H. Hodge/PA, (b)Craig Aurness/CB; **174** KS; **175** Kyodo/AP/Wide World Photos; **176** (l)Stephen J. Krasemann/DRK, (r)Brent P. Kent/ES; **177** (l)Breck P. Kent/ES, (r)Brent Turner/BLT Productions; **178** (t)Steve Kaufman/CB, (c)Galen Rowell/Mountain Light, (bl)Martin Miller, (br)David Muench/CB; **179** (t)John D. Cunningham/VU, (others)Morrison Photography; **180** Jeff Foott/DRK; **181** (l)Yann Arthus-Bertrand/CB, (r)Alfred Pasieka/SPL/PR; **182** NASA/CB; **183** (tl)Brent Turner/BLT Productions, (tr)Breck P. Kent/ES, (cl)Andrew J. Martinez/PR, (cr)Tom Pantages/PT/PQ, (bl)Runk/Schoenberger from Grant Heilman, (br)Breck P. Kent/ES; **184** (tl)Stephen J. Krasemann/DRK, (tr)PA, (bl)M. Angelo/CB, (br) Christian Sarramon/CB; **185** (t)Breck P. Kent/ES, (bl)DM, (br)Andrew J. Martinez/PR; **186** Bernhard Edmaier/SPL/PR; **187** Andrew J. Martinez/PR; **188** (t)Cliff Leight/Outside Images/PQ, (b)MM; **190** (t)Archive Photos, (bl)Stock Montage, (br)Brown Brothers; **191** (l)Arne Hodalic/CB, (r)Herbert Gehr/Timepix; **192** (t)Stephen J. Krasemann/DRK, (c)Michael Dalton/FP, (b)Tui De Roy/Minden Pictures; **193** (l)A.J. Copley/VU, (c)Barry L. Runk from Grant Heilman, (r)Breck P. Kent/ES; **194** Breck P. Kent/ES; **196** R.W. Gerling/VU; **196-197** Steve Razzetti/FPG; **197** Mark Burnett; **198** Aaron Horowitz/CB; **199** (t)Barry Sweet/AP/Wide World, (b)FI; **204** (background)National Geographic Maps, (foreground)Dorling Kindersley; **206** Dewitt Jones/CB; **208** AP; **209** (l)Chris Noble/Stone, (r)Buddy Mays/CB; **210** David Muench; **211** Mark C. Burnett; **212** AFP/CB; **213** Michael T. Sedam/CB; **214** Mark E. Gibson/VU; **216** (t)Ralph A. Clevenger/CB, (b)KS; **217** Paul Chesley/Stone; **218** John Elk III/SB/PQ; **218-219** Dale Wilson/Masterfile; **219** SS; **220** (l)Kevin Schafer/CB, (r)Charlie Ott/PR;

221 (tl)Sharon Gerig/TSA, (tr)David Muench/CB, (bl)Robert Lubeck/ES, (br)I & V/TLC/Masterfile; **223** Pat Hermansen/Stone; **224** Tom McHugh/PR; **224-225** Doug Menuez/ SB/PQ; **225** KS; **226** Jonathan Blair/CB; **227** R. & E. Thane/ES; **228** (l)DM, (r)DM/PR; **229** (t)AH, (bl)Layne Kennedy/CB, (br)Richard Cummins/CB; **232** KS; **233** USGS; **234** (background)Roger Ressmeyer/CB, (t)Martin Miller, (c)DP Schwert/ND State University, (b)Jeff Foott/Bruce Coleman, Inc.; **236** (l)Chris Rainier/CB, (r)Glenn M. Oliver/VU; **237** (tl)John Lemker/ES, (tr)Francois Gohier/PR, (b)Paul A. Souders/CB; **238** (t)Gerald & Buff Corsi/VU, (b)Dean Conger/CB; **239** (t)KS, (b)Tess & David Young/TSA; **240** (t)Gerard Lacz/ES, (b)Vanessa Vick/PR; **241** Martin G. Miller/VU; **242** Dominic Oldershaw; **243** Mark Burnett; **244** (l)Will & Deni McIntyre/PR, (r)Robert Nickelsberg/Time Magazine; **244-245** Morton Beebe, SF/CB; **246** (tl)Layne Kennedy/CB, (tr)Leonard Lee Rue III/PR, (bl)Jonathan Blair/CB, (br)C.C. Lockwood/ES; **247** (l)Martin G. Miller/VU, (r)James P. Rowan/DRK; **248** Johnny Johnson/AA; **250** Corel; **250-251** Michael T. Sedam/CB; **251** Richard Hutchings; **252** A.J. Copley/VU; **256** (t)Willard Luce/ES, (b)Fred Habegger from Grant Heilman; **258** C. Alan Chapman/VU; **259** Mark Burnett; **260** Salt Institute; **261** Stephen R. Wagner; **264** Lowell Georgia/CB; **267** Fritz Prenzel/ES; **269** James Jasek; **271** (l)Runk/Schoenberger from Grant Heilman, (c)M. L. Sinibaldi/TSM, (r)Corel; **272** (t)AP/ Wide World Photos, (b)Mammoth Cave National Park; **274** (t)Bert Krages/VU, (b)Jim Sugar Photography/CB; **275** DM; **276** (tl tr)David Muench/CB, (b)Robert Holmes/CB; **277** (l)Peter & Ann Bosted/ TSA, (r)Jean Clottes/CB Sygma; **278** (tl)Tui De Roy/Minden Pictures, (tr)Richard Thom/VU, (b)Holt Confer/DRK; **279** (l)William A. Blake/CB, (r)Michael Dwyer/SB; **281** C.C. Lockwood/ES; **282** Nuridsany et Perennou/PR; **282-283** William Mullins/PR; **283** Mark Burnett; **285** Pat & Tom Leeson/PR; **288** (background)Picture Perfect, (tl)CB, (tr)Ellis Herwig/SB/PQ, (bl br)CB; **291** Douglas Peebles/CB; **297** Timothy Fuller; **302** Jim Zuckerman/CB; **303** NOAA/NESDIS/Science Source/PR; **304** Howard Bluestein/PR; **305** MM; **306** (t)DM, (b)Mark Burnett; **307** Timothy Fuller; **308** Ron Magill/Miami Metrozoo; **309** John Berry/LA; **310** (t)Marc Muench/CB, (c)Timothy Fuller, (b)NCAR/TSA; **311** (l)Annie Griffiths Belt/CB, (r)Roger Ressmeyer/CB; **314** William E. Ferguson; **316-317** Bryan & Cherry Alexander; **317** NASA; **318** NASA; **318-319** NASA/Roger Ressmeyer/CB; **319** CB; **320** (l)Weinberg-Clark/TIB, (r)Stephen Marks/TIB; **321** (l)PE, (r)Wernher Krutein/LA; **322** Chuck Place/SB; **323** NASA; **324** (t)Roger Ressmeyer/CB, (b)Simon Fraser/SPL/PR; **325** Raphael Gaillarde/LA; **326** (t)Icon Images,

Credits

(b)Diane Graham-Henry & Kathleen Culbert-Aguilar; **327** NASA; **328** NASA/SPL/PR; **329** NASA; **330** (t tc)NASA/Science Source/PR, (bc)M. Salaber/LA; Julian Baum/SPL/ PR, **331** (tl)Dorling Kindersley Images, (tcl)TASS from Sovfoto, (tcr)NASA, (tr)NASA/JPL, (cl ccl)NASA/JPL/Caltech, (ccr)NASA, (cr)NASA/JPL, (bl)NASA, (br)NASA; **332** AFP/CB; **333** NASA; **334** NASA/Science Source/PR; **335** NASA/LA; **336** (t)NASA, (b)NASA/LA; **337** NASA/Science Source/PR; **338** NASA/JPL/Malin Space Science Systems; **339** NASA/JPL/LA; **340** (t)David Ducros/SPL/PR, (b)NASA; **341** HED Foundation/NASA; **342** Roger Ressmeyer/CB; **343** DM; **344-345** Robert McCall; **345** NASA/SPL/PR; **346** (tl)David Parker/SPL/PR, (tr)NASA/SPL/PR, (b)NASA; **347** (l)Novosti/SPL/PR, (c)Roger K. Burnard, (r)NASA; **350** David Parker/SPL/PR; **350-351** David Nunek/SPL/PR; **351** Morrison Photography; **354** Lick Observatory; **355** Francois Gohier/PR; **356** Jerry Lodriguss/PR; **357** DM; **360** (l)USGS/SPL/PR, (r)NASA/Science Source/PR; **361** (t)CB, (bl)NASA, (br)USGS/TSADO/TSA; **362** (t)JPL/TSADO/TSA, (b)CB; **363** (t)ASP/Science Source/PR, (b)NASA/JPL/Tom Stack and Associates; **364** (t)NASA/JPL, (b)Dr. R. Albrecht, ESA/ESO Space Telescope European Coordinating Facility/NASA; **365** AP/Wide World Photos; **367** Dominic Oldershaw; **369** Palomar Observatory; **370** (c)Royal Observatory, Edinburgh/SPL/PR, (others)Anglo-Australian Observatory; **371** Frank Zullo/PR; **373** R. Williams/NASA; **374** (t)Movie Still Archives, (b)NASA; **375** Mark Burnett; **376** Kevin Morris/CB; **377** NASA/Roger Ressmeyer/CB; **378** (t)Dr. Fred Espenak/PR, (b)Dr. R. Albrecht, ESA/ESO Space Telescope European Coordinating Facility/NASA; **379** (l)NASA, (r)National Optical Astronomy Observatories; **380** Rich Brommer; **382** NASA/Science Source/PR; **383** (b)NASA, others (Lick Observatory); **384-385** Diane Scullion Littler; **385** (l)Jonathan Eisenback/PT/PQ, (r)Janice M. Sheldon/Picture 20-20/PQ; **386** Oliver Meckes/ E.O.S./MPI-Tubingen/PR, Inc.; **386-387** Tim Flach/Stone; **387** MM; **389** David M. Phillips/VU; **390** (t)Don Fawcett/PR, (b)M. Schliwa/PR; **392** (t)George B. Chapman/VU, (b)P. Motta & T. Naguro/SPL/PR; **393** (t)Don Fawcett/PR, (b)Biophoto Associates/PR; **397** (l)Biophoto Associates/PR, (r)MM; **398** (background)David M. Phillips/VU, (t)Kathy Talaro/VU, (cl)Courtesy Nikon Instruments Inc., (c)Michael Gabridge/VU, (cr)Mike Abbey/VU, (b)David M. Phillips/VU; **399** (tl)James W. Evarts, (tr)Bob Krist/CB, (cl)courtesy Olympus Corporation, (cr)Mike Abbey/VU, (bl)Karl Aufderheide/VU, (br)Lawrence Migdale/SB/PQ; **402** (l)Richard J. Green/PR, (c)Dr. J.F.J.M. van der Heuvel, (r)Gelderblom/Eye of Science/PR; **405** Pam Wilson/TX Dept. of Health; **406, 407** MM; **408** (t)Quest/SPL/PR, (b)courtesy California Univ.; **409** Nancy Kedersha/Immunogen/SPL/PR; **410** (t)Oliver Meckes/PR, (b)CDC/SPL/PR; **411** (l) Keith Porter/PR, (r)NIBSC/SPL/PR; **412** Biophoto Associates/Science Source/PR; **414** Geoff Tompkinson/SPL/PR; **414-415** Inga Spence/TSA; **415** AH; **416** (l) Gary Meszaros/VU, (cr) Zig Leszczynski/ES; **419** (l)Tom J. Ulrich/VU, (r)Betty Barford/PR; **420** (t)Inga Spence/VU, (bl)Prof. P. Motta/Department of Anatomy/University "La Sapienza," Rome/SPL/PR, (br)David M. Phillips/VU; **421** Biophoto Associates/Science Source/PR; **422** (tl)Jane Hurd, (tcl)Dr. Dennis Kunkel/PT, (tcr)CNRI/PT, (tr)Jane Hurd, (bl)Dr. Dennis Kunkel/PT, (br)Paul Hirata/Stock Connection; **423** (tl)Mark Burnett/VU, (tc)William J. Weber/VU, (tr)Inga Spence/VU, (b)John D. Cunningham/VU; **424** Timothy Fuller; **425** Ariel Skelley/TSM; **426** (l)Nigel Cattlin/Holt Studios International/PR, (r)Bayard H. Brattstrom/VU; **427** Bruce Berg/VU; **430** (l)Ken Chernus/FPG, (r)Stephen Simpson/FPG; **431** Inga Spence/TSA; **432** (t)SS, (c)The Photo Works/PR, (b)The Photo Works/PR; **433** Jeff Smith/Fotosmith; **434** Arnold Zann/Black Star; **435** Charles Robinson; **436** (t) Norvia Behling, (c)Bob Langrish/AA, (b)Gil Lopez-Espina/VU; **437** (t)Betty Barford/PR, (b)Robert F. Myers/VU; **438** Alan & Linda Detrick/PR; **439** (l) Wally Eberhart/VU, (r)Richard Sheill/ES; **440** P. Parks/OSF/AA; **440-441** Sanford/Agliolo/TSM; **441** AH; **442** Wm. J. Jahoda/PR; **443** (tl)Stuart Westmorland/PR, (tr)Michael P. Gadomski/ES, (b)George Bernard/ES; **444** (t)Francis Lepine/ES; **446** (t)Roland Seitre-Bios/PA, (bl)Robert C. Gildart/PA, (br)Carr Clifton/Minden Pictures; **448** BD; **450** Dan Suzio/PR; **451** (t)Tim Davis/PR, (b)Arthur Gloor/AA; **452** Gilbert Grant/PR; **453** John Gerlach/AA; **454** Michael P. Gadomski/PR; **455** (background) Michael Boys/CB, (t)Joe McDonald/CB, (c)David A. Northcott/CB, (bl)Michael Boys/CB, (bc)Dennis Johnson/Papilio/CB, (br)Kevin Jackson/AA; **457** (tl) Ray Richardson/AA, (tr)Suzanne L. Collins/PR, (c)William E. Grenfell Jr./VU, (b)Zig Leszczynski/ES; **460** (t)GB, (b)KS; **461** KS; **462** Allen Russell/IS; **463** Courtesy Dave Grza DVM; **464** Gerald Fuehrer/VU; **465** (l)Richard Reid/ES, (r)Helga Lade/PA; **468** Gary W. Carter/VU; **468-469** Robert Mackinlay/PA; **469** Mark Burnett; **470** (l)Michel Denis-Huot/Jacana/PR, (r)Zig Leszczynski/AA; **471** (l)Jack Ballard/VU, (c)Anthony Mercieca/PR, (r)Joe McDonald/VU; **472** (t)Stephen J. Krasemann/PA, (b)Leonard Lee Rue/PR; **473** (t)The Zoological Society of San Diego, (b)Margret Miller/PR; **476** Michael Fairchild; **477** (t)Bill Bachman/PR, (b)Fateh Singh Rathore/PA; **478** Jim Brandenburg/Minden Pictures; **479** Michael Dick/AA;

Credits

480 (l)VU/Richard Thorn, (c)Arthur Morris/VU, (r)Jacana/PR; **481** (tl)T. Frank/Harbor Branch Oceanographic Institution, (bl bc)Peter J. Herring, (others)Edith Widder/Harbor Branch Oceanographic Institution; **482** Stephen Dalton/AA; **483** Richard Packwood/AA; **484** Ken Lucas/VU; **486** (t)Dave B. Fleetham/TSA, (b)Gary Carter/VU; **487** The Zoological Society of San Diego; **488** Walter Smith/CB; **488-489** Bios (Klein/Hubert)/PA Inc.; **489** Courtesy The Seeing Eye; **490** (tl)Norbert Wu/PA, (tr)Fritz Prenzel/AA; **491** (l)Valerie Giles/PR, (r)J & B Photographers/AA; **492** Alan & Sandy Carey/PR; **498** Dan Guravich/CB; **499** Ruth Dixon; **501** F. Stuart Westmorland/PR; **502** KS; **503** MM; **506** Danny Lehman/CB; **508** Harold Hofman/PR; **509** Gregory K. Scott/PR; **510** William Weber; **511** Glen M. Oliver/VU; **512** Johnny Johnson/DRK; **513** Jeff Vanuga/CB; **514** Mark Burnett; **519** Beth Davidow/VU; **520** NASA/GSFC/TSA; **521** Telegraph Colour Library/FPG; **522-523** PD; **524** file photo; **525** (t)Dan Feicht, (b)VU; **526** Jose Carillo/PE; **527** (t)AH, (b)Michael J. Howell/Rainbow/PQ; **528** (t)Garry D. McMichael/PR; **528-529** Harvey Wood/The Still Moving Picture Co.; **529** (t c)AH, (b)David Chester French/CB; **530** (t)SS, (bl)Lindsay Hebberd/CB, (br)PT/PQ; **531** (t)Ric Ergenbright/CB, (bl)Icon Images, (br)Raymond Gehman/CB; **532** NASA; **533** (t)Roger Ressmeyer/CB, (b)NASA; **534** (t)NASA/Roger Ressmeyer/CB, (c b) NASA; **535** NASA; **536** (t)PR, (b)David M. Dennis; **537** (t)Roy Morsch/TSM, (cl)Harry Rogers/PR, (cr)Donald Specker/AA, (b)Roger K. Burnard; **538** (t)Donald Specker/AA, (cl)Tom McHugh/PR, (cr)Harry Rogers/PR, (bl)Caroll W. Perkins/AA, (bc)Donald Specker/AA, (br)Patti Murray/AA; **539** (tl)Harry Rogers/PR, (tr)Ken Brate/PR, (cl)James H. Robinson/PR, (cr)Linda Bailey/AA, (bl)Ed Reschke/PA, (br)MM; **540** Timothy Fuller; **544** Roger Ball/TSM; **546** (l)GB, (r)Coco McCoy from Rainbow/PQ; **547** Dominic Oldershaw; **548** StudiOhio; **549** FI; **551** MM; **554** Paul Barton/TSM; **557** Davis Barber/PE; **569** MM; **570** (l)Dr. Richard Kessel, (c)NIBSC/SPL/PR, (r)David John/VU; **571** (t)Runk/Shoenberger from Grant Heilman, (bl)Andrew Syred/SPL/PR, (br)Rich Brommer; **572** (t)G.R. Roberts, (bl)Ralph Reinhold/ES, (br)Scott Johnson/AA; **573** Martin Harvey/DRK Photo.

Acknowledgments

"Rayona's Ride," from *A Yellow Raft in Blue Water* by Michael Dorris, copyright ©1987 by Michael Dorris. Reprinted by permission of Henry Holt and Company, LLC.

"Friends, Foes, and Working Animals," from *The Solace of Open Spaces* by Gretel Ehrlich, copyright ©1985 by Gretel Ehrlich. Used by permission of Viking Penguin, a division of Penguin Putnam Inc.

PERIODIC TABLE OF THE ELEMENTS

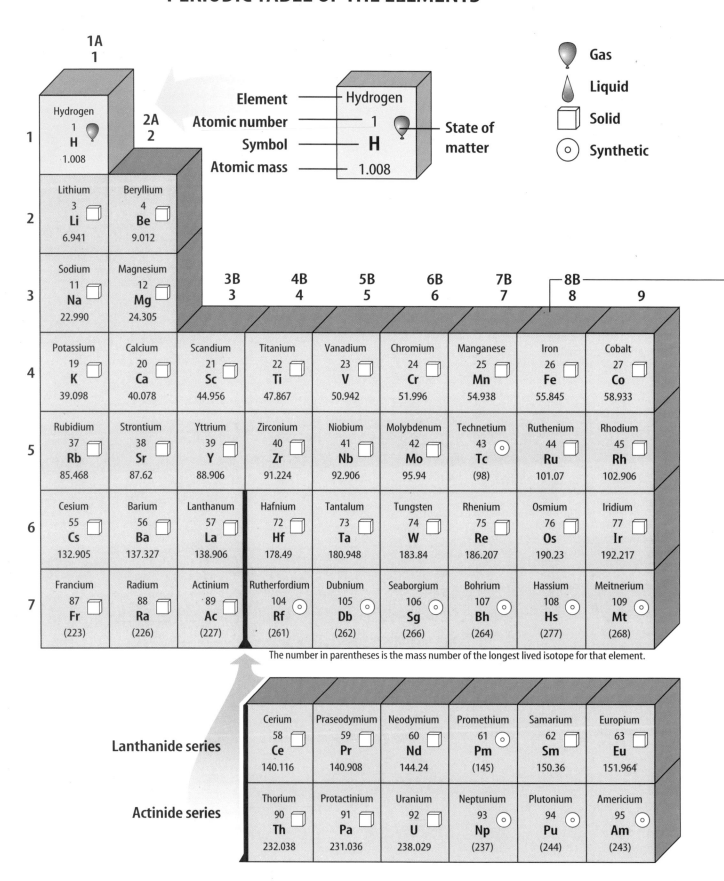